OXIDATION IN ORGANIC CHEMISTRY

Part C

This is Volume 5 of
ORGANIC CHEMISTRY
A series of monographs
Editor: HARRY H. WASSERMAN

A complete list of the books in this series appears at the end of the volume.

OXIDATION IN ORGANIC CHEMISTRY

Edited by *WALTER S. TRAHANOVSKY*

IOWA STATE UNIVERSITY OF SCIENCE
AND TECHNOLOGY
AMES, IOWA

PART C

1978

ACADEMIC PRESS New York, San Francisco, London

A Subsidiary of Harcourt Brace Jovanovich, Publishers

ACADEMIC PRESS, INC.
111 Fifth Avenue, New York, New York 10003

United Kingdom Edition published by
ACADEMIC PRESS, INC. (LONDON) LTD.
24/28 Oval Road, London NW1 7DX

Library of Congress Cataloging in Publication Data

Wiberg, Kenneth B
Oxidation in organic chemistry.

(Organic chemistry ; a series of monographs, v. 5, pts. A–)
Pts. B– edited by W. S. Trahanovsky.
Includes bibliographical references.
1. Oxidation. 2. Chemistry, Organic. I. Trahanovsky, Walter S., Date II. Title. III. Series.
QD281.09W5 547'.23 65–26047
ISBN 0–12–697252–4 (pt. C)

PRINTED IN THE UNITED STATES OF AMERICA

Contents

CHAPTER IV

Oxidations with Nitric Acid or Nitrogen Oxides

Yoshiro Ogata

CHAPTER V

Alkaline Hypohalite Oxidations

Sujit K. Chakrabartty

Contributors

Numbers in parentheses indicate the pages on which the authors' contributions begin.

Sujit K. Chakrabartty, *Alberta Research Council, Fuel Sciences Division, Edmonton, Alberta T6G 2C2, Canada* (*343*)

Roy A. Johnson, *Experimental Chemistry Branch, The Upjohn Company, Kalamazoo, Michigan 49001* (*131*)

Yoshiro Ogata, *Department of Chemistry, Faculty of Engineering, Nagoya University, Furo-cho, Chikusa-ku, Nagoya, Japan* (*295*)

Božo Plesničar, *Department of Chemistry, University of Ljubljana, 61000 Ljubljana, Yugoslavia* (*211*)

Hans J. Reich, *Department of Chemistry, University of Wisconsin, Madison, Wisconsin 53706* (*1*)

Preface

Some of the most important and common reactions in organic chemistry involve oxidation and reduction. Because of the importance of this class of reactions, numerous reagents have been developed which will bring about certain oxidations or reductions selectively. This treatise is devoted to detailed discussions of specific oxidants or topics involving oxidation of organic compounds.

This volume is concerned with oxidations which involve organoselenium intermediates, an area of chemistry which has become extremely important in organic synthesis; with oxidations which use microorganisms, a class of unusually selective oxidations which has not yet received widespread use; and with oxidations by three important classes of oxidants, peroxy acids and other peroxides, nitric acid and nitrogen oxides, and alkaline hypohalites. An attempt is made to present the main results of the most important literature which refers to these topics, organized according to oxidation of specific functional groups or types of organic compounds. The author of each review is an active researcher who has made important contributions in the area of chemistry being reviewed.

The level of all chapters is such that experts in these areas of research and students and researchers who wish a thorough and rigorous discussion of these topics should find them useful. In general, emphasis is on the scope and preparative use as well as the mechanistic aspects of the various oxidations.

The final editing of this volume was done while I was on leave at the University of Virginia. I would like to thank the faculty of the University of Virginia's Department of Chemistry for their hospitality.

WALTER S. TRAHANOVSKY

Contents of Part A

Edited by Kenneth B. Wiberg

Contents of Part B

CHAPTER I

Organoselenium Oxidations

HANS J. REICH

I. Introduction

The classical role of selenium in organic synthesis has been that of an oxidant. Selenium dioxide has been widely used for the oxidation of olefins and acetylenes to allyl and propargyl alcohols, and ketones to α-diketones

ISBN 0-12-697252-4

and enones.[1,1a] Elemental selenium has served as a dehydrogenating agent for the aromatization of hydroaromatic compounds,[2] and a number of patents have been issued for processes (including many oxidations) using selenium in various ways as a catalyst or cocatalyst.[3] These applications are outside the scope of this review although selenium dioxide chemistry pertinent to some of the organoselenium oxidations will be covered.

This chapter describes recent developments which have enlarged the scope of selenium in organic chemistry by the introduction of organoselenium reagents for performing selective oxidative transformations under unusually mild conditions. The development of these procedures has been fostered by, and has in turn promoted understanding of the role of organoselenium intermediates in various selenium dioxide oxidations. The largest group of these transformations is based on the selenoxide syn elimination, but other reactions of selenoxides such as various [2,3] sigmatropic rearrangements and Pummerer-like reactions are also becoming important. The unifying feature of these reactions is that oxidation at selenium results in cleavage of the carbon–selenium bond accompanied by an oxidative transformation in the remainder of the molecule.

Some of the methodology described here has a counterpart in organosulfur chemistry. The rationale for using either selenium or sulfur for some specific transformation is likely to be based on the peculiarities of the system. Important differences to be considered are the availability and convenience of starting reagents, toxicity hazards of selenium, relative ease of selective oxidation at selenium and sulfur, and the much greater facility of virtually all reactions involving cleavage of C–Se bonds compared to cleavage of analogous C–S bonds. These factors will be discussed in Section III, I.

When only one paper concerning a particular type of transformation has appeared, select examples have been reproduced here. Where several authors have reported examples, all have been included in a table.

A. SYNTHESIS OF ORGANOSELENIUM REAGENTS

The majority of synthetic applications of organoselenium reagents have used the phenylseleno group. Diphenyl diselenide, benzeneselenol, and benzeneselenenyl chloride are commercially available. Good experimental

[1] R. A. Jerussi, *in* "Selective Organic Transformation" (B. S. Thyagarajan, ed.), p. 301, Wiley, New York, 1970.

[1a] E. N. Trachtenberg, *in* "Oxidation" (R. L. Augustine, ed.), Vol. 1, Chapter 3. Dekker, New York, 1969; N. Rabjohn in "Organic Reactions" (W. G. Dauben, ed.) Vol. 24, p. 261. Wiley (Interscience), New York, 1978.

[2] H. A. Silverwood and M. Orchin, *J. Org. Chem.* **27**, 3401 (1962).

[3] A. V. Kollonitsch, *Ind. Eng. Chem.* **55**, (12) 18 (1963).

procedures are available for preparing these compounds in quantity as shown in Scheme 1 below.

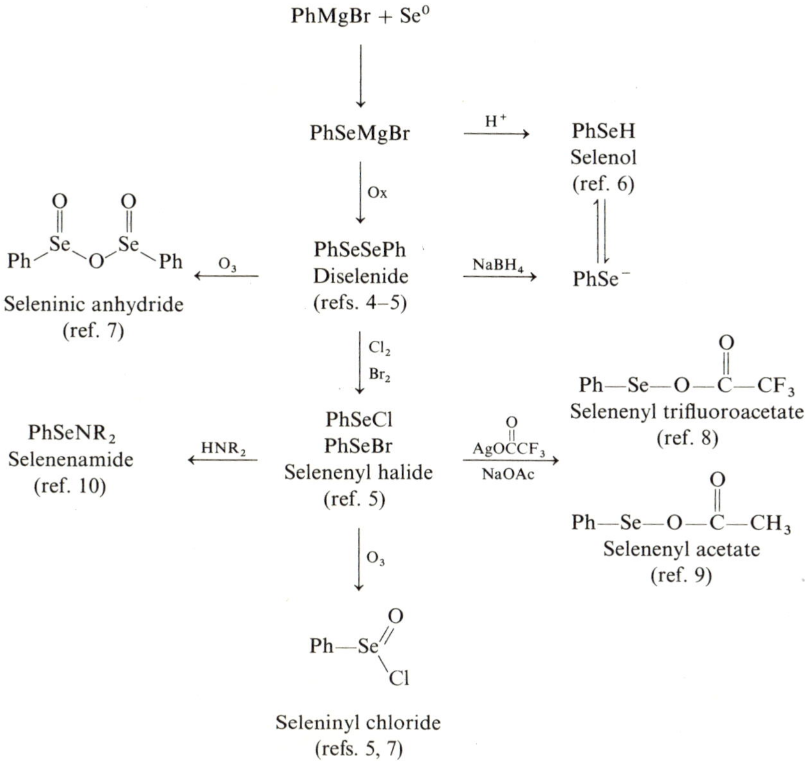

SCHEME 1

o-Nitrophenylseleno,[4,11] *p*-chlorophenylseleno,[4] and *m*-trifluoromethylseleno[4,5] compounds have also been used synthetically.

[4] K. B. Sharpless and M. W. Young, *J. Org. Chem.* **40**, 947 (1975).
[4a] K. B. Sharpless and R. F. Lauer, *J. Am. Chem. Soc.* **95**, 2697 (1973).
[5] H. J. Reich, J. M. Renga, and I. L. Reich, *J. Am. Chem. Soc.* **97**, 5434 (1975).
[6] D. G. Foster, *Org. Synth., Collect. Vol.* **3**, p. 771 (1955).
[7] G. Ayrey, D. Barnard, and D. T. Woodbridge, *J. Chem. Soc.* p. 2089 (1962).
[8] H. J. Reich, *J. Org. Chem.* **39**, 428 (1974).
[9] K. B. Sharpless and R. F. Lauer, *J. Org. Chem.* **39**, 429 (1974).
[10] H. J. Reich and J. M. Renga, *J. Org. Chem.* **40**, 3313 (1975).
[10a] H. J. Reich, J. M. Renga, and J. E. Trend, *Tetrahedron Lett.* p. 2217 (1976).
[11] P. A. Grieco, K. Hiroi, J. J. Reap, and J. A. Noguez, *J. Org. Chem.* **40**, 1450 (1975).

The selenenyl halides can be converted into selenenamides,[10,10a] or to selenenyl acetates[9] and trifluoroacetates.[8] Selenenyl chloride is converted to seleninyl chloride upon ozonization.[5,7] Seleninic anhydride is similarly prepared by ozonization of diselenide.[7]

Selenoxide syn eliminations and [2,3] sigmatropic rearrangements produce selenenic acid (PhSeOH) as the initial selenium containing fragment. While some stable aromatic selenenic acids with powerful electron-withdrawing substitutents have been isolated,[12–12c] benzeneselenenic acid itself (PhSeOH) has never been isolated or detected as an intermediate. It apparently disproportionates rapidly to diselenide and seleninic acid.[13]

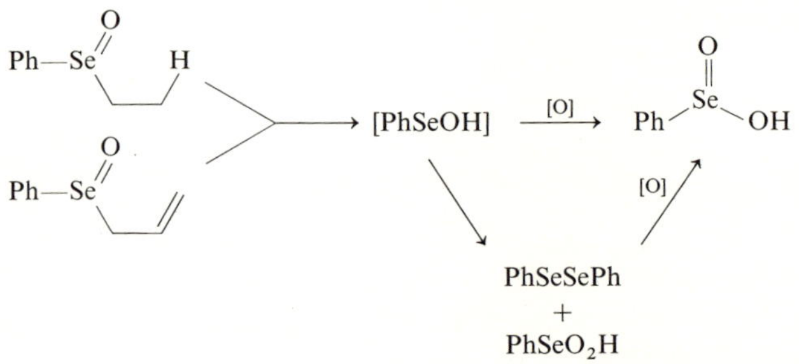

In many of the procedures described in this chapter the phenylseleno group is eventually converted to benzeneseleninic acid. This is easily converted back to diphenyl diselenide by reduction with acidic bisulfite[4] or thiosulfate.[5] The expensive selenium reagent can thus be recovered in good yield if desired.

Several different polymers containing phenylseleno groups have been prepared by Kato, Michels, and Heitz[14,14a] for the purpose of carrying out selenium based transformations. The polymers were prepared either from cross-linked polystyrene by bromination and nucleophilic displacement of bromine by NaHSe, by diazotization of poly-*p*-aminostyrene followed by treatment with KSeCN, or by copolymerization of divinylbenzene with *p*-vinylselenophenol. The polymeric reagents have the potential of alleviating the odor and toxicity problems of handling organoselenium compounds

[12] H. Rheinboldt and E. Giesbrecht, *Chem. Ber.* **89**, 631 (1956).
[12a] H. Rheinboldt and E. Giesbrecht, *Chem. Ber.* **88**, 1037 (1955).
[12b] H. Rheinboldt and E. Giesbrecht, *Chem. Ber.* **88**, 1974 (1955).
[12c] H. Rheinboldt and E. Giesbrecht, *Chem. Ber.* **88**, 666 (1955).
[13] O. Behaghel and H. Seibert, *Chem. Ber.* **66**, 708 (1933).
[14] M. Kato, R. Michels, and W. Heitz, *Polym. Lett.* **14**, 413 (1976).
[14a] R. Michels, M. Kato, and W. Heitz, *Makromol. Chem.* **177**, 2311 (1976).

as well as simplifying purification of reaction mixtures and recovery of selenium reagents.

II. Oxidation–Reduction Reactions of Organoselenium Compounds

Most of the oxides, imides, oxy acids, and their derivatives present in organosulfur chemistry are also known for selenium. Large differences do, however, exist between sulfur and selenium in the ease of various oxidation and reduction reactions, and in the stability of certain types of compounds. Particularly striking is the large preference of selenium for the Se(IV) oxidation state; only with difficulty can oxidation to Se(VI) be achieved in most cases. A second difference of some consequence is the greater stability of Se(IV) species with four σ bonds (selenuranes)[15] compared to analogous S(IV) compounds (sulfuranes).

A. SELENOLS, DISELENIDES, SELENENIC ACIDS, AND SELENINIC ACIDS

Scheme 2 illustrates some of the important oxidation–reduction relationships in this series of compounds, and gives references with provide experimental details for use of some of the key reagents. Additional references can be found in reviews.[16–16c]

Hydrogen selenide, alkyl selenols, and aryl selenols (RSeH) are easily oxidized (even by air) to diselenides. They are usually prepared *in situ* and used directly. Only a few selenenic acids stable enough for isolation are known.[12–12c,17,18] They normally disproportionate under conditions of their formation, giving diselenides and seleninic acids, or are further oxidized to the latter.

Many of the oxidizing agents are similar to those that would be effective in the sulfur series, except that oxidation to sulfonic acids is a more facile process than oxidation to selenonic acids. Many of the reducing reagents and conditions, however, are much milder than would necessary for corresponding sulfur reductions.

[15] H. J. Reich, *J. Am. Chem. Soc.* **95**, 964 (1973).

[16] T. W. Campbell, H. G. Walker, and G. M. Coppinger, *Chem. Rev.* **50**, 279 (1950).

[16a] H. Rheinboldt, *Methoden Org. Chem.* (*Houben-Weyl*), 4th Ed. **9**, 919 (1955).

[16b] D. L. Klayman and W. H. H. Günther, eds., "Organic Selenium Compounds: Their Chemistry and Biology." Wiley, New York, 1973.

[16c] K. J. Irgolic and M. V. Kudchadker, *in* "Selenium" (R. W. Zingaro and W. Cooper, eds.), p. 408. Van Nostrand-Reinhold, Princeton, New Jersey, 1974.

[17] D. L. Klayman and J. W. Lown, *J. Org. Chem.* **31**, 3396 (1966).

[18] W. Jenny, *Helv. Chim Acta* **41**, 317 and 326 (1958); G. Hölzle and W. Jenny, *Helv. Chim. Acta* **41**, 331 (1958).

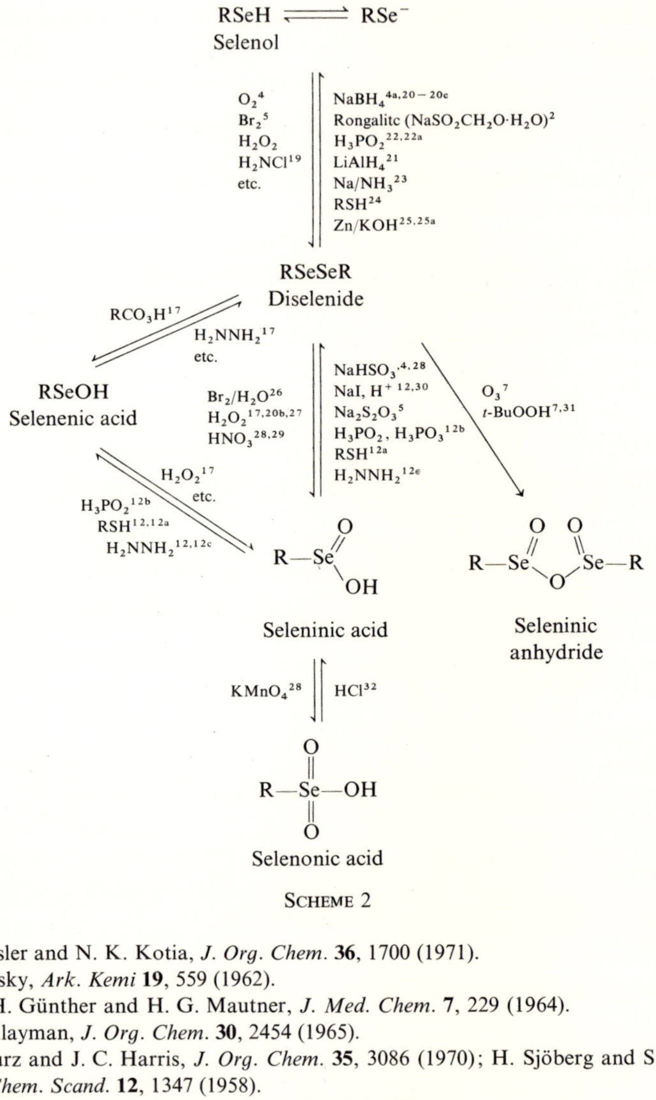

SCHEME 2

[19] H. H. Sisler and N. K. Kotia, *J. Org. Chem.* **36**, 1700 (1971).

[20] G. Zdansky, *Ark. Kemi* **19**, 559 (1962).

[20a] W. H. H. Günther and H. G. Mautner, *J. Med. Chem.* **7**, 229 (1964).

[20b] D. L. Klayman, *J. Org. Chem.* **30**, 2454 (1965).

[20c] J. L. Kurz and J. C. Harris, *J. Org. Chem.* **35**, 3086 (1970); H. Sjöberg and S. Herdevall, *Acta Chem. Scand.* **12**, 1347 (1958).

[21] G. Bergson and A.-L. Delin, *Ark. Kemi* **18**, 441 (1962).

[22] W. H. H. Günther, *J. Org. Chem.* **31**, 1202 (1966).

[22a] M. Schmidt and H. D. Block, *Chem. Ber.* **103**, 3348 (1970).

[23] G. E. Coates, *J. Chem. Soc.*, p. 2839 (1953); E. E. Aynsley, N. N. Greenwood, and J. B. Leach, *Chem. Ind.* (*London*) p. 379 (1966).

[24] W. H. H. Günther, *J. Org. Chem.* **32**, 3931 (1967).

[25] R. Lesser and R. Weiss, *Chem. Ber.* **45**, 1835 (1912); **47**, 2522 (1914); D. L. Klayman and W. H. H. Günther, eds., "Organic Selenium Compounds: Their Chemistry and Biology." p. 962. Wiley, New York, 1973.

Benzeneseleninic anhydride has been used for the oxidative hydrolysis of dithiolanes[32a] and hydrazones[32b] to ketones, for the oxidation of phenols to *o*-quinols and *o*-quinones (Section IV, A, 6) and of amines to imines (Section III, J, 1).

Areneseleninic acids in the presence of hydrogen peroxide form areneperoxyseleninic acids, which are useful oxidizing agents similar to peroxycarboxylic acids in reactivity (see Section II, B, 1).

B. SELENIDES AND SELENOXIDES

1. OXIDATION OF SELENIDES AND SELENOXIDES

Typical reagents for the oxidation and reduction of selenides, selenoxides, and selenones are shown in Scheme 3. Only the most vigorous oxidizing agents cause transformation of selenoxides to selenones.

Hydrogen peroxide in tetrahydrofuran (THF), ethanol, or dichloromethane has been by far the most commonly used oxidant. The reaction is acid catalyzed and thus a small amount of acetic acid is frequently added. A particularly effective catalyst is $PhSeO_2H$, perhaps because of the formation of perseleninic acid by the equilibration shown below.

$$\mathrm{Ph{-}\overset{\displaystyle O}{\overset{\|}{Se}}{-}OH} + H_2O_2 \rightleftharpoons \mathrm{Ph{-}\overset{\displaystyle O}{\overset{\|}{Se}}{-}OOH} + H_2O$$

In fact, selenides which cannot easily form seleninic acid are oxidized only very slowly by hydrogen peroxide, whereas oxidization of selenides to selenoxides which undergo facile syn elimination, is strongly autocatalytic. These observations suggest that perseleninic acid may be the active oxidant when seleninic acid is present.[5] This factor, together with the great exothermicity of the reaction, makes it essential that larger scale oxidations using H_2O_2 be carried out very cautiously. Cooling much below room temperature should be avoided since oxidation is very slow at 0°C and excess H_2O_2 may build up in solution. The most frequently used procedure involves addition

---

[25a] J. W. Baker, G. F. C. Barrett, and W. T. Tweed, *J. Chem. Soc.*, p. 2831 (1952).

[26] O. Behaghel and H. Seibert, *Chem. Ber.* **66**, 708 (1933).

[27] J. D. McCullough and E. S. Gould, *J. Am. Chem. Soc.* **71**, 674 (1949); E. S. Gould and J. D. McCullough, *ibid.* **73**, 1109 (1951).

[28] F. L. Pyman, *J. Chem. Soc.* **115**, 166 (1919).

[29] R. Paetzold and E. Rönsch, *Z. Anorg. Allg. Chem.* **338**, 195 (1965).

[30] J. D. McCullough, T. W. Campbell, and N. J. Krilanovich, *Anal. Chem.* **18**, 638 (1946).

[31] D. T. Woodbridge, *J. Chem. Soc., Phys. Org.* p. 50 (1966).

[32] H. W. Doughty, *Am. Chem. J.* **41**, 329 (1909).

[32a] D. H. R. Barton, N. J. Cussans, and S. V. Ley, *Chem. Commun.* p. 751 (1977).

[32b] D. H. R. Barton, D. J. Lester, and S. V. Ley, *Chem. Commun.* p. 445 (1977).

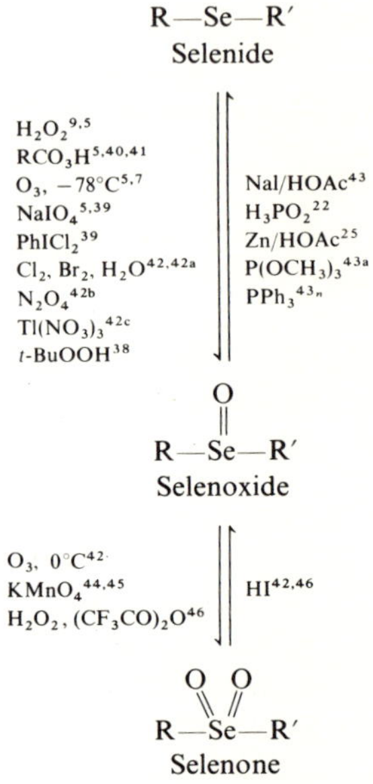

SCHEME 3

of 30% H_2O_2 to a THF or dichloromethane solution of the selenide. The two phase procedure using dichloromethane is especially recommended since the necessity for H_2O_2 phase transfer tends to moderate the rate of oxidation.[5,33] Selenoxide eliminations also proceed more rapidly in aprotic solvents than in aqueous or alcoholic media.

[33] H. J. Reich, *J. Org. Chem.* **40**, 2570 (1975).

[34] K. B. Sharpless, K. M. Gordon, R. F. Lauer, D. W. Patrick, S. P. Singer, and M. W. Young, *Chem. Scr.* **8A**, 9 (1975).

[34a] J. R. Williams and J. D. Leber, *Synthesis* p. 427 (1977).

[34b] V. Balogh, J.-C. Beloeil, and M. Fetizon, *Tetrahedron* **33**, 1321 (1977).

[35] H. J. Reich, J. M. Renga, and I. L. Reich, *J. Org. Chem.* **39**, 2133 (1974).

[36] P. A. Grieco, private communication.

[37] P. A. Grieco, Y. Yokoyama, S. Gilman, and M. Nishizawa, *J. Org. Chem.* **42**, 2034 (1977).

[37a] H. J. Reich, F. Chow, and S. L. Peake, *Synthesis* in press. (1978).

[37b] T. Kametani, H. Nemoto, and K. Kukumoto, *Heterocycles* **6**, 1365 (1977).

[38] M. Hori and K. B. Sharpless, *J. Org. Chem.* **43**, 1689 (1978).

[38a] H. J. Reich, S. Wollowitz, J. E. Trend, F. Chow, and D. T. Wendelborn, *J. Org. Chem.* **43**, 1697 (1978).

[39] M. Cinquini, S. Colonna, and G. Giovini, *Chem. Ind. (London)*, p. 1737 (1969).

[40] K. B. Sharpless, R. F. Lauer, and A. Y. Teranishi, *J. Am. Chem. Soc.* **95** 6137 (1973).

[41] H. J. Reich and S. K. Shah, *J. Am. Chem. Soc.* **97**, 3250 (1975).

[42] R. Paetzold and G. Bochmann, *Z. Anorg. Allg. Chem.* **360**, 293 (1968).

[42a] D. G. Foster, *Recl. Trav. Chim. Pays-Bas* **54**, 447 (1935).

Lactones have been isolated as products from oxidation of steroidal α-phenylselenocyclopentanones,[34–34c] and -cyclobutanones.[35] Benzeneperseleninic acid may be responsible for these Baeyer–Villiger side reactions. In fact, $PhSeO_2H$ in the presence of H_2O_2 will oxidize some ketones to lactones,[36] and some olefins to epoxides.[37–38] Under optimum conditions (30% H_2O_2, dichloromethane) the $ArSeO_2H/H_2O_2$ reagent is a cheap and selective oxidizing agent. Only a catalytic amount of seleninic acid is necessary.[37a,38] Benzeneseleninic acid and the 2-nitro- and 2,4-dinitrobenzeneseleninic acids have been used; the reactivity increases dramatically as electron attracting substituents are added. Several oxidations are shown below.

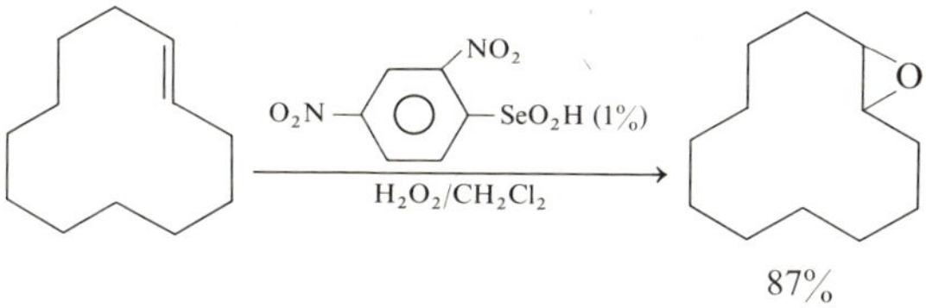

(ref. 37a)

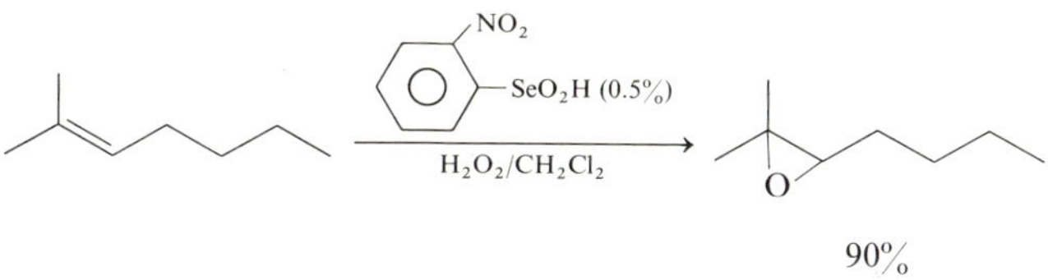

(ref. 38)

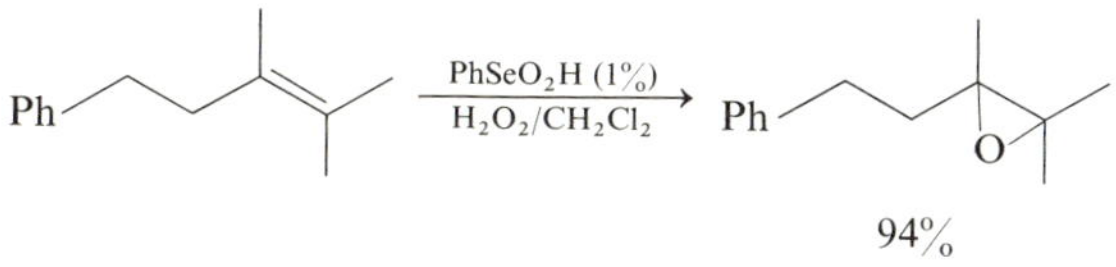

(ref. 37a)

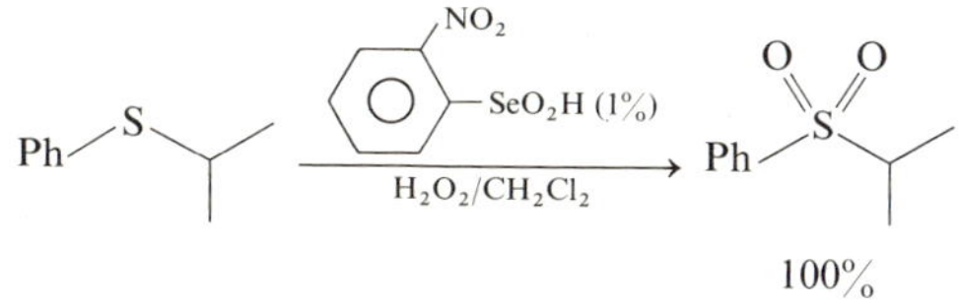

(ref. 37a)

[42b] R. Paetzold, U. Lindner, G. Bochmann, and P. Reich, *Z. Anorg. Allg. Chem.* **352**, 295 (1967).

[42c] Y. Nagao, M. Ochiai, K. Kaneko. A. Maeda, K. Watanabe, and E. Fujita, *Tetrahedron Lett.* p. 1345 (1977).

[43] H. J. Reich and F. Chow, *Chem. Commun.*, 790 (1975).

[43a] H. J. Reich and F. Chow, unpublished results.

[43b] H. Tamagaki, I. Hatanaka, and K. Tamura, *Chemistry Lett.* p. 81 (1976).

[44] J. Loevenich, H. Fremdling, and M. Föhr, *Chem. Ber.* **62**, 2856 (1929).

[45] H. Rheinboldt and E. Giesbrecht, *J. Am. Chem. Soc.* **68**, 2671 (1946).

[46] L. M. Yagupol' skii, G. P. Syrova, V. G. Voloshchuk, and V. F. Bystrov, *Zh. Obshch. Khim.* **38**, 2509 (1968).

An additional complication encountered when using hydrogen peroxide is that it is catalytically destroyed by selenoxides under some conditions, requiring the use of a large excess of oxidant.[38a]

Sodium metaperiodate is a mild reagent for the formation of selenoxides[39] and has been used for especially sensitive systems.[5,40] The necessity of working in aqueous media has disadvantages for high molecular weight selenides. Since oxidation is slow below room temperature, elimination accompanies oxidation. The reaction mixture can be buffered with sodium bicarbonate, but use of stronger base makes the oxidation inconveniently slow.

Peroxycarboxylic acids are effective oxidizing agents for selenides. They can be used in the presence of olefins or amines. Sharpless, Lauer, and Teranishi[40] found peracetic acid useful for the preparation of α,β-unsaturated esters but obtained lower yields for enones than were given by H_2O_2. *m*-Chloroperbenzoic acid (*m*-CPBA) is an especially useful reagent when it is necessary to perform oxidations at low temperatures. Oxidation of phenyl alkyl selenides to selenoxides occurs at $-78°C$ in less than 30 minutes in dichloromethane or THF solution. Low temperature oxidation allows the chemical manipulation of unstable selenoxides or the adjustment of reaction conditions for syn elimination in ways incompatible with concomitant oxidation and elimination. Under conditions where excess oxidant is not present during the syn elimination, the initially formed selenenic acid (PhSeOH) disproportionates to give a 1:1 mixture of seleninic acid ($PhSeO_2H$) and diselenide.

Ozone is another oxidant which can be used to cleanly form selenoxides at low temperature and which forms no interfering by-products during oxidation. Over-oxidation to selenones occurs only slowly at temperatures of 0°C or higher.[42] When other easily oxidizable functions such as double bonds or sulfides are absent, the blue endpoint of excess ozone can be used (ozonizations should not be carried out in THF since it is attacked by ozone at $-78°C$).

A little explored option is the use of the selenimide elimination instead of the selenoxide elimination. Sharpless *et al.*[34,46a] have caused elimination of dodecyl phenyl selenide to dodecene by treatment with Chloramine-T under phase transfer conditions. A number of *N*-tosyl selenimides have previously been prepared.[47]

[46a] M. W. Young, Ph.D. Thesis, Massachusetts Institute of Technology, Cambridge, Mass. (1975).

[47] D. Spinelli and G. Monaco, *Boll. Sci. Fac. Chim. Ind. Bologna* **20**, 56 (1962); *Chem. Abstr.* **58**, 4457f (1963); D. Hellwinkel and G. Fahrbach, *Justus Liebigs Ann. Chem.* **715**, 68 (1968); N. Derkach, *Zh. Org. Khim.* **10**, 807 (1974); S. Tamagaki, S. Oae, and K. Sakaki, *Tetrahedron Lett.* p. 649 (1975).

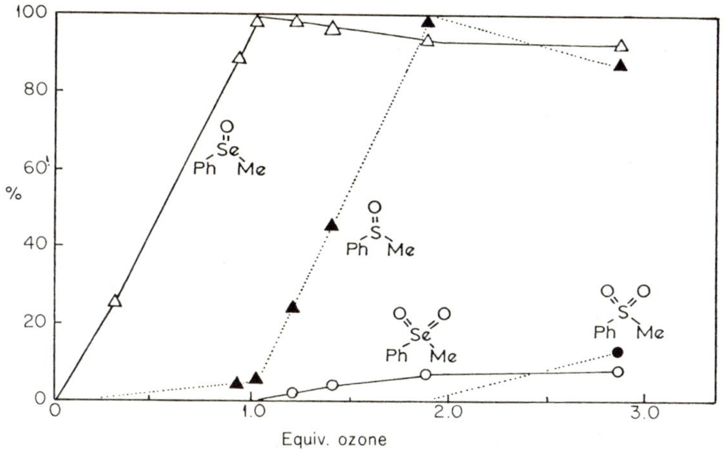

FIG. 1. Ozonization of methyl phenyl sulfide and methyl phenyl selenide in CH_2Cl_2 at 0°C.

2. Selectivity of Selenide Oxidations

Hydrogen peroxide can be used in the presence of a wide variety of potentially oxidizable functional groups, including sulfides,[48] sulfoxides,[5] amines,[48] alcohols (see Tables II and IV), ketones, and enones (Table VI). Some special problems arise when H_2O_2 reacts with $PhSeO_2H$ (see Section II, B, 1).

Sodium metaperiodate[39] also reacts with few oxidizable organic functional groups. This reagent has been used in the presence of an aldehyde.[40]

A detailed study of the ozonization of a range of selenium compounds by Ayrey, Barnard, and Woodbridge[7] showed that ozonolysis of dialkyl selenides proceeds at least 50 times as fast as the ozonolysis of olefins. Sulfides, on the other hand, are ozonized less readily than olefins.[49] Ozone is thus the reagent of choice for the oxidation of selenides in the presence of sulfides. In fact, quite selective oxidation of selenide in a sulfide–selenide mixture can be achieved at 0°C (Fig. 1).[50] At lower temperature (Fig. 2), particularly in concentrated solutions, the reaction becomes much less selective, apparently because of the intervention of peroxy sulfoxides formed by reaction of sulfide with singlet oxygen[51] or "singlet oxygenoids" in solution. This peroxysulfoxide is quite specific for the oxidation of selenoxides to selenones

[48] C. A. Wilson, II and T. A. Bryson, *J. Org. Chem.* **40**, 800 (1975).
[49] D. Barnard, *J. Chem. Soc.* p. 4547 (1957).
[50] H. J. Reich, D. Wendelborn, and S. C. Zimmerman, unpublished results.
[51] C. S. Foote and J. W. Peters, *J. Am. Chem. Soc.* **93**, 3795 (1971).

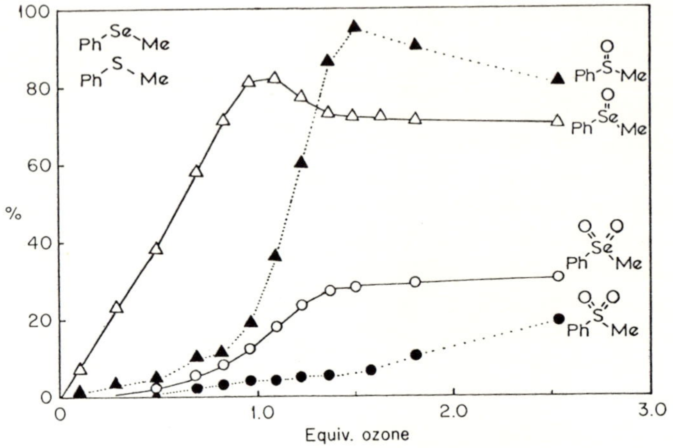

FIG. 2. Ozonization of methyl phenyl sulfide and methyl phenyl selenide in CH_2Cl_2 at $-78°C$.

(selenide or selenoxide in the absence of sulfide is not oxidized to selenone by ozone under these conditions, nor are sulfoxides appreciably converted to sulfones).

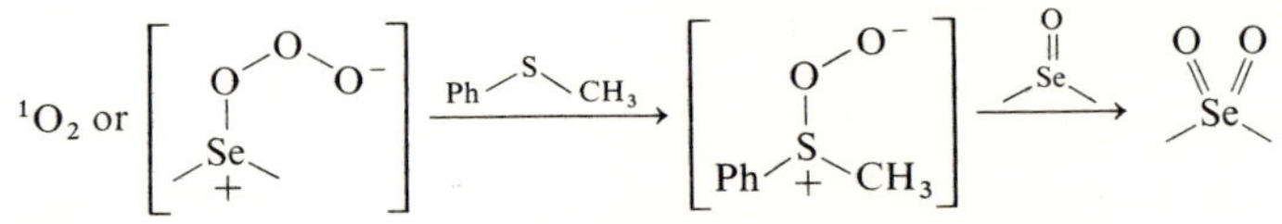

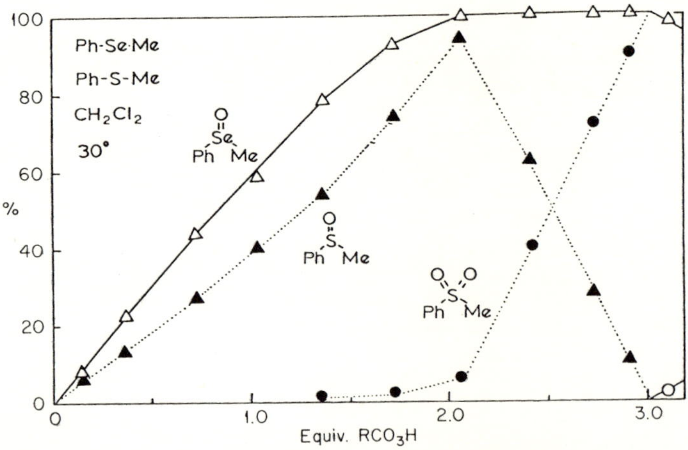

FIG. 3. Oxidation of methyl phenyl sulfide and methyl phenyl selenide with *m*-chloroperbenzoic acid.

m-Chloroperbenzoic acid has been used to oxidize selenides in the presence of olefin,[41] acetylene,[52] amino,[10,53] and alcohol functions. It was found to be superior to ozone for oxidation of an olefinic selenide.[41] Little selectivity between sulfide and selenide is shown (Fig. 3).

The oxidation of sulfide in the presence of selenide has not been achieved. In principle, however, this can be accomplished indirectly by taking advantage of the great ease of reduction of selenoxides, or of the ability of selenoxides to oxidize sulfides to sulfoxides.[54]

One example of a Jones oxidation in which an alcohol and selenide group were simultaneously oxidized to ketone and selenoxide,[55] and several examples in which allylic alcohols were oxidized to enones without affecting a phenylseleno group in the same molecule have been reported.[55,56]

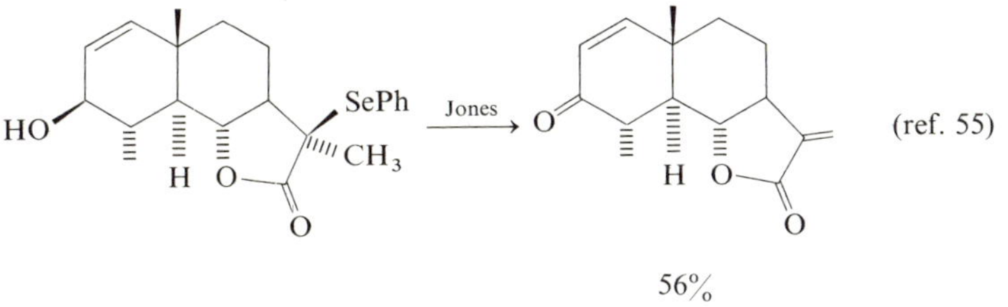

3. Selenoxides as Oxidizing Agents

Reductions of selenoxides can be readily accomplished using a large variety of reducing reagents, usually under mildly acidic conditions. In fact, it is possible to get inadvertent reductions of selenoxides during transformations in which selenenic acids or particularly selenols are generated. Such side reactions can be prevented by the presence of weak base or circumvented by using excess oxidant.

The ease with which selenoxides are reduced to selenides permits their utilization as mild and selective oxidizing agents. Such reactions are invariably performed under acidic conditions.

Barnard and Woodbridge[54] have reported that a variety of sulfides are oxidized to sulfoxides by selenoxides in acetic acid at room temperature. This reaction was considered to involve nucleophilic attack by sulfur on a protonated selenoxide. No sulfones are formed.

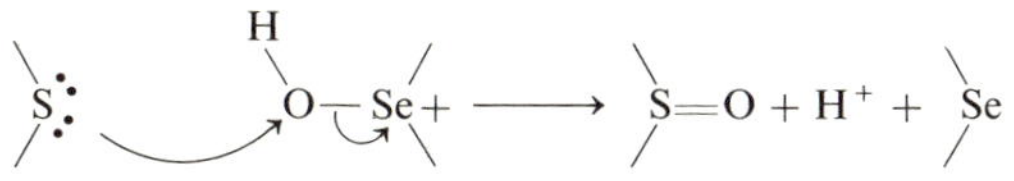

52 H. J. Reich and S. K. Shah, *J. Am. Chem. Soc.* **99**, 263 (1977).
53 E. Vedejs, J. M. Renga, M. Arco, and K. Spear, private communication.
54 D. Barnard and D. T. Woodbridge, *Chem. Ind.* (*London*) p. 1603 (1959).
55 K. Yamakawa, K. Nishitani, and T. Tominaga, *Tetrahedron Lett.* p. 2829 (1975).
56 K. Yamakawa, T. Tominaga, and K. Nishitani, *Tetrahedron Lett.* p. 4137 (1975).

The oxidation of olefin to a glycol has been reported using a polymer-bound diaryl selenoxide.[14a] It is possible that this oxidation involved a perseleninic acid (see Section II, B, 1) which could be present on the polymer in small amounts.

Balenovic and co-workers[57–57e] have published a series of papers in which diphenyl selenoxide was used to carry out oxidations. Typical examples include various oxidations of hydrazides,[57,57a] transformations of tertiary amines to amine oxides,[57c] dehydrogenation of L-ascorbic acid,[57b] and oxidation of catechols to *o*-quinones.[57d,e]

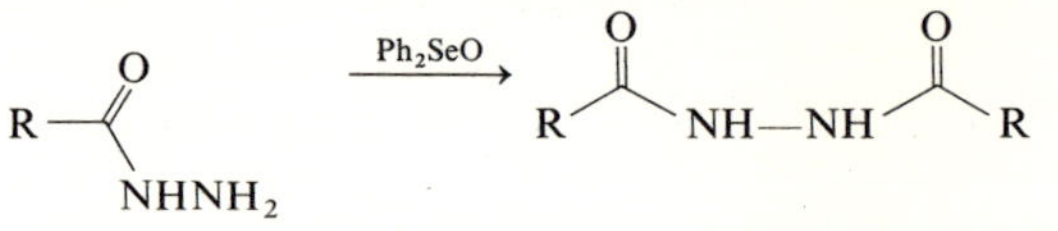

(ref. 57)

R = Ph, *m*-$NO_2C_6H_4$, PhNH, $HO_2CCHNH_2CH_2$—

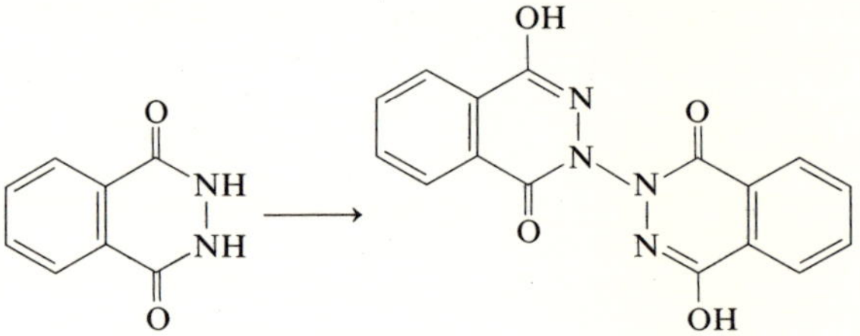

(ref. 57a)

$$R_3N \xrightarrow{R'_2SeO} R_3\overset{+}{N}-\overset{-}{O} \qquad \text{(ref. 57c)}$$

R_3N = Me_3N, Et_3N, brucine, strychnine, nicotinamide

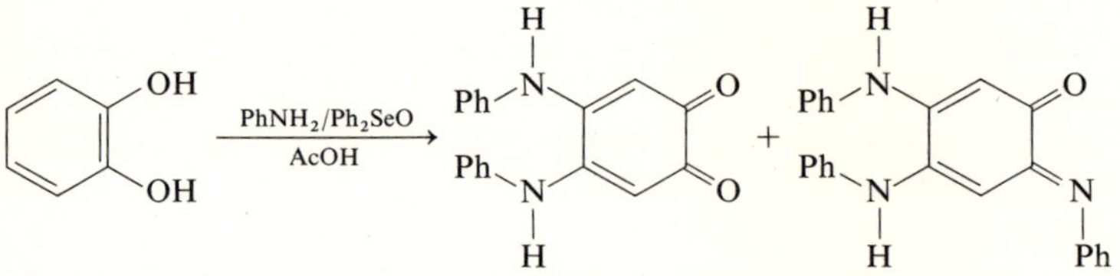

(ref. 57d)

13.5% 63%

[57] K. Balenovic, R. Lazic, V. Polak, and P. Stern, *Bull. Sci., Cons. Acad. Sci. Arts RSF Yougosl., Sect. A* **17**, 147 (1972); *Chem. Abstr.* **77**, 139499c (1972).

[57a] N. Bregant, I. Perina, and K. Balenovic, *Bull. Sci. Cons. Acad. Sci. Arts RSF Yougosl., Sect. A* **17**, 148 (1972); *Chem. Abstr.* **77**, 152092j (1972).

[57b] I. Perina, N. Bregant, and K. Balenovic, *Bull. Sci. Cons. Acad. Sci. Arts RSF Yougosl., Sect. A* **18**, 4 (1973); *Chem. Abstr.* **78**, 160,007w (1973).

[57c] M. Poje and K. Balenovic, *Bull. Sci. Cons. Acad. Sci. Arts RSF Yougosl., Sect. A*, **20**, 1 (1975); *Chem. Abstr.* **83**, 43558u (1975).

[57d] I. Perina, N. Bregant, and K. Balenovic, *Bull. Sci. Cons. Acad. Sci. Arts RSF Yougosl., Sect. A* **18**, 3 (1973); *Chem. Abstr.* **79**, 5096b (1973).

[57e] K. Balenovic, N. Bregant, and I. Perina, *Synthesis* p. 172 (1973).

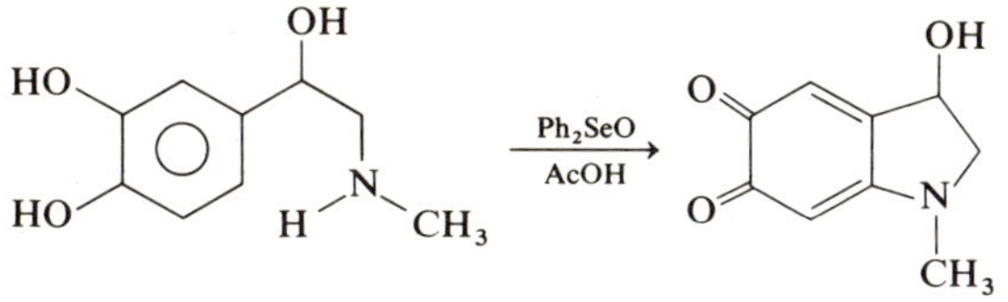
(ref. 57e)

High yields were obtained in the oxidation of adrenaline to adrenochrome.[57e]

Further work is needed to establish whether these oxidations offer any practical advantages over alternative methods, but they do emphasize the fact that selenoxides can serve as oxidizing agents for a number of functionalities under acidic conditions.

Since the metabolic role of selenium in mammals involves oxidations and reductions (at least one enzyme, glutathione peroxidase, which contains selenium at the active site, is known to catalyze hydroperoxide reduction),[58] there have been several studies of the redox properties of selenium analogs of sulfur containing amino acids.[59–59b] Thus, selenocysteine derivatives led to dehydroalanine upon oxidation but selenomethionine could be oxidized to a stable selenoxide, which spontaneously reverted to selenomethionine.[59,59a] The significance of observations such as the above,[59–59b] to the biochemical role of selenium will not be known until the mechanism of glutathione peroxidase action is better established.

III. The Selenoxide syn Elimination as a Double Bond Forming Reaction

A. INTRODUCTION

The sulfoxide syn elimination, discovered by Kingsbury and Cram[60] in 1960, has been the subject of numerous studies. The high temperature needed for elimination of alkyl aryl sulfoxides (100°–180°C) discouraged synthetic application except where special activating effects were present.

The first clear-cut case of a selenoxide elimination was provided by Huguet[61] in a largely overlooked paper published in 1967 that impinges in a number of ways on the chemistry discussed in this chapter. Oxidation of **1** gave 3-acetoxy-1-butene and a selenium compound tentatively identified as **2**.

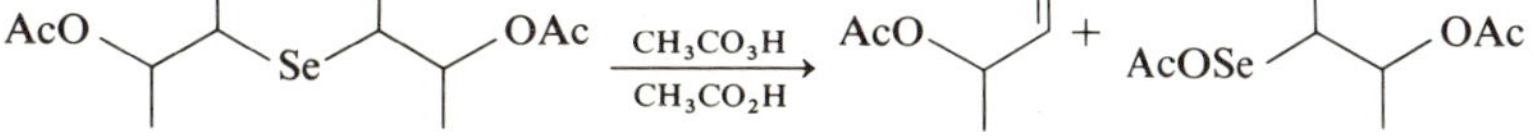
(1) **(2)**

[58] H. E. Ganther, *Chem. Scr.* **8A**, 79 (1975).
[59] R. Walter and J. Roy, *J. Org. Chem.* **36**, 2561 (1971).
[59a] R. Walter, I. L. Schwartz, and J. Roy, *Ann. N.Y. Acad. Sci.* **192**, 175 (1972).
[59b] K. A. Caldwell and A. L. Tappel, *Biochemistry* **3**, 1643 (1964).
[60] C. A. Kingsbury and D. J. Cram, *J. Am. Chem. Soc.* **82**, 1810 (1960).
[61] J. L. Huguet, *Adv. Chem. Ser.* **76**, 345 (1967).

Jones, Mundy, and Whitehouse[62] prepared and separated the diastereomeric selenoxides **3** and **4** (one of the first examples of such a separation)[63] and observed their decomposition to olefin at 0°–25°C, approximately 100°C lower than analogous sulfoxides.

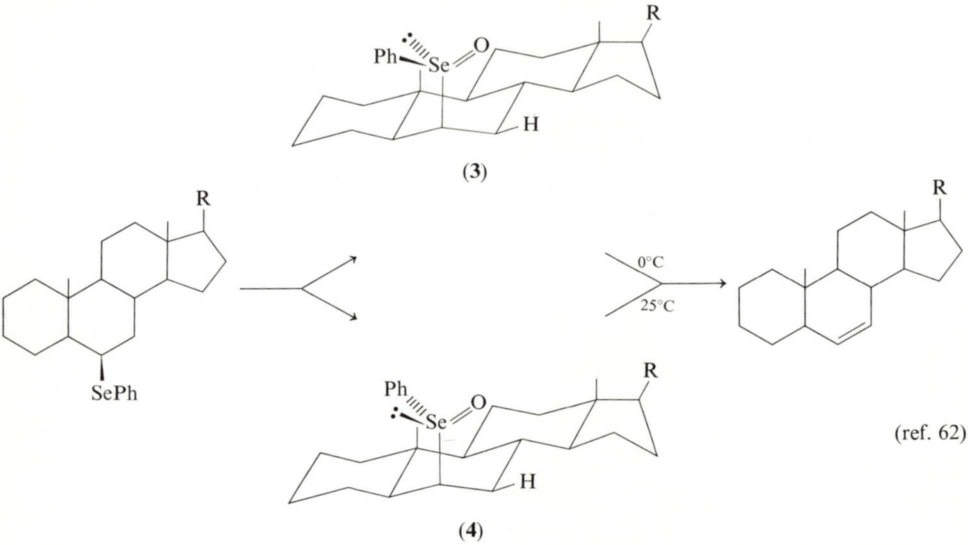

(ref. 62)

This work also suggested strong similarities between the stereochemical requirements for the sulfoxide and selenoxide eliminations. Sharpless, Young, and Lauer[64] showed that the reaction proceeded with the syn stereochemistry expected for a pericyclic reaction (Table I). These papers provided the background for the development of much of the synthetic methodology described in this chapter.

A number of alkyl selenoxides had been prepared prior to 1967 and their tendency to decompose noted.[42–42b,65,65a] In particular, dimethyl, and methyl phenyl selenoxide were easily prepared and were reasonably stable at room temperature, whereas the higher alkyl derivatives were invariably much less stable. Ethyl phenyl selenoxide, for example, decomposes on mild heating to give ethyl phenyl selenide, diphenyl diselenide, benzeneseleninic acid, and acetaldehyde.[42a] The aldehyde was identified only by Fuchsin test

[62] D. N. Jones, D. Mundy, and R. D. Whitehouse, *Chem. Commun.* p. 86 (1970).

[63] M. Cinquini, S. Colonna, and D. Landini, *Boll. Sci. Fac. Chim. Ind. Bologna* **27**, 207 (1969); *Chem. Anstr.* **72**, 89955x (1970).

[64] K. B. Sharpless, M. W. Young, and R. F. Lauer, *Tetrahedron Lett.* p. 1979 (1973).

[65] M. J. Bird and F. Challenger, *J. Chem. Soc.* p. 570 (1942).

[65a] O. K. Edwards, W. R. Gaythwaite, J. Kenyon, and H. Phillips, *J. Chem. Soc.* p.2293 (1928).

TABLE I

Olefins Prepared from Halides, Lactones, and Sulfonates by Selenoxide syn Elimination

Halide or sulfonate	$ArSe^-$ (oxidant)	Olefin	Yield[a] of olefin (%)	Ref.
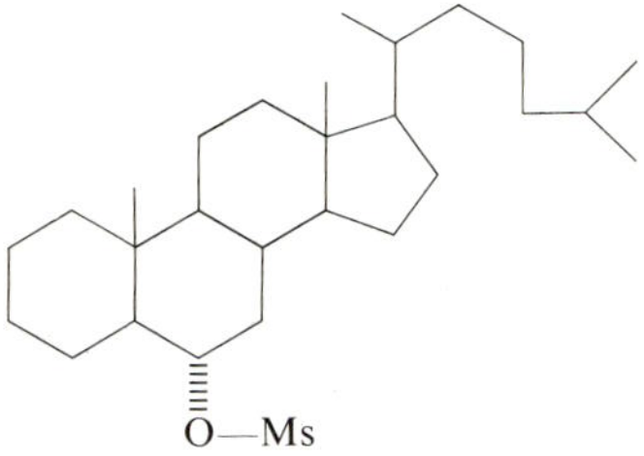	$PhSe^-$, EtOH (O_3, −78°C)	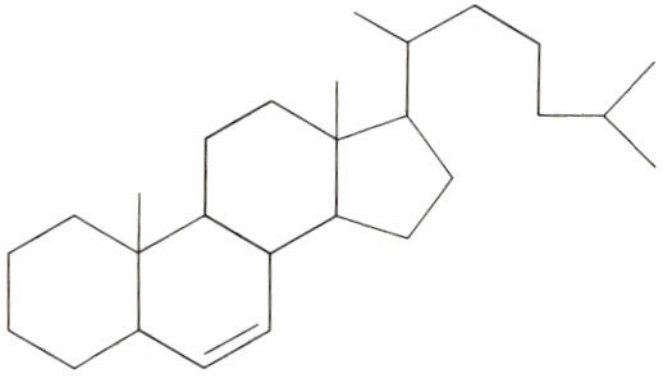	>95	62
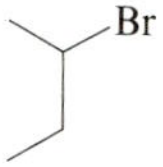	$PhSe^-$, EtOH (H_2O_2, THF, 0°C)		(43)	64
			(20)	
			(8)	
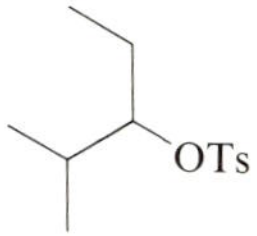	$PhSe^-$, EtOH (H_2O_2, THF, 0°C)	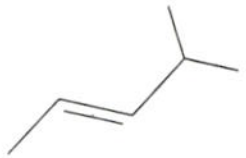	(79)	64
		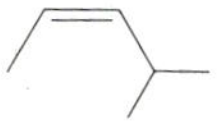	(7)	
		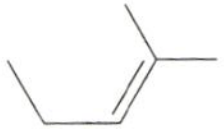	(14)	

(*cont.*)

TABLE I (*cont.*)

Halide or sulfonate	$ArSe^-$ (oxidant)	Olefin	Yield[a] of olefin (%)	Ref.
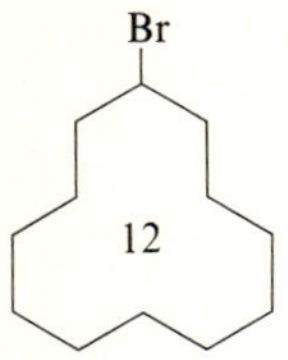	$PhSe^-$, EtOH (H_2O_2, THF)	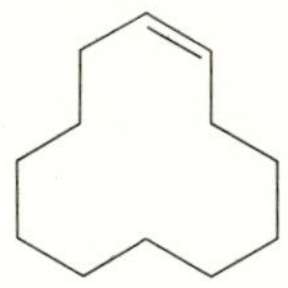	(47)	64
		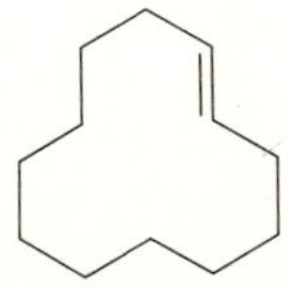	(47)	
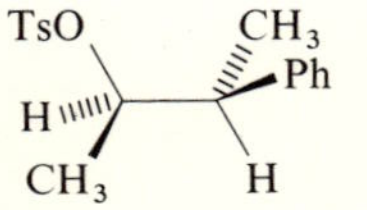	$PhSe^-$, EtOH (H_2O_2, THF, 0°C)	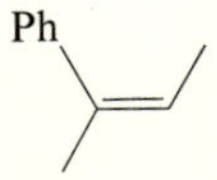	(15)	64
		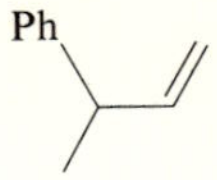	(42)	
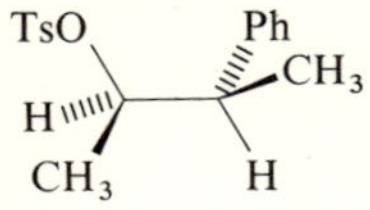	$PhSe^-$, EtOH (H_2O_2, THF, 0°C)	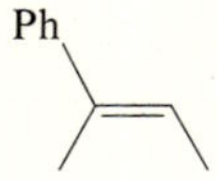	(13)	64
		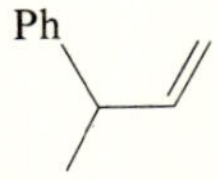	(45)	

$C_{10}H_{21}$—CH_2—CH_2—Br	$PhSe^-$, EtOH (H_2O_2, THF)	n-$C_{10}H_{21}$ 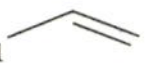	(77)	4, 64
$C_{10}H_{21}$—CH_2—CH_2—Br	$PhSe^-$, EtOH (Chloramine-T)[b]	n-$C_{10}H_{21}$ 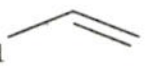	(77)	34
$C_{10}H_{21}$—CH_2—CH_2—Br	o-NO_2-$C_6H_4Se^-$ (H_2O_2, THF)	n-$C_{10}H_{21}$ 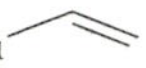	(91) 62	4
$C_{10}H_{21}$—CH_2—CH_2—Br	p-NO_2-$C_6H_4Se^-$ (H_2O_2, THF)	n-$C_{10}H_{21}$ 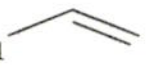	(77) 62	4
$C_{10}H_{21}$—CH_2—CH_2—Br	m-CF_3-$C_6H_4Se^-$ (H_2O_2, THF)	n-$C_{10}H_{21}$ 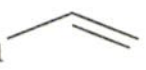	(70)	4
$C_{10}H_{21}$—CH_2—CH_2—Br	p-CH_3O-$C_6H_4Se^-$ (H_2O_2, THF)	n-$C_{10}H_{21}$ 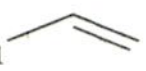	(60)	4
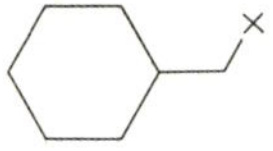	$PhSe^-$ (H_2O_2, THF)	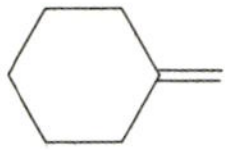	(46)	4
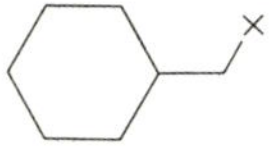	p-Cl-$C_6H_4Se^-$ (H_2O_2, THF)	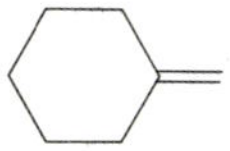	(85)	4
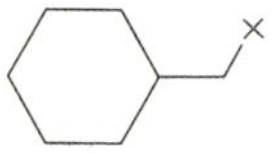	o-NO_2-$C_6H_4Se^-$ (H_2O_2, THF)	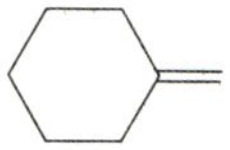	(92)	4
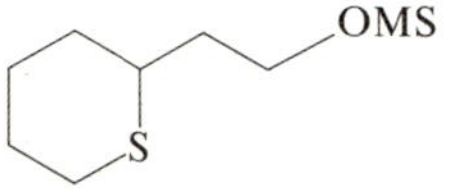	$PhSe^-$, EtOH (1. t-BuOCl, H_2O, NEt_3 2. Δ)	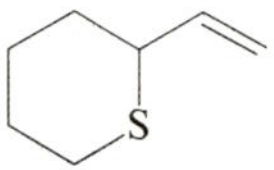	93 × 25[c]	53
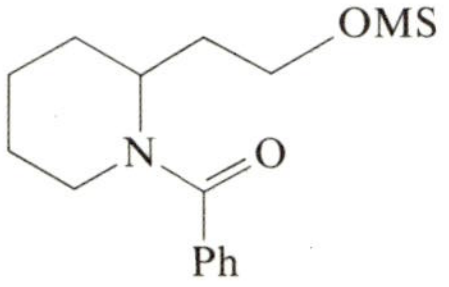	$PhSe^-$, EtOH (1. m-CPBA 2. Et_3N, Δ)	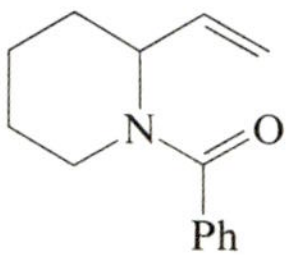	91 × 86	53

(cont.)

TABLE I (*cont.*)

Halide or sulfonate	$ArSe^-$ (oxidant)	Olefin	Yield[a] of olefin (%)	Ref.
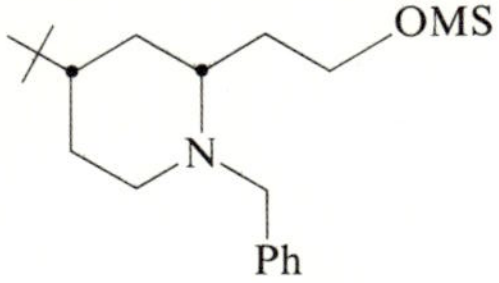	$PhSe^-$, EtOH (1. $HClO_4$, *m*-CPBA 2. Et_3N, Δ)	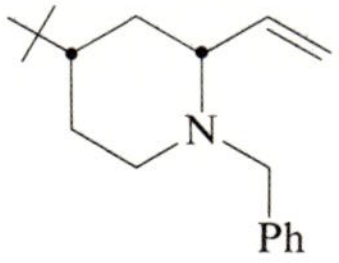	44	53
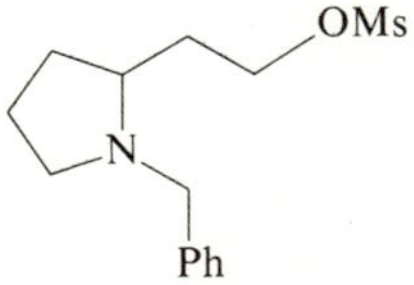	$PhSe^-$, EtOH (1. CF_3SO_3H, *m*-CPBA 2. Et_3N, Δ)	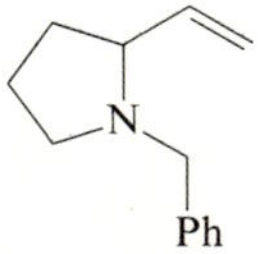	41 × 78	53
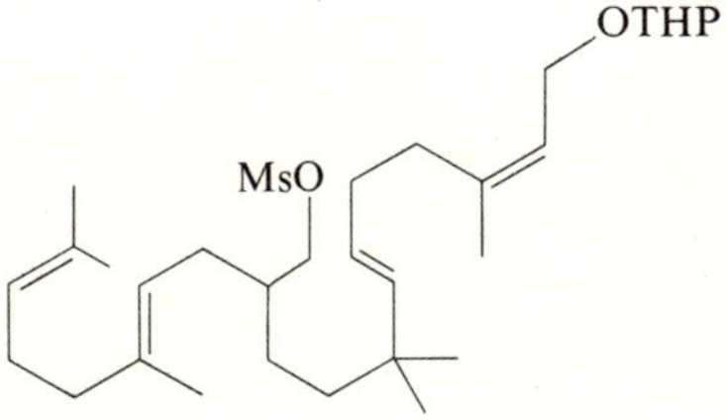	*o*-NO_2-$C_6H_4Se^-$ (H_2O_2, THF)	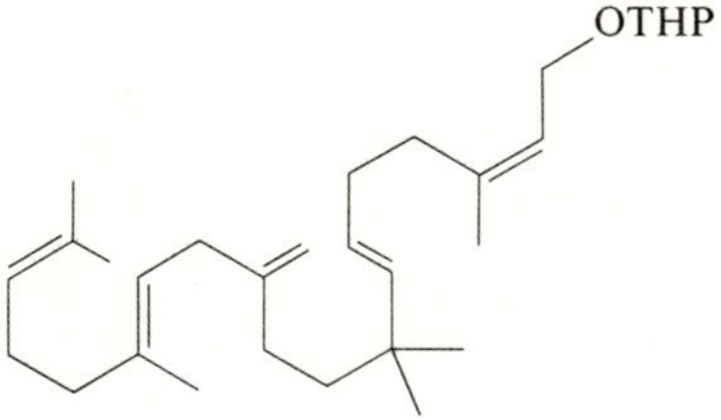	75 × 75	74
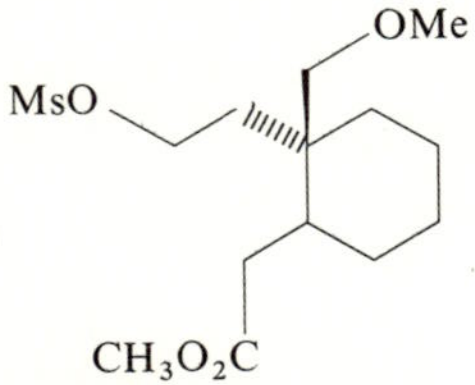	*o*-NO_2-$C_6H_4Se^-$, EtOH (H_2O_2, THF)	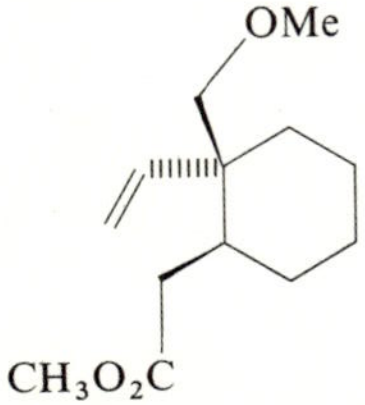	64 × 92	11

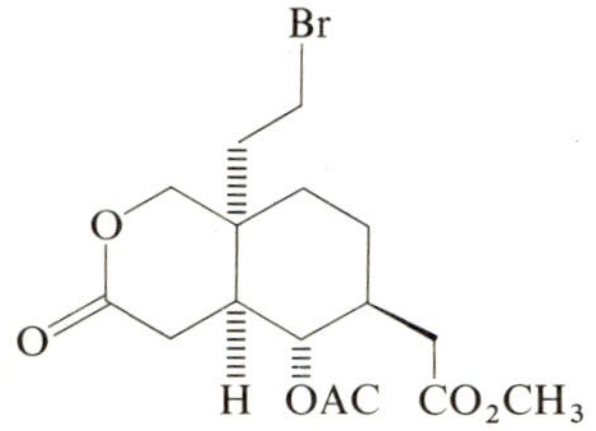	o-NO_2-$C_6H_4Se^-$, DMF (H_2O_2, THF)	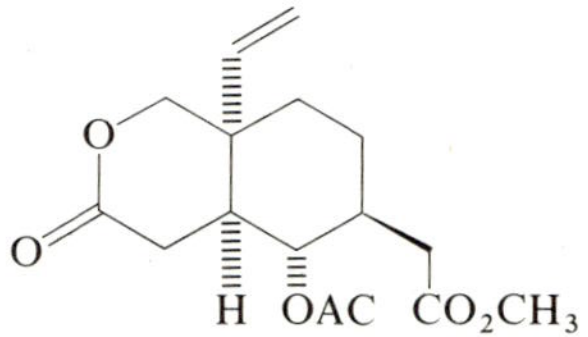	87 × >81	74a
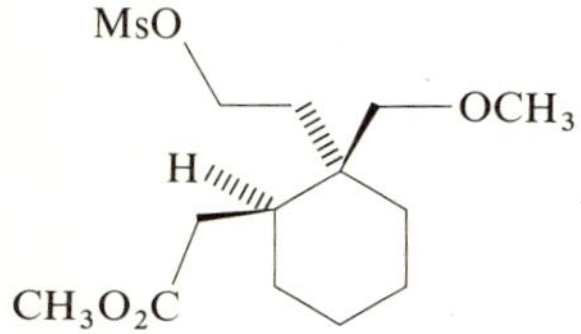	o-NO_2-$C_6H_4Se^-$ (H_2O_2, THF)	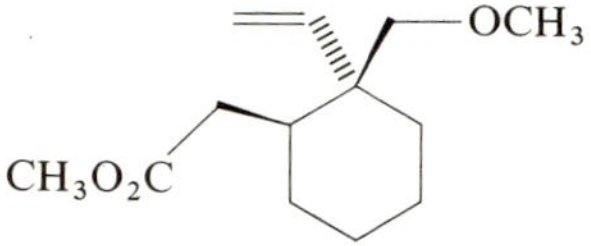	70 × 92	74b
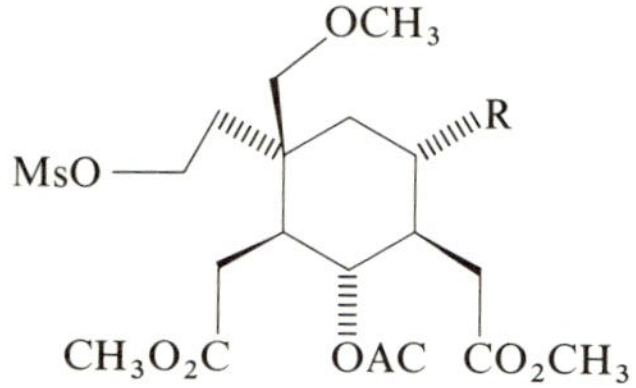	o-NO_2-$C_6H_4Se^-$, DMF (H_2O_2, THF)	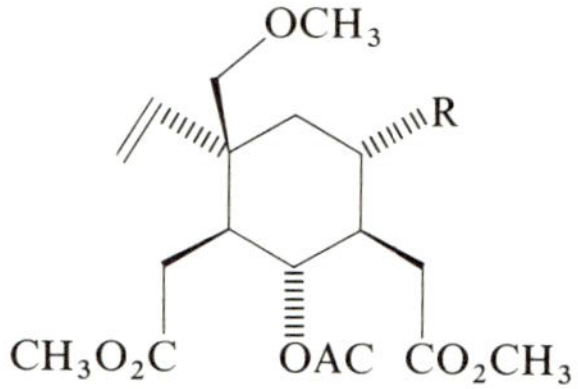	77 × 89 (R = OAc) 75 × 81 (R = H)	74c 74a
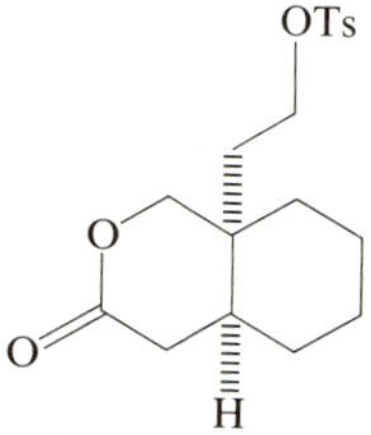	$PhSe^-$ (1. O_3, −78°C 2. Δ)	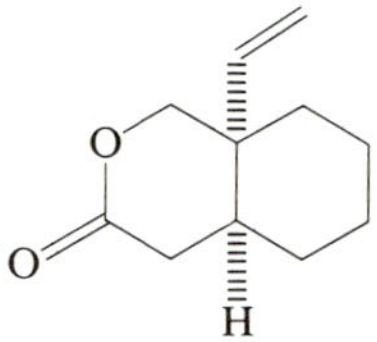	98 × 60	75

(*cont.*)

TABLE I (*cont.*)

Halide or sulfonate	$ArSe^-$ (oxidant)	Olefin	Yield[a] of olefin (%)	Ref.
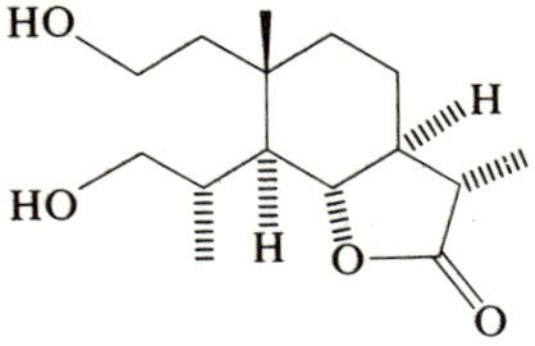	*o*-NO_2-C_6H_4SeCN, Bu_3P (50% H_2O_2, THF)	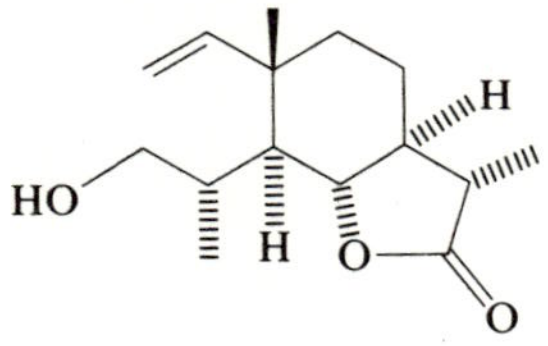	89 × 90	75a
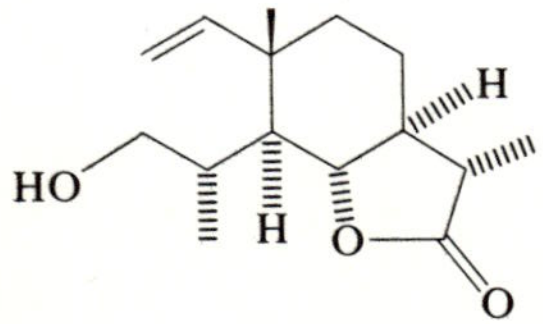	*o*-NO_2-C_6H_4SeCN, Bu_3P (50% H_2O_2, THF)	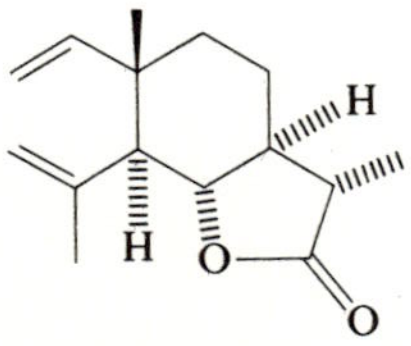	81 × 97	75a
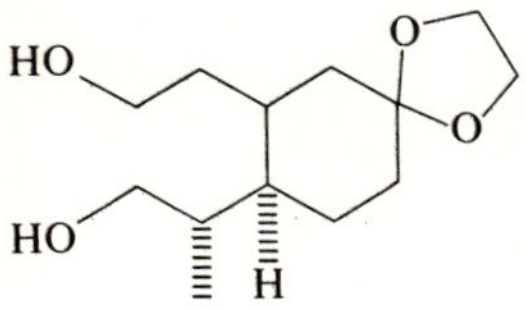	*o*-NO_2-C_6H_4SeCN, Bu_3P (50% H_2O_2, THF)	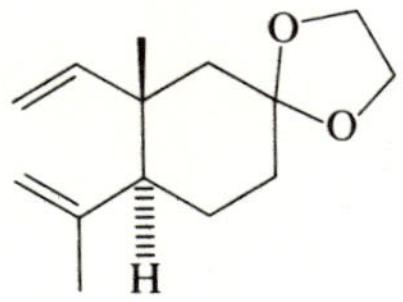	94 × 95	75b
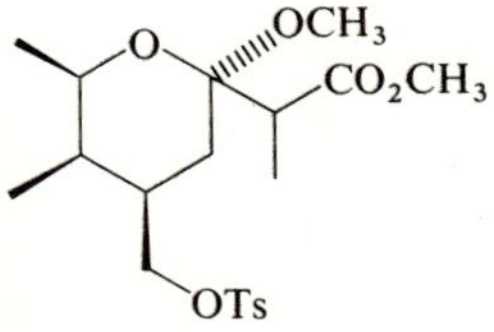	$PhSe^-$ (1. H_2O_2, THF 2. C_6H_6, NEt_3, Δ)	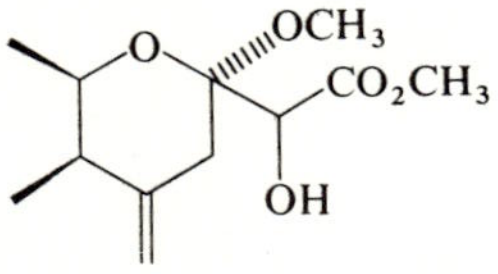	55 × 78	75c

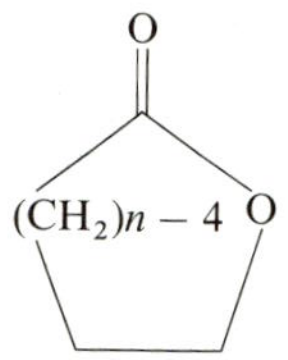	$PhSe^-$, DMF, 110°C (1. O_3 2. Δ, $CHCl_3$, Py)	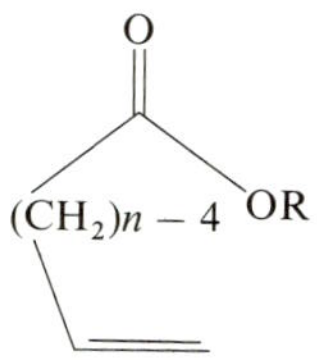		
		$n = 5$	98 × 70	75e
		$n = 6$	95 × 65	75e
		$n = 7$	90 × 71	75e
		$n = 8$	90 × 70.5	75e
		$n = 13$	55 × 60.5	75e
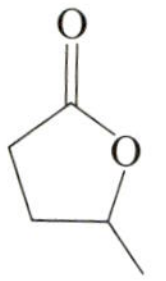	$PheSe^-$, DMF, 110°C (1. O_3 2. Δ, $CHCl_3$, Py)	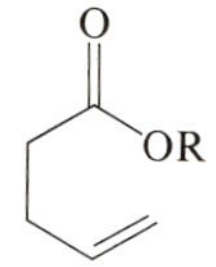	80 × 33	75e
		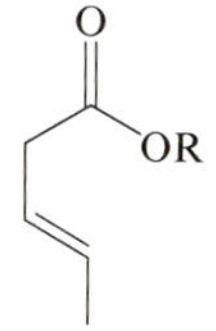	33	75e
		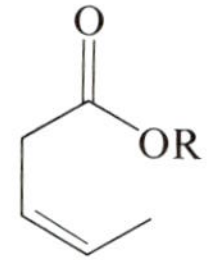	12	75e

(*cont.*)

TABLE I (*cont.*)

Halide or sulfonate	$ArSe^-$ (oxidant)	Olefin	Yield[a] of olefin (%)	Ref.
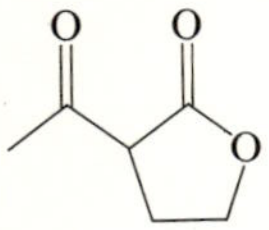	$PhSe^-$, DMF, 110°C (1. O_3 2, Δ, $CHCl_3$, Py)	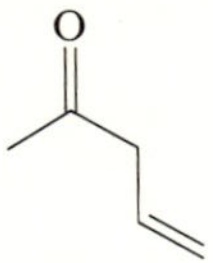	92 × 60	75e
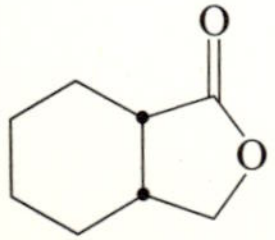	$PhSe^-$, THF, HMPA (O_3; Δ)	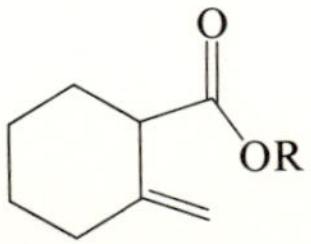	75 × 60	75d
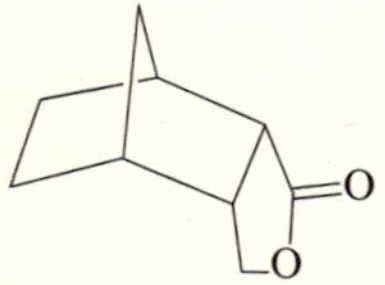	$PhSe^-$, THF, HMPA (O_3; Δ)	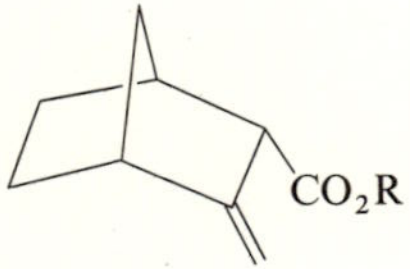	80 × 60	75d

[a] Isolated yields; gas chromatographic yields in parentheses. When two yields are given, the first is for selenide formation, the second is for the oxidation and syn elimination steps.

[b] Aliquat 336 added as phase transfer catalyst.

[c] Difficulties in selective oxidation at selenium were encountered.

and odor and thus may well have been a minor by-product. The detection of aldehyde here, together with the formation of formaldehyde from methyl selenoxides[42–42b,65,65a] and benzaldehyde from dibenzyl selenoxide,[66] led to the belief that all alkyl selenoxides decomposed in this manner[67] and hence the long delay in the recognition of the selenoxide syn elimination. In retrospect, it is surprising that the reaction was not discovered long before the sulfoxide and amine oxide eliminations were established.

The lower activation energy for selenoxide vs. sulfoxide syn eliminations as well as [2,3] sigmatropic rearrangements can reasonably be attributed to the lower C–Se bond strength[68] although reduced d–p π bonding in the selenoxide Se=O bond may also be a significant contributing factor. Weaker Se–O π bonding is indicated by the greater polarity and basicity of selenoxides (pK Me_2SO, -1.8; Me_2SeO, 2.55)[69] as well as by lower stretching force constants.[70]

B. Synthesis of Alkenes

The simplest application of the selenoxide elimination is as a replacement for conventional acid- or base-catalyzed eliminations of alcohols, epoxides, sulfonates, or halides, and these were the first applications to be developed.[4a] Nucleophilic displacement by benzeneselenolate ion gives an alkyl selenide which upon oxidation leads to selenoxide and subsequently to olefin. An oxidative step thus replaces acid or base treatment.

1. Preparation of Selenides

Selenolate anions are powerful nucleophiles, being 2–5 times more reactive than thiolate anions.[71] Additionally, they are several pK_a units less basic than thiols[72] and can therefore be used under essentially neutral conditions. The

[66] H. Rheinboldt and E. Giesbrecht, *J. Am. Chem. Soc.* **69**, 644 (1947); I. D. Entwhistle, R. A. W. Johnstone, and J. H. Varley, *Chem. Commun.* p. 61 (1976).

[67] H. Rheinboldt, *Methoden Org. Chem. Houben-Weyl, 4th Ed.* **9**, 1028 (1955); D. L. Klayman and W. H. H. Günther, eds., "Organic Selenium Compounds: Their Chemistry and Biology," p. 504. Wiley, New York, 1973.

[68] M. R. Arshadi and M. Shabrang, *J. Chem. Soc., Perkin Trans. 2* p. 1732 (1973).

[69] P. Nylen, *Z. Anorg. Allg. Chem.* **246**, 227 (1941).

[70] R. Paetzold, *Chem. Ser.* **8A**, 64 (1975).

[71] R. E. Dessy, R. L. Pohl, and R. B. King, *J. Am. Chem. Soc.* **88**, 5121 (1966); R. G. Pearson, H. Sobel, and J. Songstad, *ibid.* **90**, 319 (1968); G. Guanti, C. Dell'erba, and D. Spinelli, *Gazz. Chim. Ital.* **100**, 184 (1970).

[71a] V. Simanek and A. Klasek, *Tetrahedron Lett.* p. 3039 (1969).

[72] N. Nakamura and E. Sekido, *Talanta* **17**, 515 (1970); Y. Sogiura, Y. Hojo, Y. Tamai, and H. Tanaka, *J. Am. Chem. Soc.* **98**, 2339 (1976).

[73] H. Bauer, *Chem. Ber.* **46**, 92 (1913).

anions derived from benzeneselenol or from *o*-nitrobenzeneselenol have been used most frequently (Table I).[4,11,34,53,62,64,74–75e]

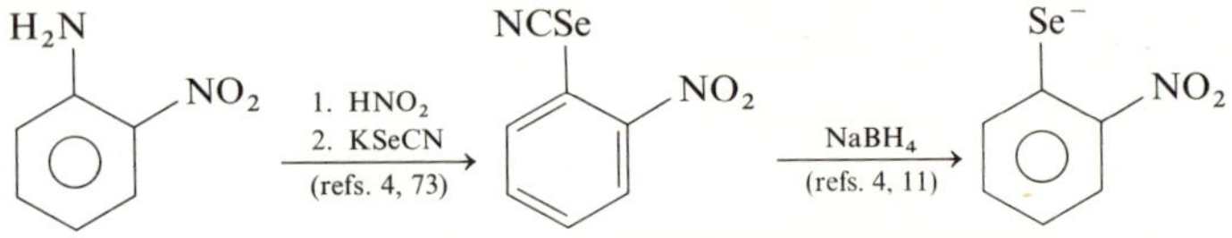

These anions are usually prepared by reduction of diselenide or selenocyanate with $NaBH_4$ in ethanol, but dimethylformamide has also been used as solvent.[74a,c] Less expensive reducing agents are available (see Section II, A).

Grieco, Gilman, and Nishizawa[76] have developed a procedure for directly converting alcohols to selenides by treatment with arylselenocyanate and tributylphosphine which is reported to give yields superior to those obtained by sulfonate displacement. This reaction is analogous to phosphine and phosphite based methods for converting alcohols to halides.

$$\text{ROH} \xrightarrow[\text{Bu}_3\text{P}]{\text{ArSeCN}} \text{RSeAr} \qquad \text{(ref. 76)}$$

The powerful nucleophilic properties of selenolate anion can be utilized in several ways. Carboxylic acid methyl esters are cleaved by PhSeNa under mild conditions, provided that borane or borate esters are not present to complex the nucleophile.[76a]

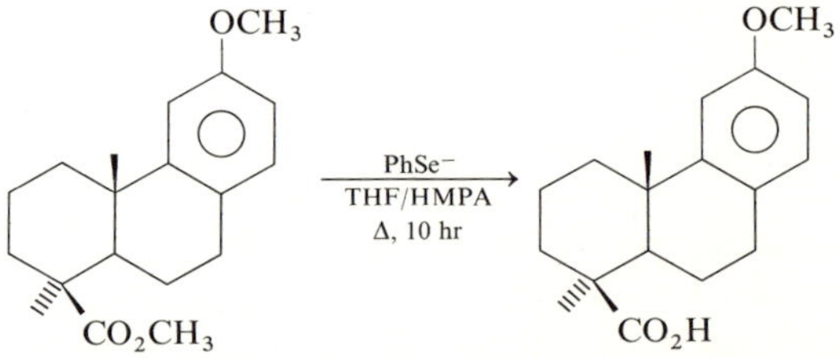

(ref. 76a)

[74] P. A. Grieco, Y. Masaki, and D. Boxler, *J. Am. Chem. Soc.* **97**, 1597 (1975).

[74a] P. A. Grieco, J. A. Noguez, and Y. Masaki, *Tetrahedron Lett.* p. 4213 (1975); *J. Org. Chem.* **42**, 495 (1977).

[74b] P. A. Grieco, J. J. Reap, and J. A. Noguez, *Synth. Commun.* **5**, 155 (1975).

[74c] P. A. Grieco, M. Nishizawa, S. D. Burke, and N. Marinovic, *J. Am. Chem. Soc.* **98**, 1612 (1976); P. A. Grieco, M. Nishizawa, T. Oguri, S. D. Burke, and N. Marinovic, *ibid.* **99**, 5773 (1977).

[75] R. D. Clark and C. H. Heathcock, *J. Org. Chem.* **41**, 1396 (1976).

[75a] P. A. Grieco and M. Nishizawa, *J. Org. Chem.* **42**, 1717 (1977).

[75b] G. Majetich, P. A. Grieco, and M. Nishizawa, *J. Org. Chem.* **42**, 2327 (1977).

[75c] K. Tsuzuki, T. Watanabe, M. Yanagiya, and T. Matsumoto, *Tetrahedron Lett.* p. 4745 (1976).

[75d] D. Liotta and H. Santiesteban, *Tetrahedron Lett.* p. 4369 (1977).

[75e] R. M. Scarborough, Jr., and A. B. Smith, III, *Tetrahedron Lett.* p. 4361 (1977).

[76] P. A. Grieco, S. Gilman, and M. Nishizawa, *J. Org. Chem.* **41**, 1485 (1976).

[76a] D. Liotta, W. Markiewicz, and H. Santiesteban, *Tetrahedron Lett.* p. 4365 (1977).

The nucleophilic opening of lactones can be carried out with uncomplexed PhSeNa in refluxing THF/HMPA,[75d] or in DMF at 110°–120°C with reagent prepared by borohydride reduction of Ph_2Se_2.[75e] The phenylseleno carboxylic acids so formed were esterified and converted to ω-olefinic esters by selenoxide syn elimination (see Table I).

Benzeneselenolate can also be used to cleave quaternary ammonium salts.[71a] *N*-Methylamines can be cleaved by treatment with PhSeH.[76b] The selenol is sufficiently acidic that the amine is protonated.

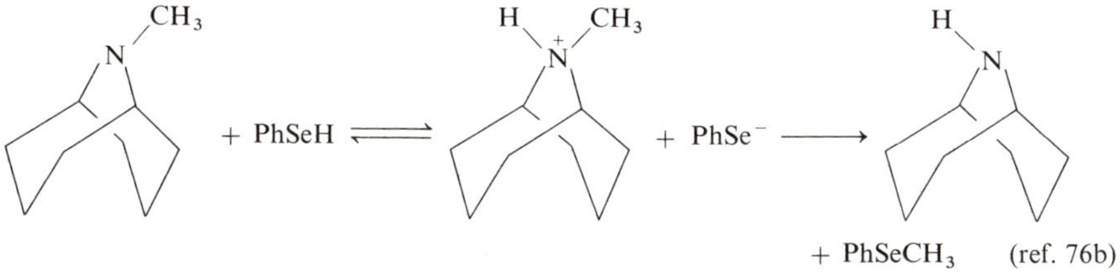

(ref. 76b)

Other useful methods for forming carbon–selenium bonds include the nucleophilic opening of epoxides (Section III, D, 1), electrophilic selenium additions to olefins (Section III, D, 2–3) enol derivatives (Section III, F, 1–2) and enolate anions (Section III, F, 3–5), reaction of electrophiles with α-lithio selenoxides and selenides (Sections III, C; III, G; IV, A, 5; IV, B), reaction of enones with selenenamides (Section III, F, 7) and the selenium dioxide oxidation of semicarbazones (Section V, C).

2. Regioselectivity of the Syn Elimination

The selenoxide syn elimination normally shows a preference for eliminations toward the less substituted carbon. The ratio of products can be roughly predicted on a statistical basis, and thus the selectivity is usually too small to be synthetically useful, e.g.

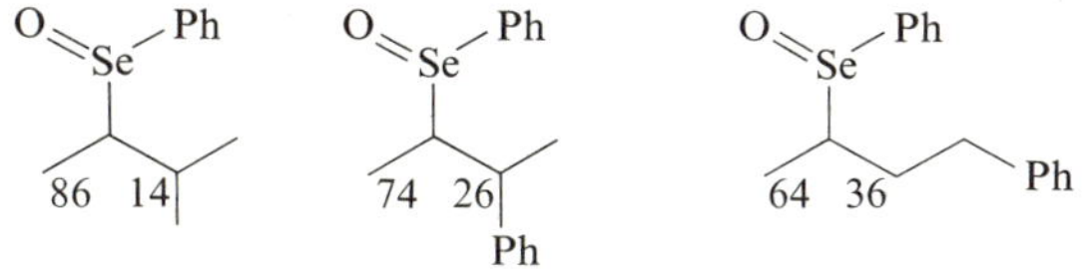

(ref. 64) (ref. 41)

The regioselectivity is similar to that observed for sulfoxide,[77,77a] amine oxide and acetate pyrolyses.

[76b] H. J. Reich and M. Cohen, unpublished results.

[77] D. W. Emerson, A. P. Craig, and I. W. Potts, Jr., *J. Org. Chem.* **32**, 102 (1967).

[77a] J. R. Shelton and K. E. Davis, *Int. J. Sulfur Chem.* **8**, 197 (1973).

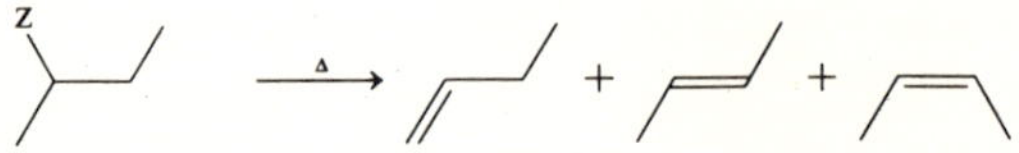

Z				
Z = Ph—Se(=O)—	61%	28%	11%	(ref. 64)
Z = Ph—S(=O)—	60%	25%	15%	(ref. 77a)
Z = $Me_2N^+(O^-)$—	67%	21%	12%	(ref. 78)
Z = $CH_3C(=O)O^-$—	57%	28%	15%	(ref. 79)

When polar or conjugative interactions are present, very high selectivity is sometimes observed. For example, homoallylic selenoxides invariably eliminate to give 1,4-dienes rather than 1,5-dienes.[41]

A particularly striking directive effect is present for β-hydroxy selenides[4a,61] which undergo selenoxide syn elimination almost exclusively away from the carbon bearing the hydroxyl group, forming allyl alcohols. The effect, studied in the cyclohexyl system (**5**) by Sharpless and Lauer,[9] holds for β-acetoxy and β-methoxy selenoxides but β-chloro selenoxides give no selectivity.

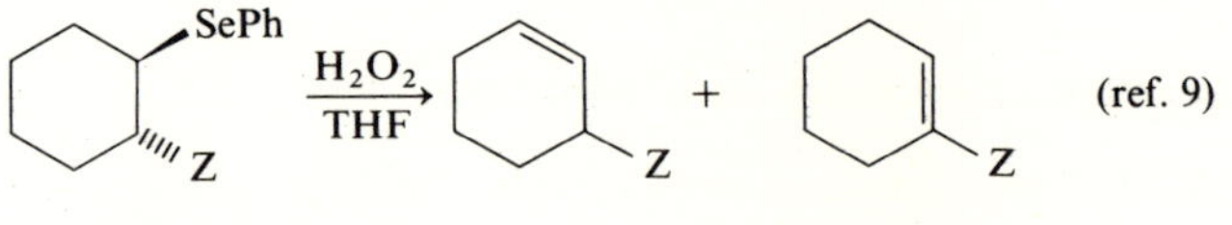

(**5**)

Z = OH	99.8%	0.2%
Z = OAc	97%	3%
Z = OCH_3	98%	2%
Z = Cl	51%	49%

A satisfactory explanation for these observations has not been presented. Dipole repulsions involving the highly polar selenoxide function may be an important factor (see Section III, H, 3).

Allylic selenoxides do not usually undergo selenoxide syn elimination to give dienes. The [2,3] sigmatropic rearrangement is faster and predominates except under special circumstances (see Section IV, A, 4).

[78] A. C. Cope, N. A. LeBel, H. H. Lee, and W. R. Moore, *J. Am. Chem. Soc.* **79**, 4720 (1957).
[79] D. H. Froemsdorf, C. H. Collins, G. S. Hammond, and C. H. DePuy, *J. Am. Chem. Soc.* **81**, 643 (1959).

3. Rate Effects

The qualitative observations available for the selenoxide syn elimination show that this reaction has many of the features demonstrated for the more thoroughly studied sulfoxide elimination. Branching at the α-carbon increases the rate (3° > 2° > 1°), while branching at the β-carbon decreases it. Scheme 4 summarizes some of the available rate data measured in $CDCl_3$ at 38°C.[38a]

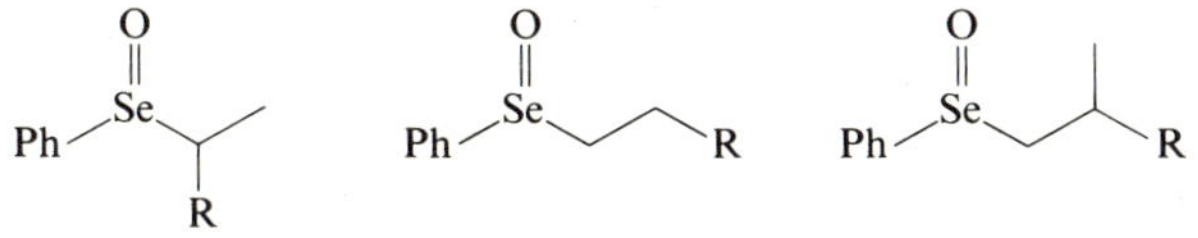

R	$t_{1/2}$(min)	R	$t_{1/2}$(min)	R	$t_{1/2}$(min)
H	360	H	360	H	675
CH_3	38	CH_3	675	CH_3	1800
Ph	<1	Ph	9	Ph	72
Cl	31	Cl	280		
		OCH_3	~4000		

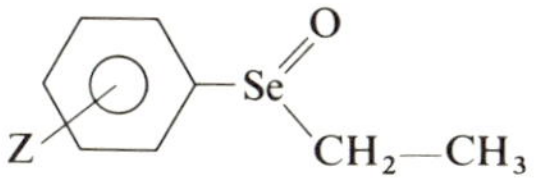

Z	$t_{1/2}$(min)
H	360
4-CH_3	550
3-CF_3	217
2-NO_2-4-CH_3	18
4-NO_2	80
4-OCH_3	890

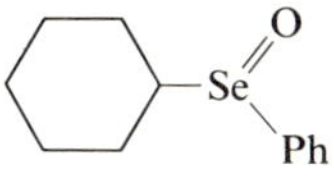

$t_{1/2} = 1820$ min

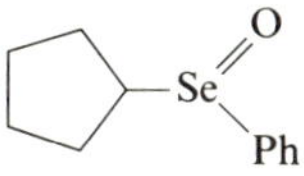

$t_{1/2} < 5$ min

Scheme 4
(All rates are in $CDCl_3$ at 38°C[38a])

Elimination is slower in protic than in aprotic solvents[38a,64,80] In fact, the presence of only 1.5 equivalents of acetic acid in chloroform results in a reduction of the rate of elimination of ethyl phenyl selenoxide by a factor of six (these conditions also result in secondary reaction of the ethylene formed, see Section III, C below).[38a] Apparently, the syn elimination is hampered when the highly polar selenoxide is hydrogen bonded.[64] Nonpolar solvents are also preferable for systems which have potential solvolytic reactivity.

[80] H. J. Reich and J. E. Trend, *J. Org. Chem.* **41**, 2503 (1976).

Electron-withdrawing substituents on the aryl group enhance the elimination rate,[4,5,74b] an effect previously documented for sulfoxide eliminations by Emerson and Korniski[81] ($\rho = 0.51$ for *p*-Z-C_6H_4SO-*n*-Pr). Thus, Sharpless and Young[6] have recommended the *p*-chloro- and *o*-nitrophenyl selenides; the latter have been used by Grieco and co-workers[74–74c,75a] in a number of natural product related syntheses for the introduction of vinyl groups (see Table I). However, Reich and Shah[41] and Clark and Heathcock[75] have found that the more readily available and more easily oxidizable phenylseleno group serves equally well if selenoxide is formed at low temperature by oxidation with ozone or *m*-chloroperbenzoic acid followed by brief thermolysis at 40°–80°C. The presence of amine during this reaction, which is conveniently carried out by adding cold selenoxide solution to refluxing hexane, dichloromethane, or carbon tetrachloride, is essential to prevent side reactions (reduction and electrophilic addition to olefins, Section III, C) which may occur.[38a]

C. SIDE REACTIONS DURING SELENOXIDE SYN ELIMINATIONS

Few side reactions have been encountered during selenoxide eliminations in simple alkyl systems. However, it appears likely that formation of carbonyl compounds can occur even in systems not specifically activated towards Pummerer-like processes since Foster, in an early study of alkyl selenoxide instability, found that ethyl phenyl selenoxide forms some acetaldehyde.[42a] Sharpless has observed that selenides undergo oxidative C–Se bond cleavage when heated with acidic hydrogen peroxide.[34]

$$C_{11}H_{23}CH_2SePH \xrightarrow[CH_3CO_2H,\ CF_3CO_2H]{H_2O_2/\Delta} C_{11}H_{23}CH_2OAc \quad (65\%) \qquad \text{(ref. 34)}$$

It was suggested that over-oxidation to selenone may be occurring under the rather vigorous reaction conditions. Little is known of the chemistry of alkyl selenones.[42] Alkyl seleninic acids, ($RSeO_2H$), are also transformed to alcohols and acetates.[34] It is possible that the acid-catalyzed SeO_2 oxidation of olefins to give *vic*-diacetates may be related to the above reactions.[82]

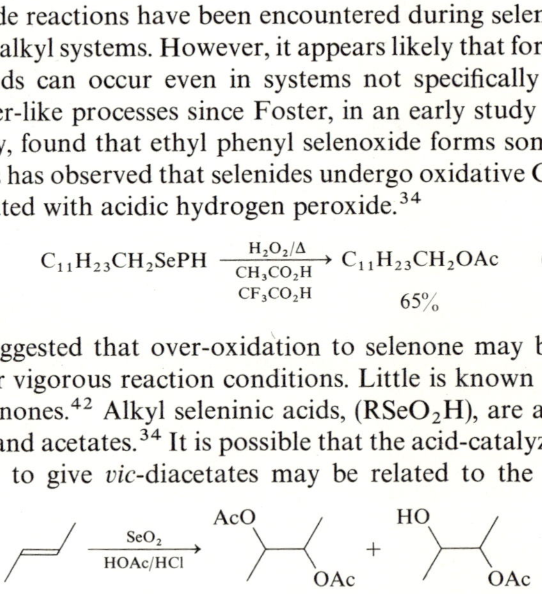

(ref. 82)

3.8:81.0 *meso*:*dl* — 15.2 — 26% total yield

[81] D. W. Emerson and T. J. Korniski, *J. Org. Chem.* **34**, 4115 (1969).

[82] K. A. Javaid, N. Sonoda, and S. Tsutsumi, *Bull. Chem. Soc. Jpn.* **43**, 3475 (1970); N. Sonoda, S. Furui, K. A. Javaid, and S. Tsutsumi, *Ann. N.Y. Acad. Sci.* **192**, 49 (1972).

These reactions are reminiscent of the conversion of alkyl iodides to alkyl acetates and *vic*-diacetates on treatment with peracetic acid, which apparently proceeds by ionic pathways.[83,84]

In systems which give stabilized carbonium ions, even selenoxides may decompose by heterolytic pathways, particularly under acidic conditions when the selenoxide is protonated. Thus, several primary and secondary alkyl phenyl selenoxides give alkyl bromides on treatment with hydrogen bromide.[85] Other examples include the formation of unrearranged cinnamyl alcohol and cinnamyl ether on oxidation of cinnamyl phenyl selenide with H_2O_2 in ethanol,[86] and the transformation of phenylseleno-substituted cyclopropanols to cyclobutanones upon oxidation (see Section III, D, 1).[87]

A final group of potential side reactions are those involving selenenic acids (RSeOH) which are formed during selenoxide syn eliminations. Normally, they disproportionate to diselenide and seleninic acid ($RSeO_2H$) or are oxidized to seleninic acid by excess oxidant. The processes in which selenenic acids may participate are listed below.

a. Oxygen transfer reactions are illustrated by the following experiment[5]:

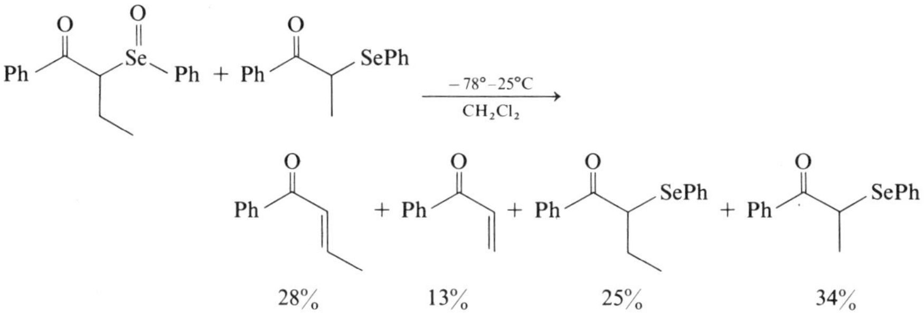

Note that little, if any, actual reduction has occurred, but that one selenide has apparently oxidized another at a rate faster than selenoxide syn elimination. In the presence of weak bases, oxygen transfer reactions are completely suppressed [reaction mixtures are ordinarily weakly acidic because of the formation of benzeneseleninic acid (pK_a 4.70)].[88]

[83] Y. Ogata and K. Aoki, *J. Org. Chem.* **34**, 3974 and 3978 (1969); Y. Ogata and I. Urasaki, *ibid.* **38**, 100 (1973).

[84] H. J. Reich and S. L. Peake, unpublished results.

[85] L. Hevesi, M. Sevrin, and A. Krief, *Tetrahedron Lett* p. 2651 (1976).

[86] H. J. Reich and I. L. Reich, unpublished results.

[87] P. H. Scudder, Ph.D. Thesis, University of Wisconsin, Madison (1977); B. M. Trost and P. H. Scudder, *J. Am. Chem. Soc.* **99**, 7601 (1977).

[88] E. S. Gould and J. D. McCullough, *J. Am. Chem. Soc.* **73**, 1109 (1951); D. DeFilippo and F. Momicchioli, *Tetrahedron* **25**, 5733 (1969).

b. Reductions can be seen when selenols are formed under acidic conditions. Pummerer-like processes, in which oxidation α to a selenoxide accompanied by reduction at selenium occurs, have been found for several types of selenides.[5,89,90] Under hydrolytic conditions benzeneselenol should be liberated, as in the example below.

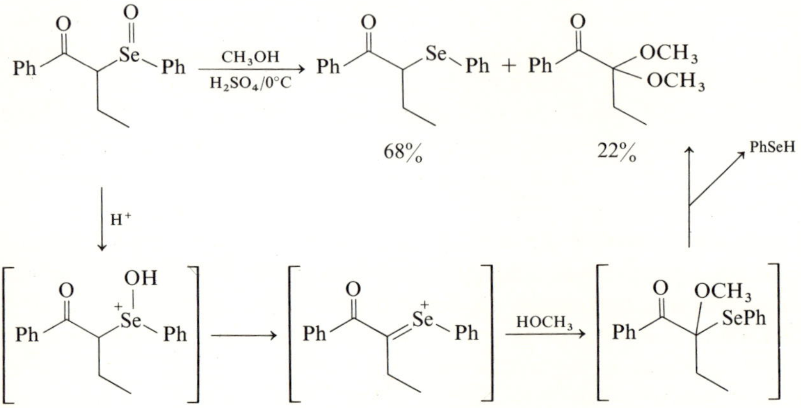

(ref. 5)

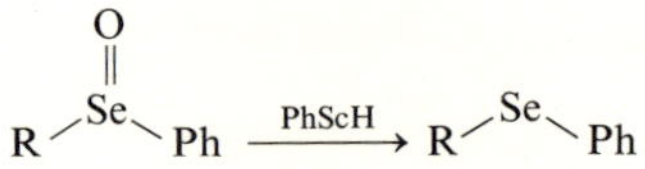

c. Selenenic acids (PhSeOH) formed by selenoxide elimination are electrophilic species and can cause side reactions. Two modes of addition are possible, one being the reversal of the syn elimination giving back selenoxide; the other, electrophilic addition to give β-hydroxy selenide.

Sulfoxide eliminations have been shown to be highly reversible in a number of systems[91] and such a process can result in loss of double bond positional specificity. The limited evidence available suggests that elimination is not reversible for alkyl selenoxides. The decomposition of selenoxide **6** in deuterated media results in no detectable deuterium incorporation in the products.[80]

[89] K. B. Sharpless and K. M. Gordon, *J. Am. Chem. Soc.* **98**, 300 (1976).

[90] Y. Okamoto, K. L. Chellappa, and R. Homsany, *J. Org. Chem.* **38**, 3172 (1973); B. E. Norcross, J. M. Lansinger, and R. L. Martin, *J. Org. Chem.* **42**, 369 (1977).

[91] J. R. Shelton and K. E. Davis, *Int. J. Sulfur Chem.* **8**, 205 (1973); Y. Makisumi, S. Takada, and Y. Matsukura, *Chem. Commun.* p. 850 (1974); R. J. Stoodley and R. B. Wilkins, *J. Chem. Soc., Perkin Trans. 1* p. 1572 (1974); D. N. Jones and D. A. Lewton, *Chem. Commun.* p. 457 (1974); D. N. Jones, *Tetrahedron Lett.* p. 2235 (1975); E. Block and J. O'Connor, *J. Am. Chem. Soc.* **96**, 3929 (1974); I. Ager, D. H. R. Barton, D. G. T. Greig, G. Lucente, P. G. Sammes, M. V. Taylor, G. H. Hewitt, B. E. Looker, A. Mowatt, C. A. Robson, and W. G. E. Underwood, *J. Chem. Soc., Perkin Trans. 1* p. 1187 (1973).

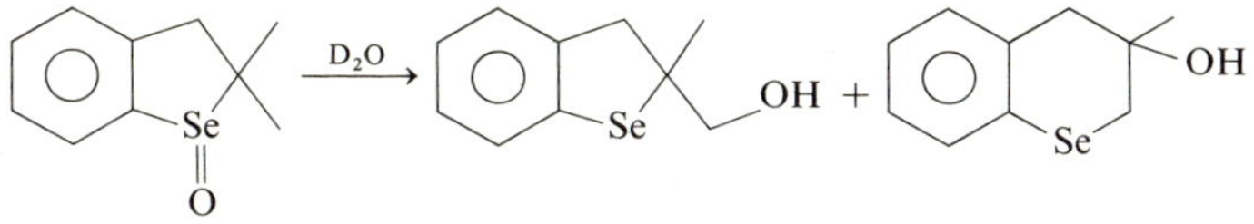

(ref. 80)

(6) (7) (8)

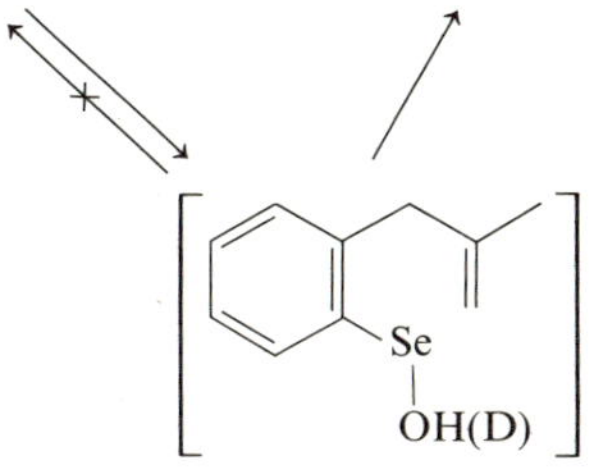

The products (**7** and **8**) illustrate a potentially more troublesome type of side reaction: electrophilic addition to a double bond to give a hydroxy selenide. Compounds **7** and **8** are formed even in basic media, suggesting that selenenic acids (or disproportionation products from them) are far more electrophilic than sulfenic acids, which give such products only under acidic conditions where sulfenyl acetates or similar species are present. Simple alkyl selenoxides also form β-hydroxy selenides on decomposition at room temperature.[38a]

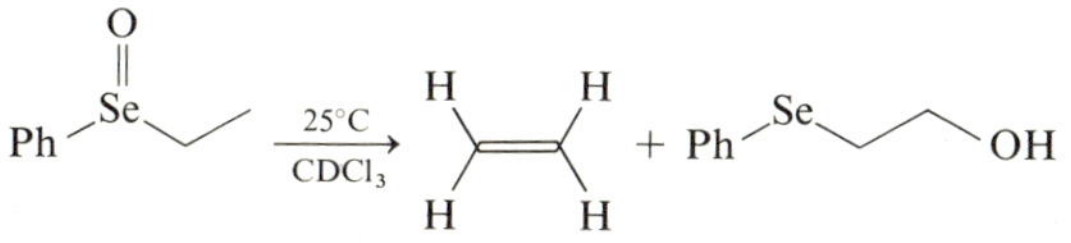

15% 85%

This side reaction becomes even more prominent under acidic conditions, such as when small amounts of acetic acid are present. PhSeOH is probably converted to PhSeOAc, which reacts with ethylene as it is formed.[38a]

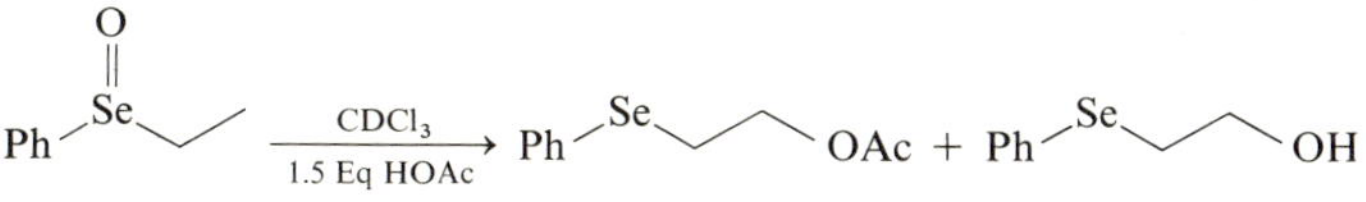

76% 10%

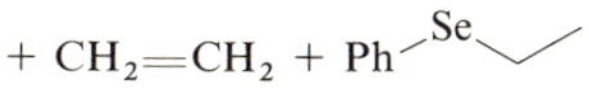

(ref. 38a)

~1% 13%

It appears that the disproportionation of selenenic acids is a reversible process, so that $PhSeO_2H$ and PhSeSePh will reform PhSeOH (or some related species such as PhSe–SeO–Ph) which can add to olefins. This has been experimentally verified.[38,38a]

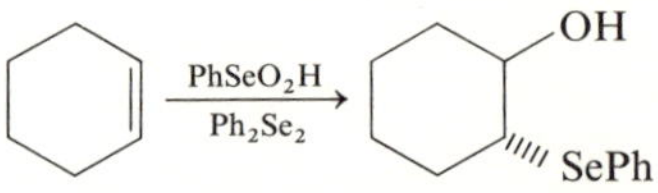

It is not known whether such electrophilic addition reactions occur in the presence of excess oxidant, but it does seem likely that poor yields of olefins, which have been reported particularly for primary selenoxides,[4,11,75] are a consequence of such reactions.

Side reactions involving electrophilic additions of selenium species should be expected whenever unusually electron-rich bonds (enols, enol ethers, enamines, etc.) are being formed or are present elsewhere in the molecule. Such problems have been encountered during selenoxide eliminations leading to enones[5] and β-amino enones[10] (see Section III, H, 2). These reactions can sometimes be prevented (including the formation of **7** and **8** during decomposition of **6**) by the presence of unhindered amines which serve to trap selenenic acids as selenenamides.[5,10,80] It is generally advantageous to carry out selenoxide elimination by a brief thermolysis (40°–80°C) rather than by slow elimination at room temperature, and also to have an amine base present ($HNEt_2$, HN(*i*-Pr)$_2$, Py) during elimination. Under these conditions, redox reactions and electrophilic additions are usually supressed.[5,38a,41,75]

Alternatively, Sharpless,[4] Grieco,[11] and Reich[5,38a] have reported that the use of electron-withdrawing substituents on the arylseleno group (such as *o*-NO_2, *p*-Cl, or *m*-CF_3) increases the rate and improves the yield of syn elimination reactions (see Scheme 4).

It should be noted that Ho and Hall[92] have successfully used syn elimination of β-alkoxy selenoxides in part of a sequence that uses β-bromo selenides as alcohol protecting groups.

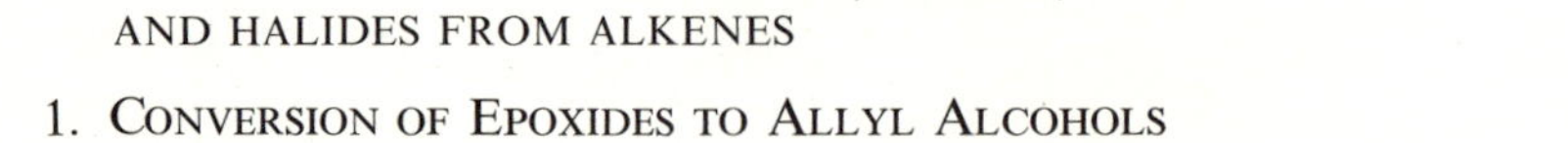

(ref. 92)

The vinyl ethers were not isolated so it is not clear whether selenenic acid addition to the enol ether occurred.

D. SYNTHESIS OF ALLYL ALCOHOLS, ETHERS, AND HALIDES FROM ALKENES

1. CONVERSION OF EPOXIDES TO ALLYL ALCOHOLS

The first synthetic methodology using the selenoxide syn elimination as the olefin-forming step was reported by Sharpless and Lauer.[4a]

[92] T.-L. Ho and T. W. Hall, *Synth. Commun.* **5**, 367 (1975).

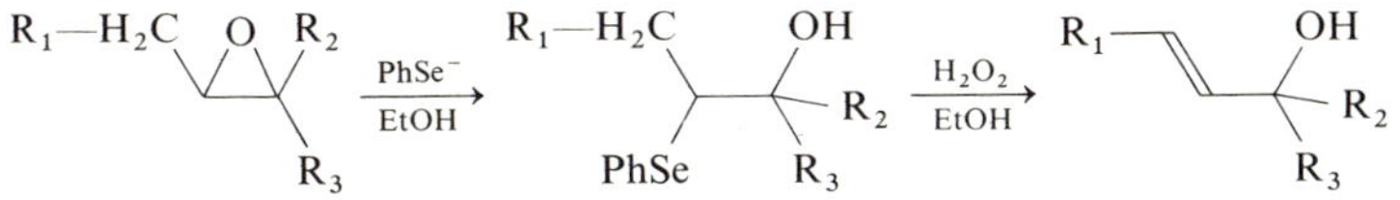

This sequence provides a mild alternative to the elimination of epoxides using lithium amide bases or diethylaluminum tetramethylpiperidide,[93] or to the singlet oxygen oxidation of olefins which also result in allylic oxidation with double bond migration. Table II lists examples of this reaction reported in the literature.[95–101]

The selenolate epoxide opening predictably occurs at the less substituted side,[4a,94] in contrast to the aluminum piperidide opening,[93] in which the more substituted C–O bond is cleaved. The reactions thus complement each other.

Trost and Scudder[87] have shown that *β*-hydroxy selenides formed by opening of oxaspiropentanes with $PhSe^-$ do not give olefins as the only products. The formation of cyclobutanones is observed, and this product becomes the only one when oxaspiropentanes derived from ketones are used.

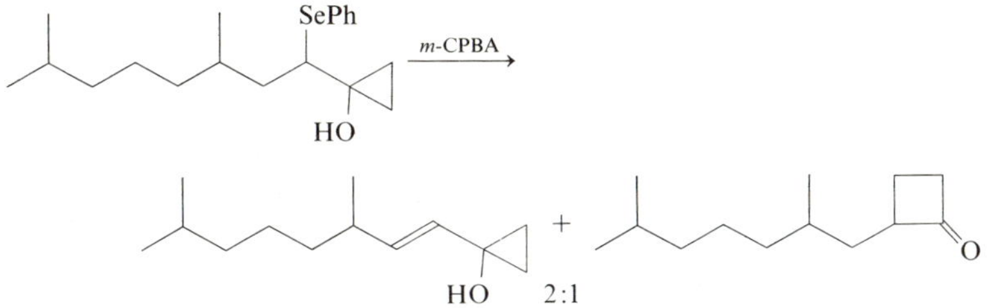

(ref. 87)

[93] A. Yasuda, S. Tanaka, K. Oshima, H. Yamamoto, and H. Nozaki, *J. Am. Chem. Soc.* **96**, 6513 (1974).

[94] G. I. Zaitseva and V. M. Al'bitskaya, *Zh. Org. Khim.* **4**, 2057 (1968); L. A. Khazemova and V. M. Al'bitskaya, *ibid.* **5**, 1926 (1969); **6**, 935 (1970).

[95] B. M. Trost and M. J. Bogdanowicz, *J. Am. Chem. Soc.* **95**, 5321 (1973); B. M. Trost, M. Preckel, and L. M. Leichter, *ibid.* **97**, 2224 (1975).

[96] B. M. Trost, *Pure Appl. Chem.* **43**, 563 (1975); B. M. Trost, *Acc. Chem. Res.* **7**, 85 (1974).

[97] K. Torssell, *Tetrahedron Lett.* p. 623 (1974).

[98] S. Ohta and S. Kimoto, *Tetrahedron Lett.* p. 2279 (1975).

[99] Y. Tsuda, T. Sano, J. Taga, K. Isobe, J. Toda, H. Irie, H. Tanaka, S. Takagi, M. Yamaki, and M. Murata, *Chem. Commun.* p. 933 (1975).

[100] S. David, A. Lubineau, and J.-M. Vatéle, *Chem. Commun.* p. 701 (1975).

[100a] S. David, A. Lubineau, and J.-M. Vatéle, *J. Chem. Soc., Perkin Trans. 1* p. 1831 (1976).

[100b] S. David and A. Lubineau, *Nouveau J. Chim.* **1**, 375 (1973).

[100c] S. Uzarewicz and E. Zientek, *Roczn. Chem.* **51**, 181 (1977).

[100d] L. A. Paquette, C. C. Liao, R. L. Burson, R. E. Wingard, Jr., C. N. Shih, J. Faylos, and J. Clardy, *J. Am. Chem. Soc.* **99**, 6935 (1977).

[101] L. A. Paquette, C. C. Liao, D. C. Liotta, and W. E. Fristad, *J. Am. Chem. Soc.* **98**, 6412 (1976).

TABLE II

CONVERSION OF EPOXIDES TO ALLYL ALCOHOLS

Epoxide	Allyl alcohol	Yield of olefin[a] (%)	Ref.
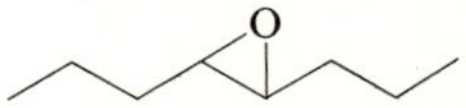	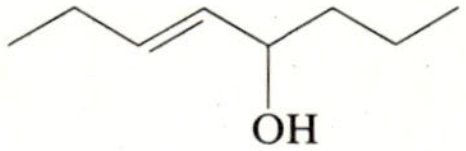	(98)	4a
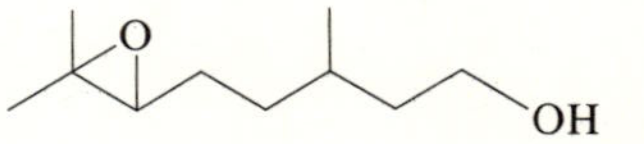	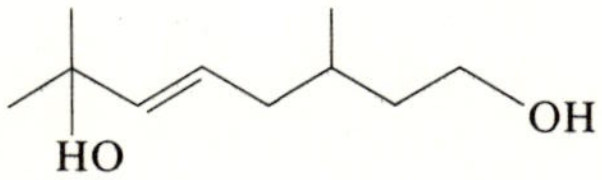	(95)	4a
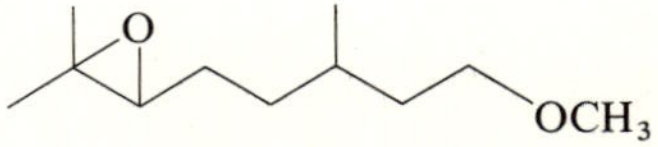	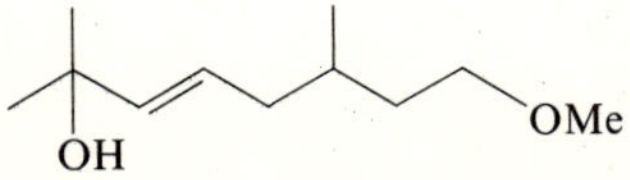	79	4a
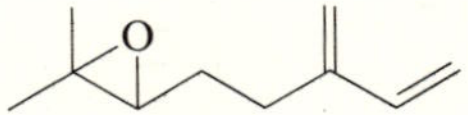	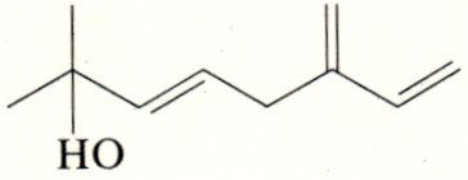	50	4a
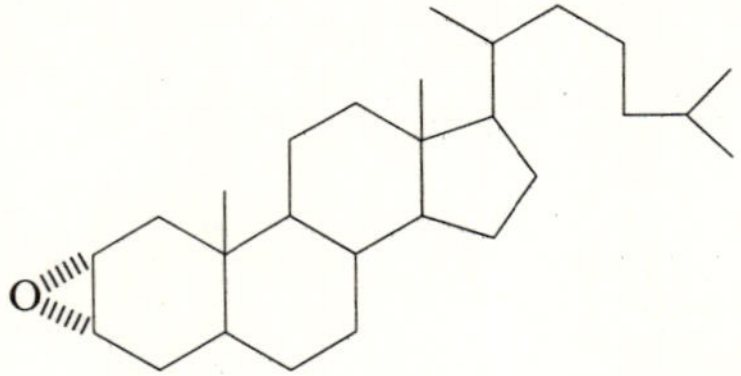	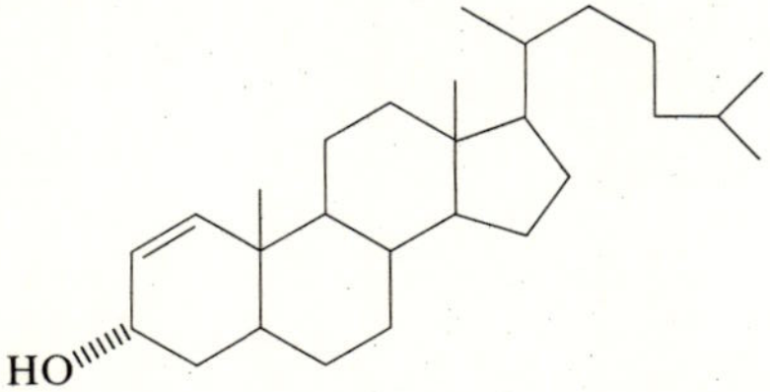	40	4a

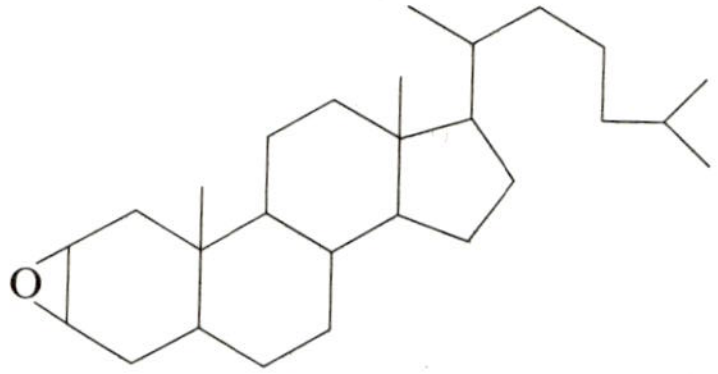	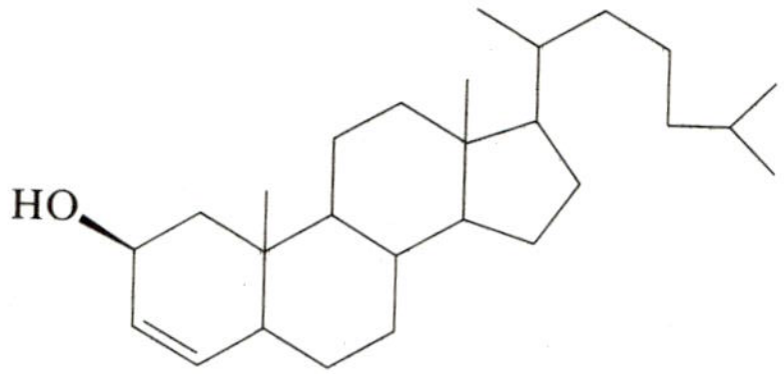	51	4a
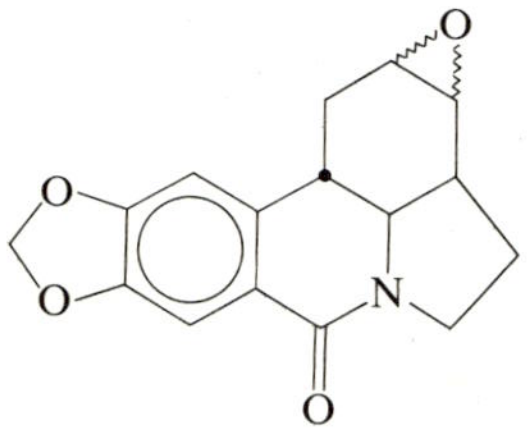	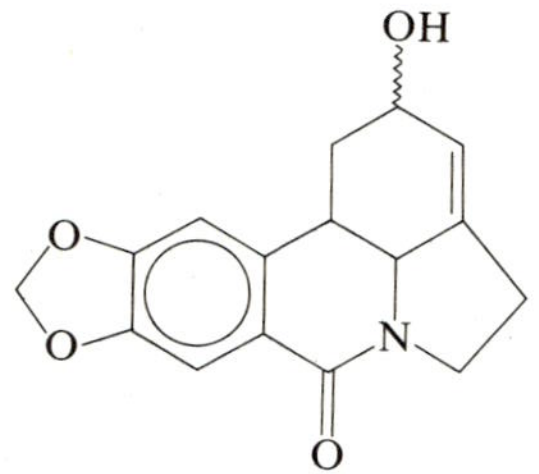	> 23[b]	97
	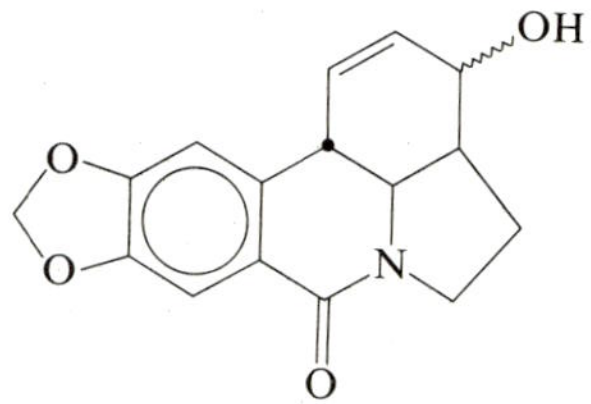	> 7[b]	
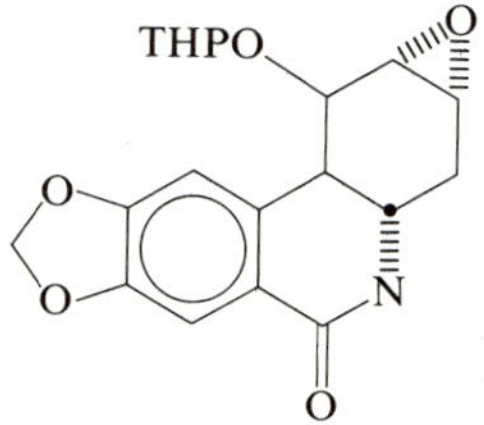	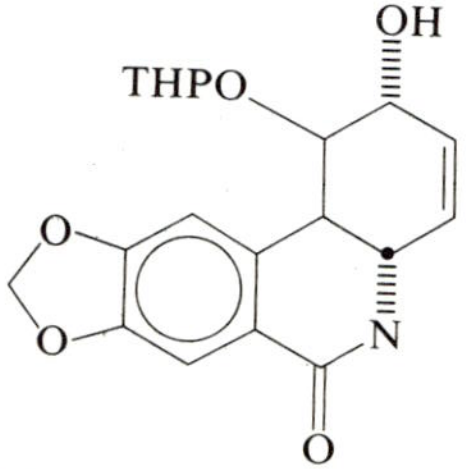	63	98

(*cont.*)

TABLE II (*cont.*)

Epoxide	Allyl alcohol	Yield of olefin[a] (%)	Ref.
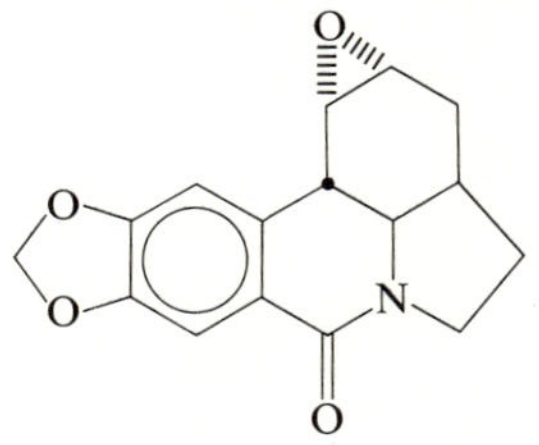	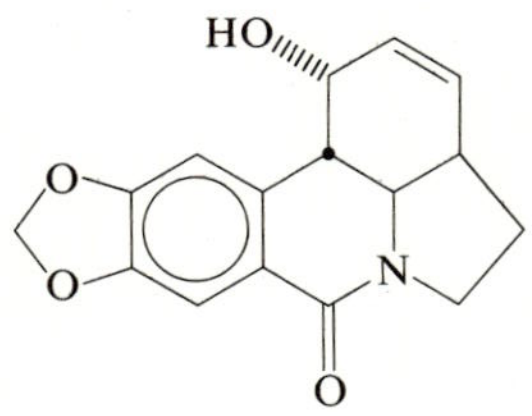	70	99
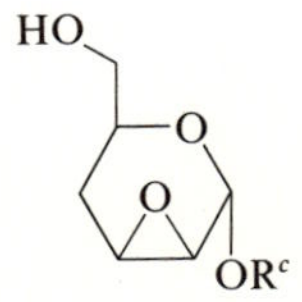	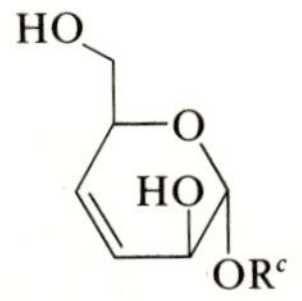	73	100
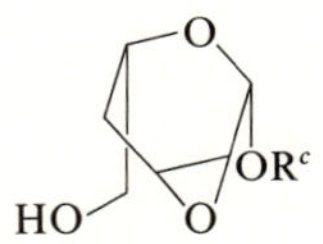	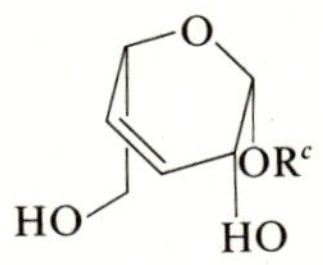	66	100a
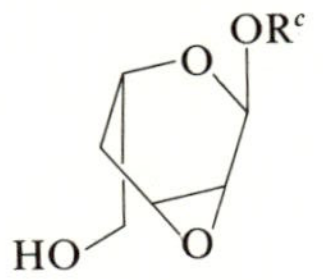	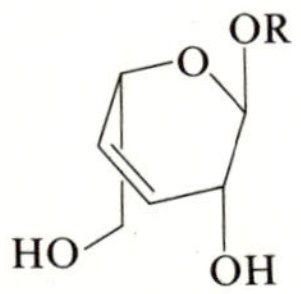	[d]	100a

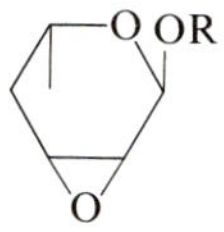	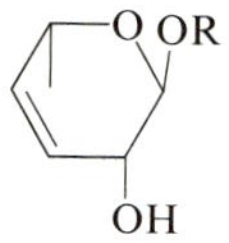	80	100b
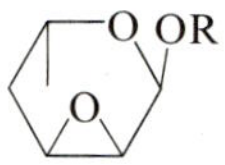	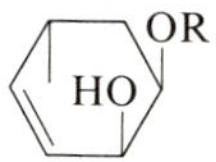	35	100b
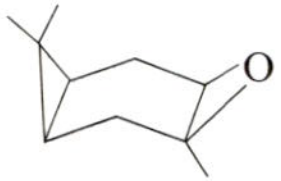	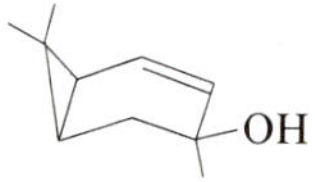	93	100c
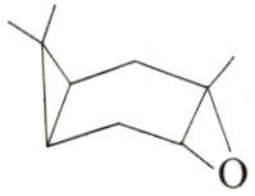	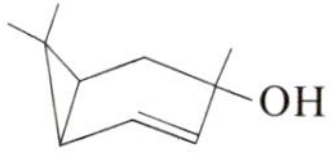	38	100c
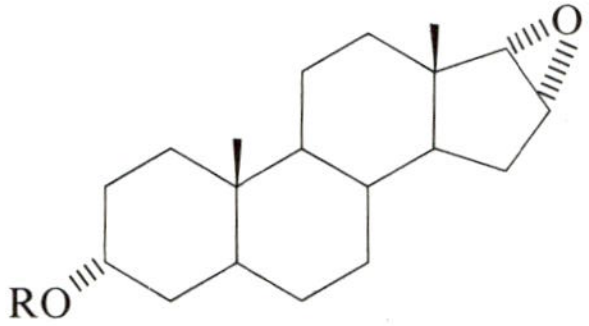	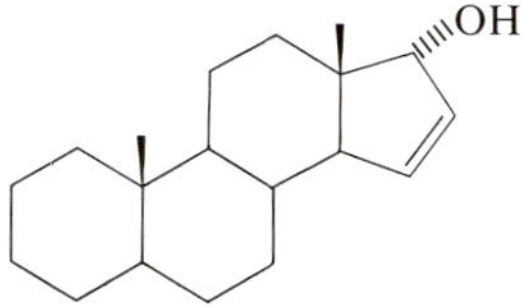	50	34b
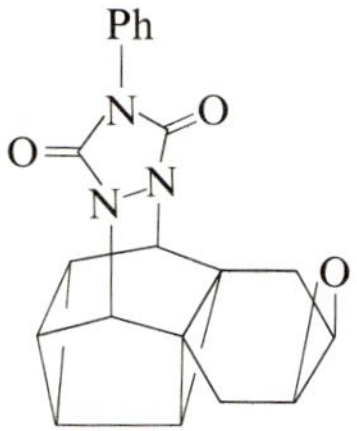	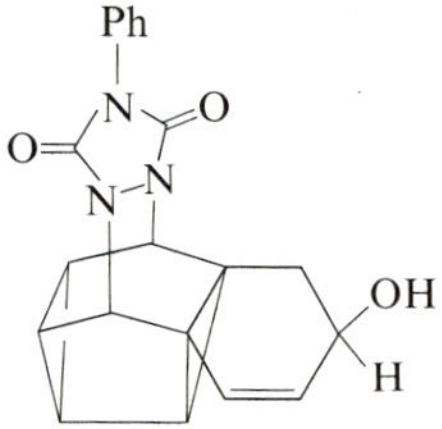	85	100d

(*cont.*)

TABLE II (*cont.*)

Epoxide	Allyl alcohol	Yield of olefin[a] (%)	Ref.
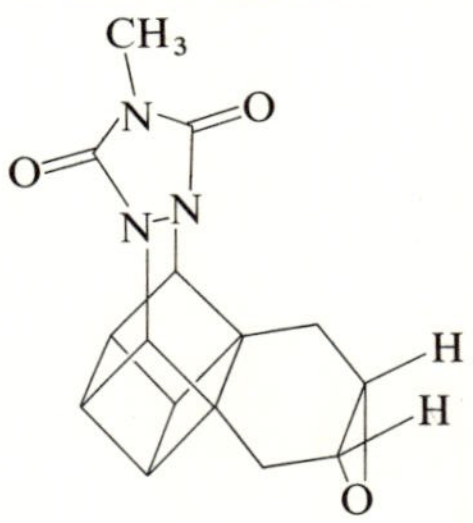	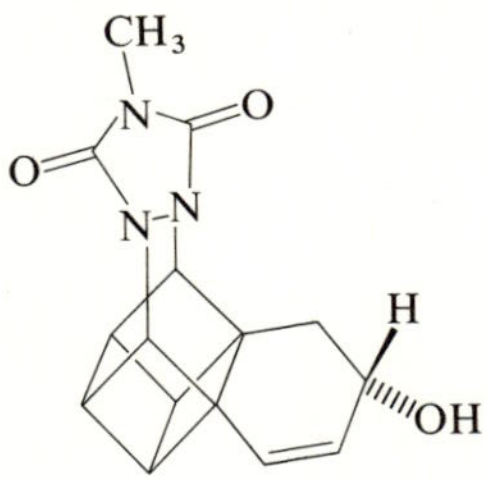	[d]	101
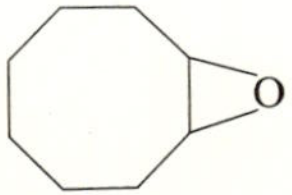	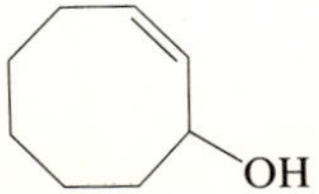	(75)	4a
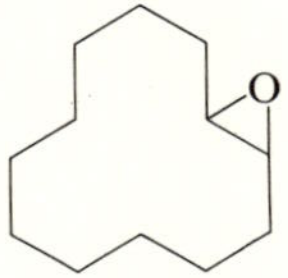	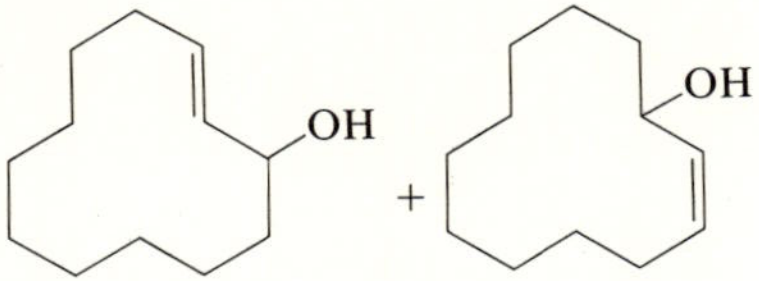	(85)	4a

[a] Isolated yields, gas chromatographic yields in brackets.
[b] Overall yield for several steps in addition to olefin formation.
[c] R = di-*O*-isopropylidene-α-D-glucofuranose.
[d] No yield reported.

This reaction is useful because it usually gives a cyclobutanone of the opposite stereochemistry from that obtained by $LiBF_4$-catalyzed opening of the oxaspiropentane.[95]

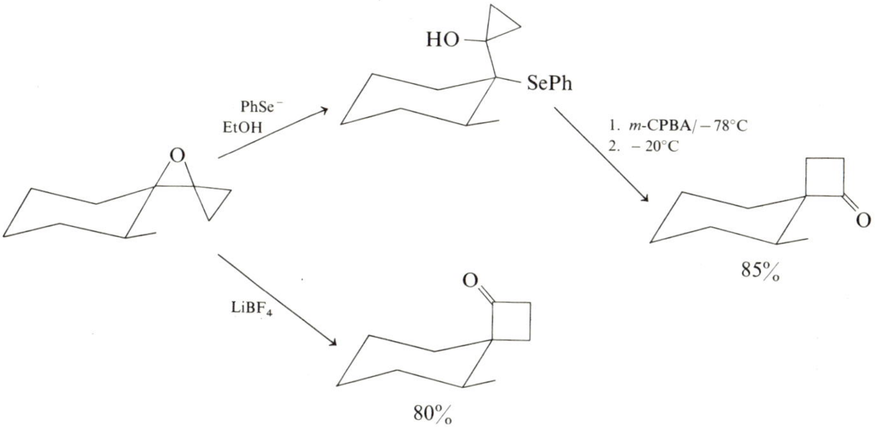

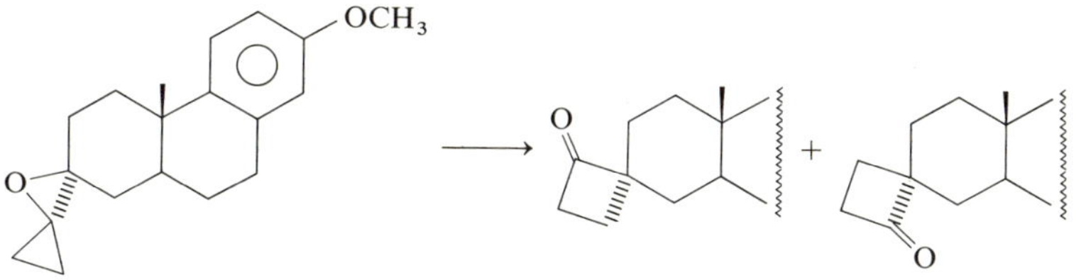

Selenide route	93:7
$LiBF_4$ route	<2:98

The cyclobutanone formation probably proceeds by ionization of the selenoxide to a highly stabilized cyclopropylcarbinyl cation. Hydrogen bonding of the cyclopropanol OH to the incipient $PhSeO^-$ group may account for the special stereochemical control observed. Cyclobutanones have served as versatile intermediates in a variety of synthetic procedures.[96]

2. Conversion of Olefins to Allyl Alcohols

A procedure for converting olefins to β-hydroxy or β-acetoxy selenides by selenenyl acetate addition was discovered by Hölzle and Jenny[102] who prepared the first stable selenenic acids and selenenic acid esters.[18] Intramolecular trapping of carboxylate during selenenyl bromide addition has been reported by Campos and Petragnani.[103]

[102] G. Hölzle and W. Jenny, *Helv. Chim. Acta* **41**, 593 (1958).
[103] M. D. M. Campos and N. Petragnani, *Chem. Ber.* **93**, 317 (1960).

Huguet[61] was able to use a selenide *catalytically* as an allylic oxidant. The sequence is illustrated below.

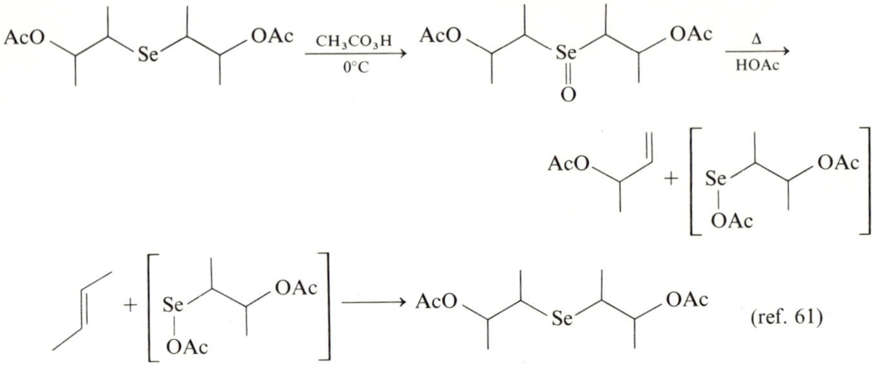

(ref. 61)

Better synthetic procedures than those used by Hölzle and Jenny were developed by Clive,[104] Reich,[8] Sharpless[9] and Toshimitsu.[104a] The additions to olefins proceed with high anti-stereospecificity but low regioselectivity so that the method is most useful as a preparative method for allyl alcohol synthesis when symmetrical olefins are used. Typical examples are shown below.

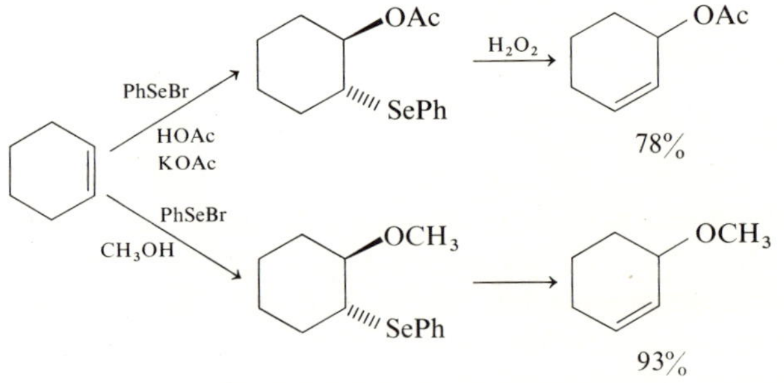

(ref. 9)

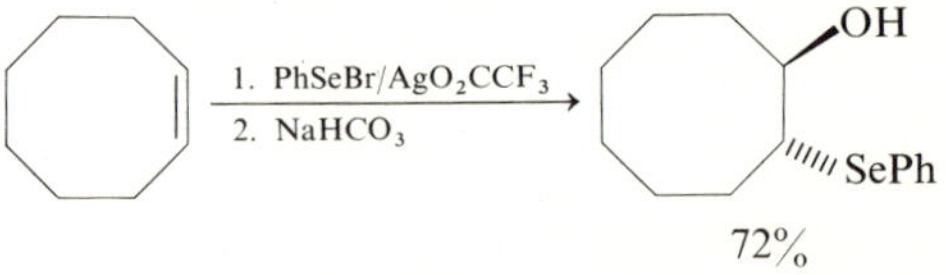

(ref. 104)

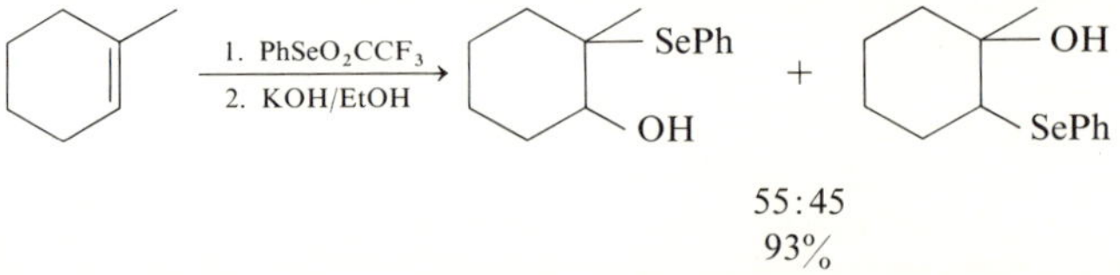

(ref. 8)

[104] D. L. J. Clive, *Chem. Commun.* p. 100 (1974).
[104a] A. Toshimitsu, S. Uemura, and M. Okano, *Chem. Commun.* p. 166 (1977).

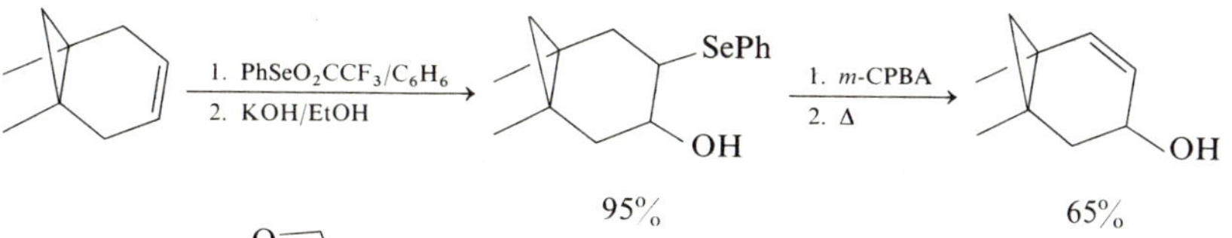

(ref. 105)

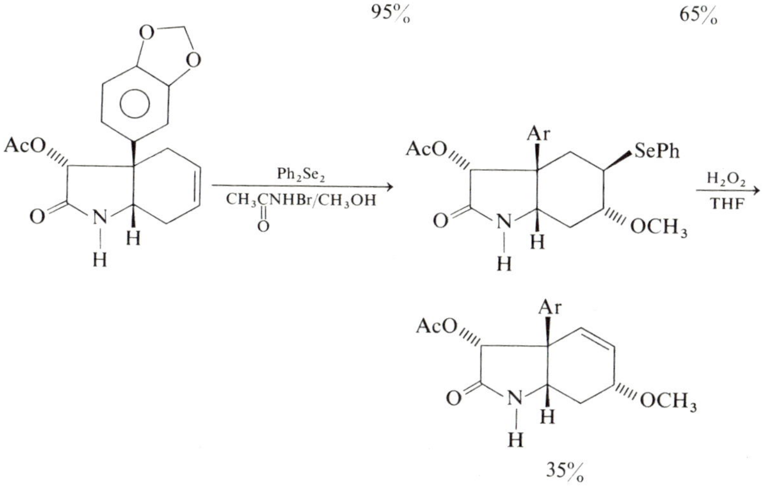

(ref. 106)

When trisubstituted olefins are converted to hydroxy selenides using $PhSeO_2/Ph_2Se_2/MgSO_4$ the tertiary alcohol is the exclusive product; presumably under these conditions thermodynamic control operates.[38]

Highly polarized olefins such as enol ethers, enol acetates, or silyl enol ethers do react with high regioselectivity to give α-phenylselenocarbonyl compounds (see Section III, F, 2).

β-Hydroxy and β-acetoxy selenides can be formed during selenoxide syn elimination by electrophilic addition of PhSeOH to the olefin (see Section III, C).

Methaneselenenyl acetate adducts are formed when dimethyl selenoxide is decomposed in acetic acid in the presence of olefin.[107–107b] Since dimethyl selenoxide apparently decomposes to give formaldehyde,[42b,65] the second equivalent of selenoxide used serves to oxidize formaldehyde selenoacylal or selenol to selenenyl acetate.[107a]

Selenenyl acetate additions also occur during selenium dioxide oxidations carried out in acetic acid. β-Acetoxy selenides **9** and **10** were isolated by Olson[108] and Huguet[61] during oxidation of ethylene and 2-butene, respectively.

[105] C. P. Casey and M. C. Cesa, private communication.
[106] K. Isobe, J. Taga, and Y. Tsuda, *Tetrahedron Lett.* p. 2331 (1976).
[107] N. Miyoshi, S. Furui, S. Murai, and N. Sonoda, *Chem. Commun.* p. 293 (1975).
[107a] N. Sonoda, N. Miyoshi, and S. Murai, *Chem. Scr.* **8A**, 109 (1975).
[107b] N. Miyoshi, S. Murai, and N. Sonoda, *Tetrahedron Lett.* p. 851 (1977).
[108] D. H. Olson, *Tetrahedron Lett.* p. 2053 (1966).

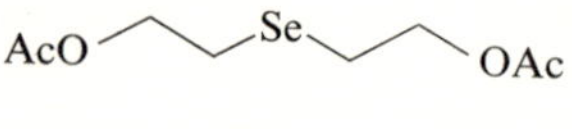

(9)

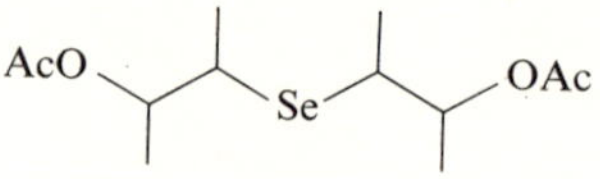

(10)

Similar adducts were isolated by Javaid, Sonoda, and Tsutsumi.[82] Although these authors assigned the selenoxide structure to their adducts, it is probable that the compounds were not selenoxides since they were eluted from silica gel with benzene (selenoxides are highly polar compounds and methanol–ether mixtures are normally necessary for chromatography on silica gel). Wiberg and Nielsen[109] and Shaefer, Horvath, and Klein[110] have also isolated selenium-containing materials from SeO_2 oxidations in acetic acid, but assigned β-acetoxy selenoxide structures to the compounds. Sharpless[34] has reinvestigated the oxidation of cyclohexene, and has concluded that β-acetoxy selenides (**11**) are formed by addition of divalent selenium intermediates to cyclohexene.

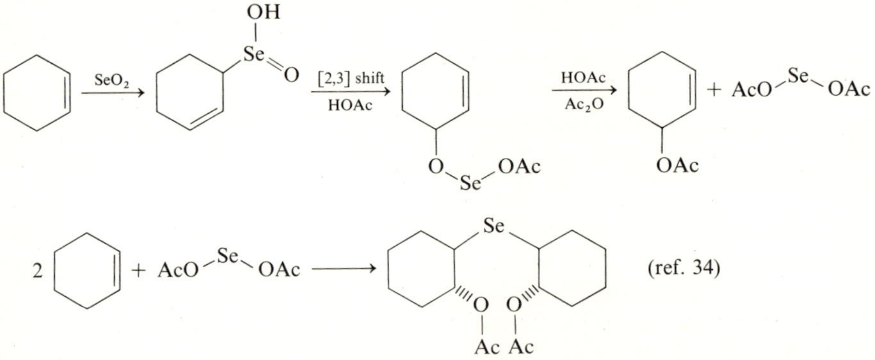

(**11**)

Note that the allyl acetates are *not* formed by selenoxides derived from β-acetoxy selenides such as **11**, since such a reaction gives allyl acetates with a rearranged double bond whereas allyl acetates formed by SeO_2 oxidation have unrearranged double bonds.

The addition of selenenyl chlorides to olefins has been studied by several groups.[111–111h] The addition proceeds with anti-stereochemistry.[111b,d] A report that the reaction leads to an isolable episelenurane[111f] has been shown to be in error.[111h]

[109] K. B. Wiberg and S. D. Nielsen, *J. Org. Chem.* **29**, 3353 (1964).
[110] J. P. Shaefer, B. Horvath, and H. P. Klein, *J. Org. Chem.* **33**, 2647 (1968).
[111] N. N. Yarovenka, V. N. Shemanina, and G. B. Gazieva, *Zh. Obshch. Khim.* **29**, 942 (1959).
[111a] N. Ya. Derkach and N. A. Pasmurteva, *Zh. Org. Khim.* **9**, 1414 (1973).
[111b] N. S. Zefirov, L. G. Gurrich, A. S. Shashkov, and V. A. Smit, *Zh. Org. Chim.* **10**, 1786 (1974).

The products of PhSeCl and PhSeBr additions to olefins can be used for the synthesis of vinyl and allyl halides, as shown by the following examples.[86,111i]

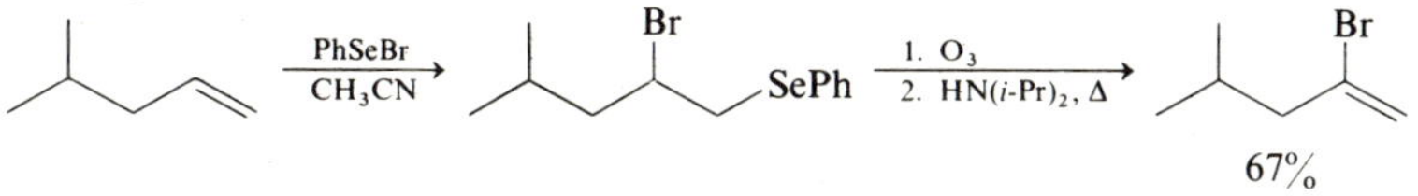

(ref. IIIi)

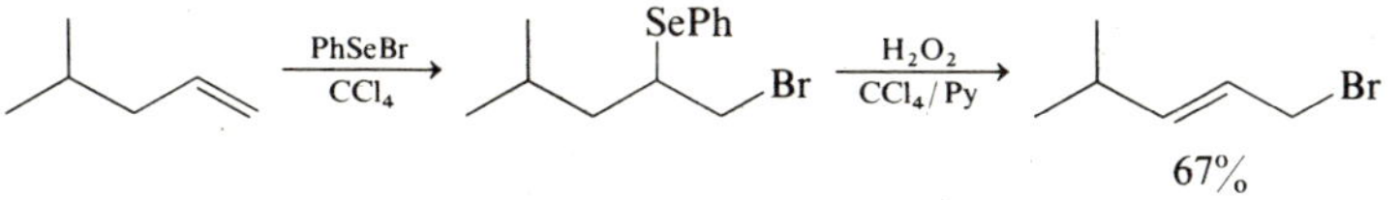

(ref. IIIi)

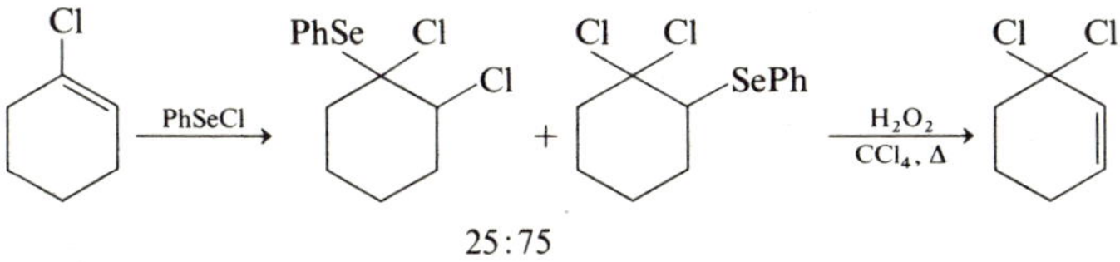

(ref. 86)

Significant variations in regiochemistry can be achieved by doing the additions under conditions of kinetic or thermodynamic control. This is also the case for reactions in which hydroxy selenides are formed.[38]

The possibility of using seleninyl derivatives to directly introduce the selenoxide function has not been systematically examined. That this is feasible at least in principle is shown by the following reaction reported by Derkach and Pasmurtseva[111a] (see Section III, F, 5).

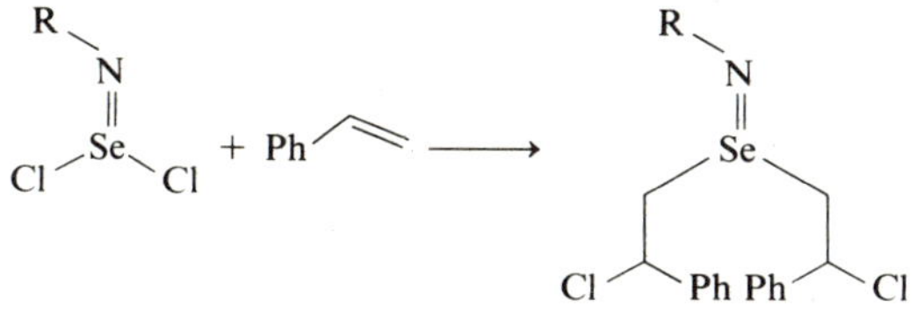

(ref. 111a)

3. Lactone and Cyclic Ether Formation during Electrophilic Selenium Additions

Campos and Petragnani[103] observed lactone formation when a γ,δ-unsaturated carboxylic acid was treated with *o*-nitrobenzeneselenenyl bromide in acetic acid.

[111c] E. G. Kataev, T. G. Mannafov, E. A. Berdnikov, and O. A. Komarovskaya, *Zh. Org Khim.* **9**, 1983 (1973).

[111d] E. G. Kataev, T. G. Mannafov, A. B. Remizov, and O. A. Komarovskaya, *Zh. Org. Khim.* **11**, 2322 (1975).

[111e] D. G. Garatt and G. H. Schmid, *Can. J. Chem.* **52**, 3599 (1974).

[111f] D. G. Garratt and G. H. Schmid, *Can. J. Chem.* **52**, 1027 (1974).

[111g] G. H. Schmid and D. G. Garratt, *Tetrahedron Lett.* p. 3991 (1975); *Chemica Scripta* **10**, 76 (1976).

[111h] H. J. Reich and J. E. Trend, *Can. J. Chem.* **53**, 1922 (1975).

[111i] S. Raucher, *Tetrahedron Lett.* p. 3909 (1977).

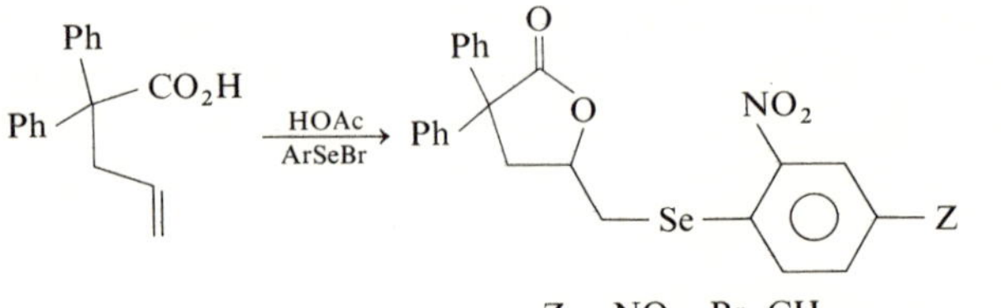

(ref. 103)

$Z = NO_2, Br, CH_3$

Improved reaction conditions for performing such lactonization using $PhSeCl/NEt_3$ have been developed by Nicolaou and Lysenko[112] and by Clive and Chittattu.[112a] The phenylseleno-substituted lactones undergo selenoxide elimination upon oxidation, or reductive cleavage upon treatment with Raney nickel.

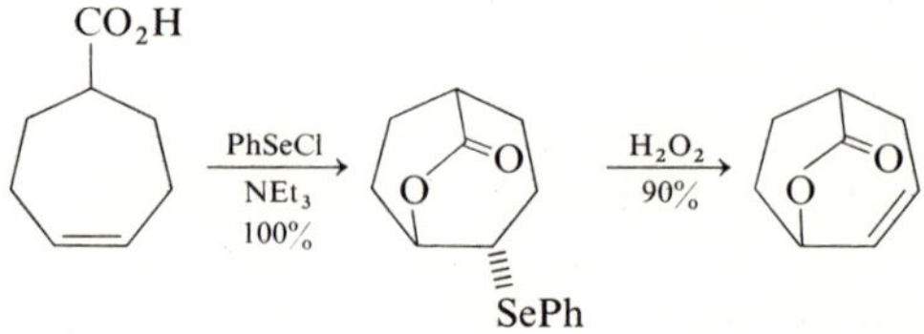

(ref. 112)

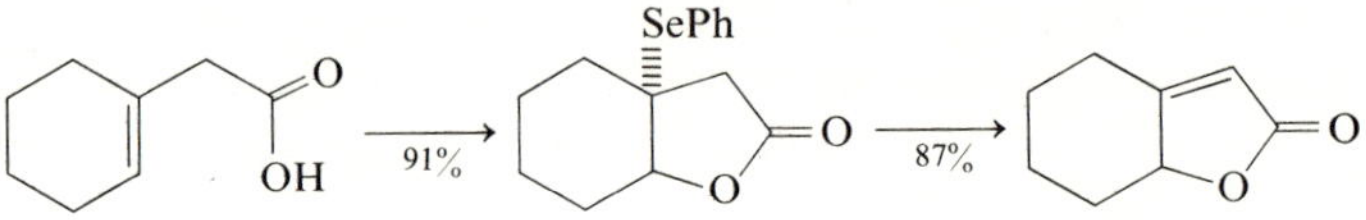

(ref. 112)

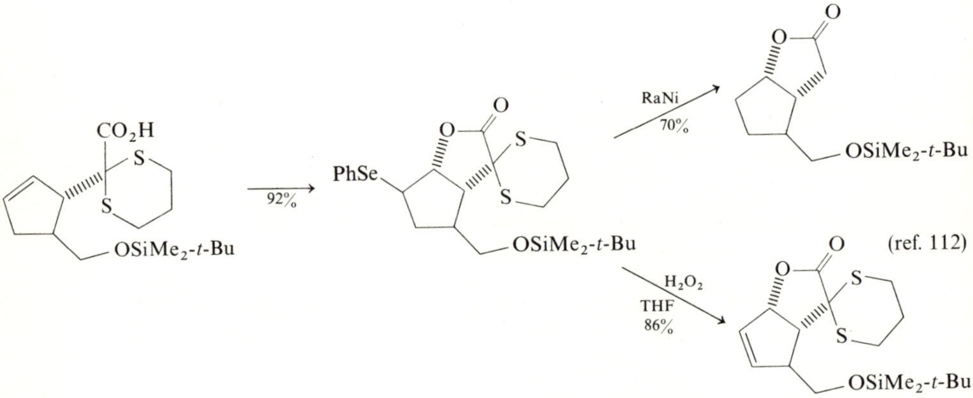

(ref. 112)

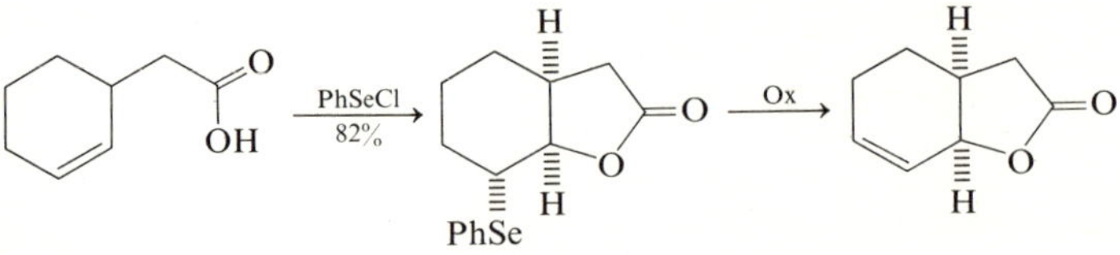

(ref. 112a)

[112] K. C. Nicolaou and Z. Lysenko, *J. Am. Chem. Soc.* **99**, 3185 (1977).
[112a] D. L. J. Clive and G. Chittattu, *Chem. Commun.* p. 484 (1977).

This phenylselenolactonization appears to be as widely applicable as the much used iodolactonization which it greatly resembles.

Cyclic ether formation can also occur during addition of PhSeCl to olefins,[112b–f] as shown by the following transformation in the prostacyclin series.

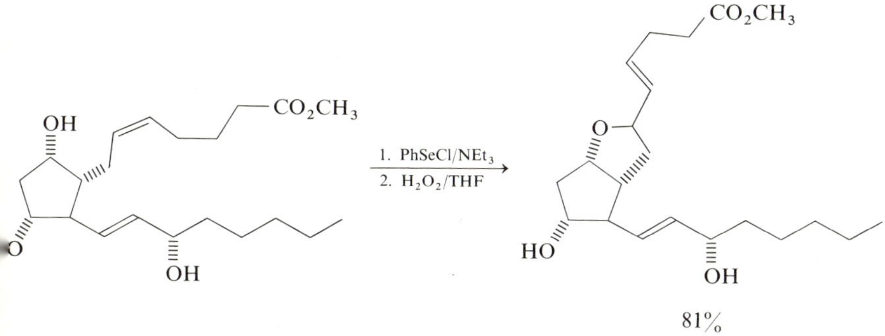

(ref. 112c)

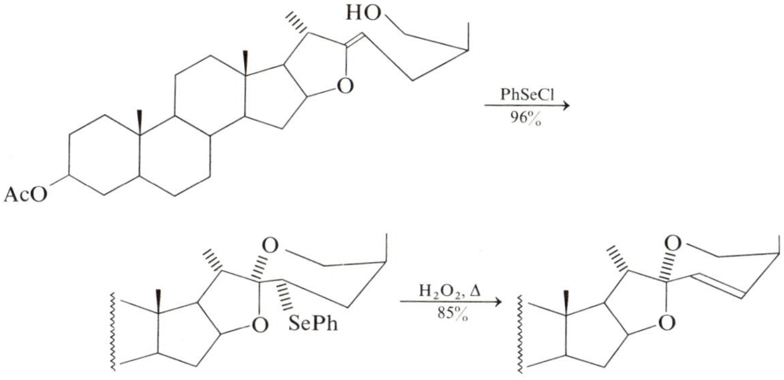

(ref. 112f)

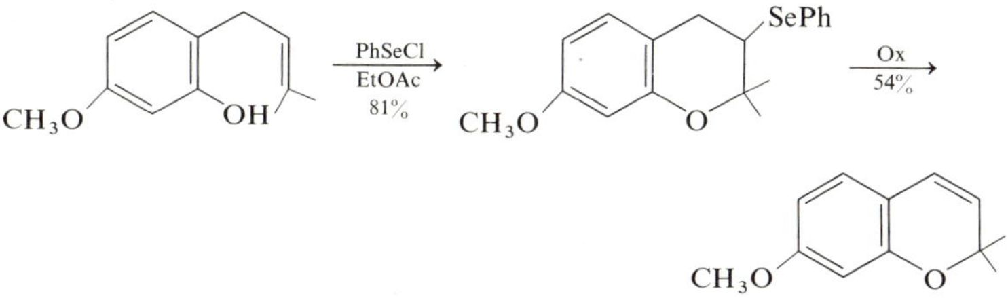

(ref. 112e)

[112b] K. C. Nicolaou and Z. Lysenko, *Tetrahedron Lett.* p. 1257 (1977).
[112c] K. C. Nicolaou and W. E. Barnette, *Chem. Commun.* 331 (1977).
[112d] E. J. Corey, G. E. Keck, and I. Szekely, *J. Am. Chem. Soc.* **99**, 2006 (1977).
[112e] D. L. J. Clive, G. Chittattu, N. J. Curtis, W. A. Kiel, and C. K. Wong, *Chem. Commun.* p. 725 (1977).
[112f] A. G. González, C. Betancor, C. G. Francisco, R. Hernández, J. A. Salazar, and E. Suárez, *Tetrahedron Lett.* p. 2959 (1977).

E. SELENIUM-STABILIZED ANIONS

The formation of selenides and selenoxides by reaction of α-lithio organoselenium compounds with alkylating agents and carbonyl compounds provides versatile synthetic routes to olefins, dienes, and allyl alcohols.

Three pathways to the required lithium reagents have been developed.

a. Deprotonation of Selenides. This procedure is limited by the necessity of using lithium amide bases since alkyllithium reagents usually attack at selenium causing Se–C bond cleavage.[41,113–113d] One of the few exceptions is the deprotonation of phenylselenotrimethylsilylmethane using *sec*-butyllithium.[41] The only simple alkyl selenide significantly deprotonated by amide bases is methyl *m*-trifluoromethylphenyl selenide using lithium tetramethylpiperidide.[114] When additional carbanion stabilizing groups are present, amide bases can be used to deprotonate a variety of selenides: $PhSeCH_2Z$ (Z = PhSe,[113b] CO_2R,[40] phenyl,[41,113d] vinyl,[33] and propargyl[52]). The very substantial acidifying effect of electron-withdrawing substituents on the phenylseleno group allows the preparation of additional lithium reagents using amide bases: *m*-CF_3–C_6H_4SeCHLi—Z (Z = H,[114] OCH_3,[43a] $SiMe_3$,[114] and ArSeCLi=CH_2[115]).

b. Cleavage of Selenoacetals and Ketals. Seebach and Peleties[113b] found that bisphenylselenomethane is cleaved to phenylselenomethyllithium on treatment with *n*-butyllithium.

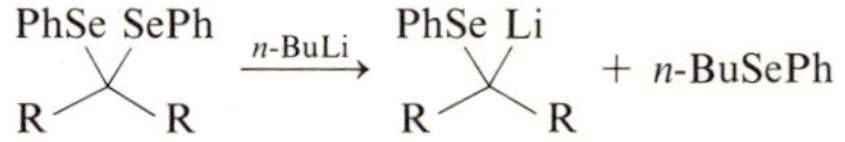

Subsequent work by Seebach[116] and Krief[113c] has extended the reaction to the preparation of secondary and tertiary α-lithio selenides. Methylseleno ketals are cleaved as well.[117] Such selenolithium compounds are now more readily available than the corresponding sulfur compounds (sulfides rarely undergo reaction with alkyllithium reagents at sulfur[118]). In fact, secondary and tertiary α-lithio sulfides are readily prepared by cleavage of mixed thio-

[113] H. Gilman and R. L. Bebb, *J. Am. Chem. Soc.* **61**, 109 (1939).
[113a] H. Gilman and F. J. Webb, *J. Am. Chem. Soc*, **71**, 4062 (1949).
[113b] D. Seebach and N. Peleties, *Angew. Chem.* **81**, 465 (1969); *Chem. Ber.* **105**, 511 (1972).
[113c] W. Dumont, P. Bayet, and A. Krief, *Angew. Chem. Int. Ed. Engl.* **13**, 804 (1974).
[113d] R. H. Mitchell, *Chem. Commun.* 990 (1975).
[114] S. K. Shah, Ph.D. Thesis, University of Wisconsin, Madison, 1977.
[115] H. J. Reich, P. D. Clark, and I. L. Reich, unpublished results.
[116] D. Seebach and A. K. Beck, *Angew. Chem., Int. Ed., Engl.* **13**, 806 (1974).
[117] J. N. Denis, W. Dumont, and A. Krief, *Tetrahedron Lett.* 453 (1976).
[118] J. F. Biellmann, J. B. Ducep, J. L. Schmitt, and J. J. Vicens, *Tetrahedron* **32**, 1061 (1976); J. Meinwald, S. Knapp, S. K. Obendorf, and R. E. Hughes, *J. Am. Chem. Soc.* **98**, 6643 (1976).

seleno acetals and ketals.[116,117] α-Lithio silanes can similarly be prepared by cleavage of α-silyl dialkyl selenides.[119] The cleavage of 1,1-bisphenylseleno ethylenes leads to the corresponding vinyllithium reagents.[115,119a]

Bisphenyltelluromethane is also cleaved by *n*-BuLi to give $PhTeCH_2Li$.[120]

c. Deprotonation of Phenyl Alkyl Selenoxides. A procedure for the *in situ* oxidation and deprotonation of alkyl selenides has been developed by Reich and Shah.[41] The product selenoxides, obtained by alkylation or reaction with carbonyl compounds, undergo selenoxide elimination directly to give olefins or allyl alcohols. (LDA = $LiN(i\text{-}Pr)_2$.

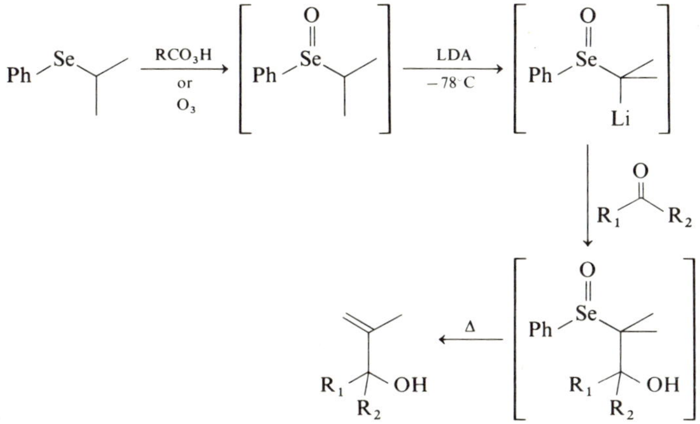

(ref. 41)

This procedure uses more readily accessible starting materials than b. above and avoids the formation of by-product butyl phenyl selenide. The selenoxide anions are less reactive, however, and may tend to cause more enolization with carbonyl compounds than selenide anions.

In the Seebach procedure involving cleavage of selenoketals, the overall synthetic transformation results in a ketone used as a vinyl anion equivalent, whereas in the Reich and Shah procedure, an alkyl halide or sulfonate is a vinyl anion equivalent.

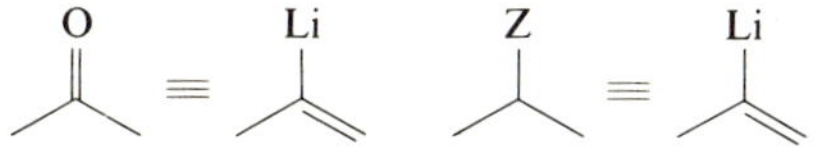

Table III lists olefins and dienes prepared by alkylation of α-lithio selenides and selenoxides followed by elimination.[41,113d,114,121] Table IV lists allyl alcohols prepared by reactions with ketones and aldehydes, with subsequent selenoxide elimination.

[119] W. Dumont and A. Krief, *Angew. Chem.* **88**, 184 (1976).
[119a] B. T. Groebel and D. Seebach, *Chem. Ber.* **110**, 867 (1977).
[120] D. Seebach and A. K. Beck, *Chem. Ber.* **108**, 314 (1975).
[121] H. J. Reich and S. K. Shah, *J. Org. Chem.* **42**, 1773 (1977).

TABLE III

OLEFINS PREPARED BY ALKYLATION OF SELENIUM-STABILIZED ANIONS

Anion	Alkyl halide	Selenide (selenoxide)	Oxidant	Olefin	Yield of olefin (%)	Ref.
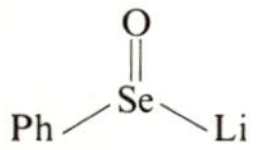	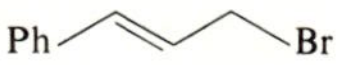	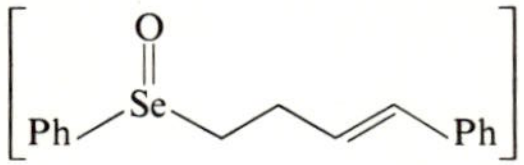	—		75	41
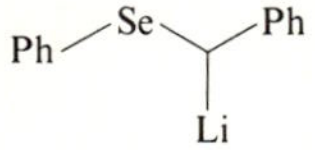	CH_3Br	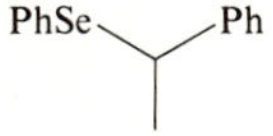	H_2O_2, THF	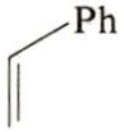	76	113d
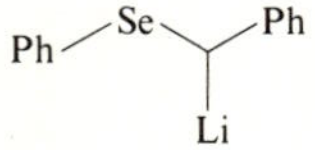	CH_3CH_2Br	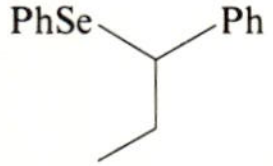	H_2O_2, THF	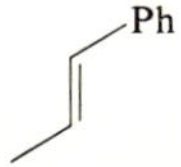	80	113d
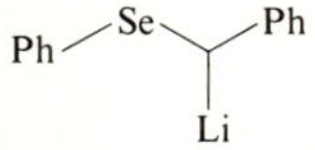	$PhCH_2Br$	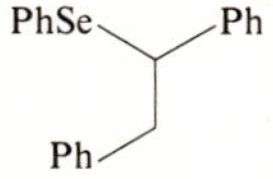	H_2O_2, THF	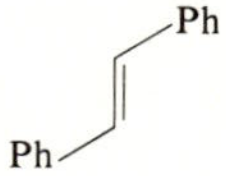	81	113d
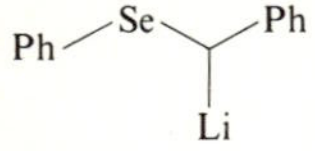	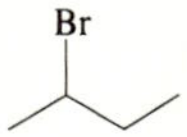	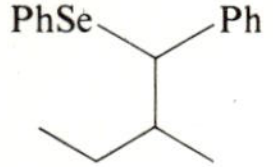	O_3, $-78°C$, Δ	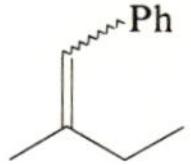	78	41
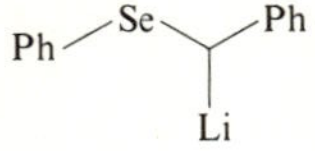		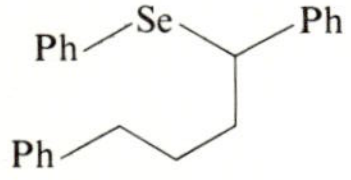	H_2O_2, CH_2Cl_2	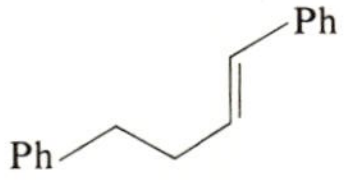	81	41

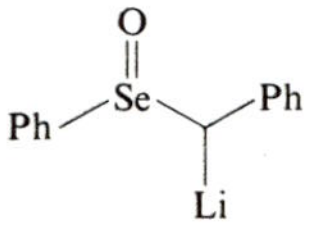	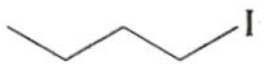	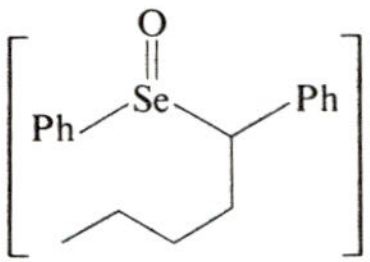	—	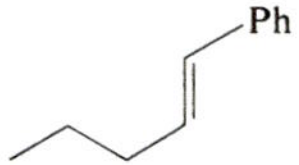	81	41
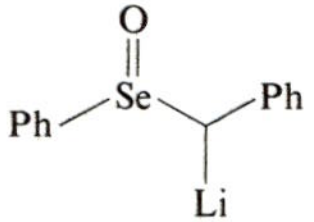	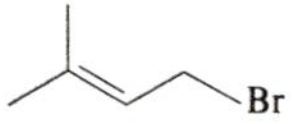	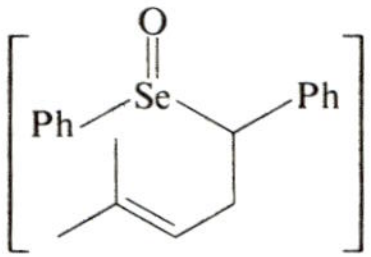	—	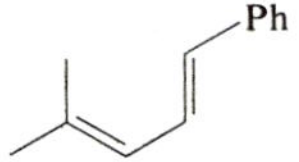	88	41
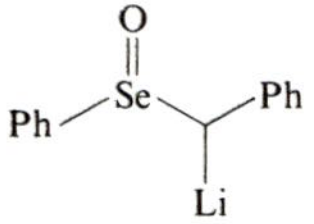	$PhCH_2Br$	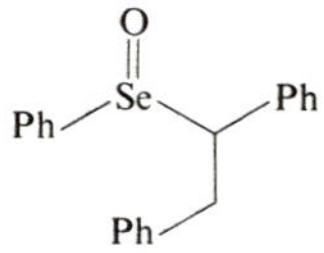	—	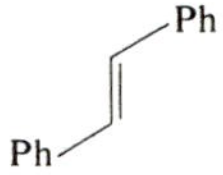	63	41
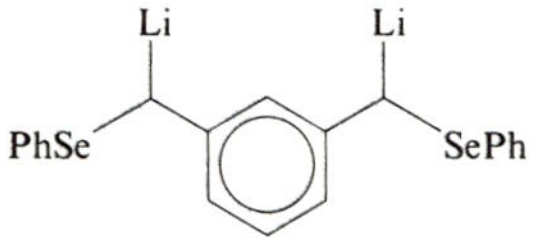	$PhCH_2Br$	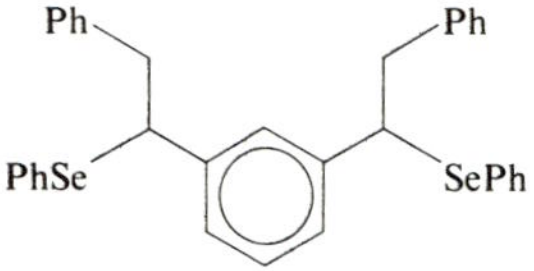	H_2O_2, THF	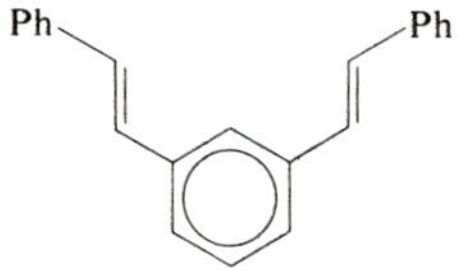	62	113d
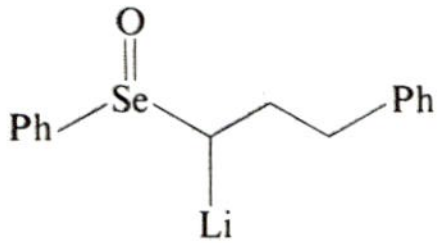	CH_3I	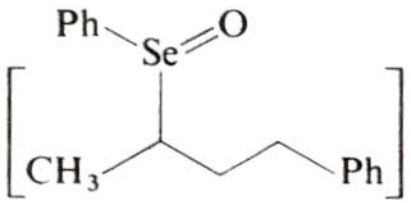	—	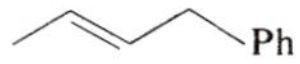	21	
					38	41

(*cont.*)

TABLE III (*cont.*)

Anion	Alkyl halide	Selenide (selenoxide)	Oxidant	Olefin	Yield of olefin (%)	Ref.
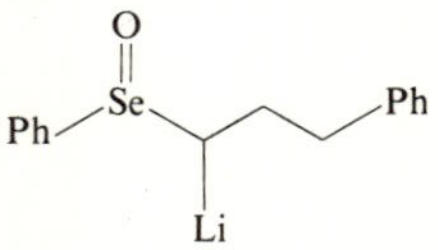	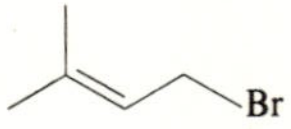	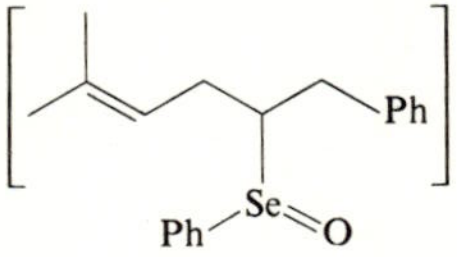	—	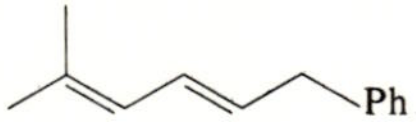	64	41
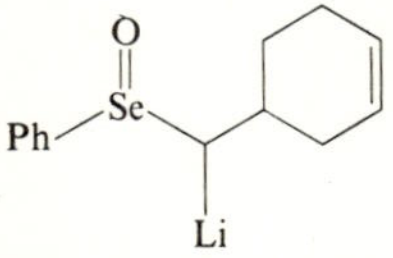	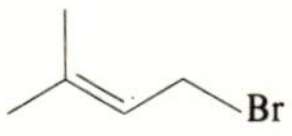	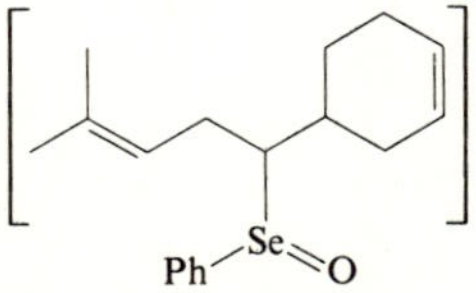	—	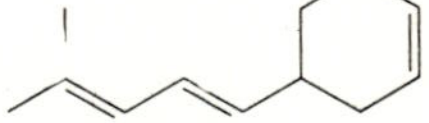	46	41
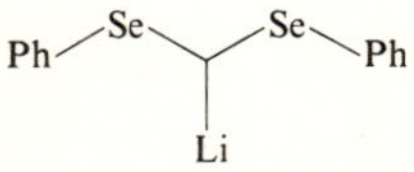	CH_3I	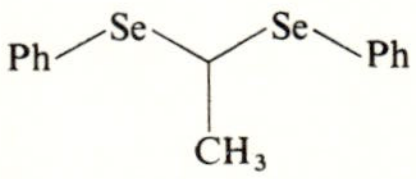	*m*-CPBA, Δ	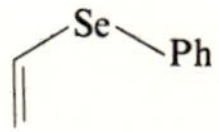	47	114
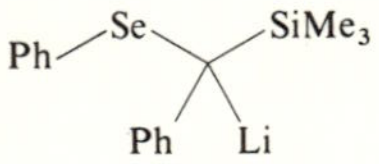	CH_3I	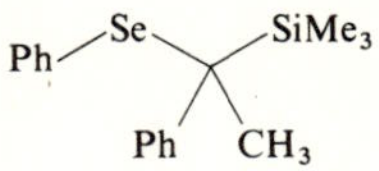	*m*-CPBA, Δ	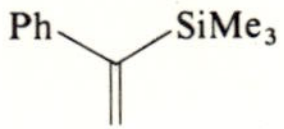	64[a]	121
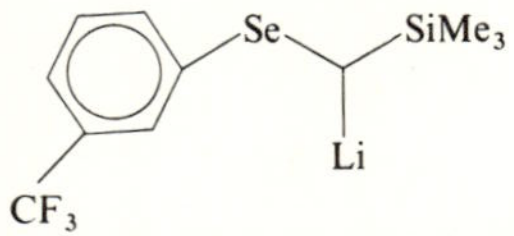		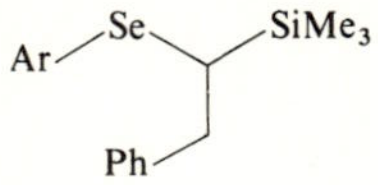	*m*-CPBA, Δ	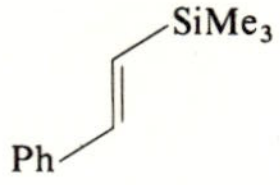	47[a]	121

[a] Other products resulting from sila-Pummerer reactions were also isolated.

TABLE IV

ALLYL ALCOHOLS PREPARED BY REACTION OF ALDEHYDES AND KETONES WITH SELENIUM STABILIZED ANIONS

Anion	Ketone or aldehyde	Selenide (selenoxide)	Oxidant	Product	Yield[a]	Ref.
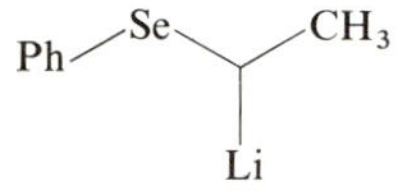	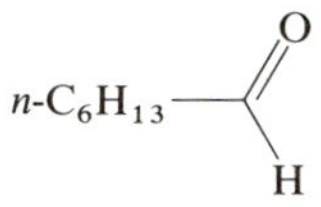	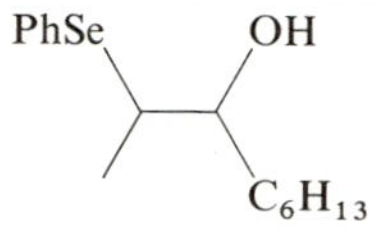	H_2O_2 EtOH	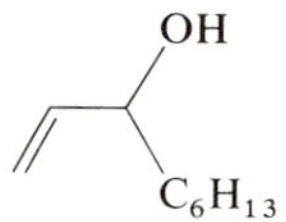	49, 78	113c
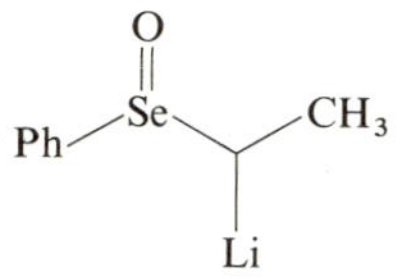	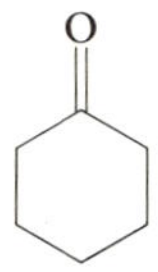	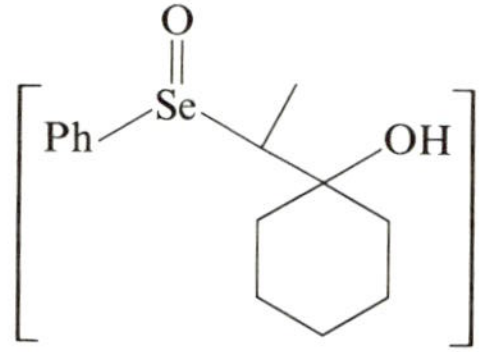	—	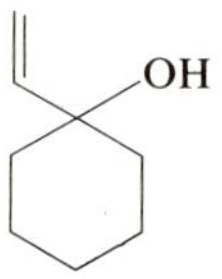	66	114
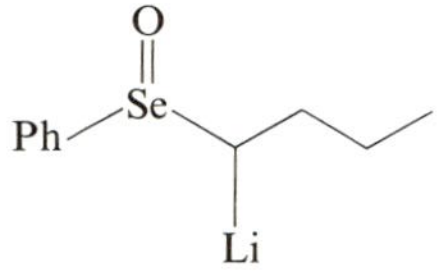	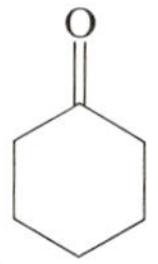	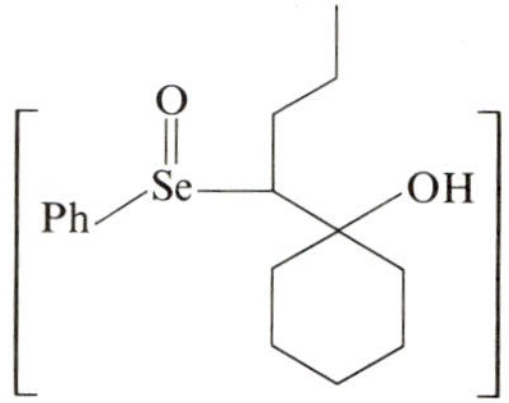	—	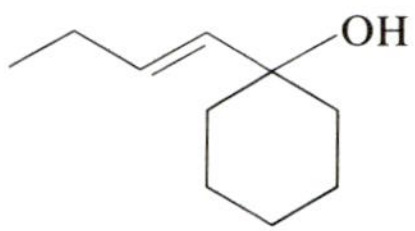	78	114
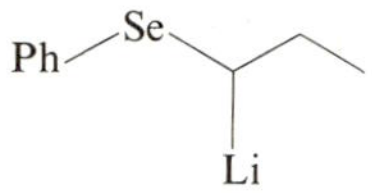	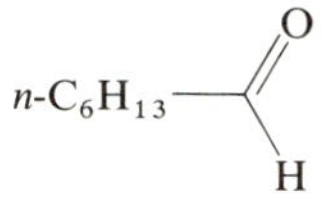	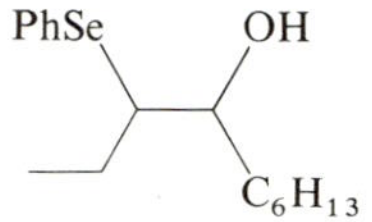	H_2O_2 EtOH	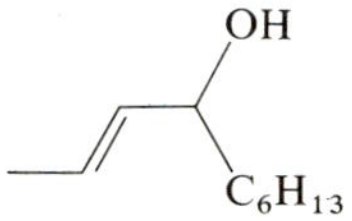	60, 74	113c

(cont.)

TABLE IV (*cont.*)

Anion	Ketone or aldehyde	Selenide (selenoxide)	Oxidant	Product	Yield[a]	Ref.
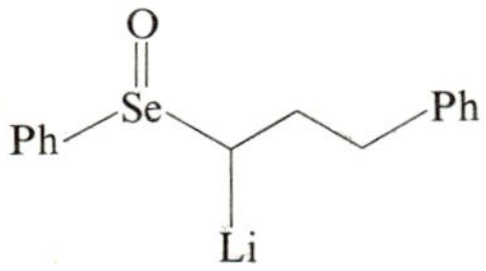		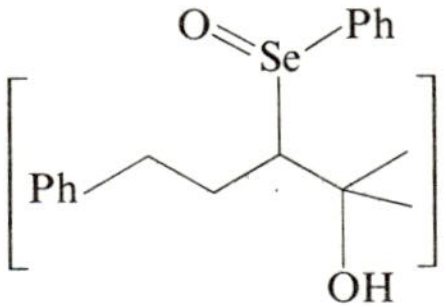	—	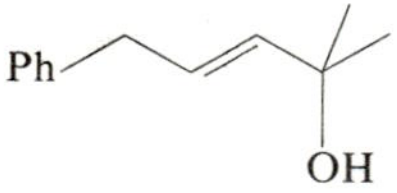	81	41
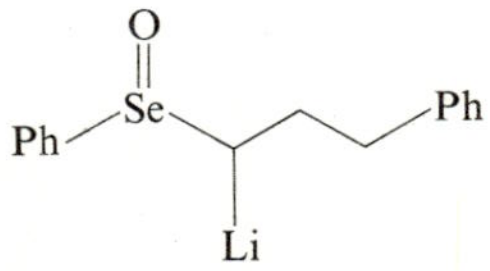	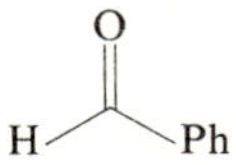	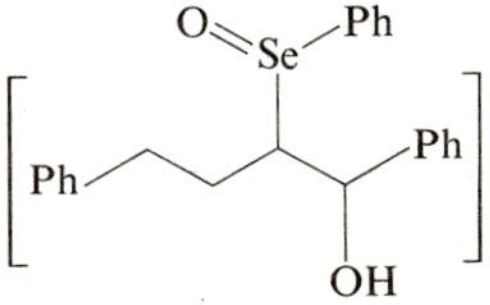	—	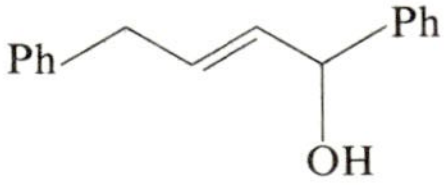	72	41
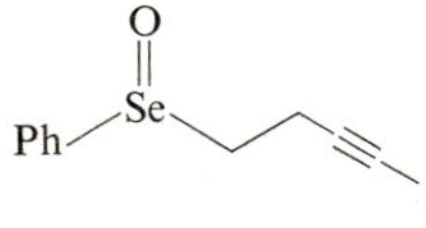	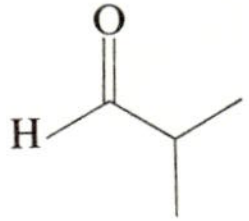	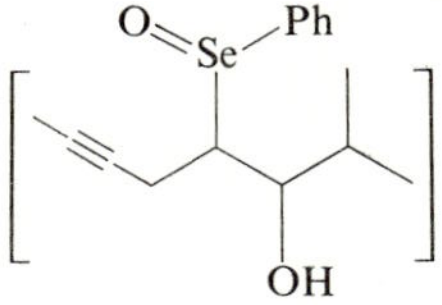	—	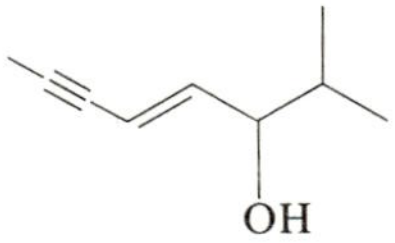	68	114
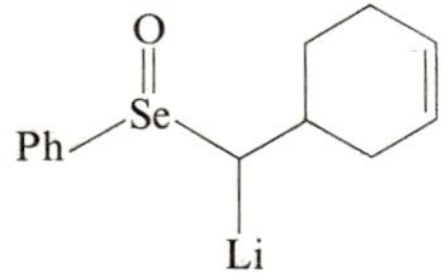	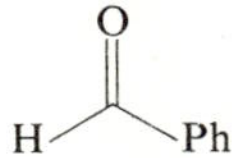	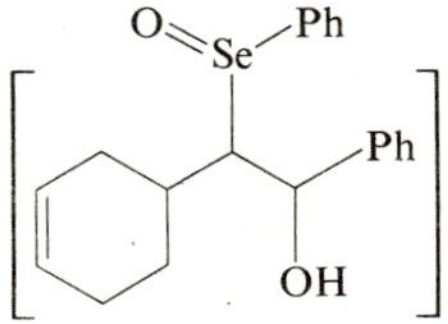	—	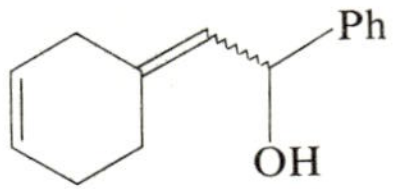	50	41

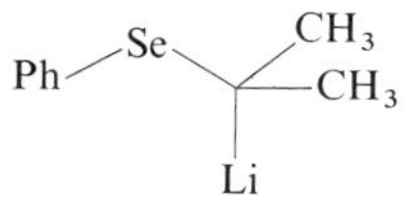	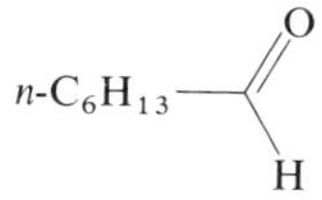	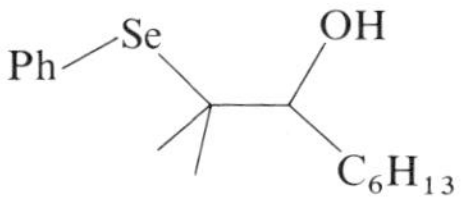	H_2O_2 EtOH	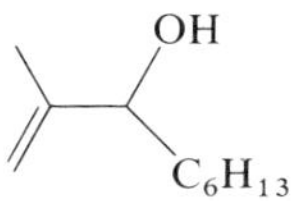	70, 80	113c
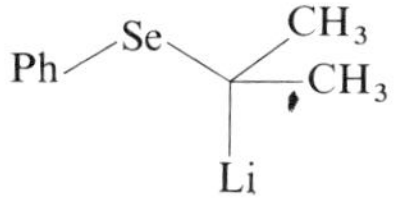	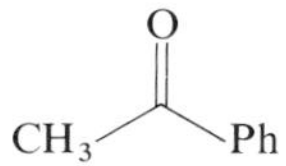	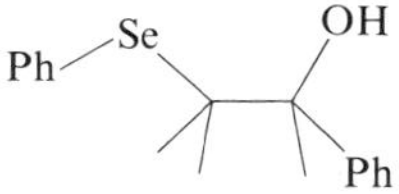	H_2O_2 EtOH	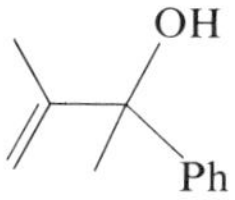	67, 80	113c
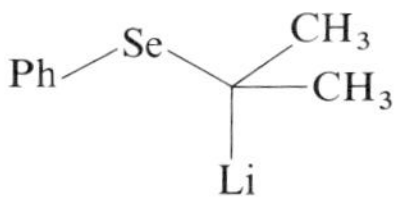	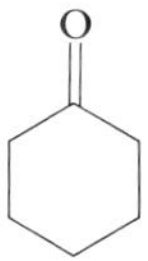	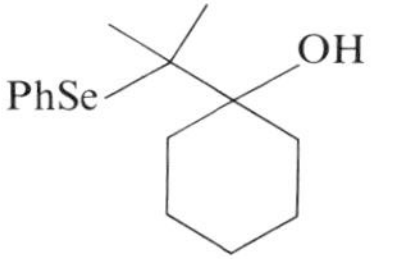	H_2O_2 EtOH	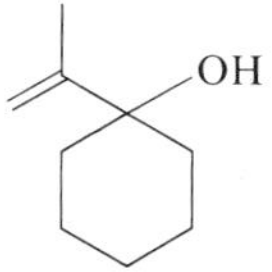	95 78, 93	116 113c
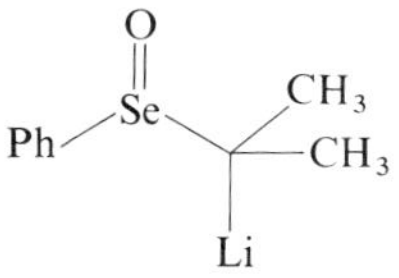	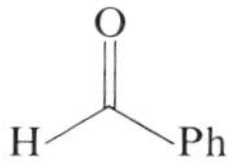	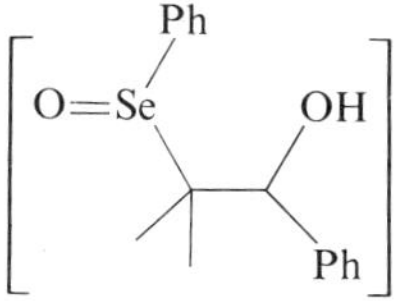	—	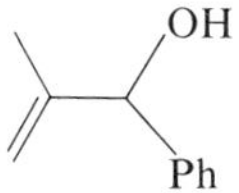	81	41
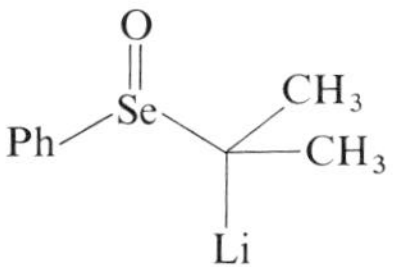	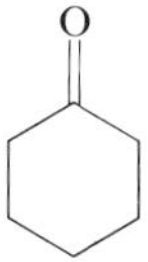	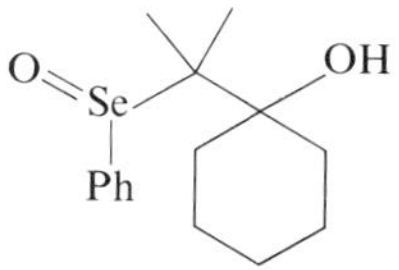	—	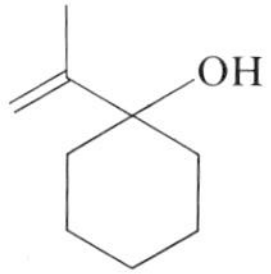	55	41
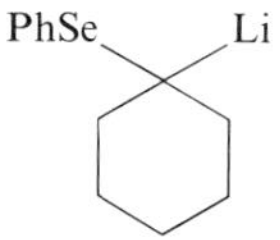	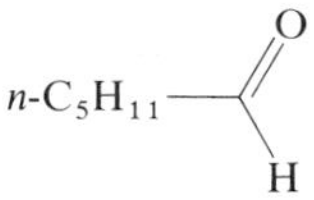	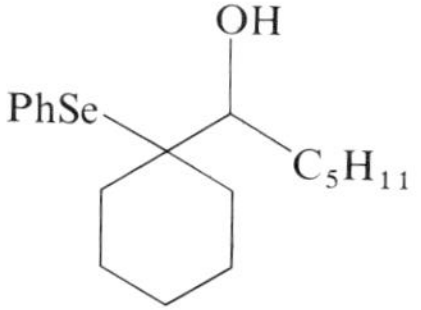	H_2O_2 EtOH	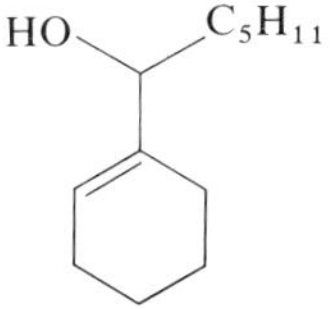	65	116

(*cont.*)

TABLE IV (*cont.*)

Anion	Ketone or aldehyde	Selenide (selenoxide)	Oxidant	Product	Yield[a]	Ref.
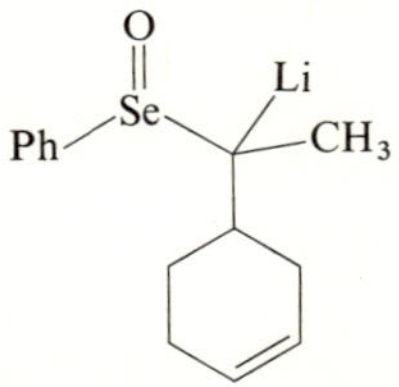	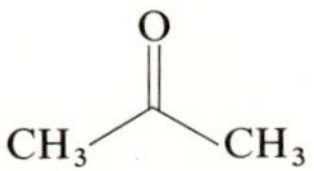	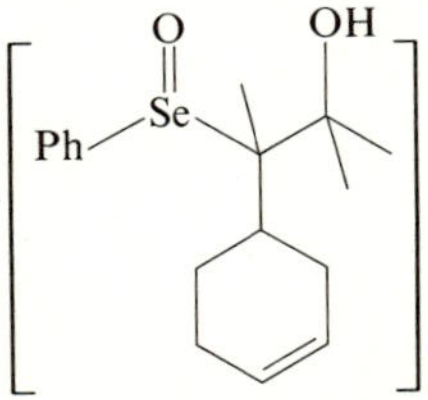	—	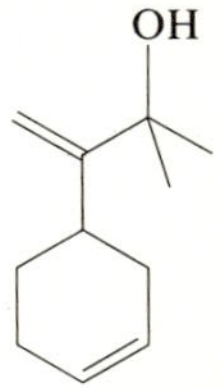	74	114
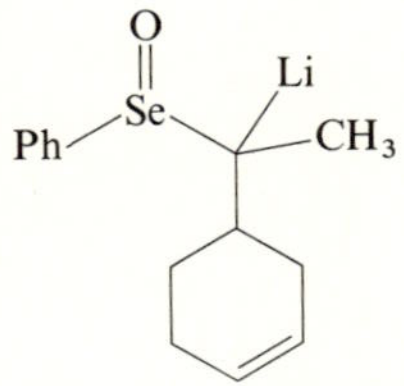	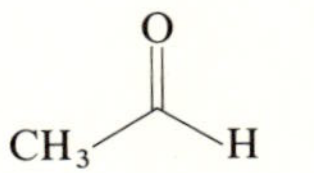	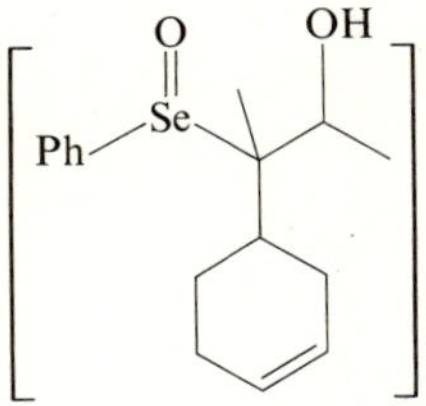	—	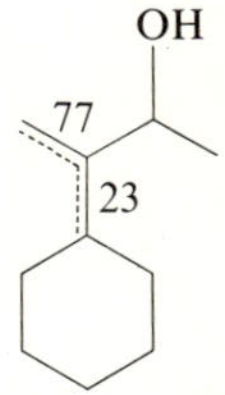	69	114

[a] Two numbers indicate isolated yield of selenide and allyl alcohol, respectively.

Allyl alcohols can also, in principle, be prepared by opening of epoxides with α-lithio selenides. Only one such example has been reported.[41,121a]

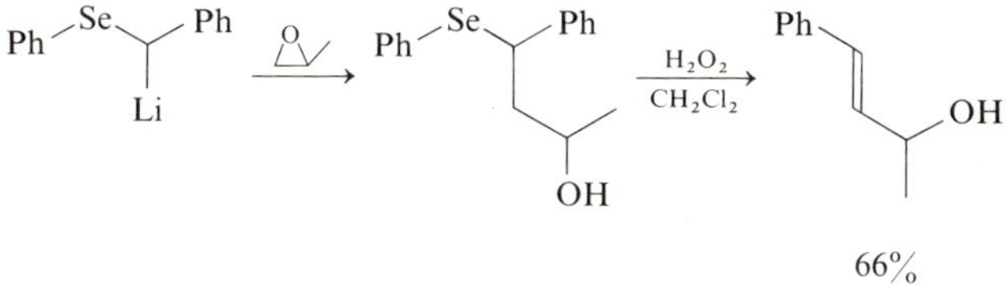

(ref. 41)

A theoretical study by Lehn, Wipff, and Demuynck[121b] has resulted in the prediction that a selenium group should be more effective at stabilizing negative charge at the α-position than should sulfur. The available evidence, however, indicates that α-protons in selenides are slightly less acidic than in sulfides.[113b,114,121c]

Reactions of selenium stabilized anions which lead to α,β-unsaturated carbonyl compounds and nitriles are discussed in Section III,G.

F. SELENENYLATION OF KETONES, ALDEHYDES, ESTERS, AND NITRILES

The dehydrogenation of ketones, esters, acids, and nitriles has been the most intensively explored area of synthetic organoselenium chemistry. These applications are based on the independent observations by Reich,[122] Sharpless,[40] and Clive[123] that α-phenylseleno carbonyl compounds are smoothly converted below room temperature to the α,β-unsaturated compounds upon oxidation to the selenoxide.

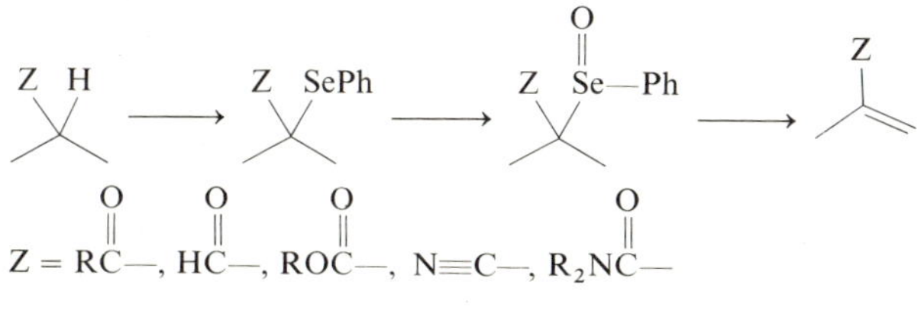

Stereochemical studies are lacking but all indications are that these reactions are pericyclic syn eliminations of the selenoxides.

[121a] M. Sevrin and A. Krief, *Tetrahedron Lett.* p. 187 (1978).
[121b] J.-M. Lehn, G. Wipff, and J. Demuynck, *Helv. Chim. Acta* **60**, 1239 (1977).
[121c] A. I. Shatenshtein and H. A. Gvozdeva, *Tetrahedron* **25**, 2749 (1969); F. G. Bordwell, *et al.*, *J. Org. Chem.* **42**, 326 (1977).
[122] H. J. Reich, I. L. Reich, and J. M. Renga, *J. Am. Chem. Soc.* **95**, 5813 (1973).
[123] D. L. J. Clive, *Chem. Commun.* p. 695 (1973).

The preparation of α,β-unsaturated carbonyl compounds using sulfoxide eliminations has been studied by Trost[124] (see Seebach and Teschner[125] for a recent review of keto sulfide chemistry). Sulfur and selenium methodology is compared in Section III, I.

A review of α-seleno ketone synthesis by Bergson and Delin[21] in 1962 provides background for the work described below. These authors explored several routes to ethylselenoacetone, including the reaction of acetone with EtSeBr, the reaction of acetone enolate with EtSeBr and Et_2Se_2, and the reaction of chloroacetone with $EtSe^-$, of which only the last gave useful yields of the selenide. It is of interest to note that the first two methods (Section III, F, 2 and 3 below) are now the most widely used procedures for the preparation of α-seleno ketones.

1. Direct Selenenylations of Ketones and Aldehydes

A number of arylseleno derivatives of acetone and acetophenone have been prepared (Table V),[111,126–128] most successfully by direct selenenylation using selenenyl thiocyanates (ArSeSCN)[126,126a] or chlorides (RSeCl). Selenenyl bromides cause mainly bromination of acetone and acetophenone.[5,21,126] Only selenenyl thiocyanates with strong electron-withdrawing groups on the aromatic ring have been prepared. These compounds are therefore not routinely available for synthetic applications and have not been used recently.

Intramolecular selenenylations of *o*-acetylselenenyl derivatives have been accomplished as illustrated in the Scheme below.[129,129a]

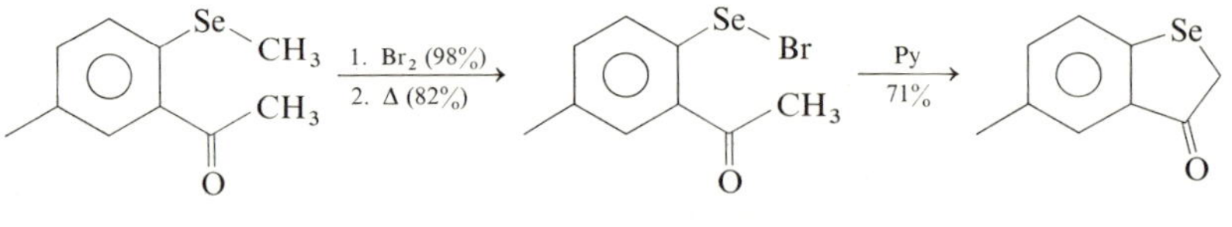

(ref. 129)

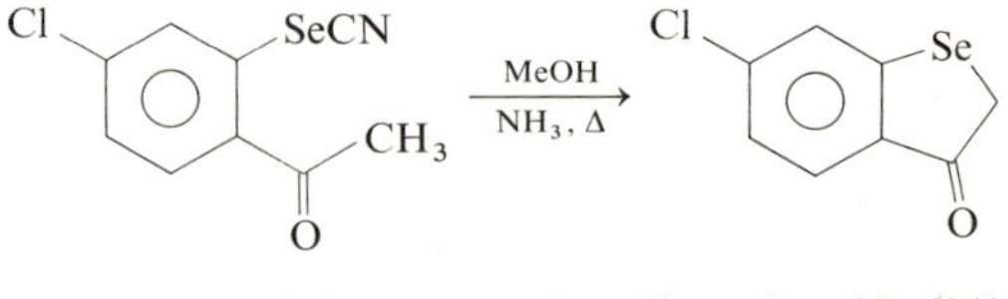

(ref. 129a)

[124] B. M. Trost and T. N. Salzmann, *J. Am. Chem. Soc.* **95**, 6840 (1973); B. M. Trost, T. N. Salzmann, and K. Hiroi, *ibid.* **98**, 4887 (1976).

[125] D. Seebach and M. Teschner, *Chem. Ber.* **109**, 1601 (1976).

[126] H. Rheinboldt and M. Perrier, *Bull. Soc. Chim. Fr.* **17**, 759 (1950).

[126a] H. Rheinboldt and M. Perrier, *Bull. Soc. Chim. Fr.* **20**, 484 (1953).

[127] D. D. Lawson and N. Kharasch, *J. Org. Chem.* **24**, 857 (1959).

[128] L. R. M. Pitombo, *Chem. Ber.* **92**, 745 (1959).

[129] J. Gosselck, *Chem. Ber.* **91**, 2345 (1958).

[129a] E. Giesbrecht and I. Mori, *An. Acad. Bras. Cienc.* **30**, 521 (1958); *Chem. Abstr.* **58**. 5611 (1963).

TABLE V

PREPARATION OF α-SELENOACETONE AND ACETOPHENONE BY DIRECT SELENENYLATION

Selenenyl compound	Derivative			Yield of selenide (%)	Ref.
	Z	R_1	R_2	$Z = CH_2{-}C(=O){-}CH_3$	
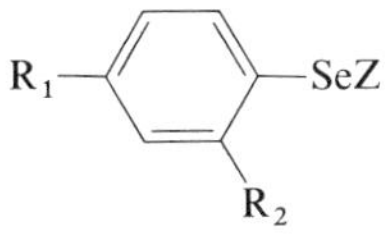	SCN	H	NO_2	99.6	126
	SCN	Cl	NO_2	99.3	126
	SCN	Br	NO_2	100	126
	SCN	CH_3	NO_2	97.4	126
	SCN	NO_2	NO_2	98.4	126
	SCN	OCH_3	NO_2	98	126a
	SCN	NO_2	OCH_3	98	126a
	Br	Cl	NO_2	34	126
	Br	Br	NO_2	43	126
	Br	CH_3	NO_2	46.5	126
	Br	NO_2	NO_2	56	126
	Cl	NO_2	NO_2	—	127
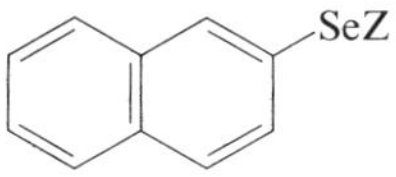	Cl			86	128
CF_3SeZ	Cl			49	111
				$Z = CH_2{-}C(=O){-}Ph$	
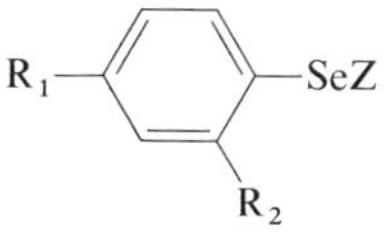	SCN	OCH_3	NO_2	87	126a
	SCN	NO_2	OCH_3	88	126a
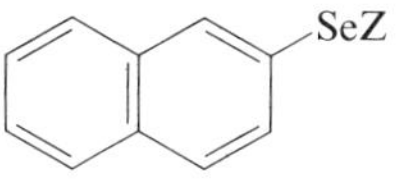	Cl			74	128

Sharpless, Lauer, and Teranishi[40] have developed a procedure for the selenenylation of ketones and even aldehydes using PhSeCl and found that the reaction is compatible with the presence of a number of functional groups including olefins and esters. Several steroidal enones have been selenenylated using this procedure and subsequently converted to dienones.[130] The reaction is acid catalyzed, analogous to the bromination of ketones, and probably involves reaction of enols with PhSeCl.

2. Reaction of Enol Derivatives and Acetylenes with Selenenylating Agents

The transformation of enol acetates, ethers, and silyl ethers to α-seleno carbonyl compounds has not been extensively studied, but all of these functional groups can probably be selenenylated. The first examples of this type of reaction were provided by Hölzle and Jenny,[102] although their reagents and conditions were not well suited for general synthetic work. These workers isolated **12** from the reaction of several selenenyl acetate derivatives with vinyl acetate.

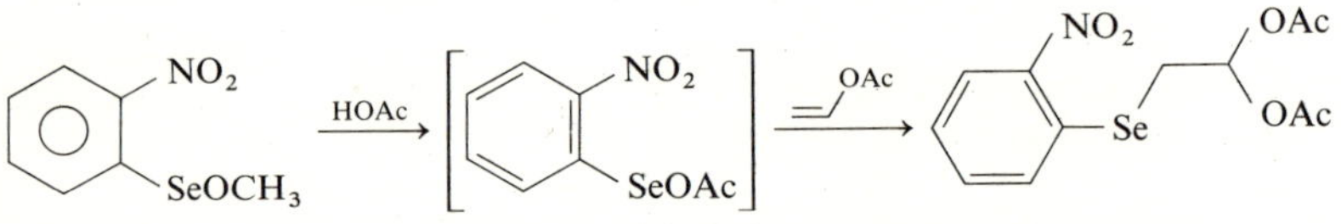

(ref. 102)

(12)

Clive[123] more recently converted the enol acetates of cyclohexanone and cyclopentanone to α-seleno ketones by addition of PhSeBr in the presence of silver trifluoroacetate. Reich[8] characterized the reaction product of PhSeBr and AgO_2CCF_3 spectroscopically as $PhSeO_2CCF_3$ and demonstrated that this reagent adds cleanly to a variety of olefins, acetylenes, and enol acetates. The reactivity of $PhSeO_2CCF_3$ appears to be greater than that of the selenenyl halides, but addition to vinyl chloride **(13)** and to the suprisingly unreactive enol acetate **(14)**[5] does not occur. The silyl ketene acetal **(15)** is converted to α-phenylseleno ester on treatment with PhSeBr.[131]

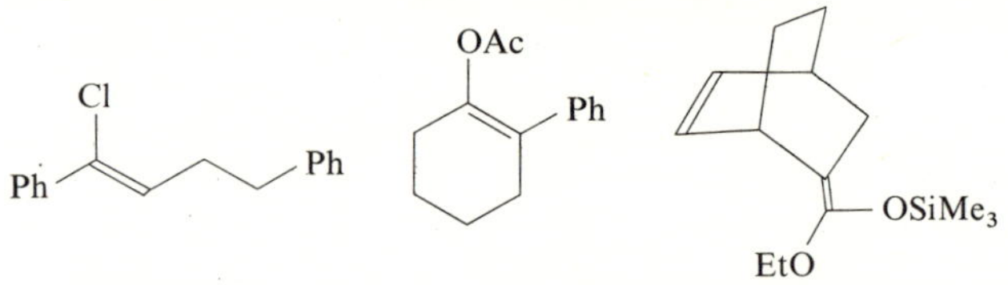

(13) **(14)** **(15)**

Several acetylenes have been converted to $PhSeO_2CCF_3$ adducts and hence to α-seleno ketones.[8]

[130] R. B. Boar, *J. Chem. Soc., Perkin Trans. 1* p. 1275 (1975); R. B. Boar, private communication.

[131] E. Vedejs and W. Wilber, private communication.

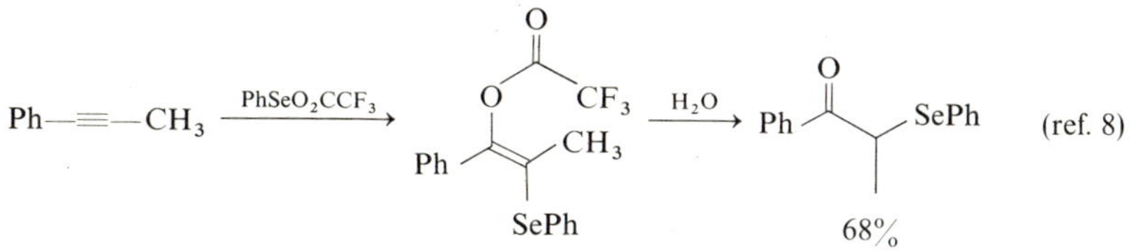

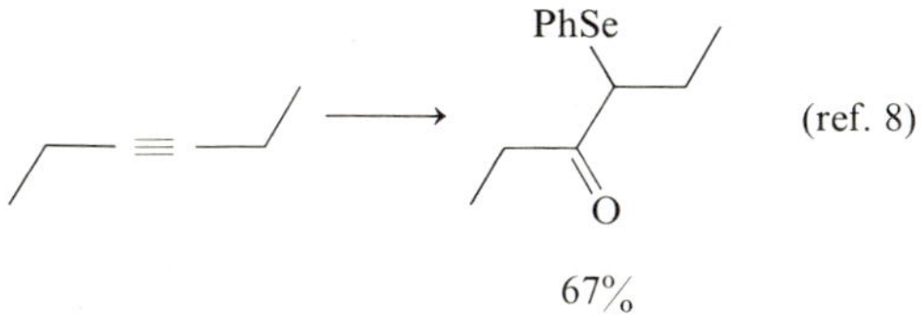

The generality and regioselectivity of this reaction is not known. A number of selenenyl chloride additions to acetylenes have been reported.[111g,132]

3. Reaction of Ketone, Ester, and Nitrile Anions with Selenenylating Agents

Bergson and Delin[21] attempted to prepare ethylselenoacetone by reaction of acetone sodium enolate with EtSeBr and Et_2Se_2. Unfortunately, acetone is particularly difficult to selenenylate using enolate chemistry and the reaction conditions chosen were not suitable. The desired product was, therefore, not obtained. However, with proper choice of reagents and reaction conditions carbanions ranging in basicity from *t*-butyllithium to sodium acetylacetonate react with PhSeSePh, PhSeBr, and/or PhSeCl to give selenides. This represents the most versatile and widely used method for the introduction of phenylseleno groups. Ketone, ester, lactone, lactam,[133] nitrile, nitrosamine,[134] and sulfone[84] anions have been selenenylated as well as more stabilized anions from β-dicarbonyl compounds, β-keto sulfoxides, and β-keto selenoxides (see Section III, F, 4). Table VI presents all reported examples of these reactions.[5,8,10,14,21,34,35,40,55,59,75a,122,123,130,135–156]

[132] L. Cherici and F. Montanari, *Gazz. Chim. Ital.* **86**, 1269 (1956); E. G. Kataev, T. G. Mannafov, and M. Kh. Mannafov, *Kinet. Katal.* **9**, 957 (1968); L. M. Kataev, N. S. Podkovyrina, T. G. Mannafov, and E. G. Kataev, *Zh. Strukt. Khim.* **10**, 1124 (1969); *Chem. Abstr.* **72**, 66229 (1970); E. G. Kataev, T. G. Mannafov, and Yu. Yu. Samitov, *Zh. Org. Khim.* **11**, 2324 (1975).

[133] P. A. Zoretic and P. Soja, *J. Org. Chem.* **41**, 3587 (1976); *J. Heterocyclic Chem.* **14**, 681 (1977).

[134] D. Seebach and D. Enders, *J. Med. Chem.* **17**, 1225 (1974).

[135] C. Giordano, *Gazz. Chim. Ital.* **105**, 1265 (1975).

[136] J. M. Renga, Ph.D. Thesis, University of Wisconsin, Madison (1975).

[137] G. Stork and S. Raucher, *J. Am. Chem. Soc.* **98**, 1583 (1976).

[138] D. N. Brattesani and C. H. Heathcock, *Tetrahedron Lett.* p. 2279 (1974).

[138a] P. A. Grieco, Y. Ohfune, and G. Majetich, *J. Am. Chem. Soc.* **99**, 7393 (1977).

Diphenyl diselenide reacts with nitrile, ester, and lactone anions, and has been recommended to be superior to the selenenyl halides for some of these systems.[40,151] Ketone enolates are not selenenylated by Ph_2Se_2 because of an unfavorable equilibrium which is established rapidly even at −78°C.[5,136] In contrast, diphenyl disulfide does sulfenylate ketone enolates.[124,157]

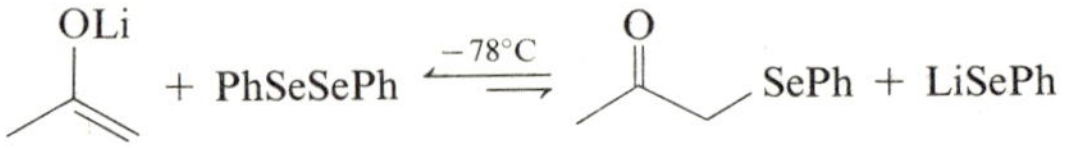

The selenenyl halides (PhSeBr and PhSeCl) react almost instantaneously with even highly stabilized anions. The kinetic enolate is trapped, as shown by conversion of 2-methyl- and 2-phenylcyclohexanone and 2-benzylcyclopentanone to the 6-and 5-phenylseleno derivatives, respectively.[5] In contrast,

---

[139] J. S. Swenton, R. M. Blankenship, and R. Sanitra, *J. Am. Chem. Soc.* **97**, 4941 (1975).
[140] P. A. Grieco and M. Nishizawa, *Chem. Commun.* p. 582 (1976).
[141] D. Caine, A. A. Boucugnani, and W. R. Pennington, *J. Org. Chem.* **41**, 3632 (1976).
[141a] J. E. McMurry and M. G. Silvestri, *J. Org. Chem.* **41**, 3953 (1976).
[141b] I. Iijima, K. C. Rice, and J. V. Silverton, *Heterocycles* **6**, 1157 (1977).
[141c] D. Caine and A. S. Frobese, *Tetrahedron Lett.* p. 3107 (1977).
[141d] T. Ishida and K. Wada, *Chem. Commun.* p. 337 (1977).
[142] T. Cynkowski, M. Kocor, *Rocz. Chem.* **50**, 257 (1976).
[142a] R. E. Ireland, P. Beslin, R. Giger, U. Hengartner, H. A. Kirst, and H. Maag, *J. Org. Chem.* **42**, 1267 (1977).
[143] S. H. Korzeniowski, D. P. Vanderbilt, and L. B. Hendry, *Org. Prep. Proced. Int.*, **8**, 81 (1976).
[144] A. G. Schultz, *J. Org. Chem.* **40**, 3466 (1975).
[145] R. F. C. Brown, F. W. Eastwood, and G. L. McMullen, *J. Am. Chem. Soc.* **98**, 7421 (1976); *Austr. J. Chem.* **30**, 179 (1977).
[145a] S. C. Welch, C. P. Hagan, D. H. White, W. P. Fleming, and J. W. Trotter, *J. Am. Chem. Soc.* **99**, 549 (1977).
[146] F. M. Dean and B. K. Park, *Chem. Commun.*, 142 (1975); F. M. Dean, private communication.
[147] T. J. Brocksom, N. Petragnani, and R. Rodrigues, *J. Org. Chem.* **39**, 2114 (1974).
[147a] T. Tsuji, K. Masaoka, T. Takahashi, A. Suzuki, and N. Miyaura, *Bull. Chem. Soc., Japan* **50**, 2507 (1977).
[147b] T. Tsuji, K. Masaoka, and T. Takahashi, *Tetrahedron Lett.* p. 2267 (1977).
[148] D. Buddhsukh and P. Magnus, *Chem. Commun.*, 952 (1975).
[148a] P. J. Kociensky, G. Cernigliaro, and G. Feldstein, *J. Org. Chem.* **42**, 353 (1977).
[149] J. Roy, W. Gordon, I. L. Schwartz, and R. Walter, *J. Org. Chem.* **35**, 510 (1970).
[150] P. A. Grieco, C. S. Pogonowski, and S. Burke, *J. Org. Chem.* **40**, 542 (1975).
[151] P. A. Grieco and M. Miyashita, *J. Org. Chem.* **39**, 120 (1974).
[152] P. A. Grieco and M. Miyashita, *Tetrahedron Lett.* p. 1869 (1974).
[153] K. Yamakawa, K. Nishitani, and A. Yamamoto, *Chem. Lett.* p. 177 (1976).
[154] J. A. Marshall and R. H. Ellison, *J. Am. Chem. Soc.* **98**, 4312 (1976).
[155] G. Ohloff, W. Giersch, K. H. Schulte-Elte, and C. Vial, *Helv. Chim. Acta* **59**, 1140 (1976).
[155a] T. Wakamatsu, K. Akasata, and Y. Ban, *Tetrahedron Lett.* p. 2755 (1977).
[156] D. N. Brattesani and C. H. Heathcock, *J. Org. Chem.* **40**, 2165 (1975).
[157] D. Seebach and M. Teschner, *Tetrahedron Lett.* p. 5113 (1973).

TABLE VI

α,β-UNSATURATED CARBONYL COMPOUNDS AND NITRILES PREPARED BY SELENENYLATION AND SELENOXIDE ELIMINATION

Starting material	Olefin	Selenide: Reagent	Selenide: Yield[a] (%)	Olefin: Oxidant	Olefin: Yield[b] (%)	Ref.
Aldehydes						
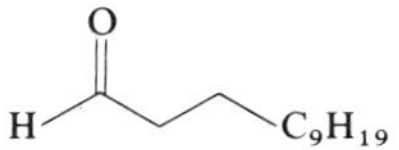	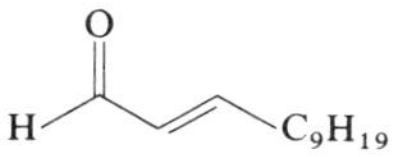	PhSeCl, EtOAc	—	$NaIO_4$	(46)	40
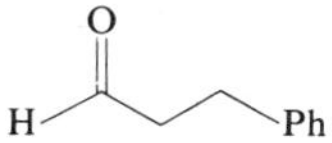	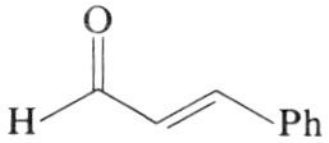	PhSeCl	—	$NaIO_4$	(67)	40
Ketones						
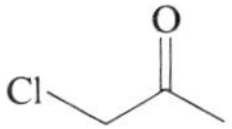	—	$C_2H_5SeMgBr$, Et_2O	19			21
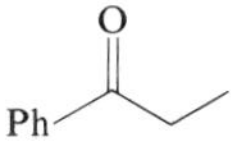	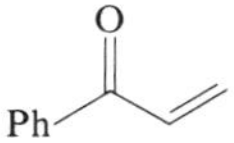	LDA, PhSeBr LDA, PhSeBr PhSeCl	88 — —	$NaIO_4$ H_2O_2, CH_2Cl_2 $NaIO_4$	89 56 (84)	5 5 40
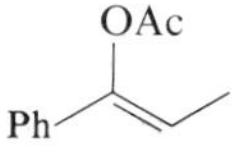	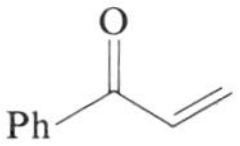	$PhSeO_2CCF_3$	83			5
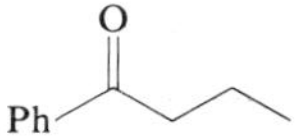	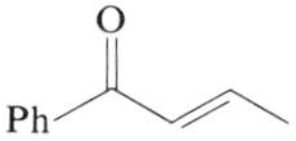	LDA, PhSeBr	83	H_2O_2, CH_2Cl_2	92	5
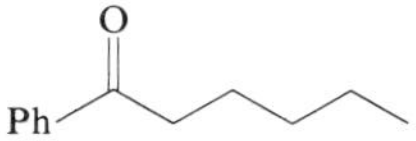	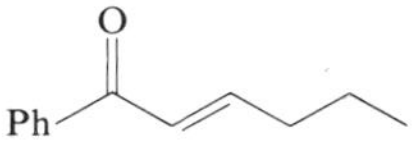	PhSeCl .	—	H_2O_2, THF	75	135

(*cont.*)

TABLE VI (*cont.*)

Starting material	Olefin	Selenide: Reagent	Selenide: Yield[a] (%)	Olefin: Oxidant	Olefin: Yield[b] (%)	Ref.
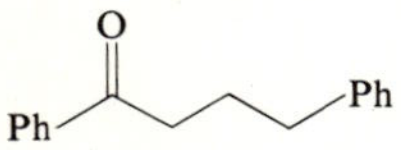	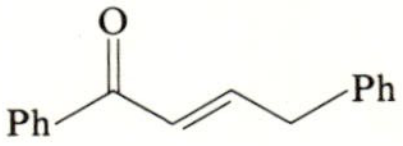	LDA, PhSeBr	—	H_2O_2, THF	84	122
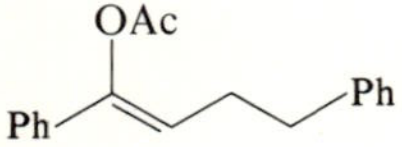	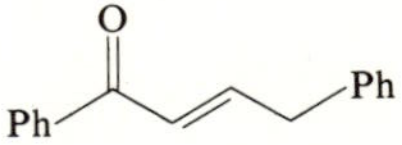	$PhSeO_2CCF_3$	96			8
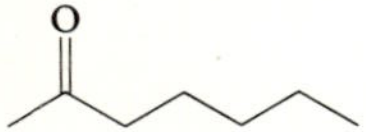	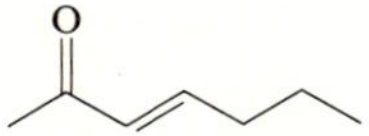	PhSeCl	—	H_2O_2, THF	(35)	40
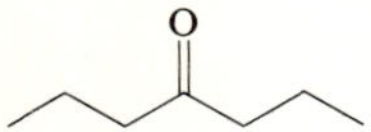	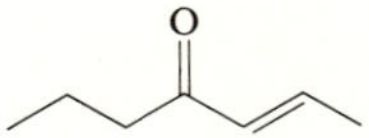	PhSeCl LDA, PhSeBr LDA, PhSeBr	— — —	H_2O_2, THF H_2O_2, CH_2Cl_2 O_3, Δ	(64) 72 80	40 5 5
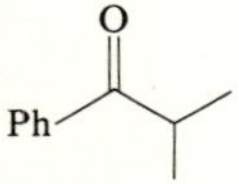	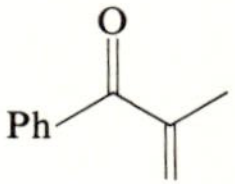	LDA, PhSeBr	33	H_2O_2, CH_2Cl_2	95	136
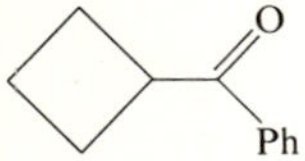	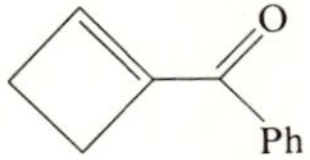	LDA, PhSeBr	87	H_2O_2, CH_2Cl_2	83	5
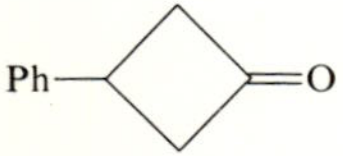	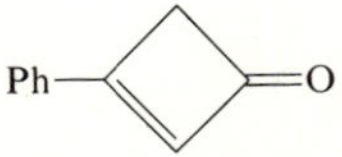	LDA, PhSeBr	50	O_3, Δ	53	5

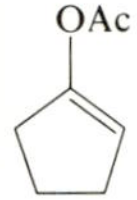	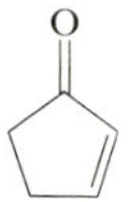	$PhSeO_2CCF_3$	62	$NaIO_4$	(95)	123
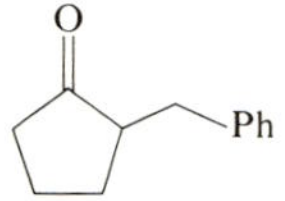	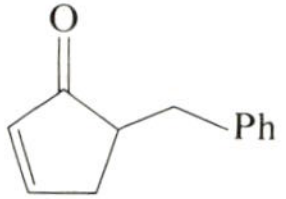	LDA, PhSeBr	—	H_2O_2, CH_2Cl_2	66	5
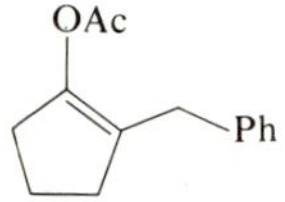	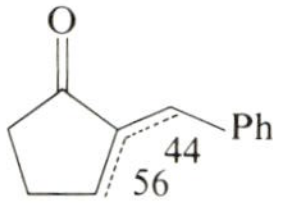	$PhSeO_2CCF_3$	96	H_2O_2, CH_2Cl_2	95	5
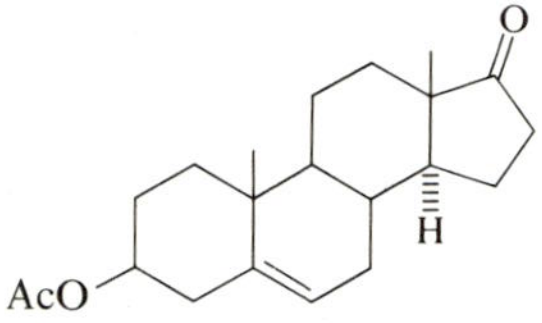	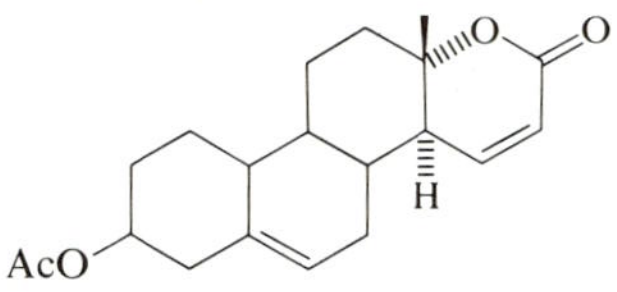	PhSeCl	—	H_2O_2, THF	39	34
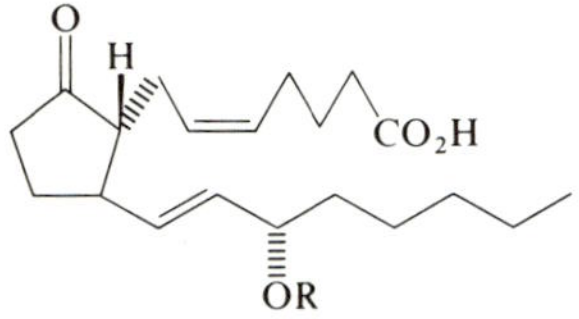	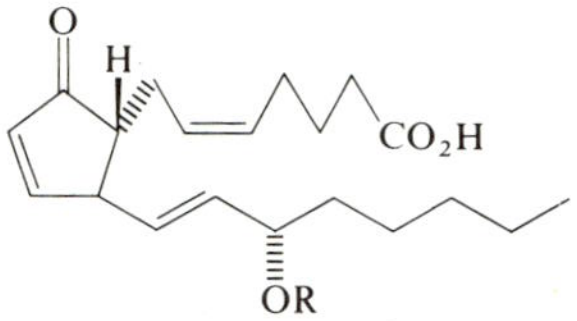	LDA, PhSeCl	—	$NaIO_4$	46[c]	137
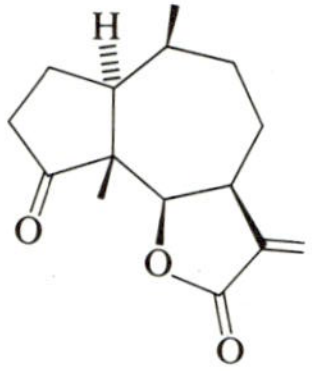	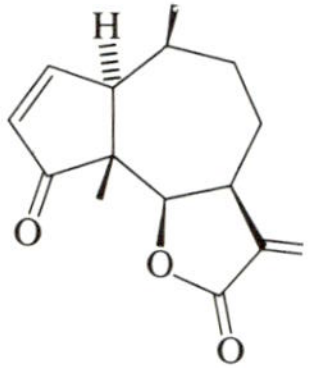	PhSeCl, EtOAc	—	$NaIO_4$, t-BuOH, H_2O	41	138a

(cont.)

TABLE VI (*cont.*)

Starting material	Olefin	Selenide: Reagent	Selenide: Yield[a] (%)	Olefin: Oxidant	Olefin: Yield[b] (%)	Ref.
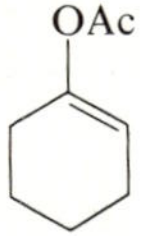	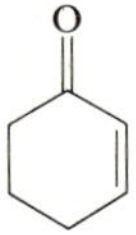	$PhSeO_2CCF_3$	70	$NaIO_4$	(83)	123
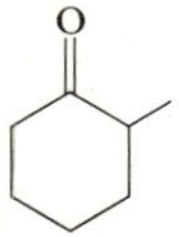	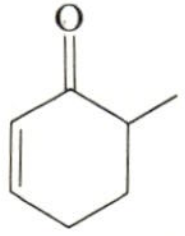	LDA, PhSeBr	89	O_3, Δ	65	5
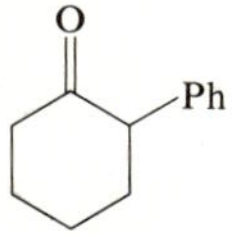	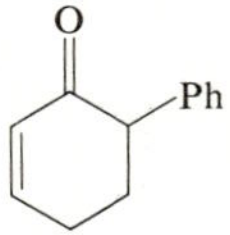	LDA, PhSeBr LDA, PhSeBr	—	H_2O_2, CH_2Cl_2 O_3, Δ	60 62	5 5
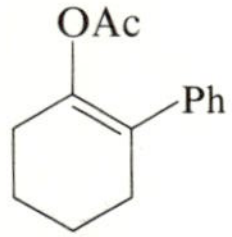	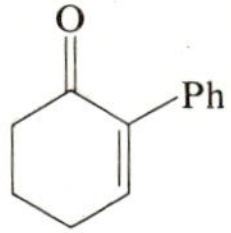	CH_3Li, PhSeBr	86	H_2O_2, CH_2Cl_2	94	5
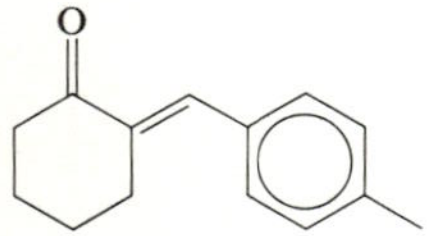	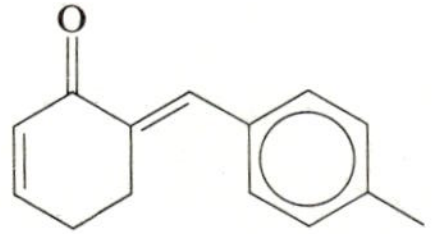	LDA, PhSeBr	68	$NaIO_4$	74	5

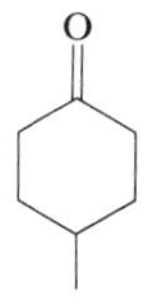	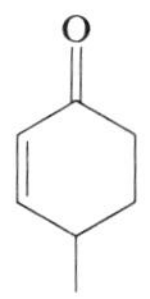	Ⓟ—SeCl	—	H_2O_2, THF	95	14a
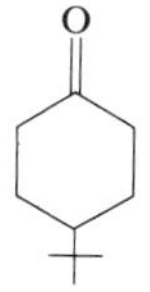	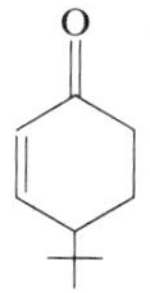	PhSeCl PhSeCl	— —	H_2O_2, THF CH_3CO_3H	(74) (45)	40 40
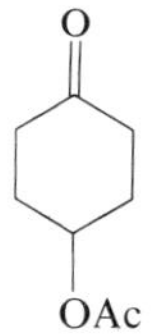	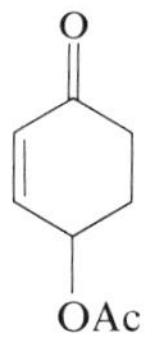	PhSeCl PhSeCl	—	H_2O_2, THF $NaIO_4$	(53) (47)	40 40
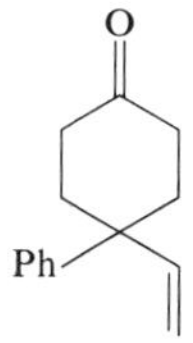	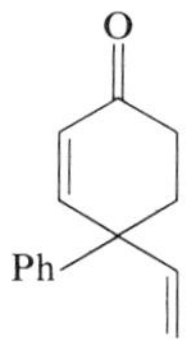	LDA, PhSeBr	—	H_2O_2	52	139
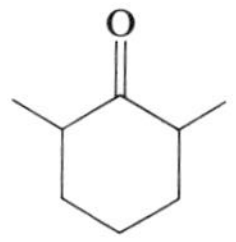	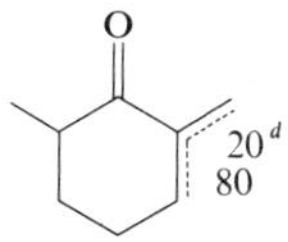	LDA, PhSeBr	—	H_2O_2, THF	(76)	122

(cont.)

TABLE VI (*cont.*)

Starting material	Olefin	Selenide: Reagent	Selenide: Yield[a] (%)	Olefin: Oxidant	Olefin: Yield[b] (%)	Ref.
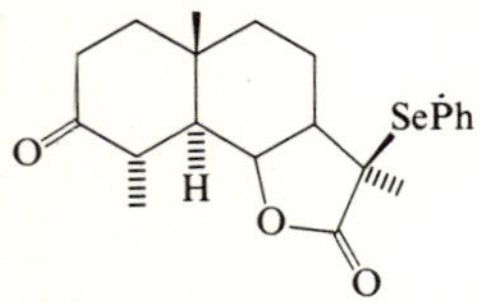	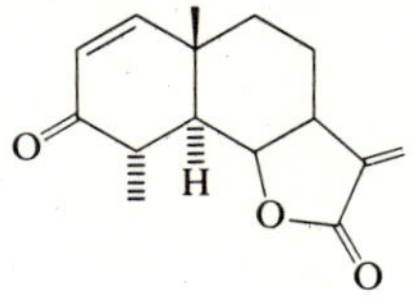	LDA, PhSeCl	76	O_3, CH_2Cl_2, Δ	60	140
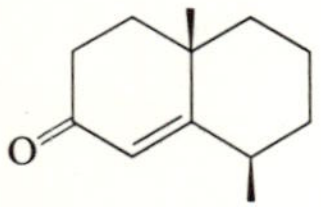	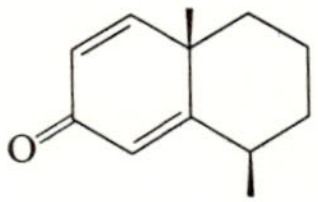	LDA, PhSeCl	—	H_2O_2, EtOAc	52	141
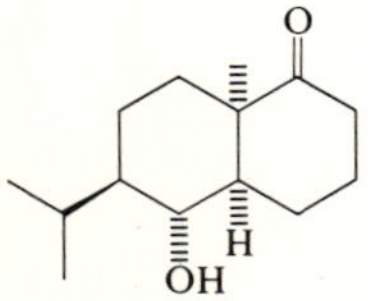	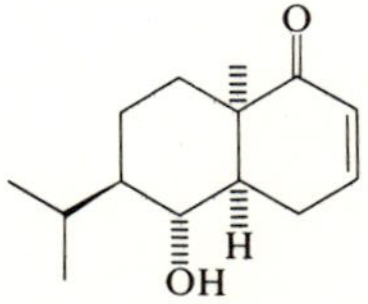	LDA, PhSeCl	—	H_2O_2, Et_2O	82	141a
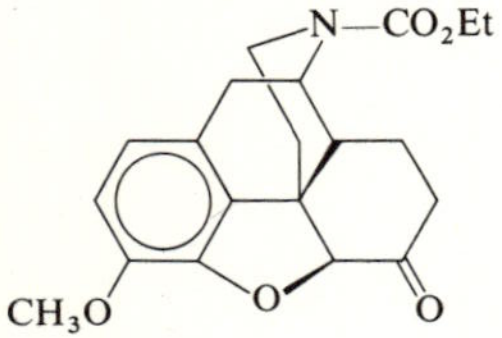	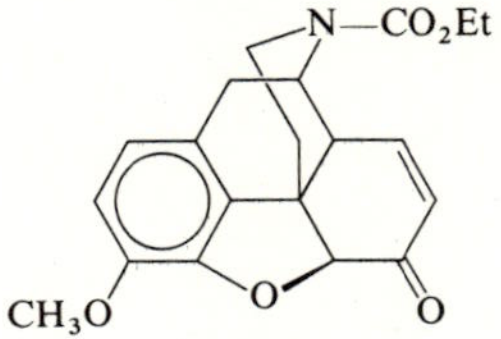	PhSeCl, EtOAc, HCl	85	$NaIO_4$, EtOAc, H_2O	71	141b
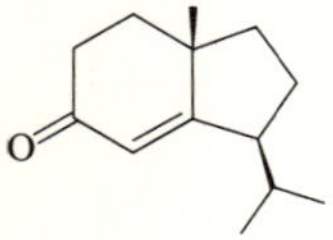	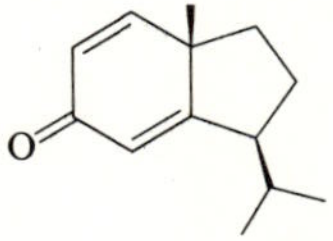	LDA, PhSeBr	—	H_2O_2, CH_2Cl_2	75	141c

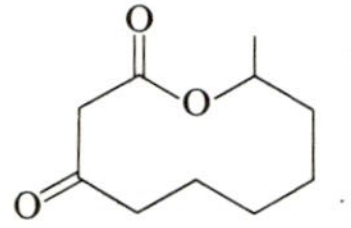	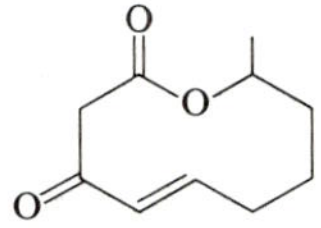	2 LDA, PhSeBr	38	$NaIO_4$	30	141d
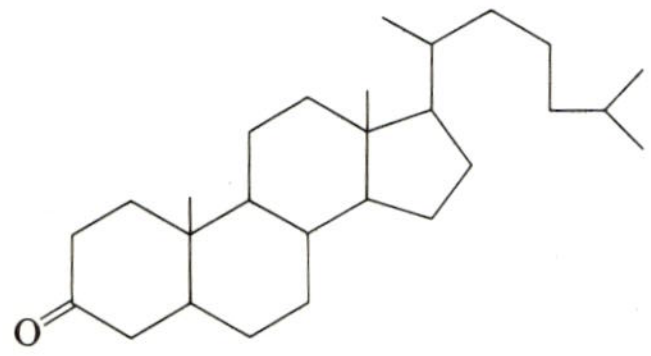	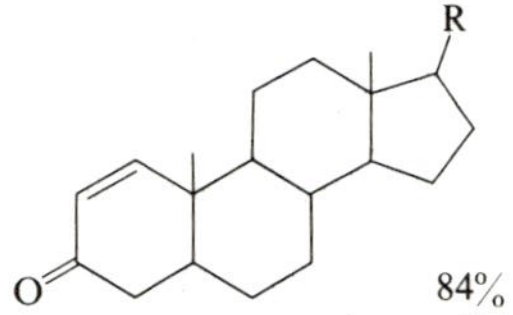 84% 4-ene 3% 1,4-diene 4%	PhSeCl, EtOAc	—	H_2O_2	(91) 45	40
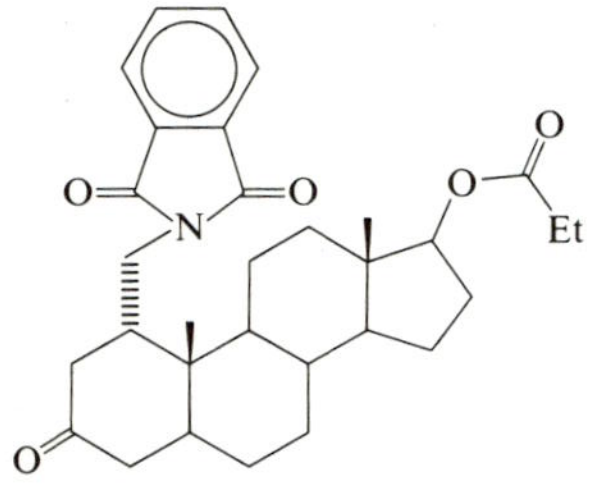	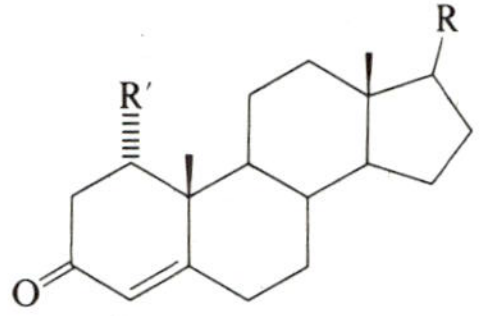	PhSeCl, EtOAc	—	H_2O_2, THF	68	142
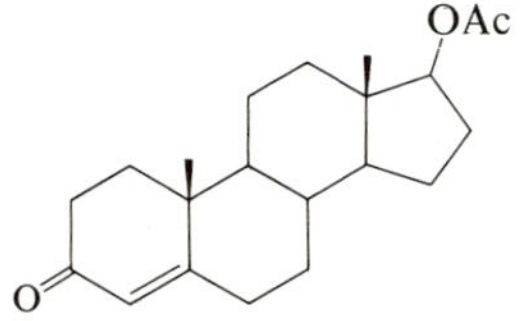	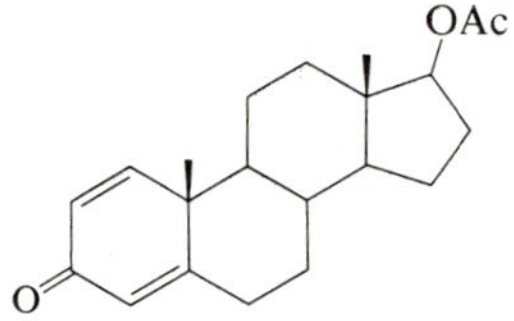	PhSeCl	—	H_2O_2	50	130

(*cont.*)

TABLE VI (*cont.*)

Starting material	Olefin	Selenide		Olefin		Ref.
		Reagent	Yield[a] (%)	Oxidant	Yield[b] (%)	
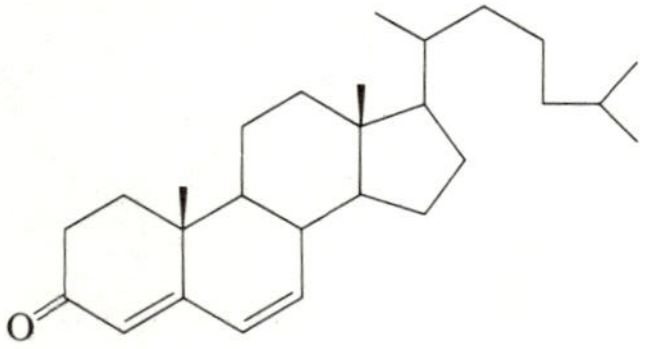	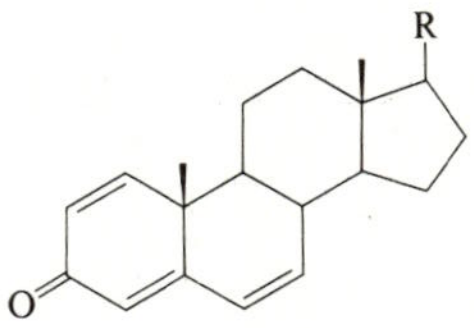	PhSeCl	—	H_2O_2	50	130
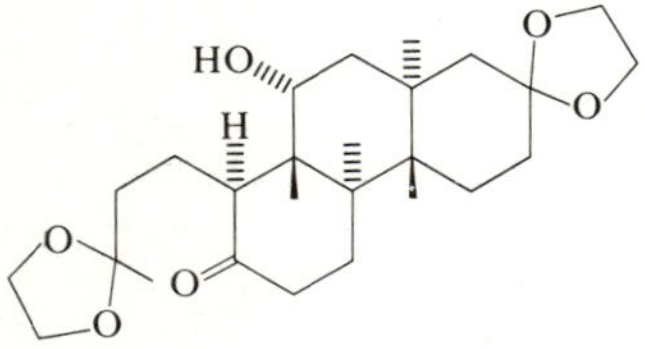	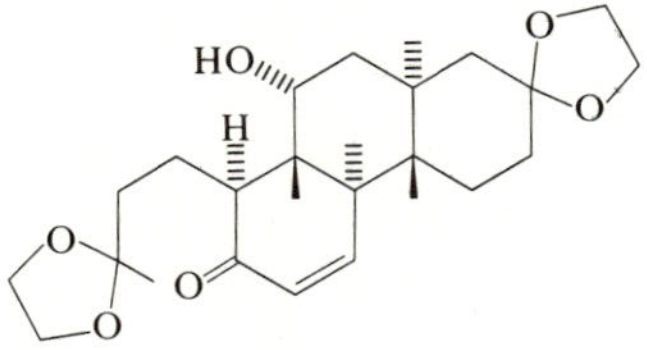	LDA, PhSeBr	—	H_2O_2, CH_2Cl_2	~100	142a
		LDA, PhSeBr	80	H_2O_2, CH_2Cl_2	70	5
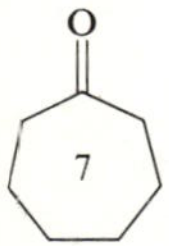	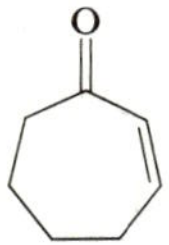	LDA, PhSeBr	—	O_3, Δ	55	5

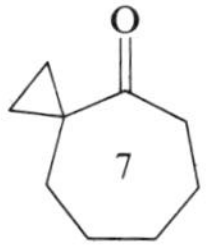	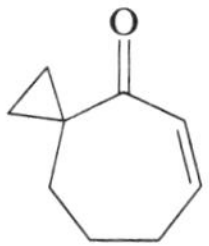	LDA, PhSeBr	85	O_3, Δ	85	5
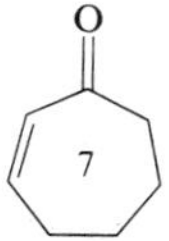	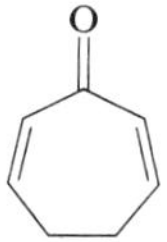	LDA, PhSeBr	81	H_2O_2, CH_2Cl_2	61	5
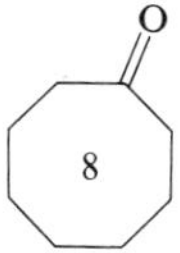	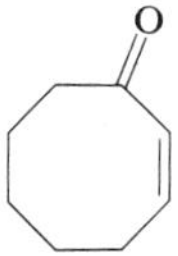	LDA, PhSeBr	70	O_3, Δ	74	5
				TsNClNa, Δ	54	46a
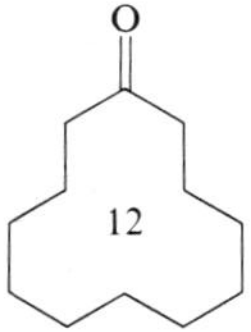	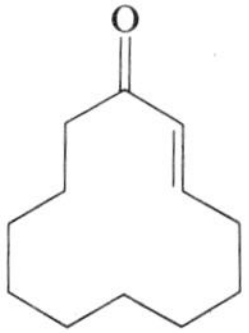	PhSeCl	—	H_2O_2, THF	77	40
		LDA, PhSeBr	—	H_2O_2	72	5
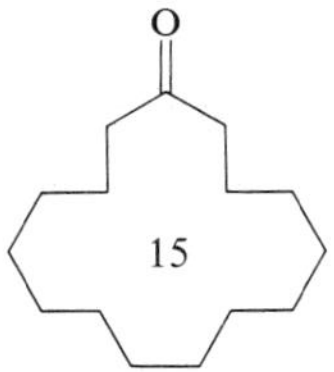	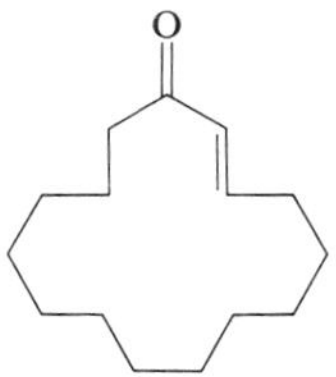	LDA, PhSeCl	—	H_2O_2	> 79[c]	143
		PhSeCl, EtOAc	—	H_2O_2	60	143

(*cont.*)

TABLE VI (*cont.*)

Starting material	Olefin	Selenide: Reagent	Selenide: Yield[a] (%)	Olefin: Oxidant	Olefin: Yield[b] (%)	Ref.
β-Amino and β-Hydroxy Ketones (Section III, F, 9)						
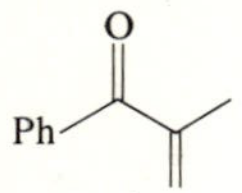	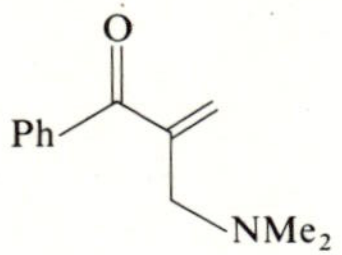	$PhSeNMe_2$, $CDCl_3$	93	*m*-CPBA, Δ	84	10
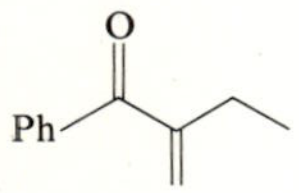	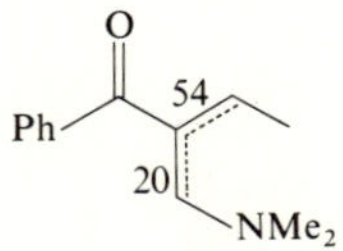	$PhSeNMe_2$, $CDCl_3$	94	*m*-CPBA, Δ	74	10
	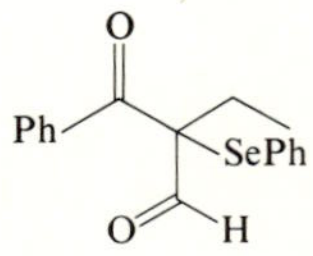				10	
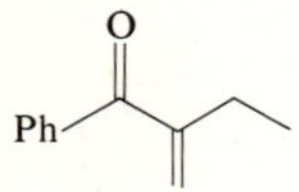	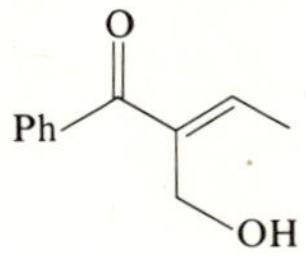	$PhSeO\overset{O}{\overset{\|}{C}}CF_3$, NH_4OH	80	*m*-CPBA, Δ	74	136
	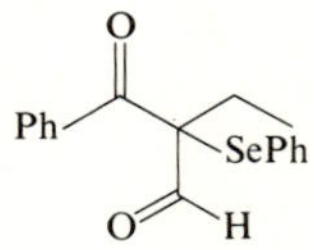				14	

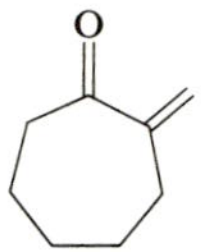	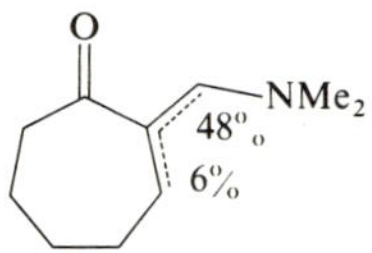	$PhSeNMe_2$, C_6H_6	87	*m*-CPBA, Δ	54	10
	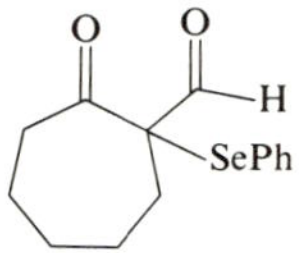				22	
β-Dicarbonyl Compounds (*Section III, C, 4*)						
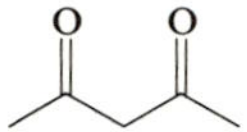						144
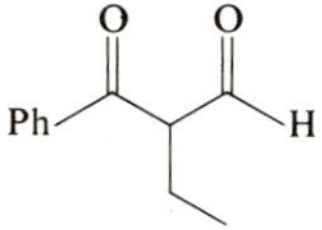	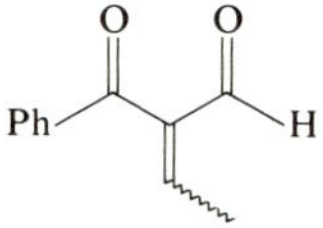	$PhSeNEt_2$	89	H_2O_2, CH_2Cl_2	55	10,136
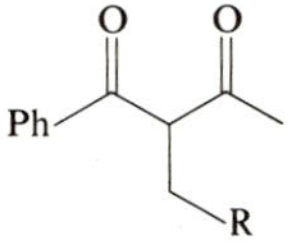	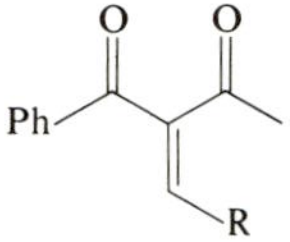					
	R = H	NaH, PhSeBr	93	H_2O_2, CH_2Cl_2	80	5
	R = *n*-Bu	NaH, PhSeBr	96	H_2O_2, CH_2Cl_2	93	5
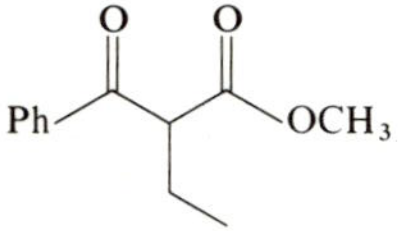	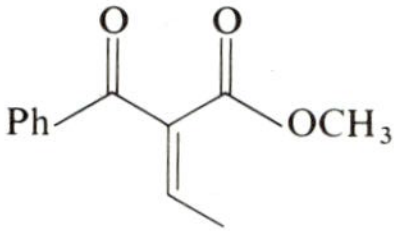	$PhSeNMe_2$	92	H_2O_2, CH_2Cl_2	92	10,136

(*cont.*)

TABLE VI (*cont.*)

Starting material	Olefin	Selenide: Reagent	Selenide: Yield[a] (%)	Olefin: Oxidant	Olefin: Yield[b] (%)	Ref.
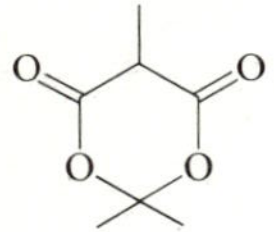	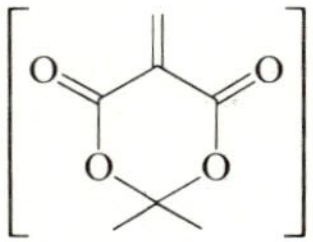	NaH, PhSeBr	67	*m*-CPBA, $-10^\circ C$	53[h]	145
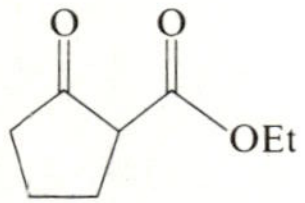	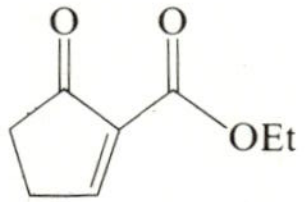	NaH, PhSeBr	—	H_2O_2, CH_2Cl_2	81	5
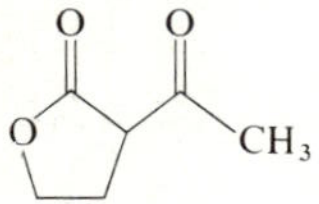	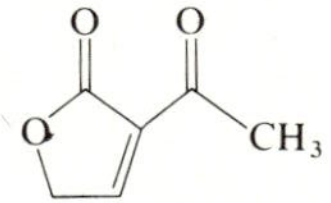	$PhSeNMe_2$	90	H_2O_2, CH_2Cl_2	<4	136
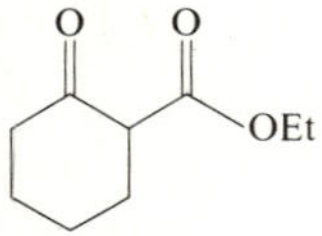	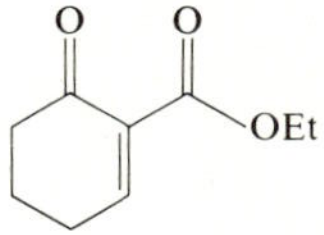	NaH, PhSeBr	—	H_2O_2, CH_2Cl_2	89	5
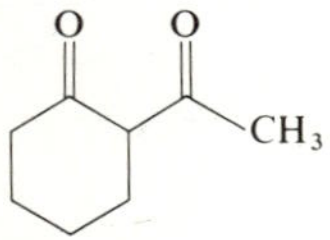	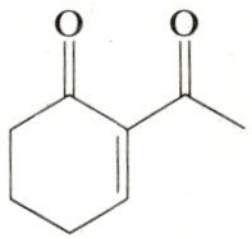	NaH, PhSeBr $PhSeNMe_2$	— 92	H_2O_2, CH_2Cl_2 —	84 —	5 136

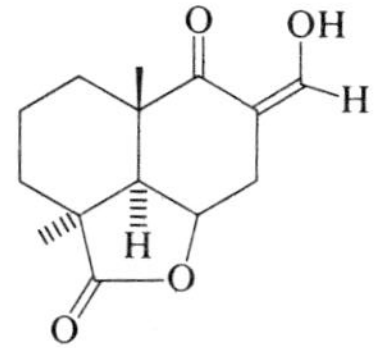	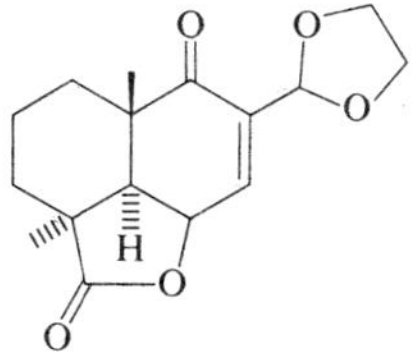	PhSeCl, NEt_3	—	1. *m*-CPBA 2. $(CH_2OH)_2$, H^+	54	145a
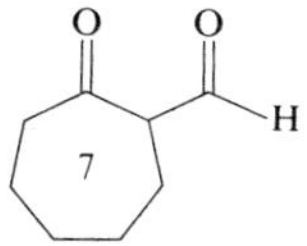	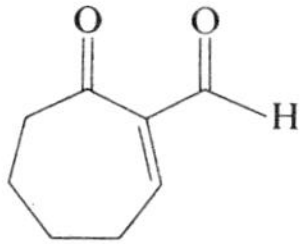	$PhSeNEt_2$	90	H_2O_2, CH_2Cl_2	74	10
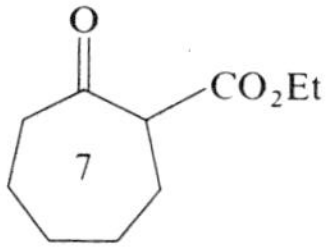	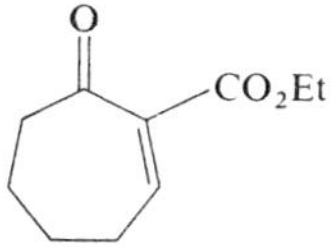	NaH, PhSeBr	90	H_2O_2, CH_2Cl_2	93	5
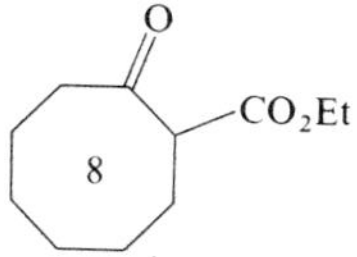	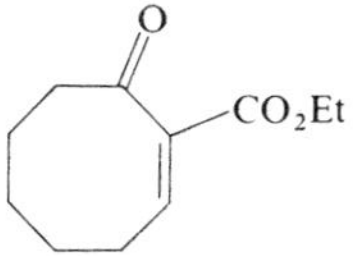	NaH, PhSeBr	96	H_2O_2, CH_2Cl_2	93	5
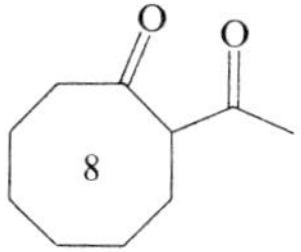	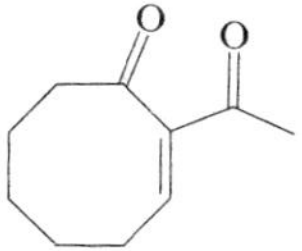	NaH, PhSeBr	—	H_2O_2, CH_2Cl_2	94	146

(*cont.*)

TABLE VI (*cont.*)

Starting material	Olefin	Selenide: Reagent	Selenide: Yield[a] (%)	Olefin: Oxidant	Olefin: Yield[b] (%)	Ref.
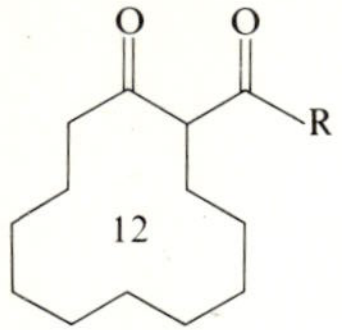	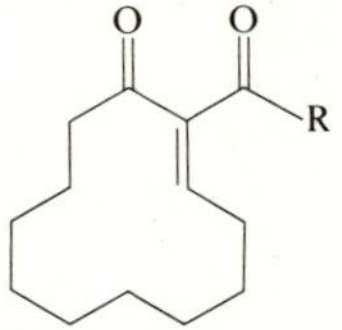					
	$R = CH_3$	NaH, PhSeCl	—	H_2O_2	96	146
	$R = t$-Bu	NaH, PhSeCl	—	H_2O_2	66	146
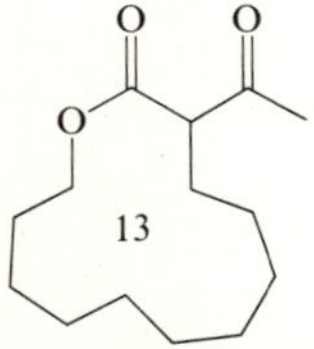	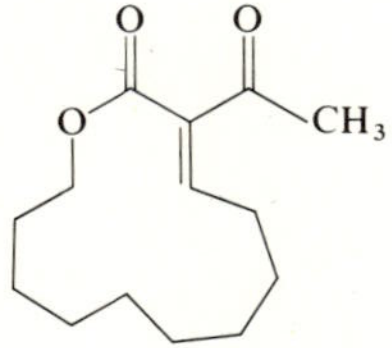	NaH, PhSeCl	—	H_2O_2	[e]	146
β-Keto Sulfoxides, Selenoxides						
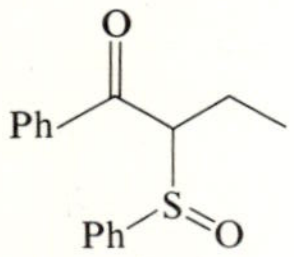	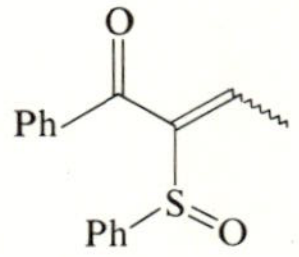	LDA, PhSeBr	78	H_2O_2, CH_2Cl_2	71	5
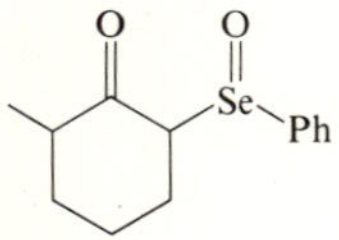	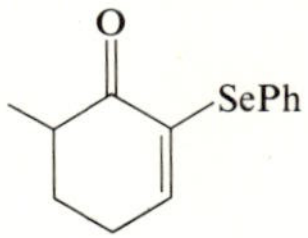	LDA, PhSeCl	—	Δ	57	35

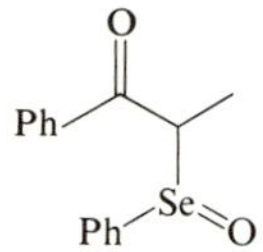	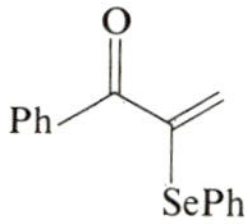	LDA, PhSeCl	—	Δ	57	136
Carboxylic Acids and Esters						
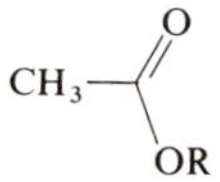	—	LDA, PhSeBr	85 (R = *t*-Bu) 80 (R = Et)			147
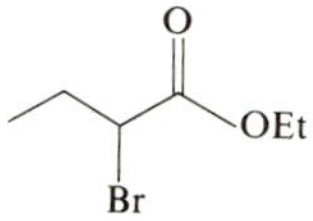	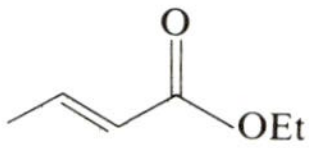	Ⓟ—Se^-		H_2O_2	75	14
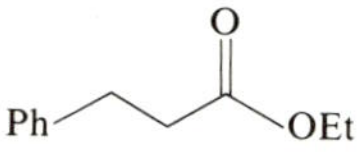	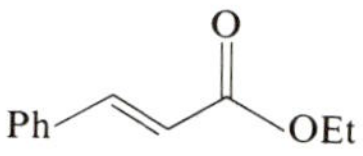	$LiNR_2$,[g] PhSeBr	—	CH_3CO_3H, THF	(80)	40
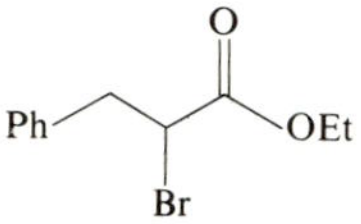	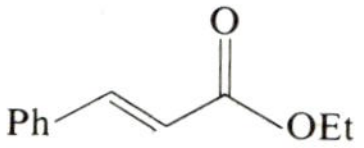	$PhSe^-$, EtOH	—	CH_3CO_3H, THF	(78)	40
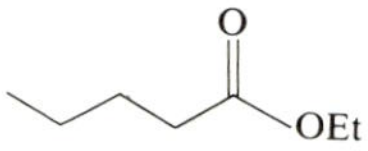	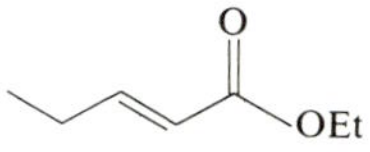	$LiNR_2$,[g] PhSeBr	—	CH_3CO_3H, THF	(83)	40
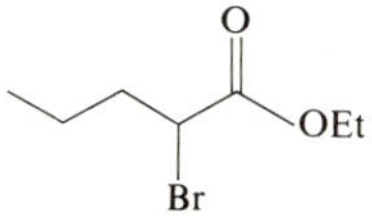	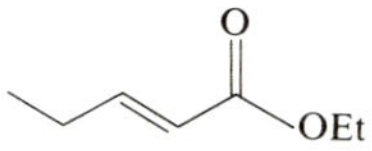	$PhSe^-$, EtOH	—	CH_3CO_3H, THF	(89)	40

(*cont.*)

TABLE VI (*cont.*)

Starting material	Olefin	Selenide: Reagent	Selenide: Yield[a] (%)	Olefin: Oxidant	Olefin: Yield[b] (%)	Ref.
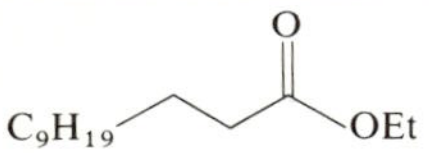	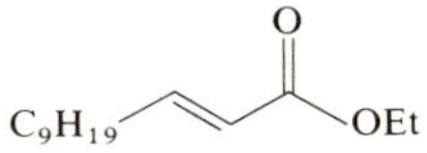	$LiNR_2$,[g] PhSeBr	—	CH_3CO_3H, THF	(79) 83	40
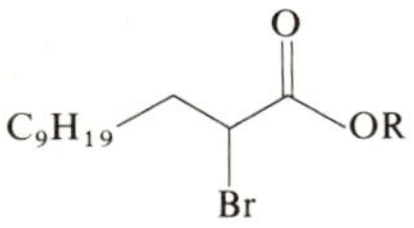	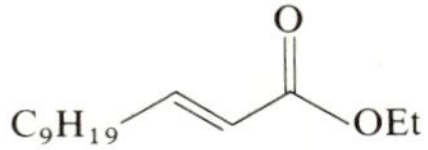	$PhSe^-$, EtOH	—	CH_3CO_3H, THF (R = Et) (R = H)	(82) 91	40 34
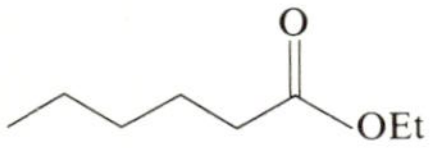	—	LDA, PhSeBr	80	—	—	147
	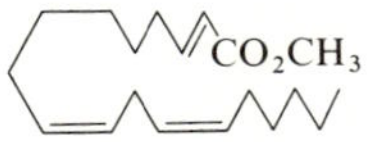	$LiNR_2$,[g] Ph_2Se_2	—	$NaIO_4$	80	40
		$LiNR_2$, PhSeBr	47	$NaIO_4$, MeOH	81	147a
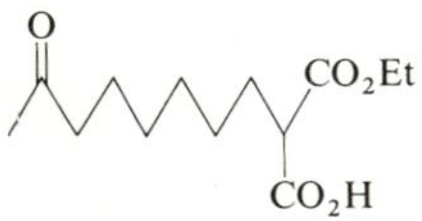		NaH, Ph_2Se_2, Δ	47	$NaIO_4$	91	147b
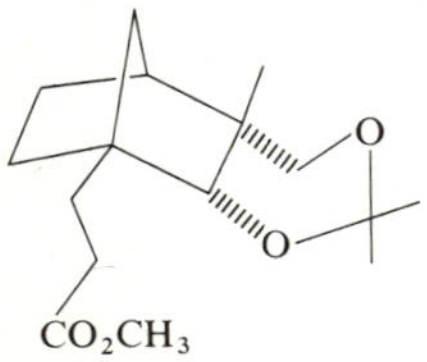	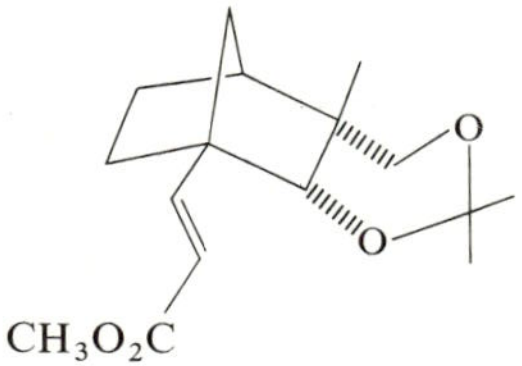	LDA, Ph_2Se_2	—	$NaIO_4$	90	148

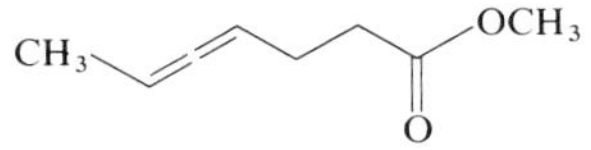	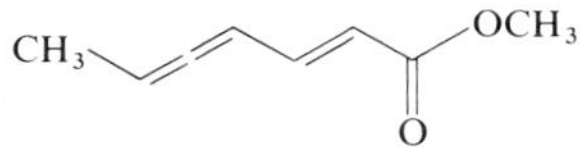	LDA, Ph_2Se_2	—	$NaIO_4$	85	148a
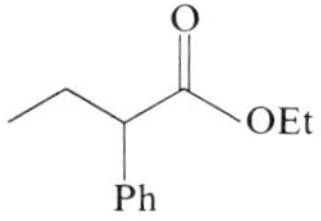	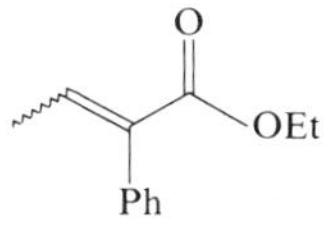 *E*/*Z*: 46/54	LDA, PhSeBr	—	H_2O_2	79	5
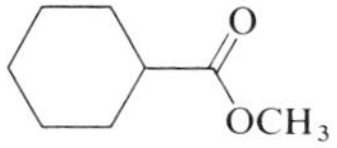	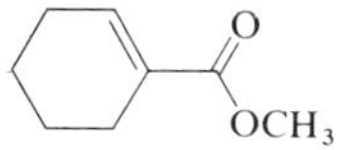	LDA, PhSeBr	—	H_2O_2	96	5
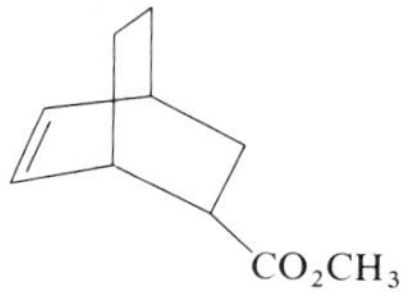	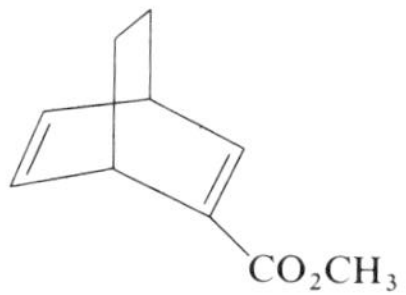	LDA, PhSeBr	—	H_2O_2	68	5
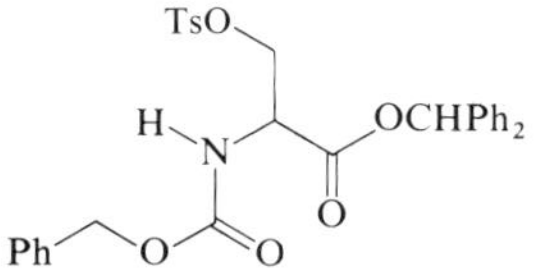	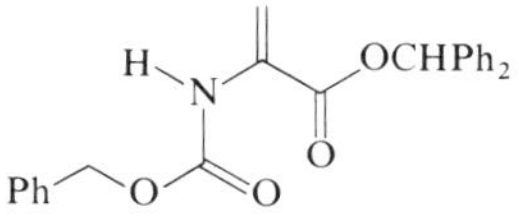	NaHSe, Ph_2CHBr, KOH	79%	H_2O_2	~100	59,149
Lactones						
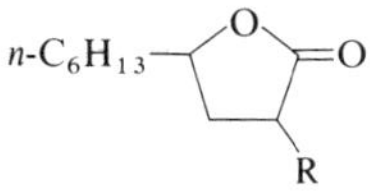	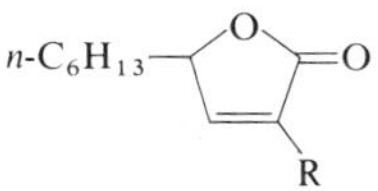					
	R = H	LDA, PhSeBr	—	CH_3CO_3H, EtOAc	56	40
	R = Et	LDA, PhSeCl	70	H_2O_2, THF	80	150
	R = *n*-Bu	LDA, PhSeCl	80	H_2O_2, THF	98	150
	R = CH_2Ph	LDA, PhSeCl	72	H_2O_2, THF	85	150

(*cont.*)

TABLE VI (*cont.*)

Starting material	Olefin	Selenide: Reagent	Selenide: Yield[a] (%)	Olefin: Oxidant	Olefin: Yield[b] (%)	Ref.
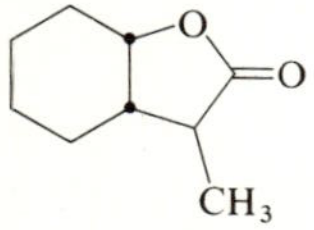	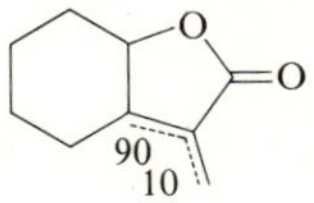 *exo*/*endo* = 10/90	LDA, PhSeCl	88	H_2O_2, THF	95	150,151
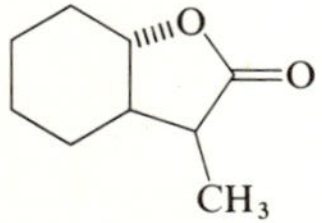	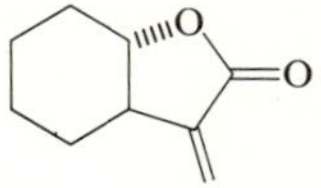	LDA, Ph_2Se_2	85	H_2O_2, THF	96	151
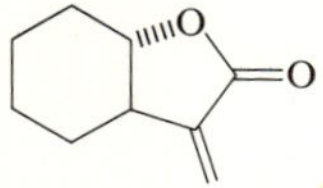	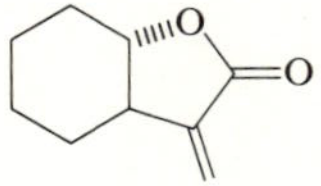	$PhSe^-$, EtOH	60	H_2O_2, THF	97	152
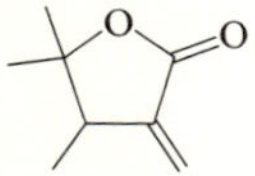	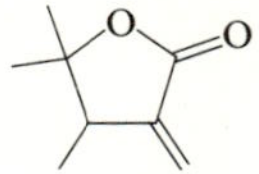	$PhSe^-$, EtOH	54	H_2O_2, THF	99	152
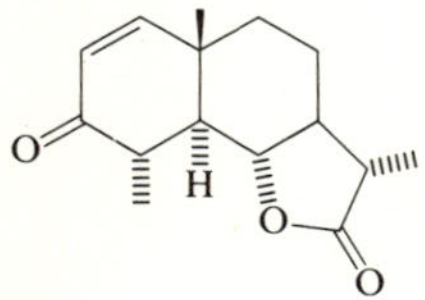	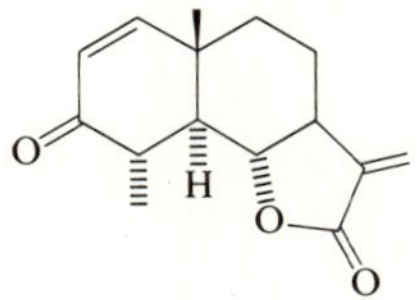	LDA, Ph_2Se_2	15	H_2O_2, THF	~100	55

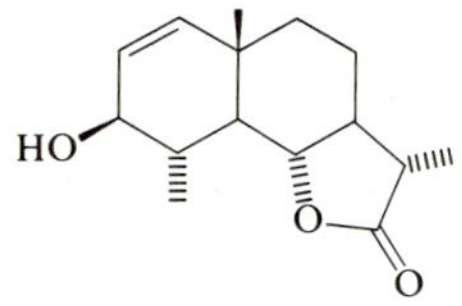	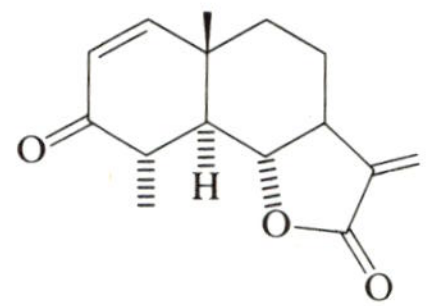	LDA, Ph_2Se_2	66	H_2CrO_4 (Jones)	56	55
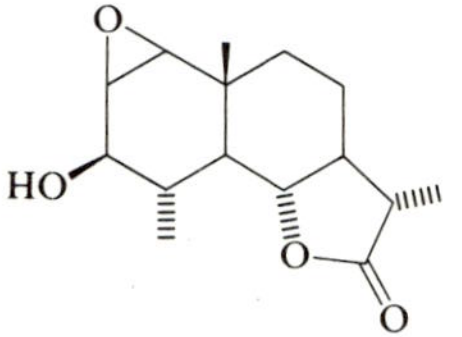	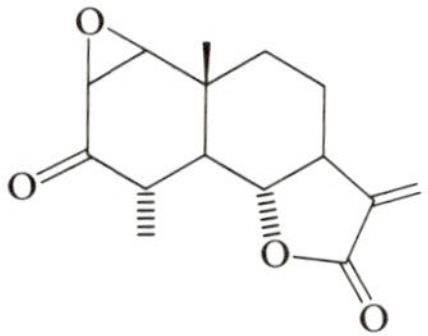	LDA, Ph_2Se_2	81	MnO_2, H_2O_2	60,75	55
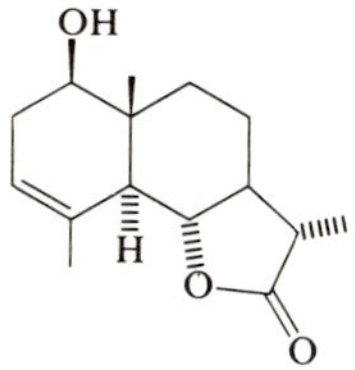	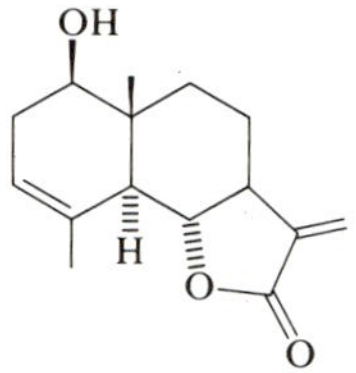	LDA, Ph_2Se_2	—	H_2O_2, THF	Good	55
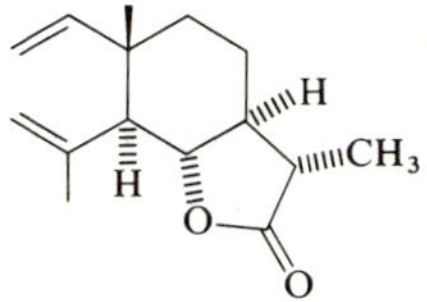	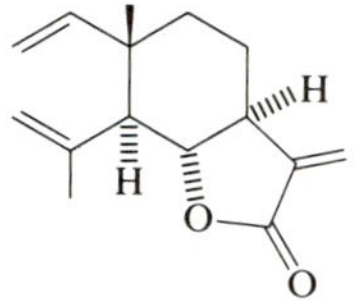	LDA, Ph_2Se_2, HMPA	97	H_2O_2, THF	93	75a
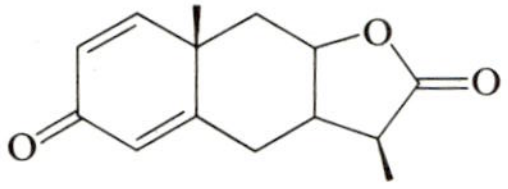	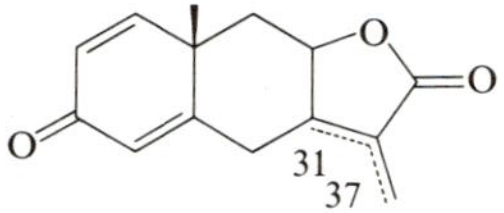	LDA, Ph_2Se_2	42	H_2O_2, THF	68	153

(*cont.*)

TABLE VI (*cont.*)

Starting material	Olefin	Selenide: Reagent	Selenide: Yield[a] (%)	Olefin: Oxidant	Olefin: Yield[b] (%)	Ref.
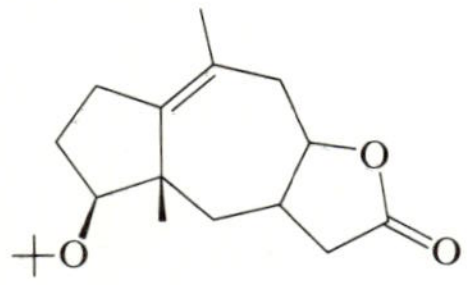	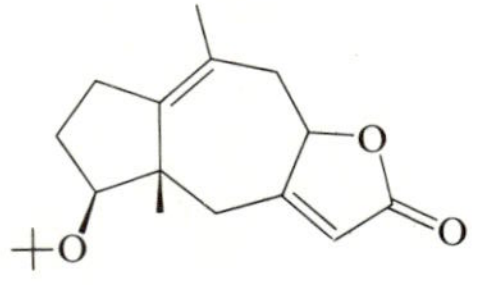	LDA, Ph_2Se_2	—	H_2O_2	> 70[c]	154
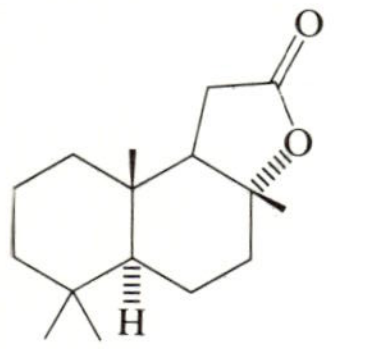	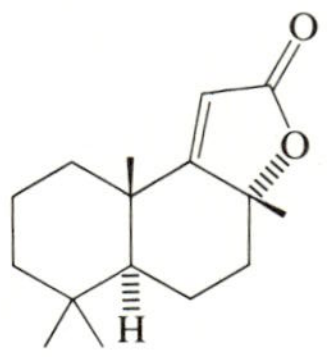	$LiNR_2$,[g] PhSeBr	—	CH_3CO_3H, EtOAc	[e]	155
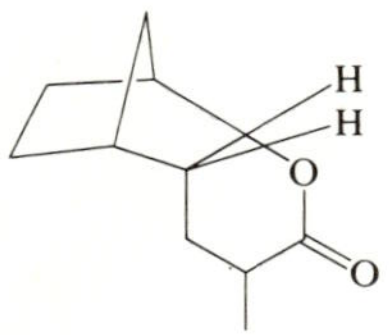	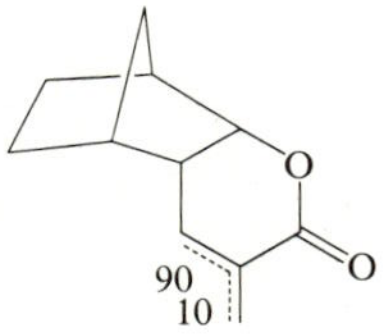	LDA, Ph_2Se_2	62	H_2O_2, THF	99	151
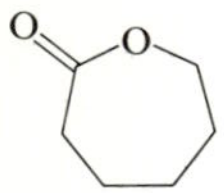	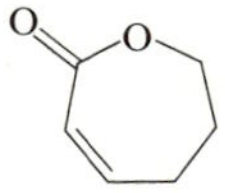	LDA, PhSeBr	70	O_3, Δ	70	5

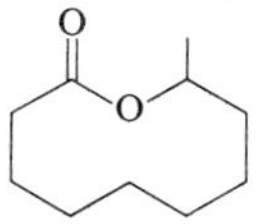	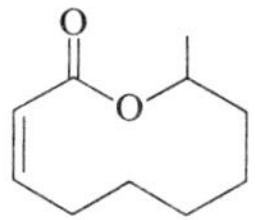	LDA, PhSeBr	—	H_2O_2, AcOH	82	155a
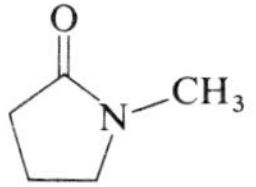	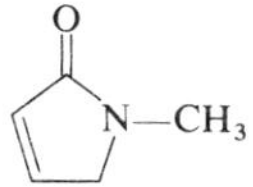	LDA, PhSeCl	55	H_2O_2, MeOH	57	133
Nitriles						
	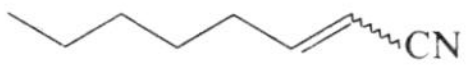 *E*/*Z* 54/46	$LiNR_2$,[g] Ph_2Se_2	—	H_2O_2, THF	96	138
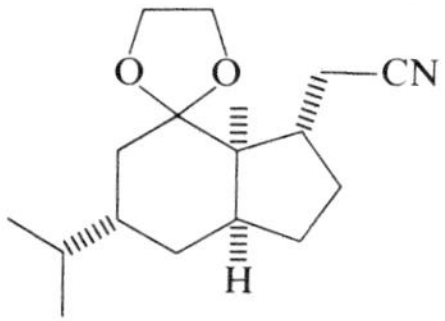	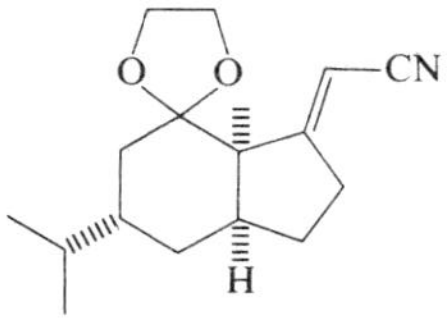	$LiNR_2$,[g] Ph_2Se_2	—	H_2O_2, THF	42	156

[a] If no yield is given, olefin yield is overall.
[b] Isolated yields, gas chromatographic yields in parentheses.
[c] Overall yield for several steps including olefin formation.
[d] See Section III, H, 3 for regioselectivity.
[e] Yield not reported.
[f] Selenoxide prepared *in situ* by oxidation of α-phenylseleno ketone with ozone.
[g] Lithium isopropylcyclohexylamide.
[h] Isolated as the Diels–Alder adduct with cyclopentadiene.

sulfenylation of 2-phenylcyclohexanone [$LiN(i\text{-}Pr)_2$, Ph_2S_2] occurs to give the 2-phenylthio compounds,[158] as a consequence of the enolate equilibration. The more reactive sulfenylating reagent $PhSO_2$–SPh results in trapping of the kinetic enolate prepared from 2-methylcyclohexanone.[124]

The major side reaction encountered during selenenylation of anions is a consequence of the substantial acidifying effect of the phenylseleno group. Thus, compounds which have an additional hydrogen at the position being selenenylated are themselves quite acidic and may be deprotonated by other bases in solution. This problem is particularly severe during selenenylation of nitriles. Brattesani and Heathcock[138] obtained good yields of monoseleno nitriles only when two equivalents of lithium amide base were used, so that monoselenated material is deprotonated as soon as it is formed. The lower reactivity of the monoseleno anion compared to the unsubstituted anion ensures that little bisseleno compound is formed. A similar technique has been used for the selenenylation of lactams.[133]

Proton transfers are probably responsible for minor contamination by starting material usually encountered during selenenylation of methylene groups α to ketones and esters.

Copper enolates prepared by cuprate additions to enones have been selenenylated. It is necessary to use sufficient PhSeBr or PhSeCl to react with all organometallics present. The overall transformation corresponds to a β-alkylation of an enone.

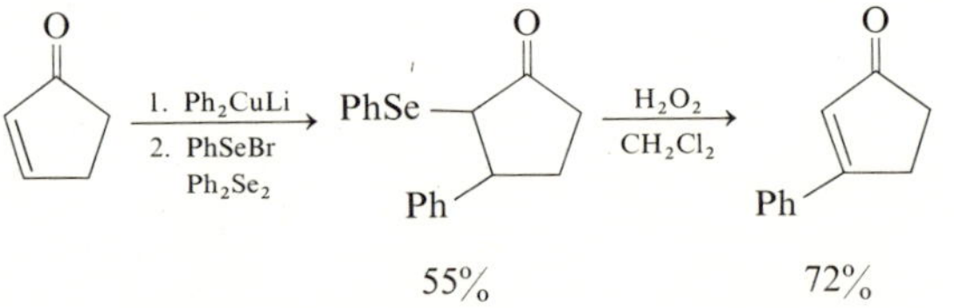

(ref. 35)

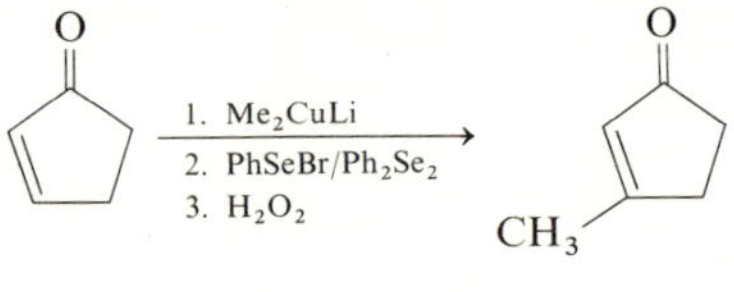

(ref. 136)

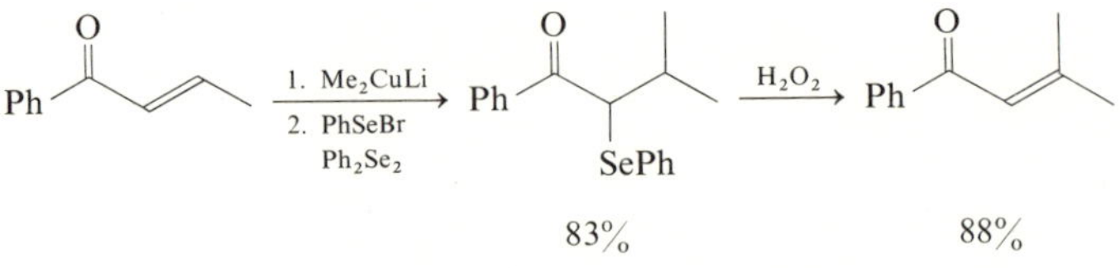

(ref. 5)

[158] F. Cooke and P. Magnus, *Chem. Commun.* p. 519 (1976).

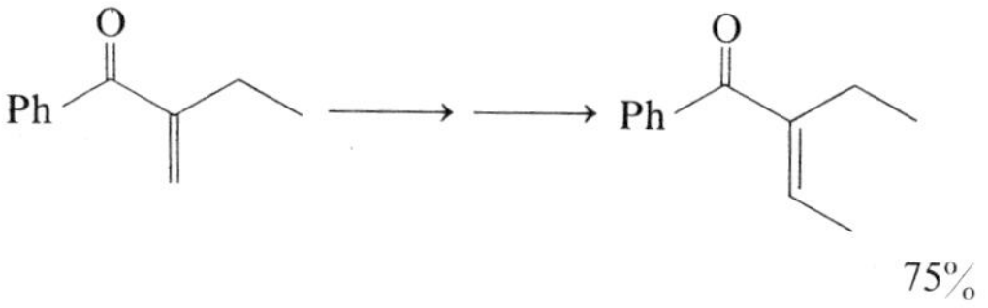

(ref. 136)

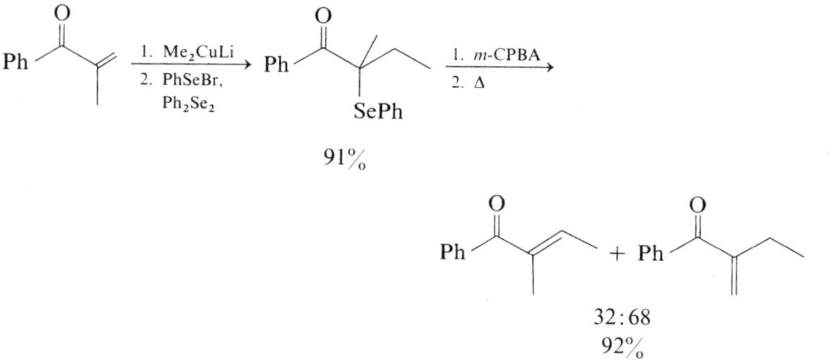

(ref. 136)

α,β-Unsaturated ester anions are selenenylated in the γ position.[48] It is not known whether these are the kinetically-controlled products or whether they are formed by [1,3] sigmatropic rearrangement of α-seleno derivatives. Allyl selenides have been reported to undergo facile [1,3] rearrangements.[159,159a]

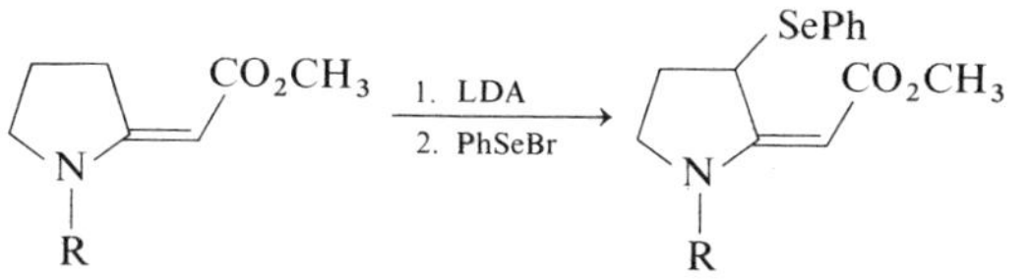

(ref. 48)

(16)

Oxidation and acid-catalyzed rearrangement transforms **16** to a pyrrole derivative.

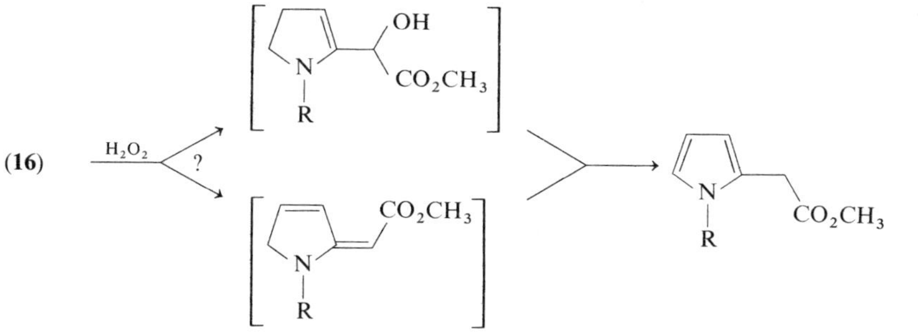

(ref. 48)

[159] E. G. Kataev, G. A. Chmutova, A. A. Musina, and E. G. Yarkova, *Dokl. Akad. Nauk. SSSR* **187**, 1308 (1969).

It was not established whether the transformation occurred by [2,3] sigmatropic rearrangement or by selenoxide syn elimination (see Section IV, A, 4). Furans and thiophenes can also be prepared in this way.

4. β-DICARBONYL COMPOUNDS

Stabilized enolates are best selenenylated using PhSeCl(Br) and the sodium salt of the carbonyl compound (prepared with NaH in THF). Stabilized enolates which have been selenenylated include those derived from β-diketones, β-keto esters, β-keto sulfoxides, and β-keto selenoxides.[5,35] Reich and Renga[10] have developed an alternative procedure for selenenylation of β-dicarbonyl compounds involving direct reaction with selenenamides ($PhSeNMe_2$, $PhSeNEt_2$) which proceeds in the absence of strong base. Reactions of this sort have precedent in sulfur chemistry.[160]

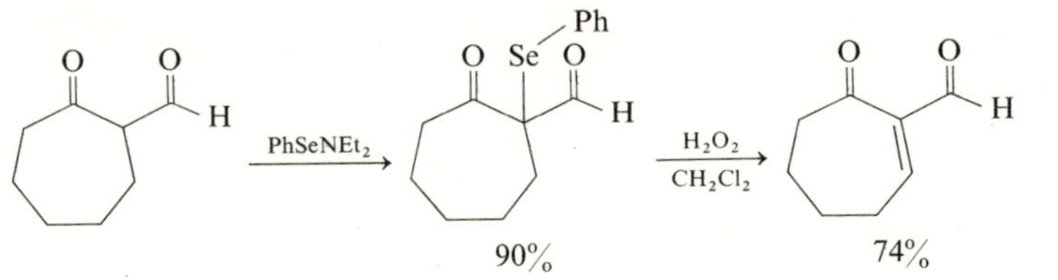

(ref. 10)

The dehydrogenation of β-dicarbonyl compounds using the selenoxide elimination promises to be one of the more useful applications of this reaction. In most cases conditions are mild enough so that enolization does not occur (a notable exception is 2-formylcyclohept-2-enone, which enolizes very rapidly). The yields are frequently almost quantitative.

No good alternative procedures for carrying out these dehydrogenations are available although 2,3-dichloro-5,6-dicyano-1,4-benzoquinone[161] and SeO_2[162] have worked in some cases. Successful acetoxylation–acetate pyrolyses[163] or halogenation–dehydrohalogenations[164] have been reported occasionally, but the latter procedure usually fails because competing Favorskii reactions occur, the halo compounds rearrange, or the products rearrange under the often vigorous reaction conditions.[161,162,164] The other

159a K. B. Sharpless and R. F. Lauer, *J. Org. Chem.* **37**, 3973 (1972).

160 T. Mukaiyama, S. Kobayashi, and T. Kumamoto, *Tetrahedron Lett.* p. 5115 (1970).

161 J. A. Edwards, M. C. Calzada, L. C. Ibáñez, M. E. C. Rivera, R. Urquiza, L. Cardona, J. C. Orr, and A. Bowers, *J. Org. Chem.* **29**, 3481 (1964).

162 J. N. Marx, J. H. Cox, and L. R. Norman, *J. Org. Chem.* **37**, 4489 (1972); J. N. Marx and L. R. Norman, *ibid.* **40**, 1602 (1975).

163 D. Gorenstein and F. H. Westheimer, *J. Am. Chem. Soc.* **92**, 634 (1970).

164 J. A. Brenner, J. Org. Chem. **26**, 22 (1962); C. W. T. Hussey and A. R. Pinder, *J. Chem. Soc.* p. 3525 (1961); p. 1517 (1962).

main route is the aldol (Knoevenagel) condensation, although specialized procedures such as the vacuum pyrolysis of 3-phenylpropargyl acetate to give 1-acetyl-1-benzoylethene[165] provide efficient routes to specific compounds.

5. Seleninylation of Ketone and Ester Enolates

The direct introduction of the selenoxide function by reaction of enolates with seleninylating reagents has been explored to some extent. Reich, Renga, and Reich[5] treated several enolates with benzeneseleninyl chloride and found that reasonable yields could be obtained in some systems.

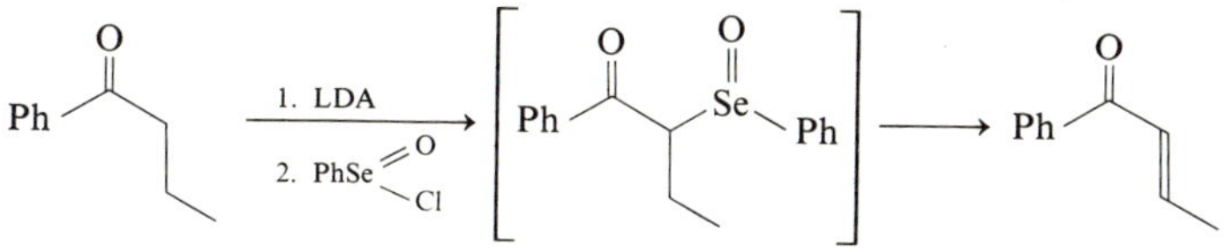

80%

(ref. 5)

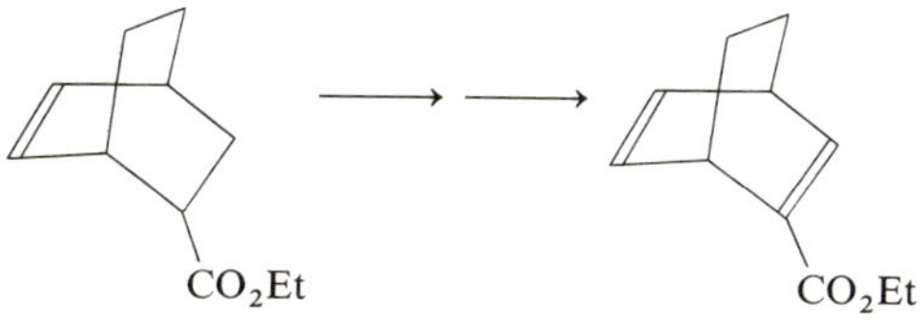

68%

(ref. 5)

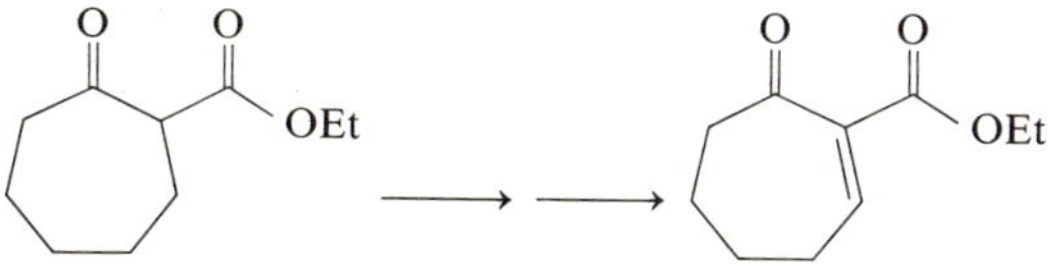

70%

(ref. 136)

The great sensitivity of keto selenoxides toward base and the enhanced acidity of any α-protons present make it necessary that special care be taken with this method. The two step procedure using PhSeCl or PhSeBr followed by oxidation usually gives better yields.

The use of benzeneseleninic anhydride as a seleninylating agent for ester enolates has been mentioned in a footnote.[166]

[165] W. S. Trahanovsky and P. W. Mullen, *J. Am. Chem. Soc.* **94**, 5086 (1972).

[166] D. H. R. Barton, P. D. Magnus, and M. N. Rosenfeld, *Chem. Commun.* p. 301 (1975); D. H. R. Barton, S. V. Ley, P. D. Magnus, and M. N. Rosenfeld, *J. Chem. Soc., Perk. I*, p. 567 (1977).

6. Reaction of α-Halo Carbonyl Compounds with Selenolate Anions

The reaction of α-halo ketones with selenolate anions has been rarely used to prepare α-seleno ketones. The reaction gives mainly dehalogenation, but improved yields are obtained in *aprotic* solvents.[5,21]

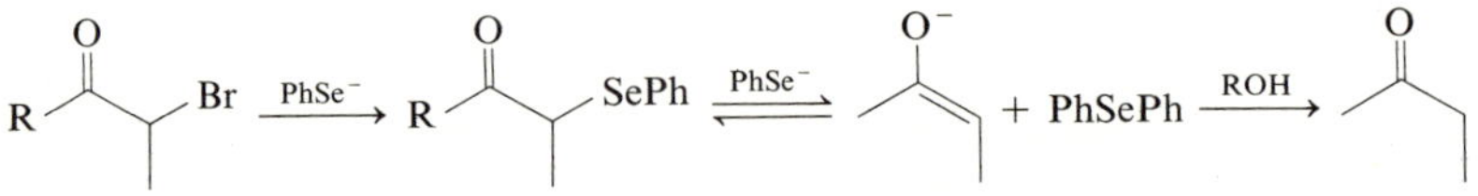

This could be due to several factors. One is that α-seleno ketones are known to be rapidly deselenated by selenolate anions in protic solvents and this reaction may be occurring during attempted synthesis of selenides. In hydroxylic solvents the enolate would be protonated. Another possibility is that attack of the soft nucleophile occurs at the halogen to give enolate and selenenyl halide. Under protic conditions the enolate is discharged; under aprotic conditions seleno ketone is formed.

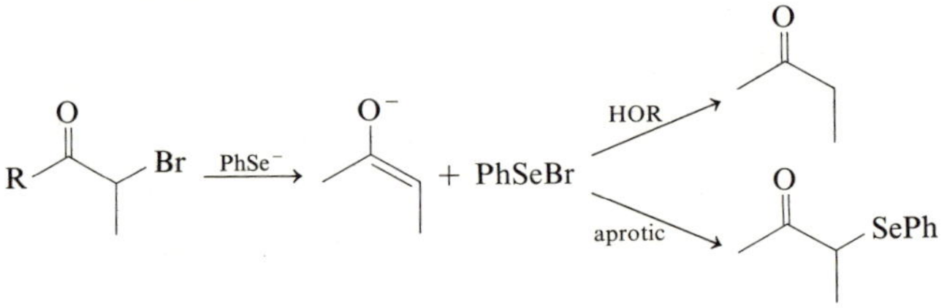

α-Halo carboxylic acids and esters smoothly react with selenols to give α-seleno acids[20,25,34,167] and esters[40] using ethanol as solvent. The greater basicity of ester anions apparently slows down the side reactions outlined above for ketones.

The diethylacetal of bromoacetaldehyde has been converted to the selenide by nucleophilic displacement.[168]

7. Reaction of α,β-Unsaturated Carbonyl Compounds with Selenenamides

α-Phenylseleno β-dialkylamino carbonyl compounds can, in certain systems, be prepared by reaction of enones with *N,N*-dialkylbenzeneselenenamides.[10] The reaction involves a Michael addition followed by an intramolecular selenenylation.[10,10a] The synthesis is limited to ketones which are reactive Michael acceptors. Only one reaction of this type is known for

[167] G. T. Morgan and W. H. Porritt, *J. Chem. Soc.* p. 1755 (1925).

[168] F. C. McIntire and E. P. Painter, *J. Am. Chem. Soc.* **69**, 1834 (1947).

sulfenamides: *N,N*-dimethylbenzenesulfenamide adds to dimethyl acetylenedicarboxylate under extreme conditions.[10a]

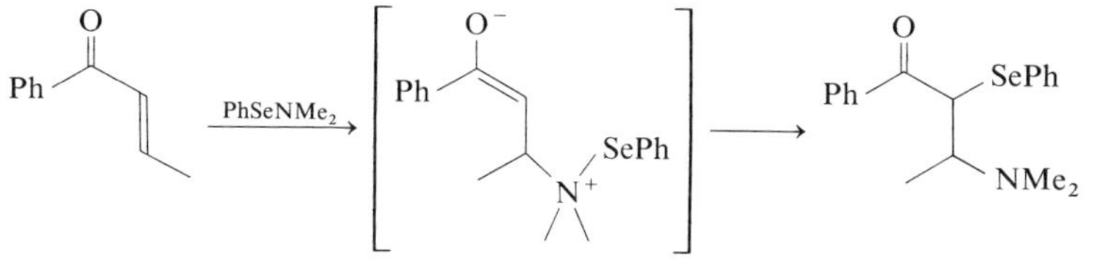

(ref. 10)

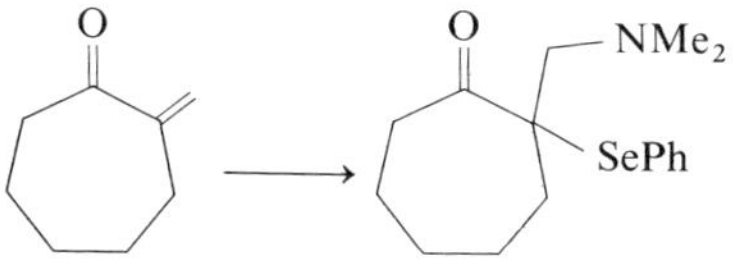

(ref. 10)

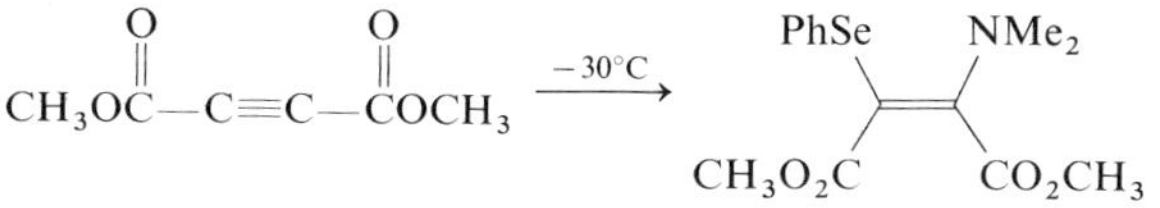

(ref. 10a)

G. Synthesis of α-Phenylseleno Carbonyl Compounds Using Selenium-Stabilized Anions

1. Reaction of Enolates Derived from α-Phenylseleno Carbonyl Compounds

Sharpless, Lauer, and Teranishi[40] have shown that ethyl α-phenylselenoacetate can be deprotonated using amide bases and that the anion can be alkylated.

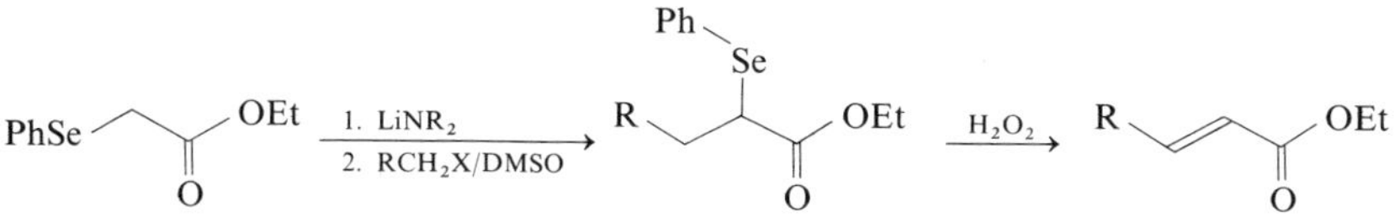

(ref. 40)

R = CH_3 60%
R = Ph 65%

An α-phenylseleno lactone has been methylated by Grieco and Miyashita[151] who were in this way able to achieve stereochemical control and thus regiospecific formation of α-methylene lactones.

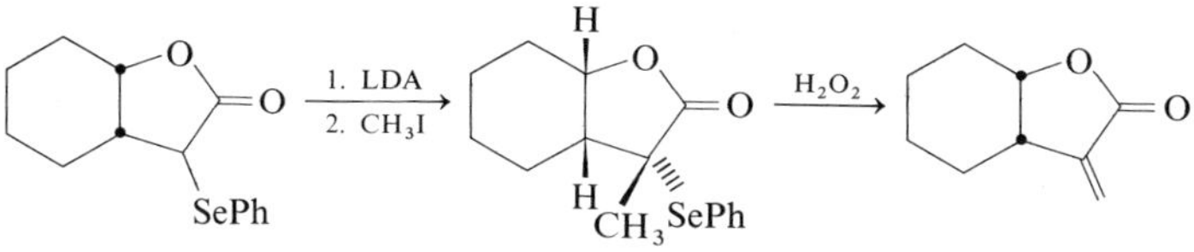

(ref. 151)

Introduction of the phenylseleno group after the methylation step resulted in a predominance of endocyclic elimination due to opposite stereochemical orientation of the methyl and seleno group.

α-Phenylseleno ketones can also be deprotonated under some conditions, but deselenation occurs very readily, particularly if the hydrogen is tertiary. α-Phenylselenoacetophenone can therefore be deprotonated quite cleanly.[169]

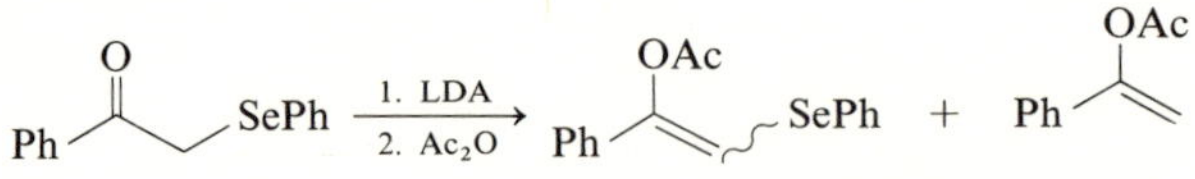

(ref. 169)

94:6

α-Phenylselenopropiophenone gives mainly deselenation with lithium diisopropylamide in THF,[169] whereas potassium hydride[170,171] does give the enolate; methylation can be achieved in 70% yield.[136]

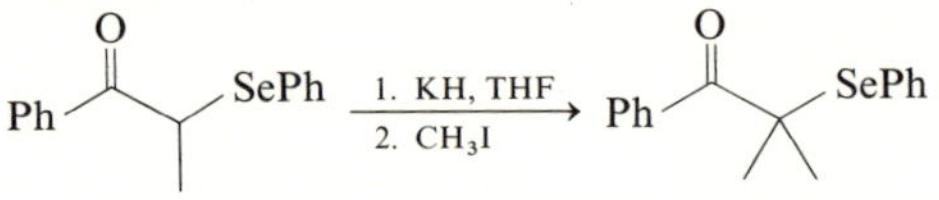

(ref. 136)

70%

Alkylation of an α-phenylselenocyclohexanone was used by Grieco, Nishizawa, Oguri, Burke, and Marinovic[74c] in a recent synthesis of vernolepin.

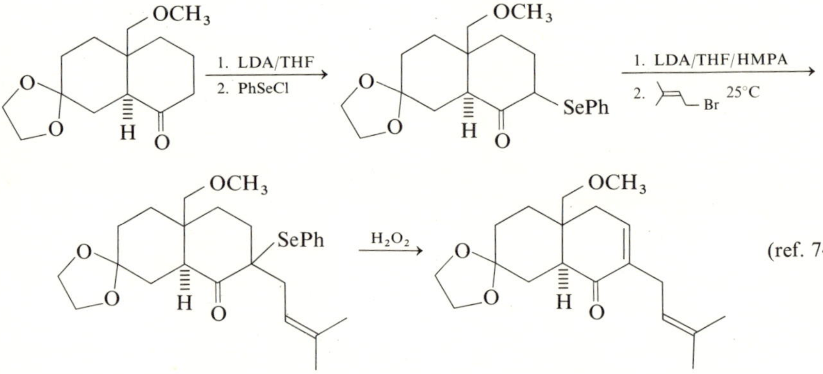

83% 65%

(ref. 74c)

A surprising feature of this sequence is the very high preference for endocyclic elimination (<20% exocyclic). Other cyclohexanone systems have

[169] H. J. Reich and M. Cohen, unpublished results.
[170] C. A. Brown, *J. Org. Chem.*, **39**, 1324 and 3913 (1974).
[171] C. A. Brown, *Chem. Commun.* p. **680** (1974).

shown a preference for endocyclic elimination but the above system has an additional feature in that the exocyclic elimination leads to a dienone system (see Section III, H, 3). The high reactivity of prenyl bromide leads to an unusual side reaction during alkylation of α-phenylselenoacetophenone and -propiophenone.

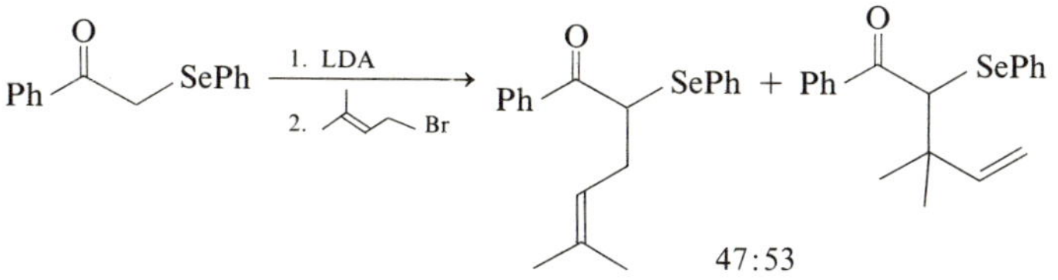

(ref. 169)

The major product apparently results from Se-alkylation, followed by a [2,3] sigmatropic rearrangement of the ylid.[169]

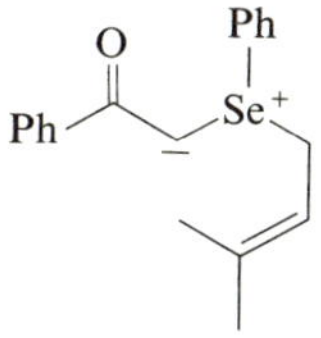

Alkylation at sulfur in a 1-methylthiovinylcopper system has been postulated to account for the formation of cumulenes.[172]

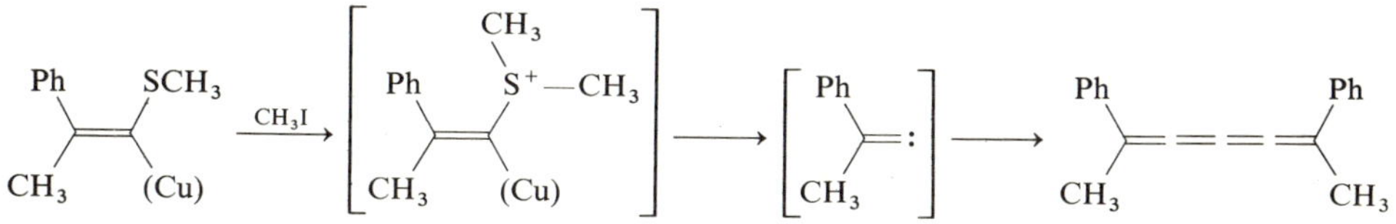

Several examples of α-selenino ketone enolate reactions have been reported.

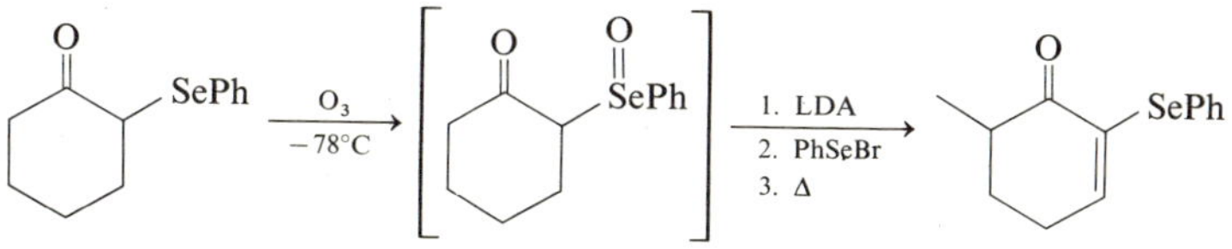

(ref. 35)

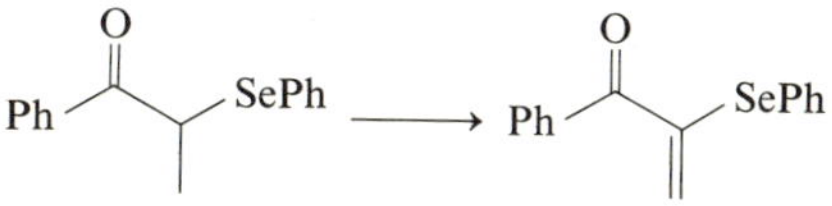

(ref. 136)

[172] H. Westmijze, J. Meijer, and P. Vermeer, *Tetrahedron Lett.* p. 2923 (1975).

α-Phenylseleno nitriles can be alkylated by normal enolate[173] or by phase transfer techniques.[173a] The latter procedure provides a convenient synthesis of 1-phenylseleno-1-cyanocyclopropane, which can be converted to 1-cyanocyclopropene. Cyclopropenes have not been prepared previously by syn elimination.

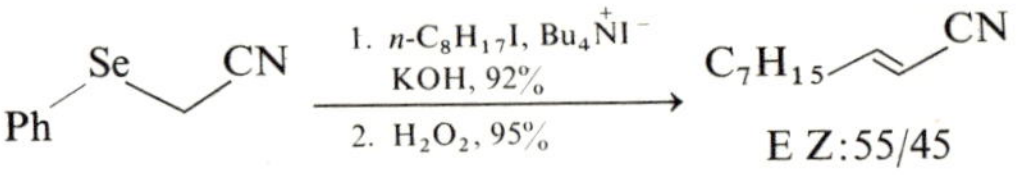

(ref. 173a)

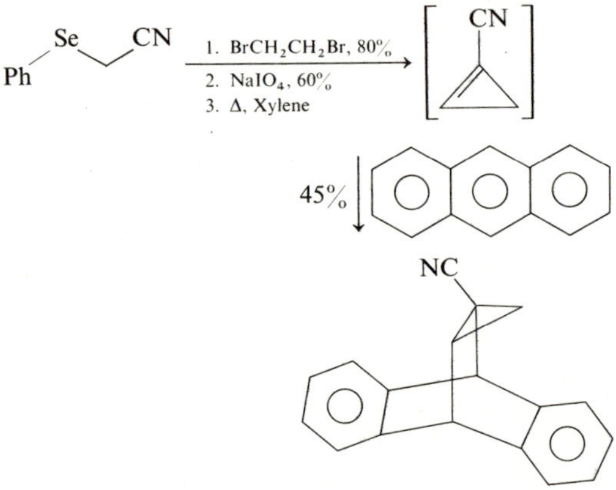

(ref. 173a)

Grieco and Yokoyama[173] have developed a procedure for the formation of α-phenylseleno nitriles from aldehydes by treatment with ArSeCN/Bu_3P.

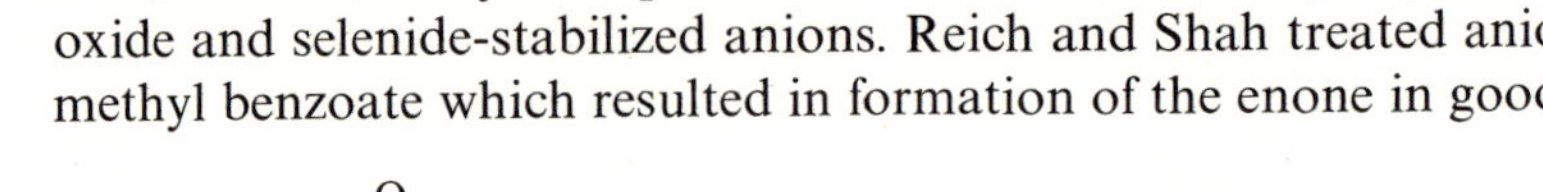

(ref. 173)

2. Acylation and Carboxylation of Selenium-Stabilized Anions

α-Seleno carbonyl compounds are also accessible by acylation of selenoxide and selenide-stabilized anions. Reich and Shah treated anion **17** with methyl benzoate which resulted in formation of the enone in good yield.

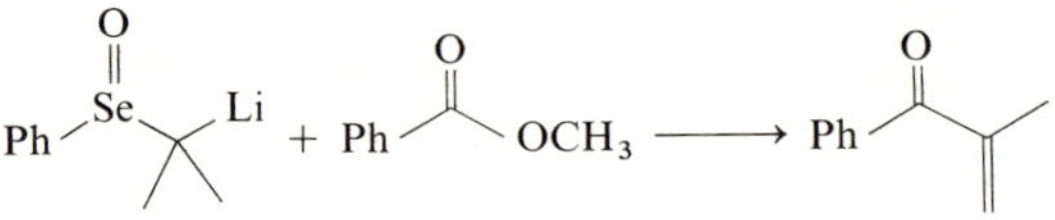

(ref. 41)

(**17**) 81%

[173] P. A. Grieco and Y. Yokoyama, *J. Am. Chem. Soc.* **99**, 5211 (1977).
[173a] Y. Masuyama, Y. Ueno, and M. Okawara, *Chemistry Lett.* p. 835 (1977).

The reaction gives low yields of enones when the lithium reagent is not tertiary or when the ester is aliphatic.[41]

Denis, Dumont, and Krief[117] have formylated, acylated, benzoylated, carbomethoxylated, and carboxylated several anions prepared using the selenoacetal cleavage (see Section III, E). α-Seleno aldehydes, ketones, esters, and acids were prepared in yields of 25–80%.

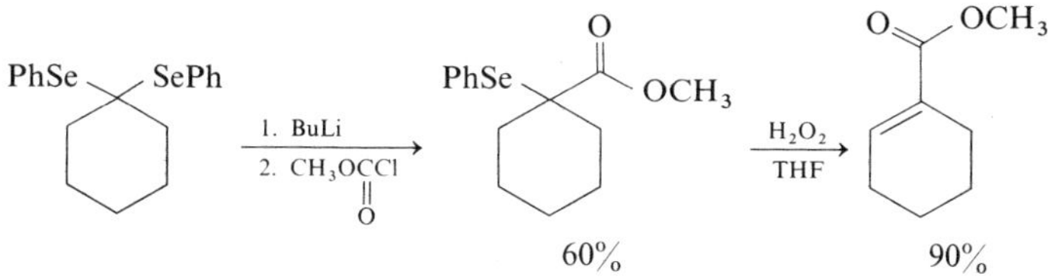

(ref. 117)

H. Conversion of α-Phenylseleno Carbonyl Compounds to Olefins

1. Oxidation of α-Phenylseleno Carbonyl Compounds (see Section II, B, for a discussion of selenide oxidation)

In the majority of cases where α,β-unsaturated carbonyl compounds have been prepared by selenoxide elimination, the introduction of selenium is a separate step, but the oxidation and syn elimination are frequently carried out together. The keto selenoxides undergo syn elimination well below room temperature (the β-dicarbonyl selenoxides eliminate below −10°C) so oxidants compatible with the products are desirable. Two equivalents of oxidant are necessary under these conditions, since the initially formed selenenic acid (PhSeOH) is oxidized (directly, or via disproportion to diselenide) to seleninic acid ($PhSeO_2H$). Hydrogen peroxide has been by far the most commonly used reagent for oxidation. Two side reactions have been reported: the oxidation of some β-dicarbonyl enones to epoxides[35] with excess H_2O_2 (prevented by addition of exactly two equivalents of H_2O_2) and the Baeyer–Villiger oxidation of certain cyclobutenones (ozone was successfully used here[35]) and D-ring steroidal enones[34–34b] to lactones. It is not known whether the use of stoichiometric amounts of peroxide would prevent these over-oxidations.

Examination of Table VI shows several other oxidants which have been found useful for α-phenylseleno carbonyl compounds. Sodium metaperiodate has been used for several particularly sensitive systems, while ozone and *m*-chloroperbenzoic acid have been used to perform oxidations at low temperature where syn elimination is slow.

The possibility of using Chloramine-T to form selenimides has been examined for α-phenylseleno carbonyl compounds; in one case: 2-phenylselenocyclooctanone was converted to cyclooctenone in 54% yield.[46a]

2. Side Reactions

There have been few reports of failures in achieving the introduction of phenylseleno groups alpha to carbonyl functions. The most common side reaction, either by direct selenenylation using PhSeCl, or by reaction of enolates with PhSeCl is the formation of α,α- or α,α′-disubstituted compounds. Weller and Rapoport[174] have reported that an intermediate in a synthesis of codeine could not be selenenylated using either reaction of enolate with PhSeBr, ketone with PhSeCl, or enol acetate or enol ether with PhSeBr.

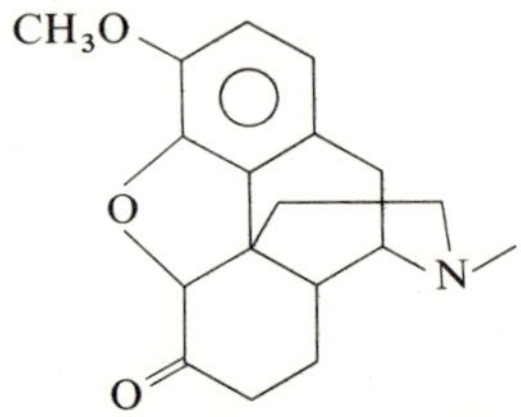

The reaction of enol derivatives with $PhSeO_2CCF_3$ was not attempted. Failure to achieve selenenylation was ascribed to interaction of PhSeCl and PhSeBr with the tertiary amine. This explanation is supported by the observation of Iijima, Rice, and Silverton[141b] that the *N*-carboethoxy derivative can be smoothly selenenylated and converted to the enone (see Table VI).

Aside from processes involving reaction of products with oxidant, which can usually be predicted from the known reactivity of the various oxidizing agents, there are several side reactions which can occur during selenoxide elimination of α-phenylseleno ketones (esters and lactones have not been found to give any side reactions).

The conversion of *cyclic* α-phenylseleno ketones (6-, 7-, and 8-membered rings) where α-hydrogens are present frequently give low yields during the oxidation–syn elimination step if ordinary oxidation conditions (H_2O_2, $NaIO_4$) are used.[35] Esters, lactones, acyclic ketones, and cyclic ketones, where the selenide has no α-hydrogens, are almost invariably converted to unsaturated compounds in high yield under ordinary oxidation conditions.

Careful study of reaction conditions for the 2-phenylselenocyclooctanone system by Reich, Renga, and Reich[5] has revealed that two major side reactions occur: an acid-catalyzed seleno-Pummerer reaction (prevented by having weak base present) which leads to α-diketone, and a more complex pathway leading to α-phenylseleno enone involving selenenylation of intermediate keto selenoxide.

[174] D. D. Weller and H. Rapoport, *J. Med. Chem.* **10**, 1171 (1976).

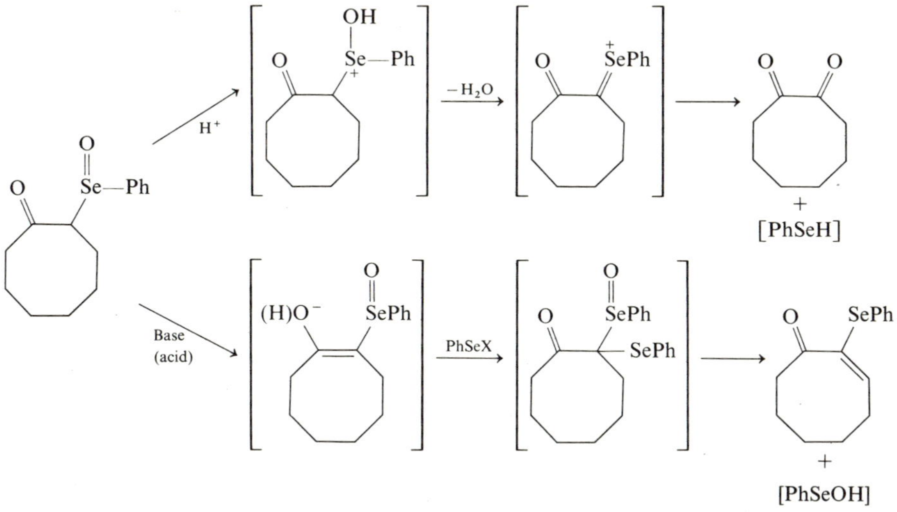

Table VII lists some of the experiments illustrating the changes in yields caused by changes in reaction conditions. The suberic acid is apparently formed by further oxidation of α-diketone and is the major product when H_2O_2 is used as oxidant. Dramatic improvements in yield are observed when elimination is carried out by a short pyrolysis in the presence of base, particularly when the base is diethylamine (trapping of PhSeX to give $PhSeNEt_2$).[10]

TABLE VII

PRODUCTS FROM OXIDATION–ELIMINATION OF 2-PHENYLSELENOCYCLOOCTANONE (%)[5]

Conditions	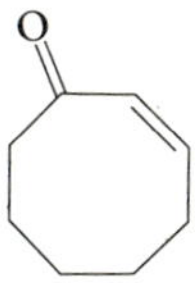		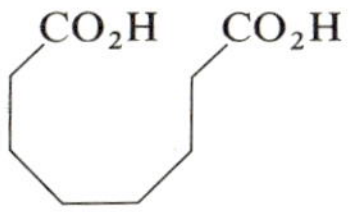	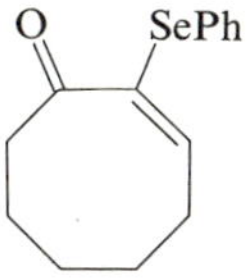
H_2O_2/CH_2Cl_2	16	13	39	2
O_3, HN(*i*-Pr)$_2$, Δ	67	<2		14
O_3, $HNEt_2$, Δ	74	<2		<2

Electrophilic reactions of PhSeOH, or similar species derived from it, can also be a problem when alkyl selenoxides undergo syn elimination (see Section III, C). Suitably activated enones react with selenenylating species formed during selenoxide syn elimination as well.[136]

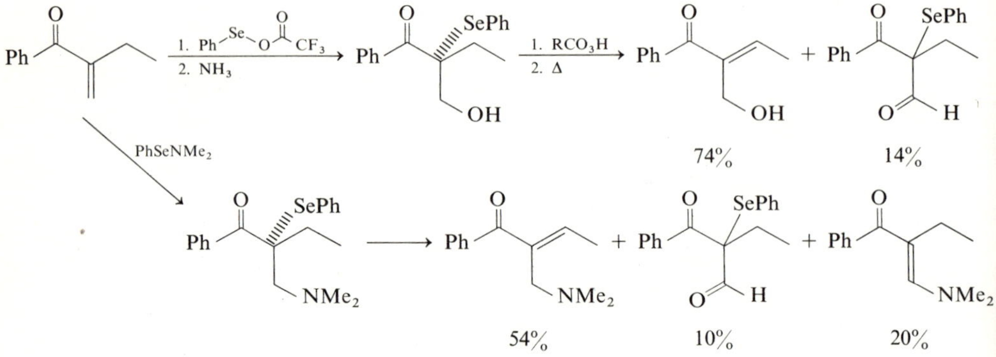

The α-phenylseleno-β-dicarbonyl compounds above are thought to be formed by reaction of PhSeX with enol or enamine.

3. Regioselectivity

The α-phenylseleno carbonyl compounds capable of two modes of elimination are listed below together with product ratios obtained. The systems are too disparate to arrive at any real generalizations.

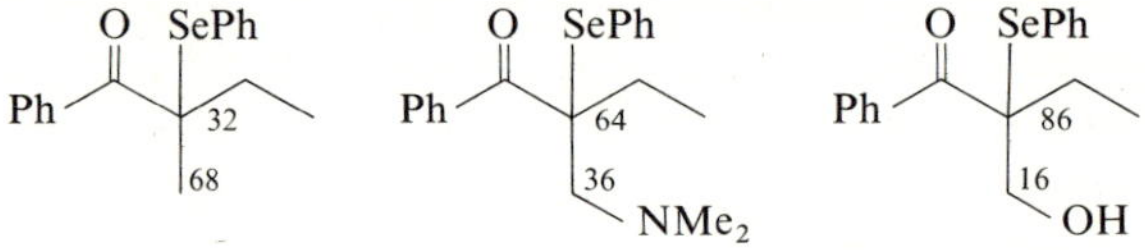

(ref. 136)

CH_2Cl_2 CH_2Cl_2 CH_2Cl_2

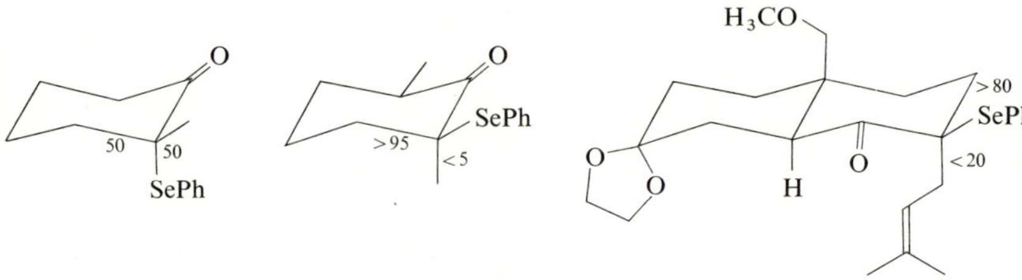

EtOH (ref. 122) EtOH (ref. 122) THF/H_2O (ref. 74c)

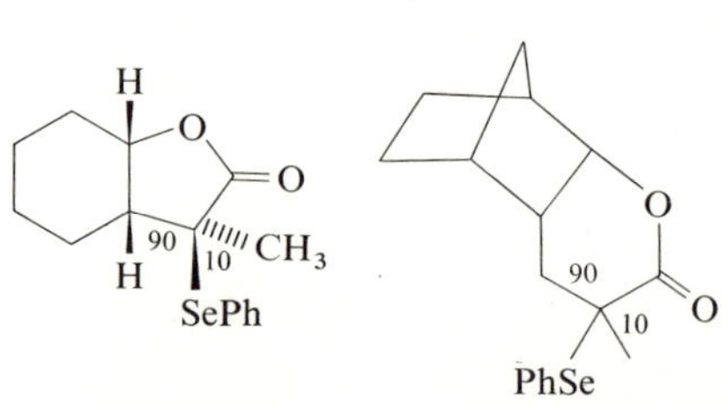

(ref. 151)

THF/H_2O THF/H_2O

Trost and Leung[175] have made the important observation in the sulfur series that although α-phenylsulfino γ-lactones preferentially undergo endocyclic elimination, the ring-opened hydroxy acid gives elimination toward the methyl group.

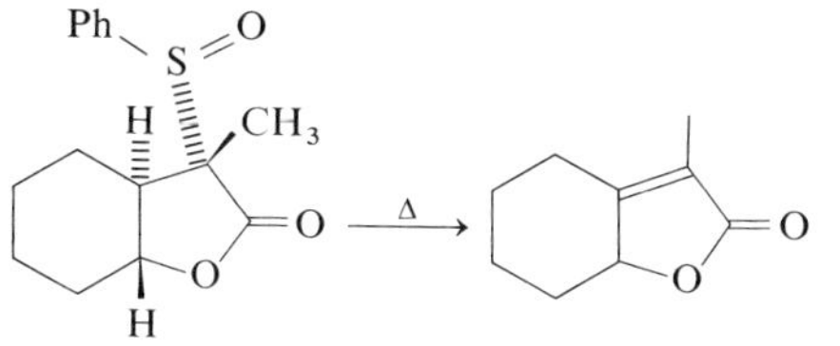

(ref. 175)

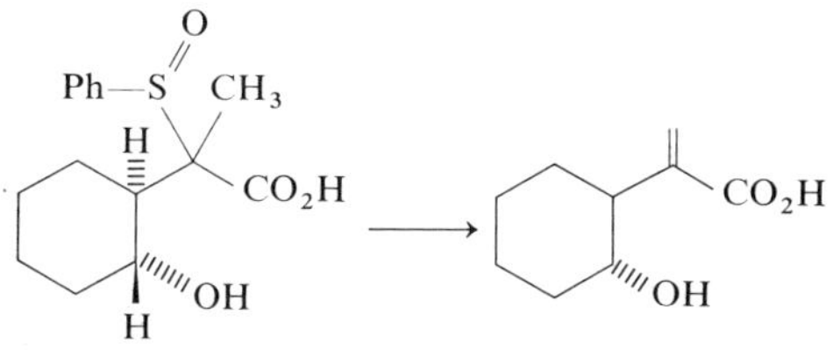

(ref. 175)

This effect was ascribed to dipole repulsions between the carbonyl and sulfinyl groups, which are minimized for the endocyclic transition state in the lactone, overcoming the normal preference for elimination toward methyl groups. Dipole repulsions may also be a factor in the greater rate of elimination of keto sulfoxides compared to alkyl sulfoxides. Since selenoxides are even more polar then sulfoxides such effects should be equally important in selenoxides.

A complicating feature is that diastereomeric selenoxides are present in most of the compounds of interest. These are likely to eliminate at different rates[62] and with different regioselectivities (high positional control has been achieved in elimination of hindered diastereomeric steroidal sulfoxides).[176] Under acidic conditions, selenoxides are probably undergoing epimerization at selenium with rates comparable to selenoxide elimination.[177]

I. COMPARISON OF SELENIUM AND SULFUR METHODOLOGY

The major advantage of the selenium based procedure is the much milder conditions under which elimination occurs, usually at about 100°C lower temperature than sulfoxide eliminations.[62] Sulfoxide eliminations are thus conveniently carried out only for activated systems where temperatures

[175] B. M. Trost and K. K. Leung, *Tetrahedron Lett.* p. 4197 (1975).

[176] D. N. Jones, A. C. F. Edmonds, and S. D. Knox, *J. Chem. Soc., Perkin Trans. 1* p. 459 (1976).

below 100°C are usually sufficient. Another factor of particular importance is the greater ease of selenenylation of ketones because of the convenience and stability of PhSeCl and PhSeBr as contrasted with the instability of PhSCl. A wider range of reaction conditions and cheap oxidizing agents (particularly H_2O_2) are available for selenide oxidations since problems with over-oxidation to selenones are absent. The ability to use excess oxidant also results in cleaner crude reaction products since all selenium is converted to $PhSeO_2H$, which is base extractable and recoverable. The oxidation of sulfide specifically to sulfoxide requires selective and expensive oxidants such as sodium metaperiodate or *m*-chloroperbenzoic acid. The greater nucleophilicity and lower basicity of $PhSe^-$ also has advantages when nucleophilic displacements are to be carried out. The ease of reduction of selenoxides can provide synthetic flexibility in some systems.[43]

Disadvantages of the selenium method include the probable greater toxicity of organoselenium compounds as well as their greater cost. The latter factor, however, is frequently compensated for by the efficiency of selenenylation with PhSeCl, as compared to using Ph_2S_2 (2 equivalents of Ph_2S_2 are sometimes used during enolate sulfenylations, and $PhSO_2SPh$ must be used in some situations). The instability of most selenoxides prevents their easy purification; even keto sulfoxides are stable to well above room temperature and hence can be purified chromatographically or otherwise. The enhanced electrophilicity of PhSeOH compared to PhSOH sometimes results in side reactions which require special elimination conditions for selenoxides (Sections III, C and III, H, 2). The potential of translating chirality at sulfinyl sulfur into chiral, diasteromeric, or constitutional differences in the remainder of the molecule is a promising one and a number of examples are known.[176] The vastly greater configurational lability of selenoxides (as much as 10^{10} greater racemization rate[177]) will make applications of this sort much more difficult in the selenium series, although not necessarily impossible as shown by the work of Jones, Mundy, and Whitehouse.[62]

In the area of anion chemistry, the greater stabilizing effect of sulfur and greater resistance of sulfides and sulfoxides to nucleophilic attack at sulfur make many types of α-lithio sulfides and sulfoxides more accessible and more stable than corresponding selenium compounds. Notable exceptions are the simple α-lithio selenides which are readily prepared by cleavage of selenoacetals and ketals (Section III, E).

There are some reactions for which the difference between sulfur and selenium is such that entirely different products are obtained. One such case

[177] M. Oki and H. Iwamura, *Tetrahedron Lett.* p. 2917 (1966); J. E. Trend, Ph.D. Thesis, University of Wisconsin, Madison (1976).

is the cleavage of selenoacetals using alkyllithium reagents as mentioned above. Others are the addition of selenenamide to enones (Section III, F, 7), the rearrangement of propargyl selenoxides (Section IV, B, 1), the reductive elimination of β-hydroxy selenides using reagents such as CH_3SO_2Cl–NEt_3,[43] $SOCl_2$–NEt_3,[178] $SnCl_2$–$POCl_3$,[43] TsOH, or $HClO_4$,[178a,b] the reaction of selenenic acids with olefins to give β-hydroxy selenides (sulfenic acids give sulfoxides) (Section III, C), or the pyrolysis of 1,2,3-selenadiazoles to give acetylenes (Section V, C).

J. Formation of Carbon–Heteroatom Double Bonds

The potential of using selenoxide eliminations to form C=Z double bonds has been explored to only a very limited extent and with limited success.

1. Carbonyl Compounds and Imines

The decomposition of alkyl benzeneseleninates appears not to have been studied, but the few known esters are stable well above room temperature.

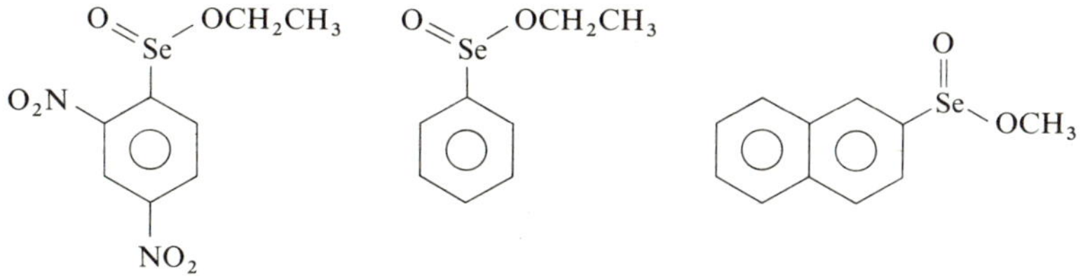

m.p. 78–80°C[7] m.p. 35°–41°C[7] (distilled at 45°C) 106°C[44]

In particular the 2,4-dinitrophenyl derivative should be prone toward decomposition to acetaldehyde if a syn elimination process was to occur. Anthraquinoneseleninic acid has been reported to oxidize alcohols to aldehydes.[179]

Czarny has isolated good yields of carbonyl compounds on treatment of certain amines with benzeneseleninic anhydride[180] or benzeneseleninyl chloride.[180a]

[178] J. Rémion and A. Krief, *Tetrahedron Lett.* p. 3743 (1976).
[178a] J. Rémion, W. Dumont, and A. Krief, *Tetrahedron Lett.* p. 1385 (1976).
[178b] A. M. Léonard-Coppens and A. Krief, *Tetrahedron Lett.* p. 3227 (1976).
[179] O. Behaghel and W. Müller, *Chem. Ber.* **68**, 1549 (1935).
[180] M. R. Czarny, *Chem. Commun.* p. **81** (**1976**).
[180a] M. R. Czarny, *Synth. Commun.* **6**, 285 (1976).

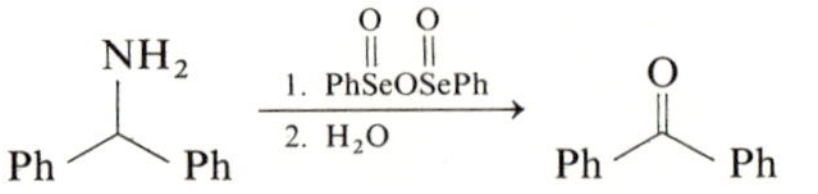

(ref. 180)

88%

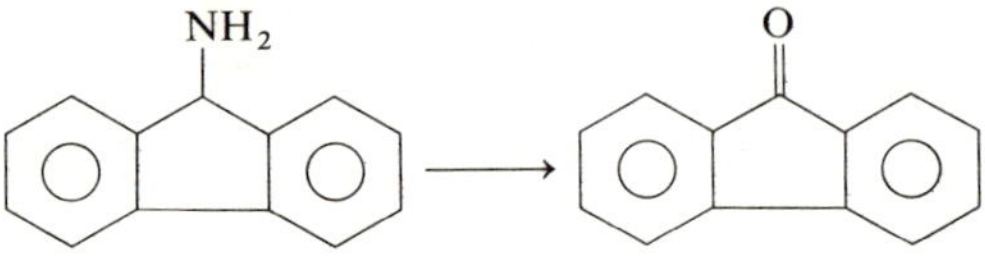

(ref. 180)

84%

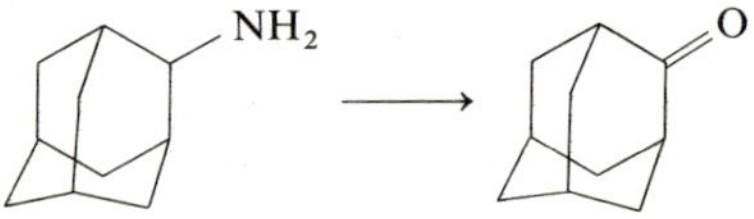

(ref. 180)

91%

Benzylamine was converted to benzonitrile in good yield.

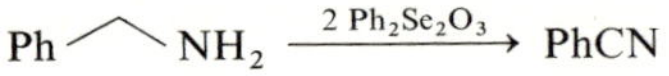

(ref. 180)

96%

A mechanism involving intermediate formation of a seleninamide, followed by elimination was suggested. The reaction gave unidentified products with 2-phenylethylamine and 1,2-diphenylethylamine.

2. Thiocarbonyl

Baldwin, Haber, and Kitchin[181] prepared the phenylseleno sulfide (**18**). Oxidation with *m*-chloroperbenzoic acid led to cyclization to **19** instead of formation of thione.

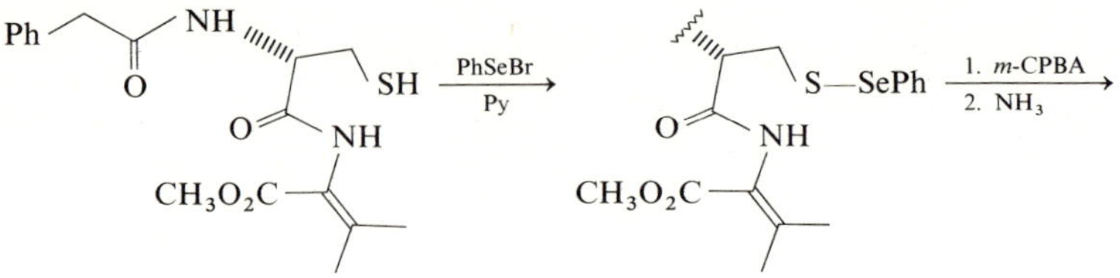

(**18**)

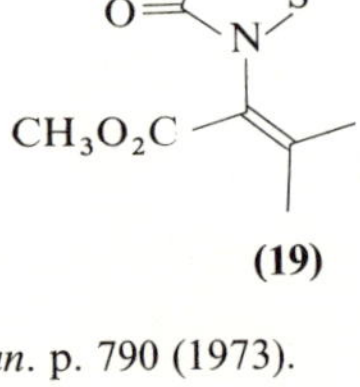

(ref. 182)

(**19**)

[181] J. E. Baldwin, S. B. Haber, and J. Kitchin, *Chem. Commun.* p. 790 (1973).

A β-lactam thione has been prepared by decomposition of a thiolsulfinate, possibly by a syn elimination pathway.[182]

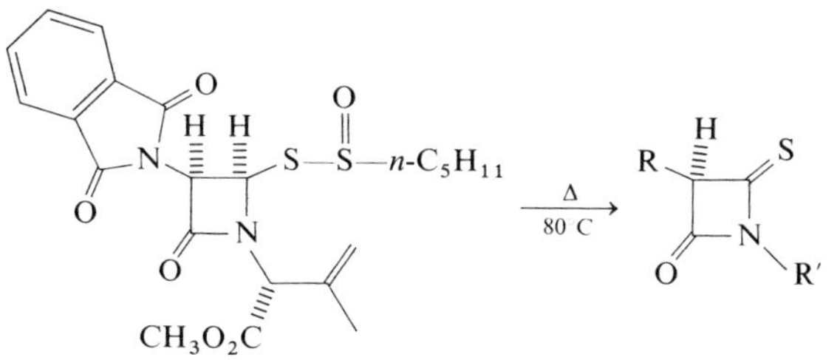

(ref. 182)

3. SILAALKENE

The possibility of forming silaalkenes by selenoxide syn elimination was explored by Reich and Shah.[121] Oxidation of the silane (**20**) using various oxidants gave only products of the sila-Pummerer reaction (benzaldehyde) but no products (**21**) expected for the silaalkene.

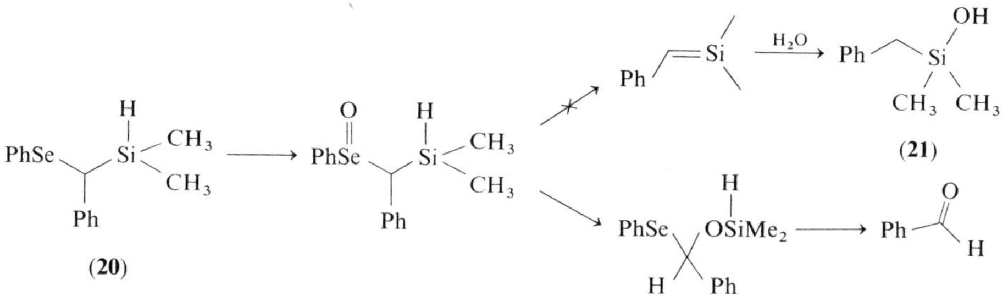

K. FORMATION OF TRIPLE BONDS

No examples of the syn elimination of vinyl selenoxides to give acetylenes or allenes have been reported. The selenoxides below are stable above 80°–100°C[115,182a]

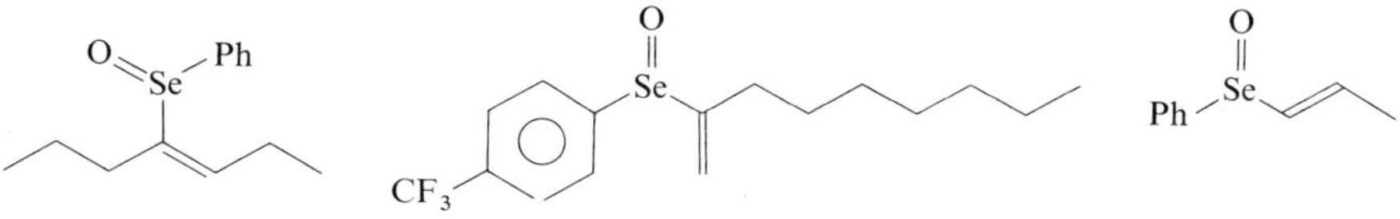

At higher temperatures decomposition to uncharacterized products occurs. Significant amounts of the expected allenes or acetylenes are not formed.[115]

[182] T. S. Chou, G. A. Koppel, D. E. Dorman, and J. W. Paschal, *J. Am. Chem. Soc.* **98**, 7864 (1976).

[182a] K. B. Sharpless and M. W. Young, private communication.

IV. Sigmatropic Rearrangements of Organoselenium Compounds

A. SIGMATROPIC REARRANGEMENTS OF ALLYL SYSTEMS

1. ALLYL SELENIDES

The intra- and intermolecular 1,3-rearrangements of allyl sulfides[183] apparently also occur for allyl selenides[159,159a] although at considerably lower temperatures than the sulfur analogs.

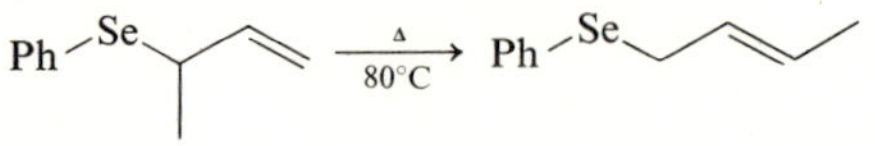

(ref. 159)

2. ALLYL DISELENIDES

Geranyl diselenide isomerizes to geranyl linalyl selenide in the presence of triphenylphosphine.[159a] The process is again more facile than in the sulfur system.[184]

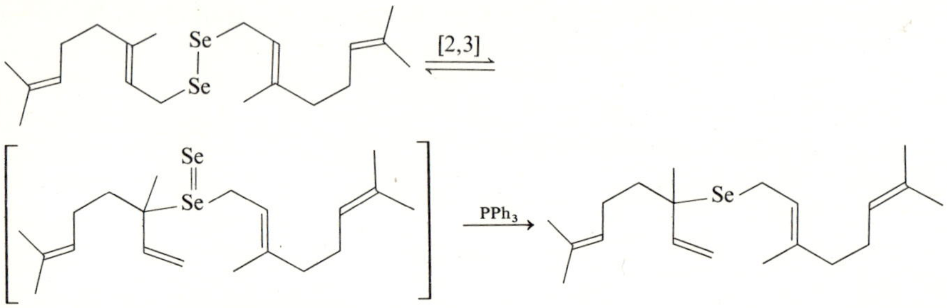

$t_{1/2}$ = 2.5 hours, 25 °C (ref. 159a)

3. ALLYL SELENINIC ACIDS

Hydrogen peroxide oxidation of allyl diselenides (as well as allyl selenols) gave allyl alcohols with allylic rearrangement.[185] The reaction probably involves the rearrangement of allyl seleninic acids although other species may be involved.

4. ALLYL SELENOXIDES

Although benzyl phenyl and dibenzyl selenoxides are stable, no stable allyl selenoxides are known. Allyl phenyl selenoxides, prepared of ozonolysis of allyl selenides, can be observed at low temperature by NMR spectroscopy; they rearrange between −45°C and −30°C. The allyl selenenates are not usually observed since they readily hydrolyze or decompose to allyl alcohols

[183] H. Kwart and N. Johnson, *J. Am. Chem. Soc.* **92**, 6064 (1970).

[184] G. M. Blackburn and W. D. Ollis, *Chem. Commun.* p. 1261 (1968), G. Höfle and J. E. Baldwin, *J. Am. Chem. Soc.* **93**, 6307 (1971).

[185] K. B. Sharpless and R. F. Lauer, *J. Am. Chem. Soc.* **94**, 7154 (1972).

and other products. Clear-cut evidence is lacking, but indications are that the selenenate is more stable than selenoxide.[86]

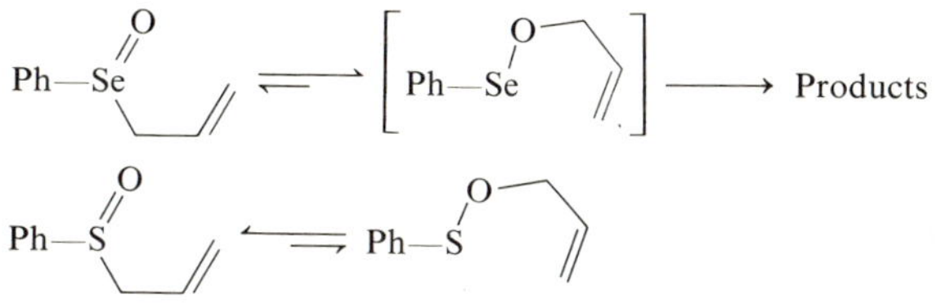

For the majority of allyl sulfoxides only trace amounts of sulfenate isomer is present at equilibrium.[186]

The first [2,3] sigmatropic rearrangement of allyl selenoxides was reported by Sharpless and Lauer[185] (Table VIII). No trapping agents are required for these rearrangements; the selenenate ester suffers rapid hydrolysis under the oxidation conditions. No mechanistic or stereochemical studies have been reported.

Carbonium ion processes can intervene in some cases. Thus, when cinnamyl phenyl selenide is oxidized in ethanol using H_2O_2, a mixture of materials is obtained including unrearranged alcohol and ethers, whereas oxidation in dichloromethane (H_2O_2, two phase) gives only the rearranged alcohol in high yield.[86]

A [2,3] sigmatropic shift is part of a conversion of α-phenylseleno enones to γ-hydroxy enones.[52]

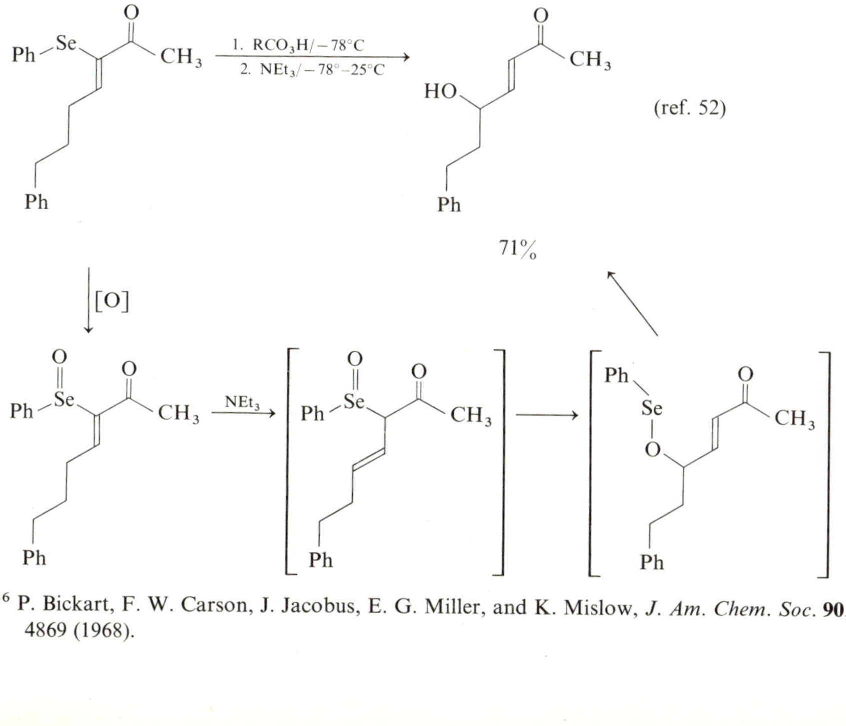

[186] P. Bickart, F. W. Carson, J. Jacobus, E. G. Miller, and K. Mislow, *J. Am. Chem. Soc.* **90**, 4869 (1968).

TABLE VIII

ALLYL ALCOHOLS PREPARED BY OXIDATION OF ALLYL SELENIDES

Selenide	Oxidant	Allyl alcohols	Yield (%)	Ref.
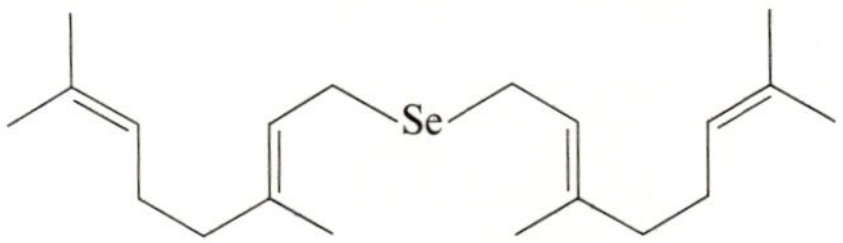	H_2O_2 THF	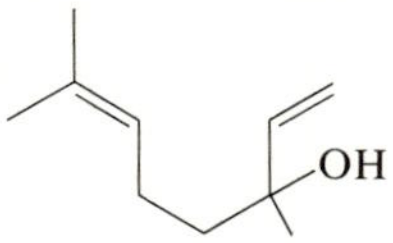	70	185
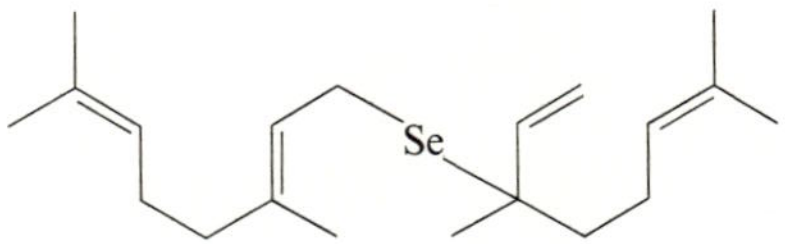	H_2O_2 THF	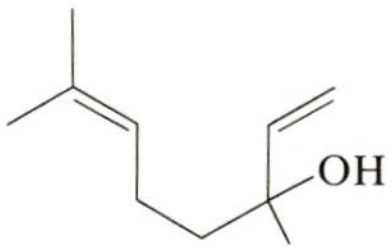	36	185
		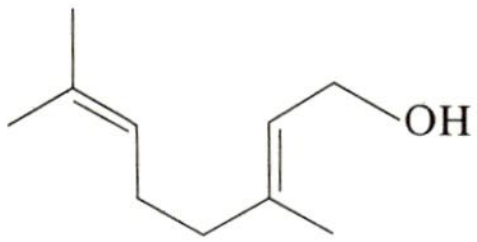	22	
		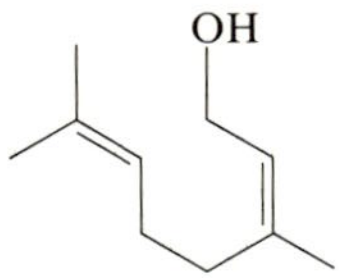	14	
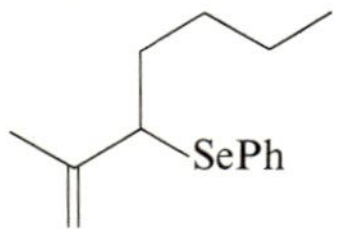	H_2O_2 THF	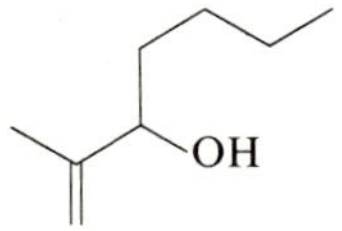	14	185
		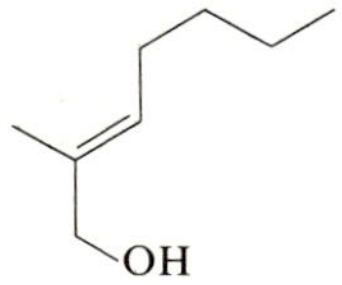	71	

		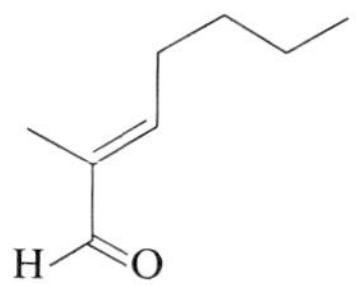	5	
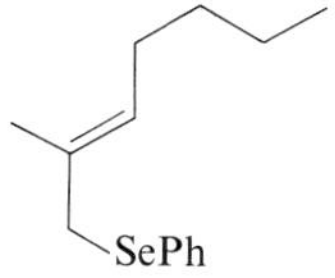	H_2O_2 THF	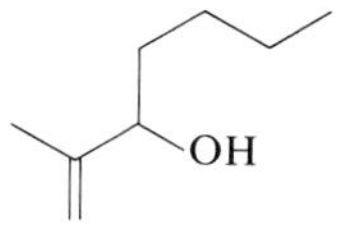	100	185
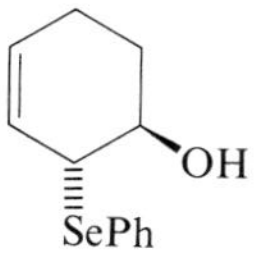	H_2O_2 THF	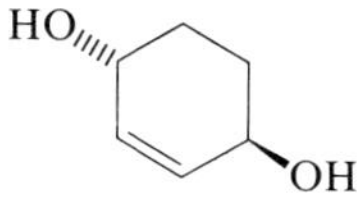	50[a]	4a
	H_2O_2 CH_2Cl_2	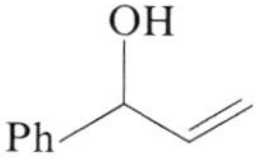	79	86
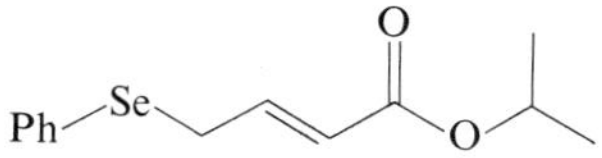	H_2O_2 CH_2Cl_2	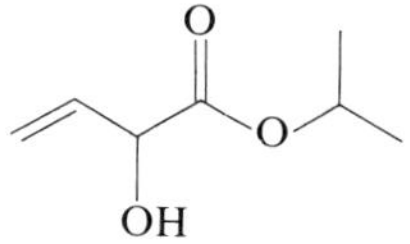	86	84
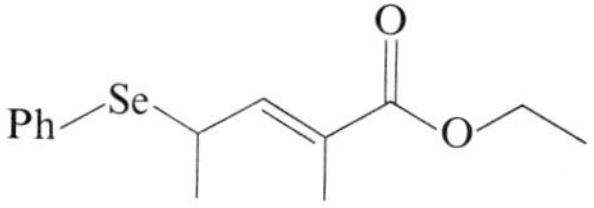	H_2O_2 CH_2Cl_2	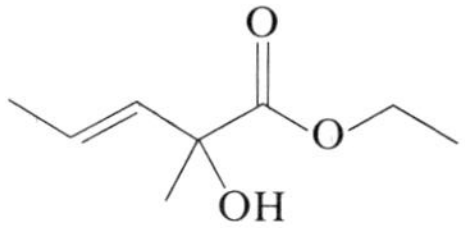	84	84

[a] Overall yield for selenide formation and oxidation.

Similar transformations have been observed for sulfoxides.[187]

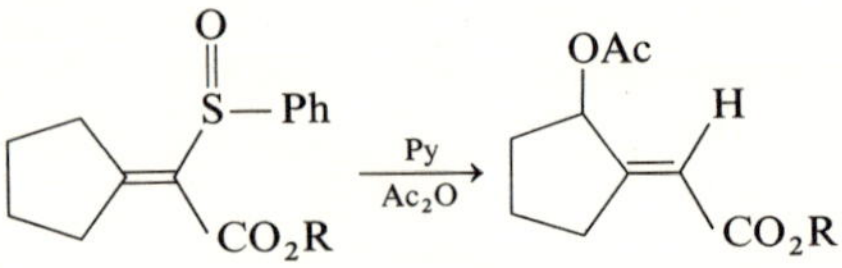

(ref. 187)

The [2,3] sigmatropic rearrangement of allyl selenoxides is faster than syn elimination, so dienes are not usually formed. Several exceptions are listed below.

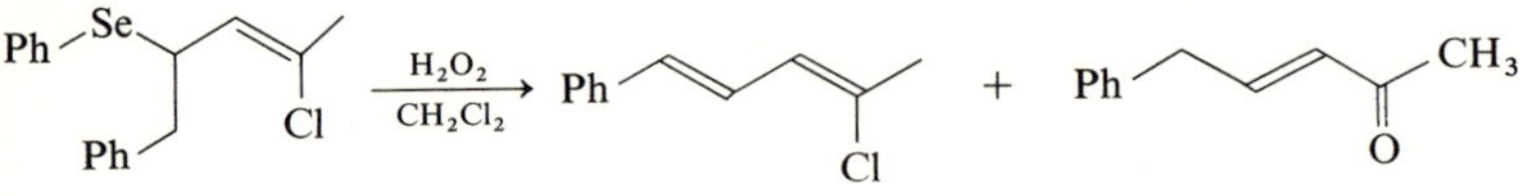

(ref. 33)

15:85

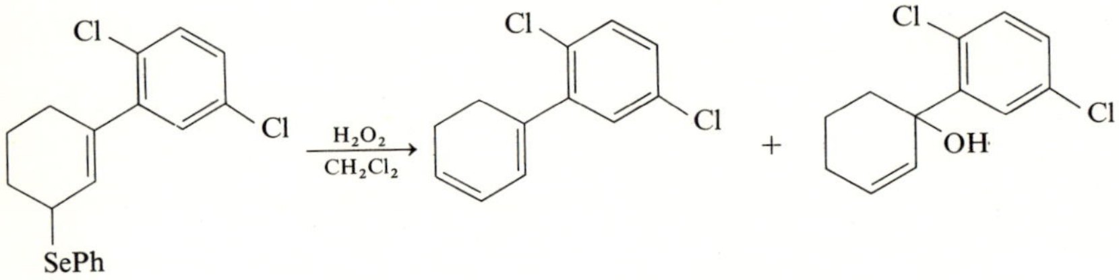

(ref. 86)

40:60

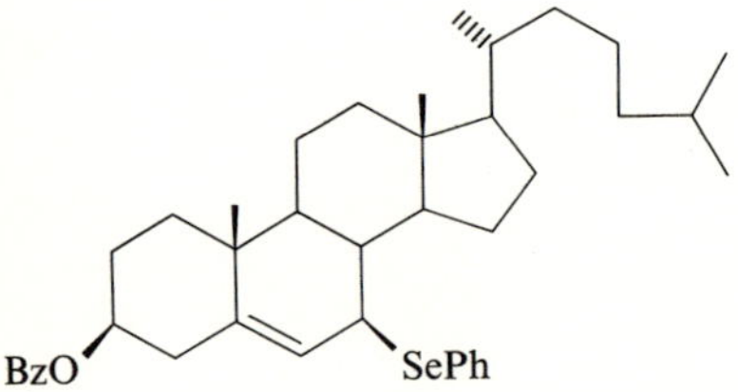

(22)

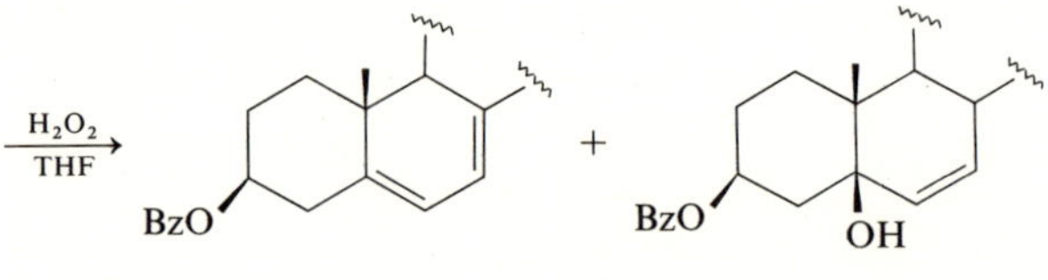

(ref. 187a)

50:50

[187] H. Kosugi, H. Uda, and S. Yamagiwa, *Chem. Commun.* p. 192 (1975).

[187a] W. G. Salmond, M. A. Barta, A. M. Cain, and M. C. Sobala, *Tetrahedron Lett.* p. 1683 (1977).

Under favorable circumstances dienes can be prepared by treatment of allyl alcohols with ArSeCl/NEt_3. This reaction represents the first example of an allyl selenenate ester rearranging to an allyl selenoxide.[187b]

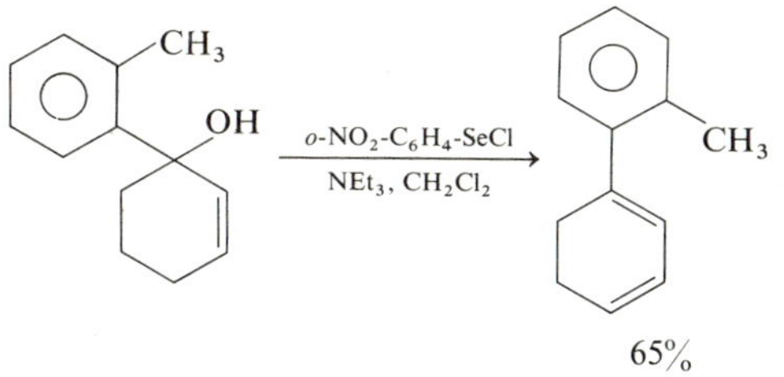

(ref. 187b)

A more widely applicable diene synthesis utilizes 2,4-dinitrobenzenesulfenyl chloride instead of the selenenyl chloride above.[187b]

The diene and allyl alcohol products from oxidation of **22** are thought to be formed stereospecifically, one from each selenoxide diastereomer.[187a] Oxidation of a γ-phenylseleno α,β-unsaturated ester (**16**) was part of a sequence in a furan, thiophene, and pyrrole synthesis (see Section III, F, 3).[48] It was not established whether selenoxide syn elimination or [2,3] sigmatropic shift occurred. From the observation that the ester below gave only rearrangement,[84] it can be assumed that the latter was the case.

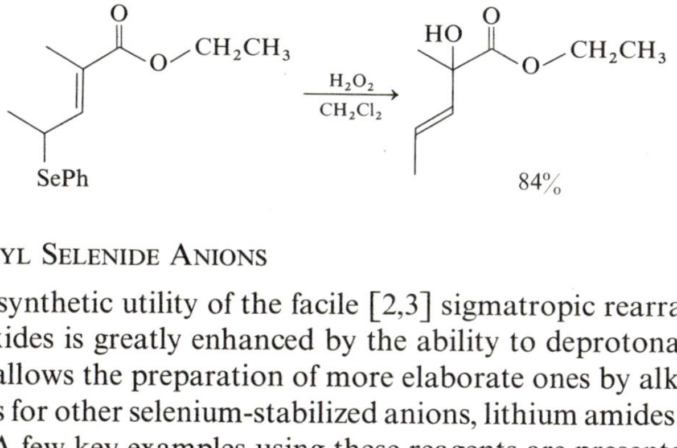

(ref. 84)

5. Allyl Selenide Anions

The synthetic utility of the facile [2,3] sigmatropic rearrangement of allyl selenoxides is greatly enhanced by the ability to deprotonate allyl selenides which allows the preparation of more elaborate ones by alkylation or silylation. As for other selenium-stabilized anions, lithium amides are the preferred bases. A few key examples using these reagents are presented below.[33]

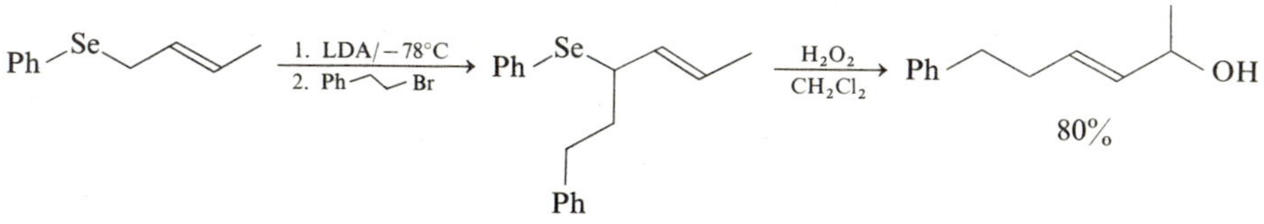

[187b] H. J. Reich, I. L. Reich, and S. Wollowitz, unpublished results.

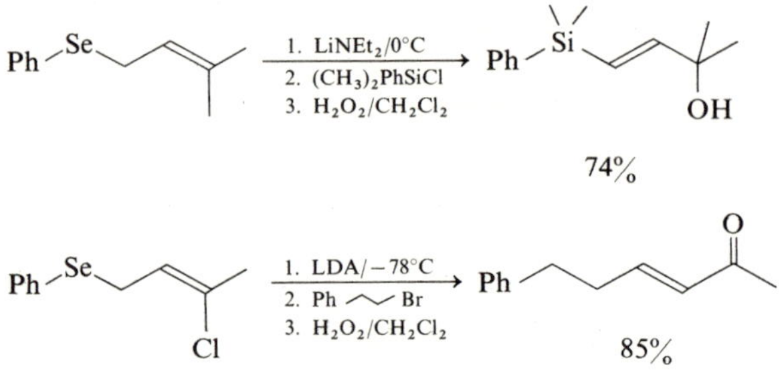

α-Alkylation predominates for most substituted allyl systems; about 20% γ-substitution occurs for allyl selenide, less for crotyl or prenyl. The α/γ selectivity is quite similar to that observed for the sulfur analogs.[188,188a]

Allyl sulfides, including γ-chloroallyl sulfides, can be alkylated and the derived sulfoxides undergo [2,3] sigmatropic rearrangement to enones,[188] similar to the third example above.

6. Enol and Phenol Seleninates

An attempt to α-hydroxylate ketones by reaction of enolates with benzeneseleninyl chloride (*O*-seleninylation followed by [2,3] sigmatropic shift) led to enone formation by *C*-seleninylation and selenoxide syn elimination.[5]

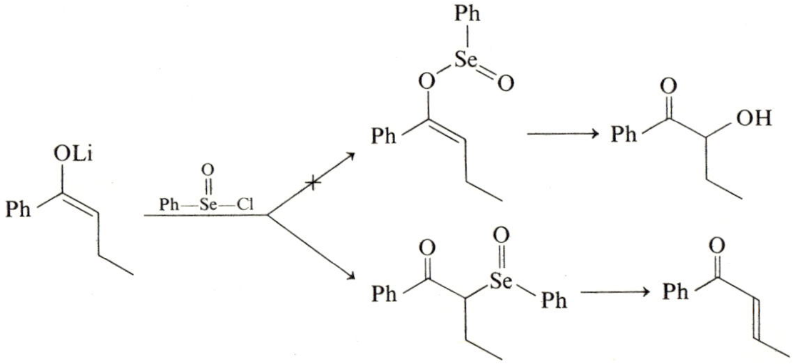

o-Hydroxylations of phenols were carried out by Barton, Magnus, and Rosenfeld[166] using benzeneseleninic anhydride as the oxidizing agent. The oxidations leading to *o*-quinol dimers were postulated to proceed via [2,3] sigmatropic rearrangement of aryl benzeneseleninates (**23**) and were best carried out by reaction of phenoxide ion with the anyhydride.

[188] P. T. Lansbury and R. W. Britt, *J. Am. Chem. Soc.* **98**, 4577 (1976).
[188a] D. A. Evans, *Acc. Chem. Res.* **7**, 147 (1974).

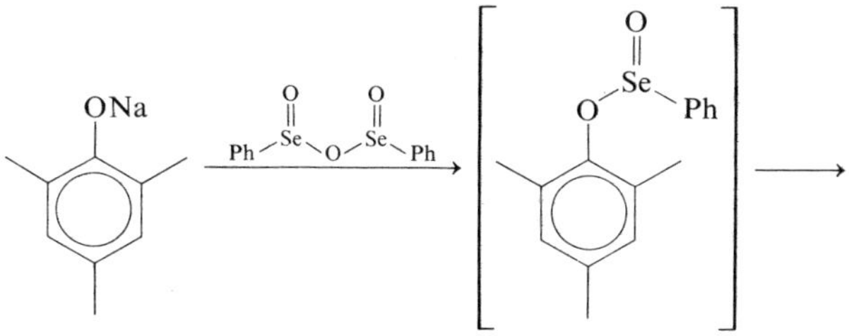

(23)

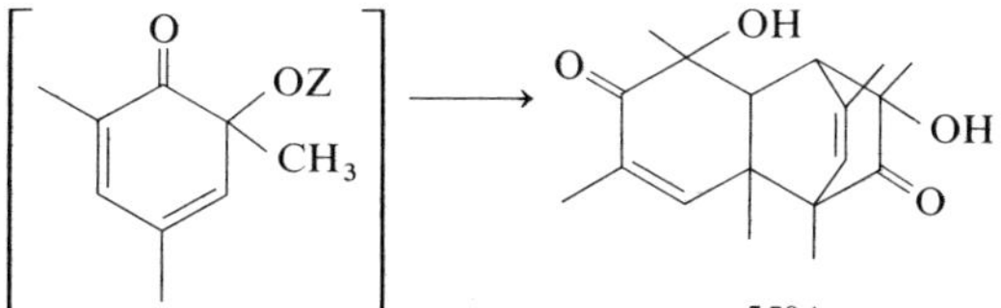

55%

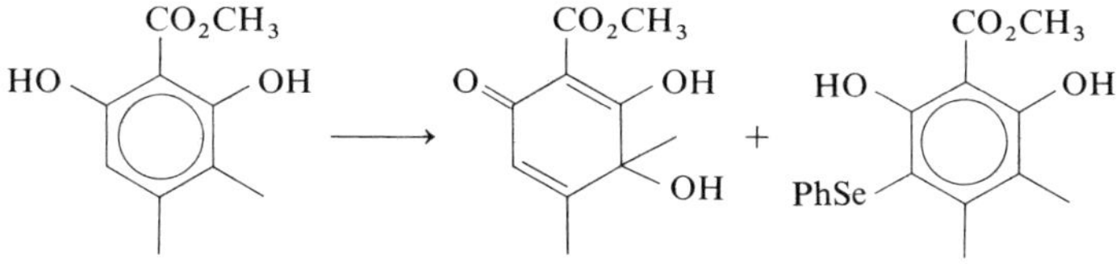

(ref. 166)

75% 17%

Phenols lacking *o*-substituents are oxidized to *o*-quinones by seleninic anhydride.[189]

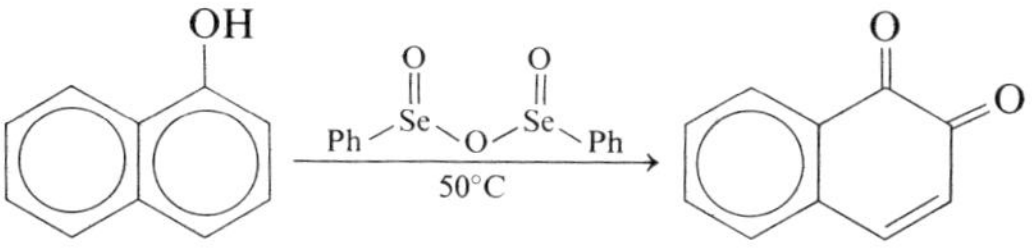

(ref. 189)

62%

B. SIGMATROPIC REARRANGEMENTS OF PROPARGYL AND ALLENYL SYSTEMS

1. PROPARGYL SELENOXIDES

Some interesting chemistry which is strikingly different from that shown by analogous sulfur systems has been found in the chemistry of propargyl selenoxides.[52] The utility of these reactions is again based on the capability of preparing elaborate propargyl systems through the agency of propargyl anions. Treatment of propargyl selenide with one or two equivalents of LDA

[189] D. H. R. Barton, A. G. Brewster, S. V. Ley, and M. N. Rosenfeld, *Chem. Commun.* p. 985 (1976).

at $-78°C$ leads to the mono or dianion. These anions can be treated with a variety of electrophiles. The products, when oxidized, are transformed to disubstituted acrylyl systems. Selenium-free compounds are obtained under some conditions (H_2O_2, MeOH), although in poorer yield.

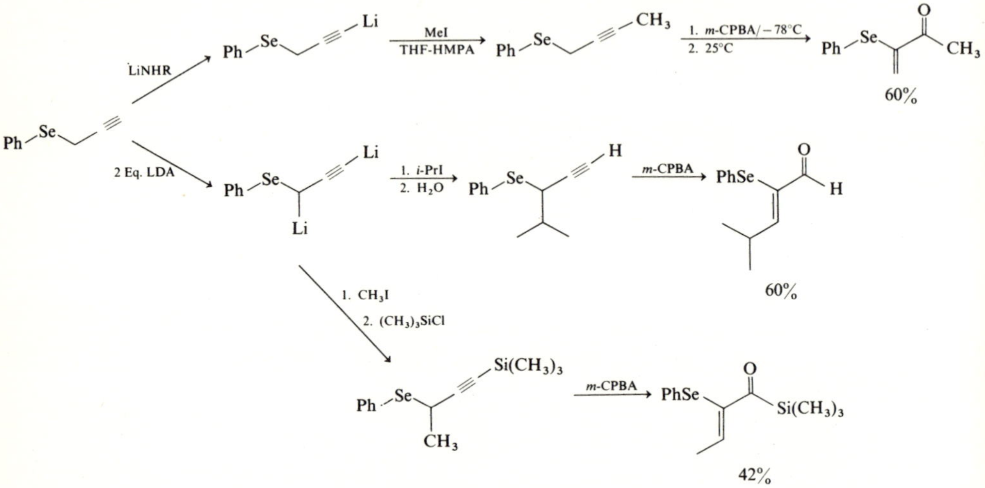

(ref. 52)

Phenyl propargyl selenoxide itself could be observed by low temperature NMR; it rearranges cleanly to α-phenylseleno acrolein at $-31°C$ ($k_1 = 1.1 \times 10^{-4}\ sec^{-1}$). The rearrangement is intermolecular as shown by appropriate cross-over experiments.

Phenyl propargyl sulfoxides also undergo [2,3] sigmatropic rearrangements, but the intermediate allenol sulfenate (**24**) suffers a [3,3] shift (Claisen-type rearrangement) to eventually give substituted benzothiophenes.[190]

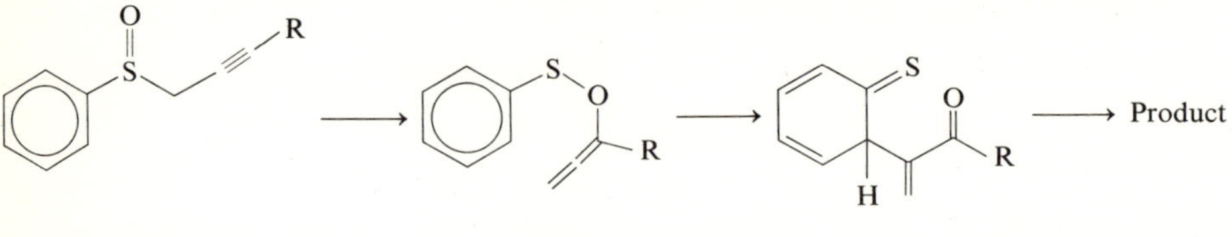

(ref. 1

(**24**)

No products analogous to those observed for the selenium system were isolated. Propargyl amine oxides behave similarly, giving substituted indoles.[191]

[190] K. C. Majumdar and B. S. Thyagarajan, *Chem. Commun.* p. 83 (1972); Y. Makisumi and S. Takada, *Chem. Commun.* p. 848 (1972).

[191] B. S. Thyagarajan, J. B. Hillard, K. V. Reddy, and K. C. Majumdar, *Tetrahedron Lett.* p. 1999 (1974).

2. Allenyl Selenoxide

The oxidation of allenyl selenides also leads to products of sigmatropic rearrangement. Substituted systems can be prepared by alkylation of α-lithioallenyl selenide (**25**).[114]

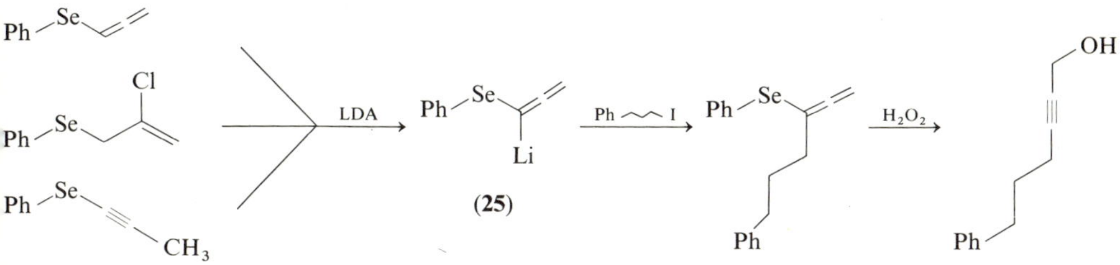

(**25**)

V. Organoselenium Intermediates in Selenium Dioxide Oxidations and Related Reactions

The extensive chemistry of selenium dioxide olefin and carbonyl compound oxidations has been reviewed recently,[1,1a] and no effort will be made here to duplicate the coverage provided by these reviews. There have, however, been a number of recent developments which bear on the mechanism of the SeO_2 oxidation and which complement the chemistry described in earlier sections of this review.

A. Olefin Oxidations

1. Mechanism

Perhaps the most widely studied SeO_2 oxidation is the transformation of olefins to allyl alcohols. Key features of the reaction are

a. Oxidation usually occurs without double bond rearrangement.

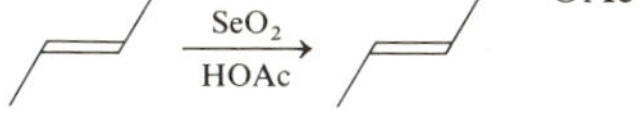

b. 2-Methyl-2-alkenes undergo oxidation of the *E*-methyl group.

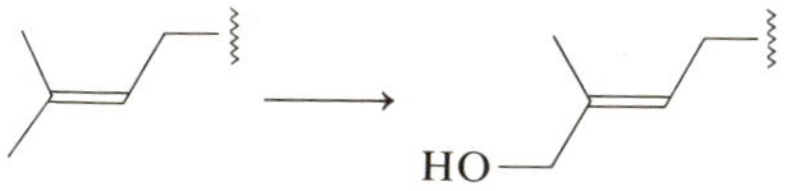

c. 1-Substituted cyclohexenes give oxidation at the 6-position (this rules out direct formation of *allylic* free radicals, carbonium ions, or carbanions.

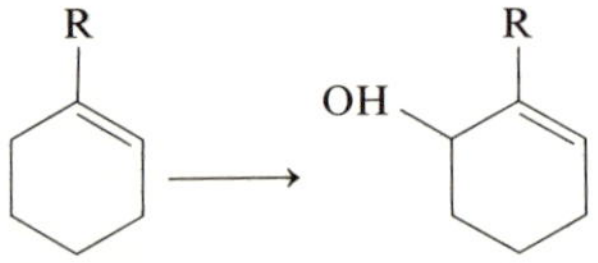

Highly-substituted systems may lose positional specificity, and carbonium ion intermediates are probably involved in such systems.

Wiberg and Nielsen[109] had proposed allylic seleninic acids as intermediates on the basis of stereochemical studies and on the observation that organoselenium compounds were frequently formed as by-products (Section III, D, 2). This suggestion was rejected repeatedly by other workers who used the argument that benzyl seleninic acid was a stable compound and allyl seleninic acids should be equally so.

Sharpless and Lauer[185] suggested that allyl seleninic acids had a decomposition pathway available that was not easily accessible to the benzyl derivative (i.e., a [2,3] sigmatropic rearrangement) and they showed that allyl diselenides, allyl selenols, as well as allyl selenides underwent smooth oxidation to rearranged allyl systems presumably by a [2,3] sigmatropic shift of an allyl selenium oxide species (Section IV, A, 3 and 5). These authors furthermore demonstrated that the stereospecificity (point b above) was also a feature of these rearrangements.

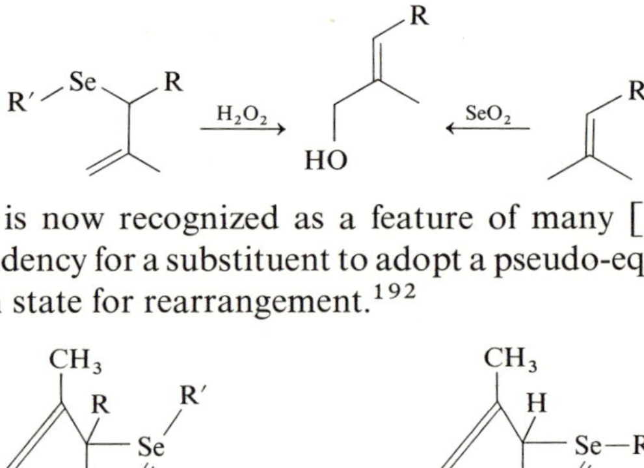

This specificity is now recognized as a feature of many [2,3] sigmatropic shifts, i.e. the tendency for a substituent to adopt a pseudo-equatorial position in the transition state for rearrangement.[192]

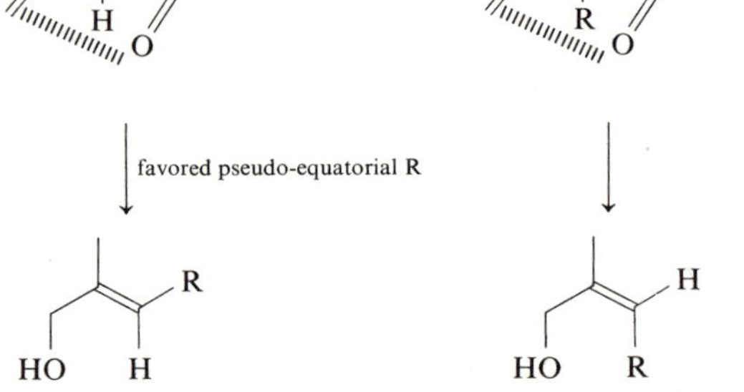

[192] J. E. Baldwin and J. E. Patrick, *J. Am. Chem. Soc.* **93**, 3556 (1971); P. A. Grieco and R. S. Finkelhor, *J. Org. Chem.* **38**, 2245 (1973); P. A. Grieco, *Chem. Commun.* p. 702 (1972).

Even more direct evidence is provided by the trapping of intermediate seleninic acids as selenino lactones (with rearranged double bond) by judiciously located hydroxyl groups.[193]

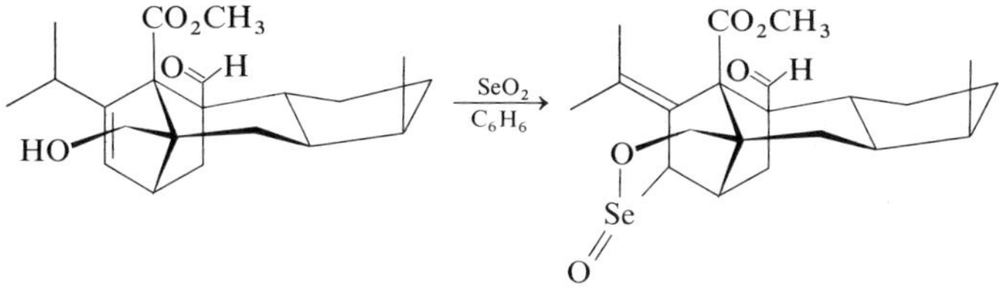

(ref. 193)

34%

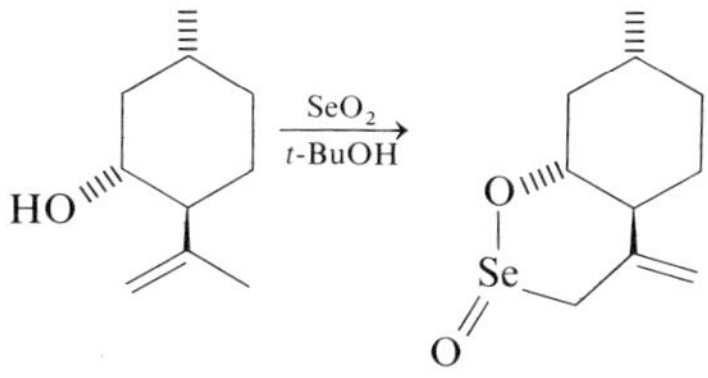

39%

The mechanism for SeO_2 oxidation proposed by Sharpless[34] which satisfactorily accounts for the observed properties of SeO_2–olefin reactions (a.–c. above) is depicted below.

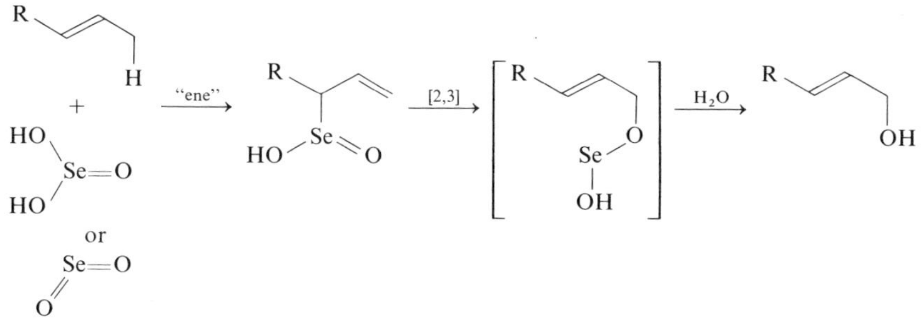

2. Selenium Diimide Aminations

Sharpless, Hori, Truesdale, and Dietrich[194] have discovered that aminating reagents can be prepared by reaction of $SeCl_4$ with $TsNH_2/NEt_3$ and by reaction of Se° with anhydrous Chloramine-T. The two reagents differ

[193] D. Arigoni, A. Vasella, K. B. Sharpless, and H. P. Jensen, *J. Am. Chem. Soc.* **95**, 7917 (1973).

[194] K. B. Sharpless, T. Hori, L. K. Truesdale, and C. O. Dietrich, *J. Am. Chem. Soc.* **98**, 269 (1976).

in regioselectivity and reactivity, but both react with olefins and acetylenes to give allylic *N*-tosylamines without double bond rearrangement. The reactions occur at room temperature and appear to proceed more cleanly than SeO_2 oxidations. It provides the first method for allylic amination of olefins.

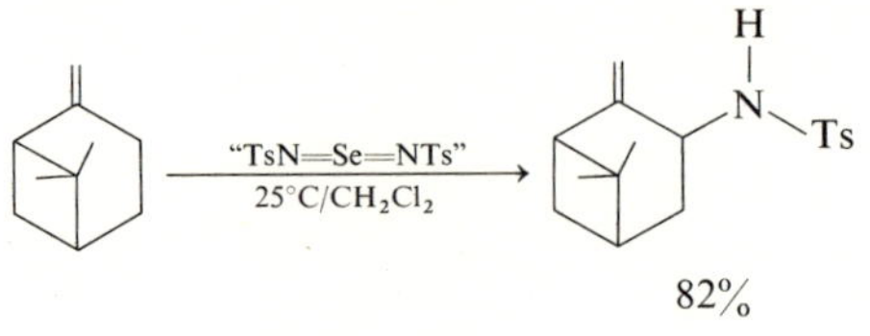

(ref. 194)

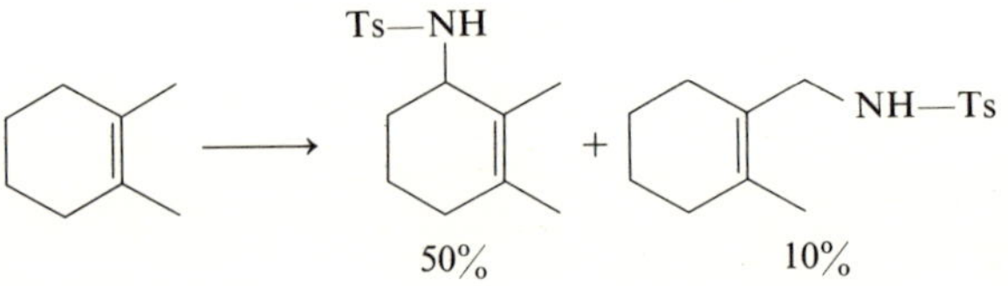

(ref. 194)

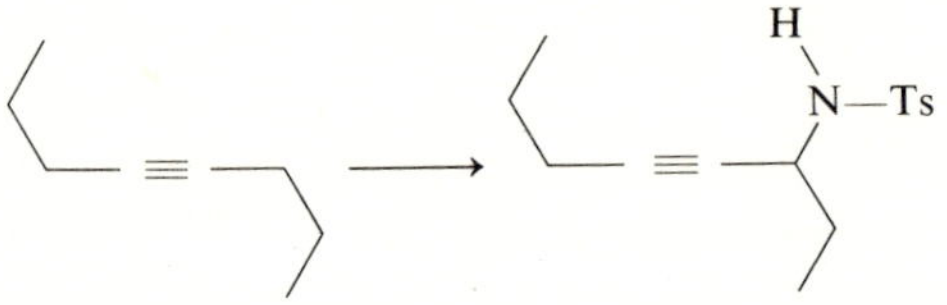

(ref. 194)

Schönberger and Kresze[195,195a] and Sharpless and Hori[196] have shown that sulfur diimides also aminate olefins. The former group has succeeded in isolating the intermediate rearranged iminosulfinamide (**26**), as well as the product of its [2,3] sigmatropic rearrangement.

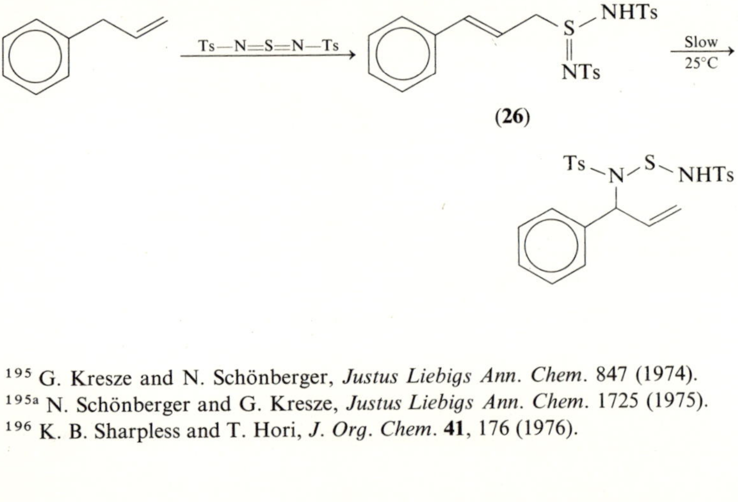

(**26**)

(ref. 195a)

[195] G. Kresze and N. Schönberger, *Justus Liebigs Ann. Chem.* 847 (1974).
[195a] N. Schönberger and G. Kresze, *Justus Liebigs Ann. Chem.* 1725 (1975).
[196] K. B. Sharpless and T. Hori, *J. Org. Chem.* **41**, 176 (1976).

The reaction of dienes with selenium diimide reagents leads to 1,2-bisamino derivatives presumably by a sequence involving [2 + 4] cycloaddition, Se–N bond cleavage, and [2,3] sigmatropic rearrangement.[197]

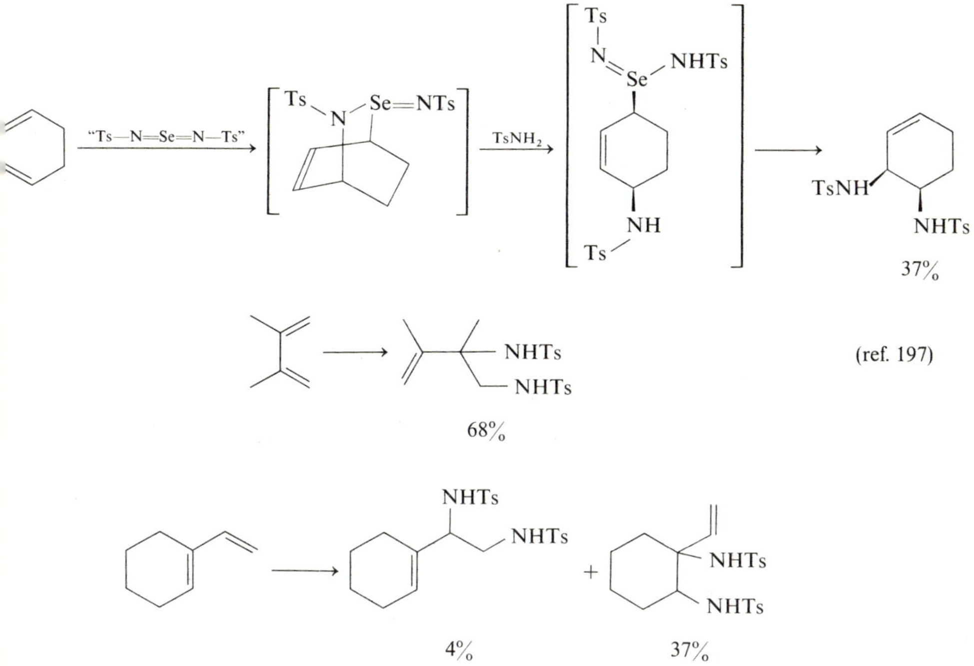

B. KETONE OXIDATIONS

The oxidation of ketones and aldehydes to α-dicarbonyl compounds using SeO_2 is one of the most important applications of this reagent. Some systems give α,β-unsaturated carbonyl compounds as by-products or even as the major product.[162,198–199a]

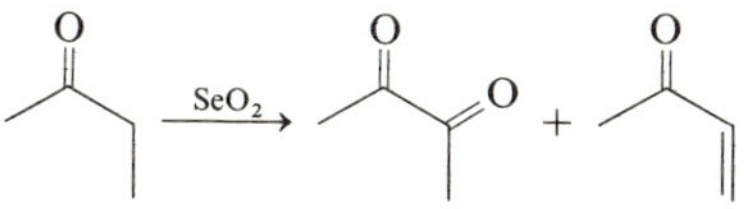

[197] K. B. Sharpless and S. P. Singer, *J. Org. Chem.* **41**, 2504 (1976).

[198] See Trachtenberg, [1a] p. 166, for a detailed discussion of dehydrogenation using SeO_2.

[199] J. N. Marx and L. R. Norman, *Tetrahedron Lett.* p. 2867 (1973).

[199a] J. N. Marx, J. C. Argyle, and L. R. Norman, *J. Am. Chem. Soc.* **96**, 2121 (1974); D. Caine and F. N. Tuller, *J. Org. Chem.* **38**, 3663 (1973).

The Corey–Shaefer mechanism[200] for SeO_2 oxidation, based on a careful study of the oxidation of deoxybenzoin to benzil, is shown below.

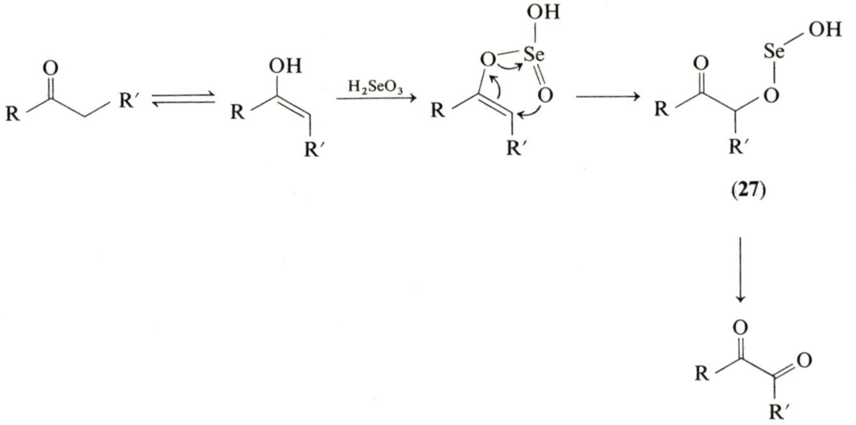

(27)

Sharpless and Gordon[89] have pointed out several troublesome features of this mechanism.

1. The intermediate Se(II) species **27** should hydrolyze readily to a ketol, yet ketols are not intermediates in the formation of α-dicarbonyl compounds. It is pertinent to note that the γ-phenylselenocrotonate (**28**) leads to an α-hydroxy compound **30**.[84] The intermediate selenenate ester (**29**), closely related to the Corey–Shaefer intermediate **27**, does not give an α-dicarbonyl compound.

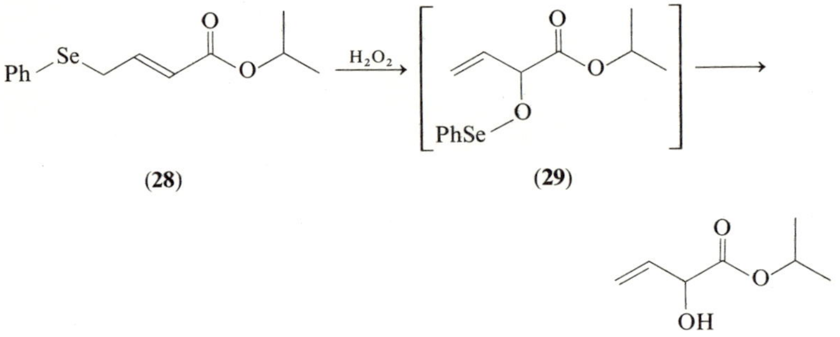

(28) (29)

(ref. 84)

86%

(30)

2. The formation of α,β-unsaturated carbonyl compounds, sometimes in excellent yield, would not appear to be easily reconciled with intermediate **27**.

[200] E. J. Corey and J. P. Schaefer, *J. Am. Chem. Soc.* **82**, 918 (1960).

3. Organoselenium compounds are frequently isolated as by-products from selenium dioxide oxidations.[199,201–201b]

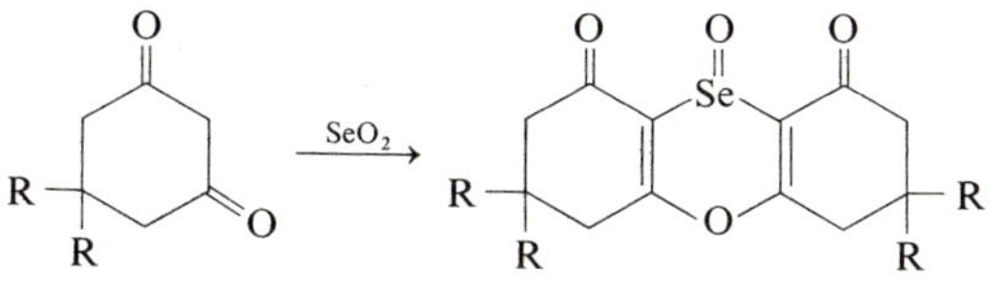

$R = CH_3$ (ref. 201)
$R = H$ (ref. 201a)

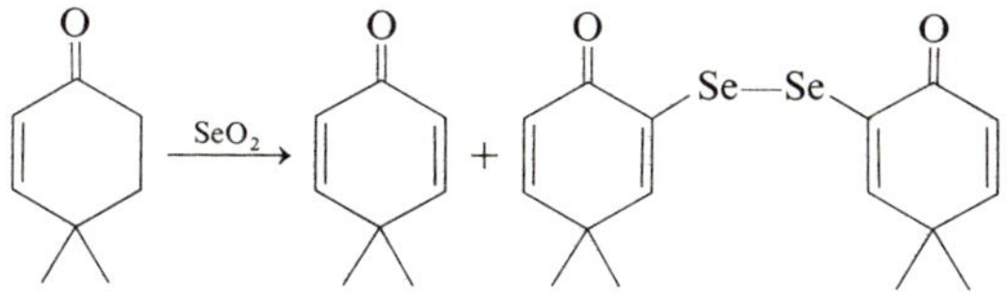

(ref. 199)

70% 14%

These must be explained by other processes than the Corey–Shaefer mechanism.

Sharpless and Gordon[89] have proposed a mechanism involving an organoselenium intermediate: the *β*-ketoseleninic acid (**31**).

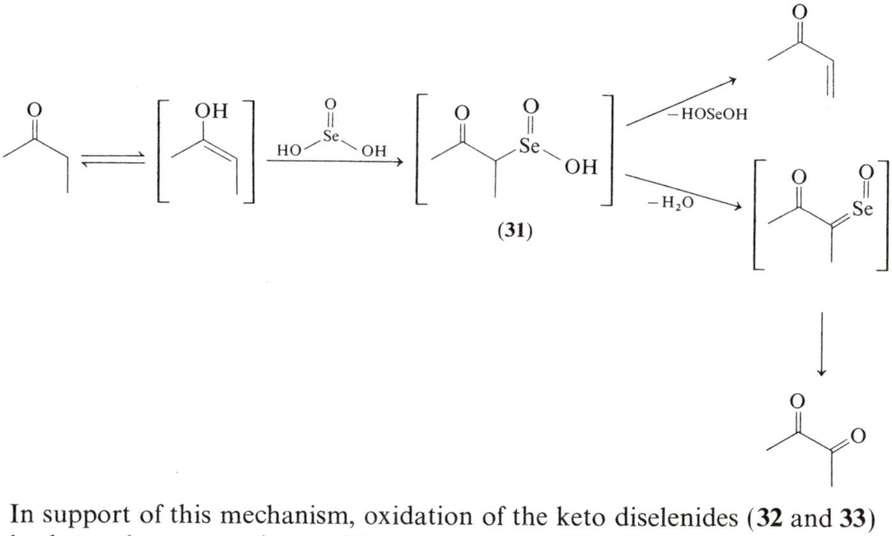

In support of this mechanism, oxidation of the keto diselenides (**32** and **33**) leads to the appropriate *α*-diketones, presumably via the corresponding *β*-keto seleninic acids.

[201] W. Borsche and H. Hartmann, *Chem. Ber.* **73**, 839 (1940); H. Stamm and K. Gossrau, *ibid.* **66**, 1558 (1933).
[201a] F. E. King and D. G. I. Felton, *J. Chem. Soc.* p. 274 (1949).
[201b] D. Paquer, M. Perrier, and J. Vialle, *Bull. Soc. Chim. Fr.* p. 4517 (1970).

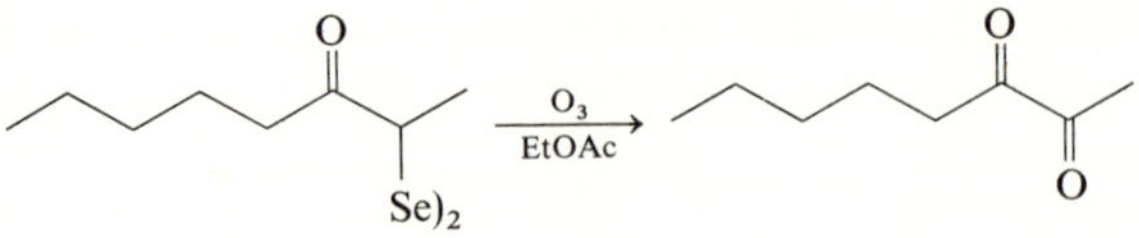

(32) 34% (ref. 89)

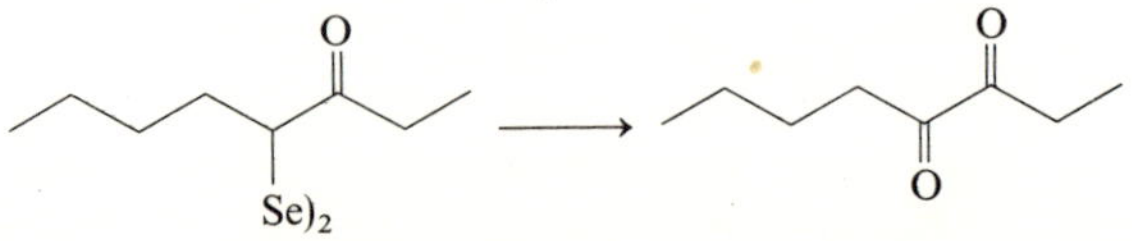

(33) 28%

Substantial precedent exists for the Pummerer-like decomposition of selenium oxides: Selenoxides give carbonyl compounds under appropriate conditions (see Section III, C)[5,34,90] and methaneseleninic acid (CH_3SeO_2H) gives formaldehyde upon thermolysis.[202]

It is important to note the great similarity between this mechanism for selenium dioxide oxidation, and the mechanisms involved in various reactions of keto selenoxides (Section III, H, 2), even to the extent that seleno enones are formed, which are similar to the by-products found for SeO_2 oxidation (see above).

It is also quite reasonable to postulate keto seleninic acids as intermediates in the formation of 1,4-oxaselenin oxides during SeO_2 oxidation of dimedone.[201–201b]

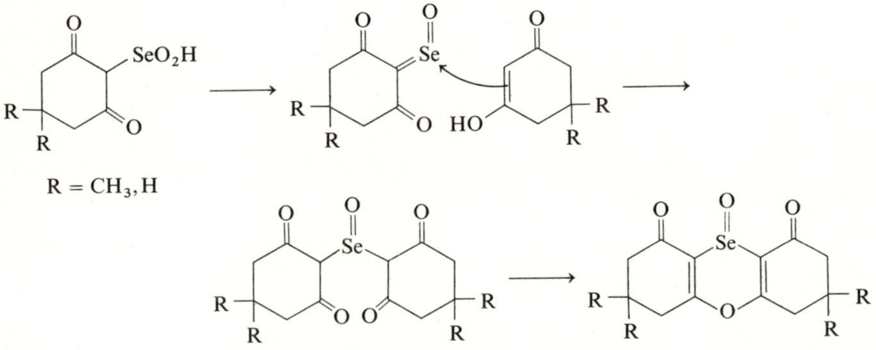

In fact, the selenium heterocycle isolated during SeO_2 oxidation of dimedone bisoxime can be considered a trapped selenine.[201a,b]

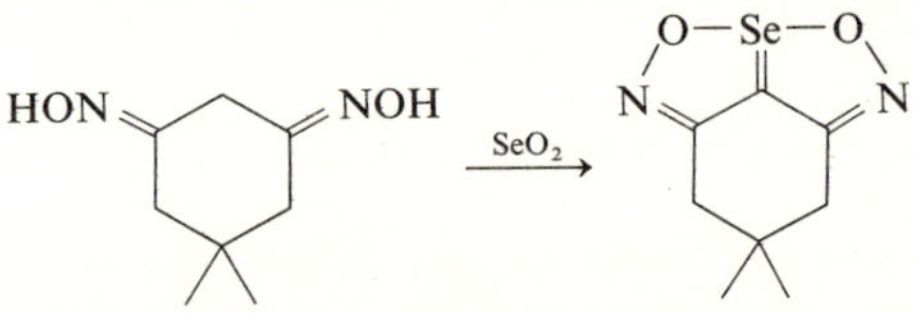

[202] M. L. Bird and F. Challenger, *J. Chem. Soc.* p. 570 (1942).

The formation of 1,2,3-selenadiazoles during SeO_2 oxidation of semicarbazones (Section V, C below) is consistent with the mechanism postulated for ketone oxidations.

C. OXIDATION OF SEMICARBAZONES WITH SeO_2—SYNTHESIS OF ACETYLENES

In 1969, Lalezari, Shafiee, and Yalpani[203] reported that the SeO_2 oxidation of acetophenone semicarbazone gave 4-phenyl-1,2,3-selenadiazole (**34**) instead of the expected phenylglyoxal ketosemicarbazone.[204] This reaction became synthetically important when the above workers found that pyrolysis of **34** gave phenylacetylene in good yield.[205]

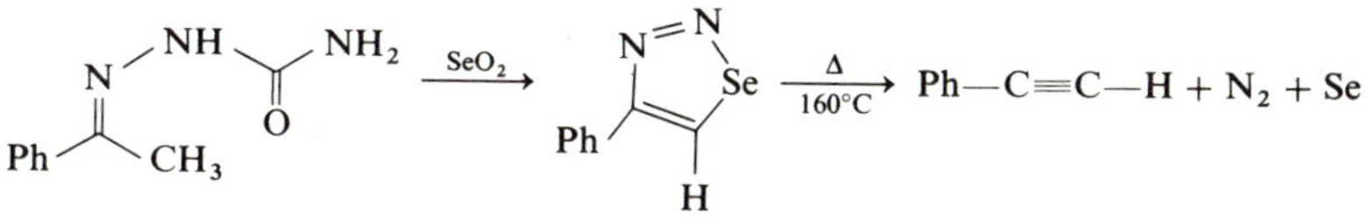

(ref. 205)

(**34**)

The formation of 1,2,3-selenadiazoles by SeO_2 oxidation has been carried out with a wide variety of ketone and aldehyde semicarbazones. For unsymmetrical ketones both isomeric products are formed.[206] The ratio is probably controlled by the acidity of the hydrazone α-hydrogens.

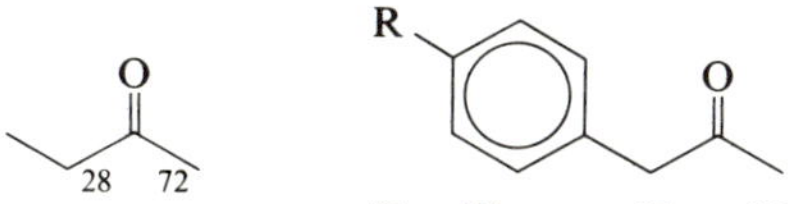

R			
R = H	67	33	(ref. 206)
CH_3	71	39	
OCH_3	69	31	
F	91.1	8.9	

The mechanism can be written[204,207] as either an intra- or intermolecular reaction of an electrophilic selenium (IV) species with the enamine tautomer of the hydrazone.

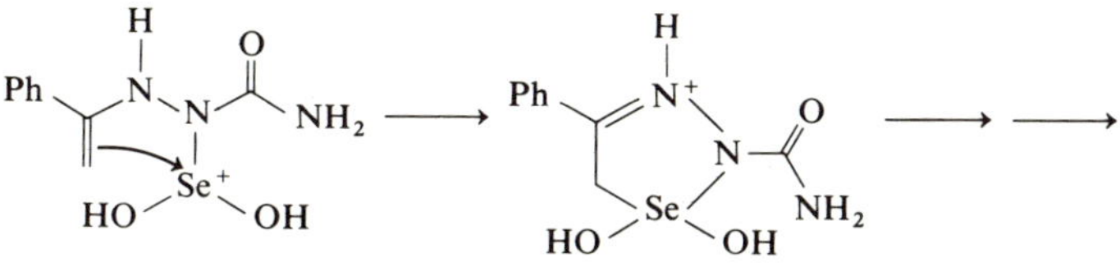

(**34**)

[203] I. Lalezari, A. Shafiee, and M. Yalpani, *Tetrahedron Lett.* 5105 (1969).
[204] I. Lalezari, N. Sharghi, A. Shafiee, and M. Yalpani, *J. Heterocycl. Chem.* **6**, 403 (1969).
[205] I. Lalezari, A. Shafiee, and M. Yalpani, *Angew. Chem., Int. Ed. Engl.* **9**, 464 (1970).
[206] I. Lalezari, A. Shafiee, and M. Yalpani, *J. Org. Chem.* **36**, 2836 (1971).
[207] H. Meier and E. Voight, *Tetrahedron* **28**, 187 (1972).

Such a mechanism is closely analogous to the mechanism of SeO_2 oxidation of ketones proposed by Sharpless and Gordon[89] (Section V, B, 1).

The pyrolysis of 1,2,3-selenadiazoles formed from acyclic ketones generally gives excellent yields of acetylenes (Table IX).[205,206,208–212] The reaction is usually carried out by mixing the heterocycle with 5–10 times its weight of sand or glass beads and heating to 160°–180°C. If temperatures much higher than this are used further reactions occur, giving substituted selenophenes and diselenins as products.[213]

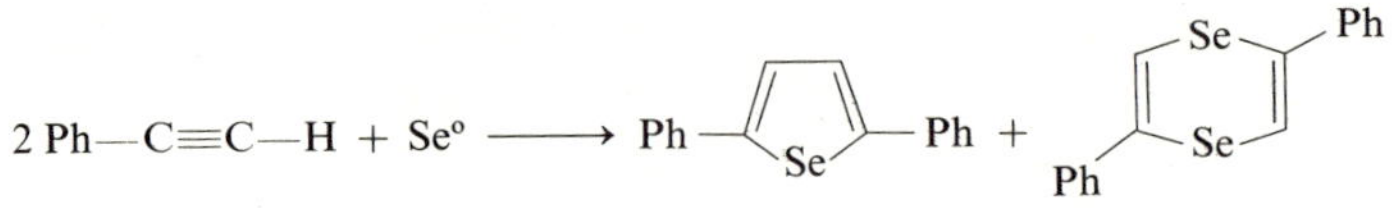

Strained cyclic acetylenes can also be prepared by pyrolysis of selenadiazoles, a route extensively examined by Meier and co-workers.[207,209a,212,213a,b] Various cyclooctynes can be isolated from selenadiazole pyrolyses (Table X),[203,205,211,213a,214] whereas cycloheptyne and cyclohexyne can only be trapped as decarbonylated tetraphenylcyclopentadienone adducts.

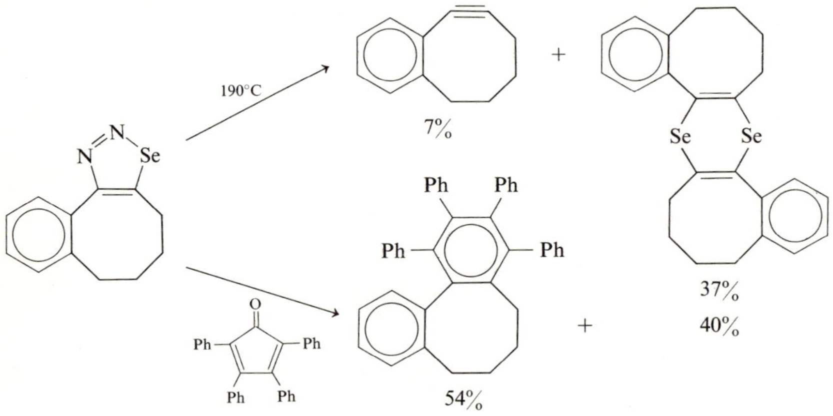

(ref. 209a)

[208] H. Meier and I. Menzel, *Tetrahedron Lett.* p. 445 (1972).
[209] A. Caplin, *J. Chem. Soc., Perkin Trans. 1* p. 30 (1974).
[209a] H. Meier, M. Layer, W. Combrink, and S. Schniepp, *Chem. Ber.* **109**, 1650 (1976).
[210] I. Lalezari, A. Shafiee, and H. Golgolab, *J. Heterocycl. Chem.* **10**, 655 (1973).
[211] H. Golgolab and I. Lalezari, *J. Heterocycl. Chem.* **12**, 801 (1975).
[212] H. Meier, S. Schniepp, and W. Combrink, *Chem. Ztg.* **99**, 461 (1975).
[213] I. Lalezari, A. Shafiee, and A. Rashidbaigi, *J. Heterocycl. Chem.* **13**, 57 (1976).
[213a] H. Meier and H. Gugel, *Synthesis* p. 338 (1976).
[213b] H. Meier, M. Layer, and A. Zetzsche, *Chem.-Ztg.* **98**, 460 (1974).
[214] I. Lalezari, A. Shafiee, and M. Yalpani, *J. Heterocycl. Chem.* **9**, 1411 (1972).

TABLE IX

FORMATION OF 1,2,3-SELENADIAZOLES FROM ACYCLIC SEMICARBAZONES AND PYROLYSIS OF THESE TO ACETYLENES

Carbonyl[a] compound	Acetylene	Yield (%)		Ref.
		1,2,3-Selenadiazole	Acetylene	
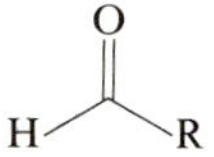				
$R = CH_3$	—	25	—	206
$R = (CH_2)_nCH_3$				
$n = 1$	—	27	—	206
$n = 2$		22	—	206
$n = 3$		28	>80	206
$n = 4$	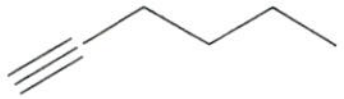	27	>80	206
$n = 6$		25	>80	206
$R = CH_2CH_2Ph$		28	>80	206
$R = CH_2CN$	$\equiv$—CN	50	>80	206
$R = CH_2CO_2Et$	$\equiv$—CO_2Et	80	>80	206
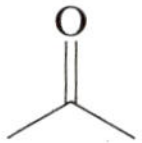	—	30	—	

(*cont.*)

TABLE IX (*cont.*)

Carbonyl[a] compound	Acetylene	Yield (%) 1,2,3-Selenadiazole	Yield (%) Acetylene	Ref.
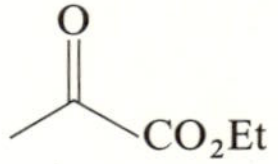	H—≡—CO_2Et	33	65	208
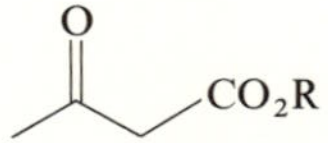	CH_3—≡—CO_2R			
R = Et		90	>80	206
R = H		80	>80	206
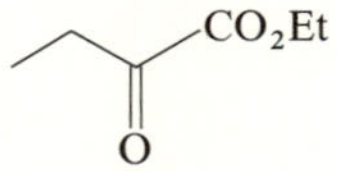	CH_3—≡—CO_2Et	17	[b]	208
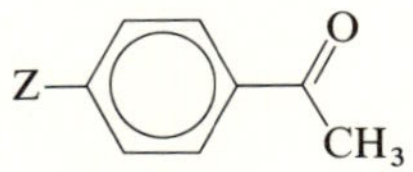	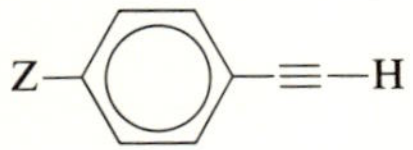			
Z = H		70[c]	84	205
Z = F		73	11	205
Z = Cl		65	80	205
Z = Br		70	85	205
Z = NO_2		65	80	205
Z = CH_3		65	70	205
Z = OCH_3		50	60	205
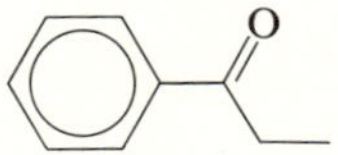	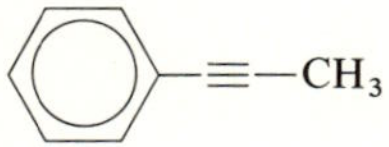	62	—	209
		68	72	205

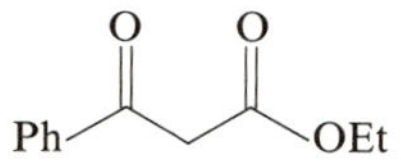	Ph—≡—CO_2Et	80	>80	206
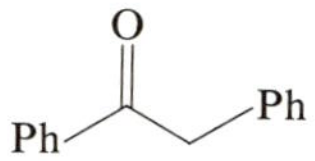	Ph—≡—Ph	61	67[d]	209a
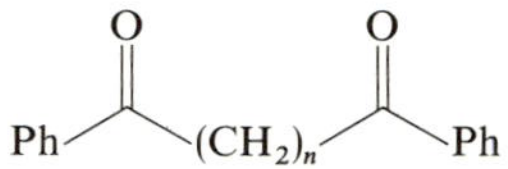	Ph—C≡C—$(CH_2)_{n-2}$—C≡C—Ph			
$n = 2$		48	96	210
$n = 4$		60	98	
$n = 5$		56	93	
$n = 6$		55	95	
$n = 7$		70	92	
$n = 8$		62	98	
$n = 10$		60	93	
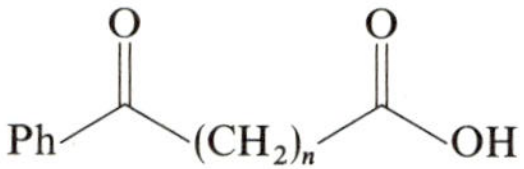	Ph—C≡C—$(CH_2)_{n-1}$—CO_2H			
$n = 3$		62	98	210
$n = 5$		56	97	
$n = 6$		58	99	
$n = 7$		60	94	
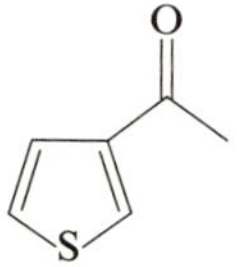	—	50	—	209

(*cont.*)

TABLE IX (*cont.*)

Carbonyl[a] compound	Acetylene	Yield (%)		Ref.
		1,2,3-Selenadiazole	Acetylene	
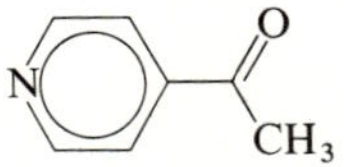	—	19	—	209
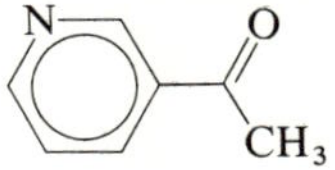	—	42	—	209
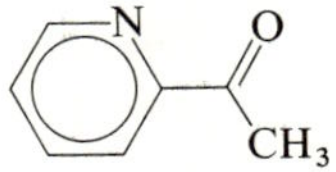	—	37	—	209
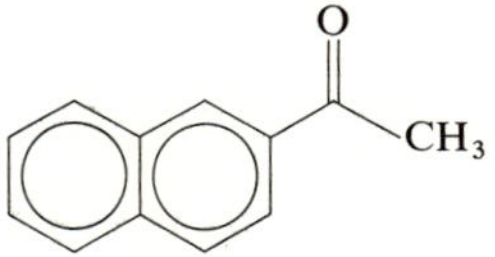	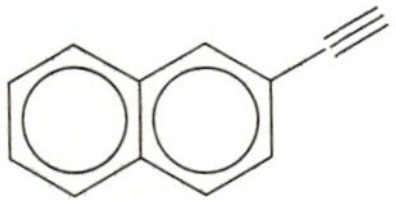	80 87	60 —	211 209
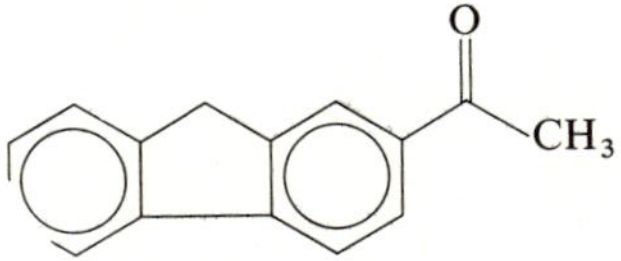	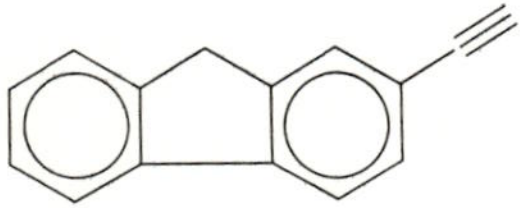	16	4	211

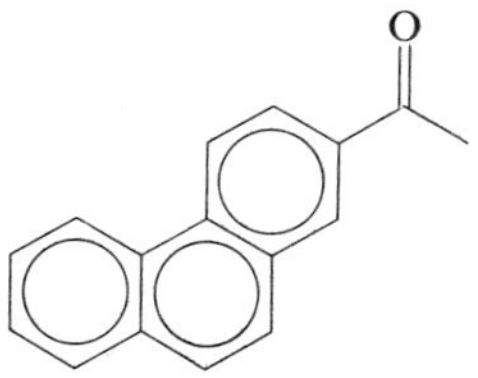	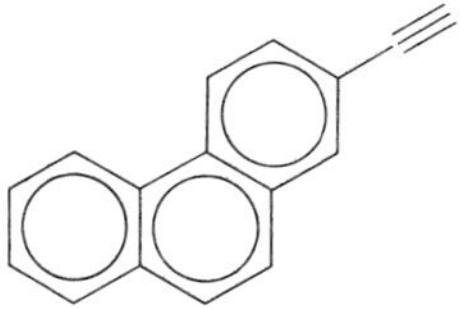	55	35	211
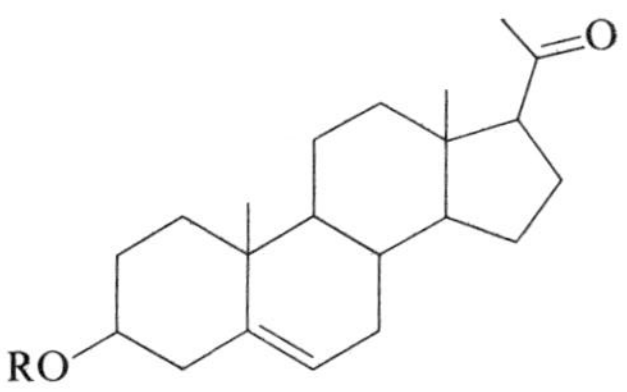 R = H R = Ac	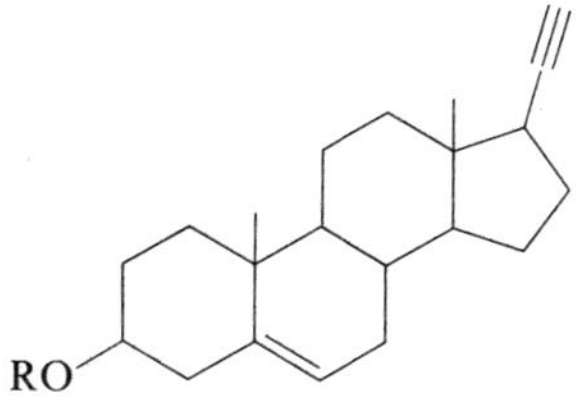	12 17	47 55	211 211
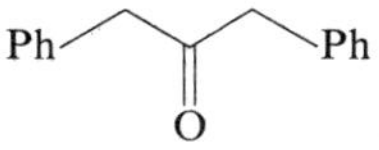		55	—	212

[a] The corresponding semicarbazone was the starting material for selenadiazole preparation.
[b] Yield not given.
[c] See Caplih[209] for the preparation of a large number of 1,2,3-selenadiazoles.
[d] Treatment of semicarbazone with SeO_2 in HOAc gave diphenylacetylene directly.

TABLE X

FORMATION AND PYROLYSES OF 1,2,3-SELENADIAZOLES DERIVED FROM CYCLIC KETONES

Carbonyl[a] compound	Yield (%)			Ref.
	Selenadiazole	Acetylene	Diselenin	
	22	0	30	214
	18	0[b]	27	207
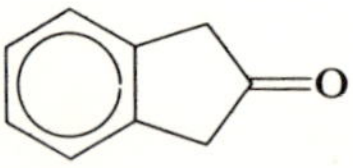	26	—	—	205
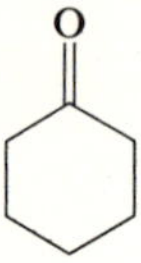	60	—	78	214
	65	6[b]	77	207
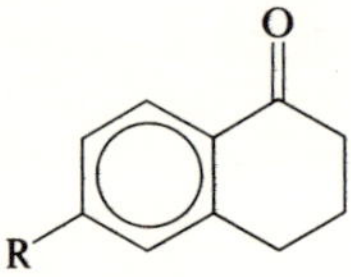				
R = H	33	0[b]	60	205
R = OCH_3	75	—	60	211
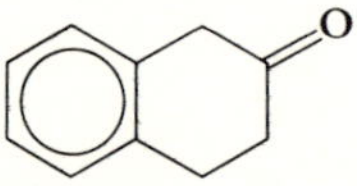	77	0[b]	20	205
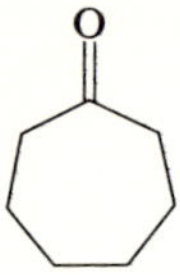	50	—	60	214
	45	29[b]	66	207
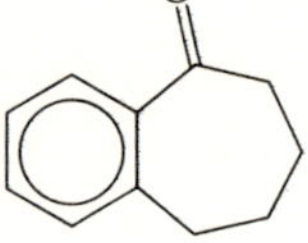	44	—	40	205
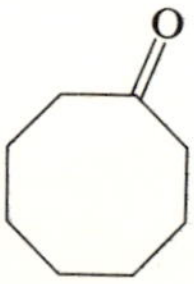	80	55	75	214
	81	49,51[b]	71	207

TABLE X (*cont.*)

Carbonyl[a] compound	Yield (%)			Ref.
	Selenadiazole	Acetylene	Diselenin	
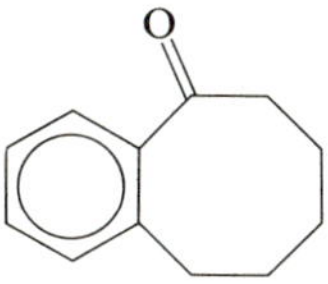	50	11	37	205
		54[b]	30	205
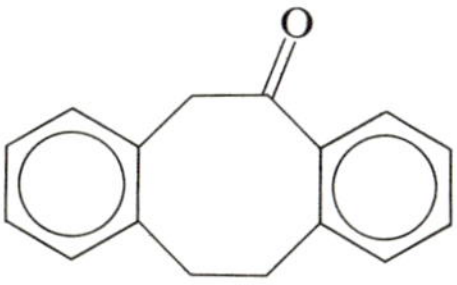	74	51	11,40[b]	213a

[a] The corresponding semicarbazone was the starting material for selenadiazole preparation.

[b] Isolated as the decarbonylated tetracyclopentadiene adduct.

Cycloalkynes generated by selenadiazole pyrolysis have also been trapped by reaction with the rhodium complex (**35**).[215] Thermolysis of compounds **36** and **37** gave no products derived from cyclohexyne or cyclopentyne.[209a]

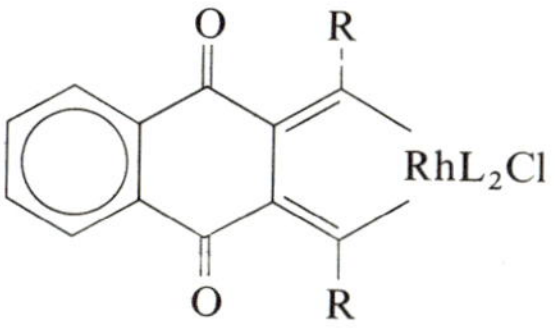

(**35**)

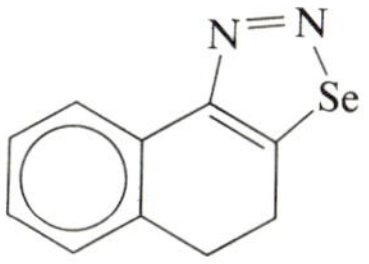

(**36**)

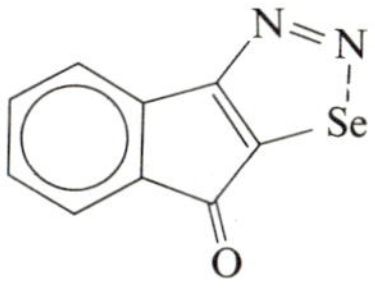

(**37**)

Selenadiazoles are apparently also decomposed to acetylenes on treatment with *n*-butyllithium as shown by the following reaction.[213b]

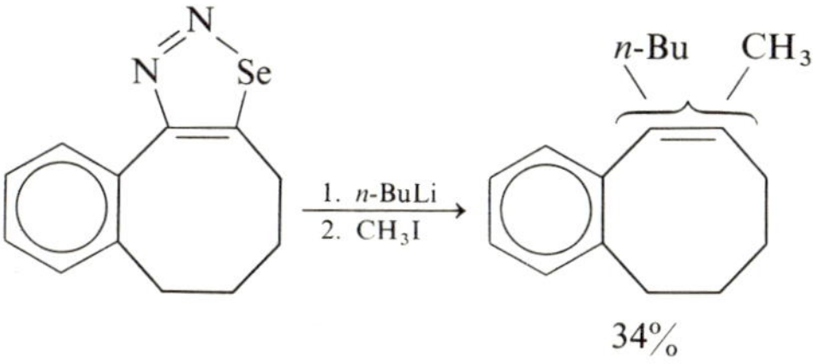

(ref. 213b)

[215] E. Muller and G. Odenigbo, *Chem.-Ztg.* **97**, 662 (1973); *Justus Liebigs Ann. Chem.* 1435 (1975).

Major by-products during selenadiazole pyrolysis are 1,4-diselenins perhaps formed by dimerization of a selenirene or related species (**38**).

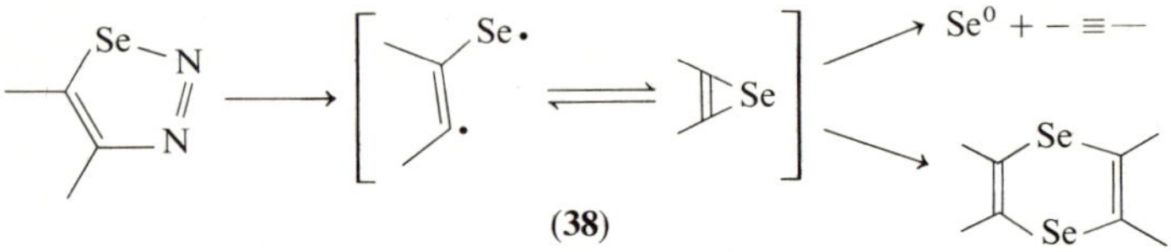

(**38**)

The formation of diselenin becomes the major pathway when the preparation of strained acetylenes is attempted.

Selenadiazoles are sources of a selinirene fragment in several reactions. Thus, treatment of **39** with iron carbonyl gives a compound assigned structure **40**.[216]

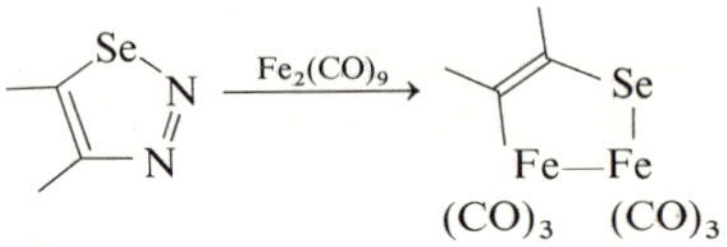

(**39**) (**40**)

Pyrolysis of selenadiazole **41** in the presence of CSe_2[217] or CS_2[218] gives **42** and **43**.

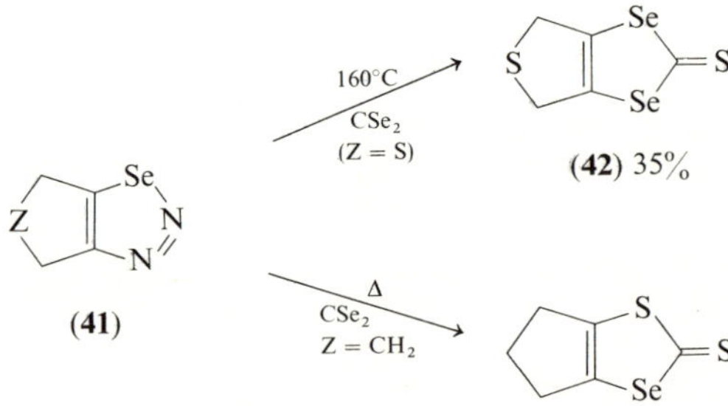

(**41**) (**42**) 35% (ref. 217) (**43**) 53% (ref. 218)

These compounds serve as precursors for tetraselenofulvalenes and dithiodiselenofulvalenes which are of interest as components of potential organic electrical conductors.

[216] T. L. Gilchrist, P. G. Mente, and C. W. Rees, *J. Chem. Soc., Perkin Trans. 1* p.2165 (1972).
[217] C. Berg, K. Bechaard, J. R. Andersen, and C. S. Jacobsen, *Tetrahedron Lett.* p. 1719 (1976).
[218] H. K. Spencer, M. V. Lakshmikantham, M. P. Cava, and A. F. Garito, *Chem. Commun.* p. 867 (1975).

VI. Miscellaneous Selenium-Based Transformations

A number of reactions involving organoselenium reagents or intermediates which do not fall within the scope of this review, but which may be of interest to organic chemists are briefly described below.

Reich and Chow[43] have shown that β-hydroxy selenides undergo "reductive elimination" under a variety of conditions, of which the best was treatment with methanesulfonyl chloride-triethylamine. Krief and coworkers[178–178b] have reported several other reagents for this transformation. Since β-hydroxy selenides can be prepared by addition of α-lithio selenides and selenoxides to carbonyl compounds,[43,178–178b] by nucleophilic opening of epoxides,[42,178a] by addition of organometallics to α-phenylseleno aldehydes[178b] (α-phenylseleno ketones are largely deselenated[43a]), or by reduction of α-phenylseleno ketones with borane,[43a]

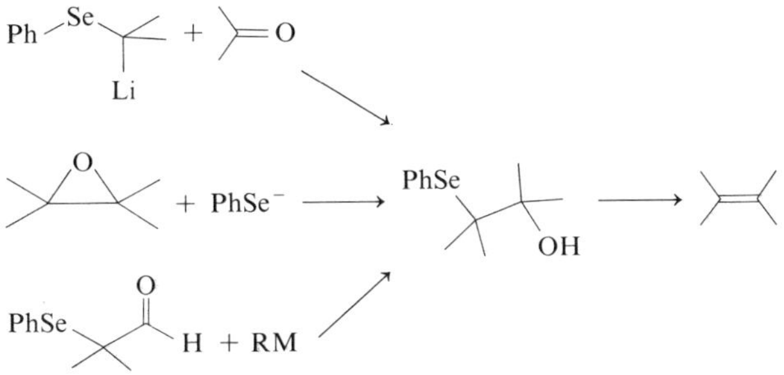

the reaction provides a procedure for the synthesis of olefins, vinyl selenides,[41] and vinyl ethers,[43a] and for the deoxygenation of epoxides.[178a]

β-Hydroxy selenides can also serve as precursors for epoxides via the selenonium salt.[219]

The Raney nickel or Li/RNH_2 hydrogenolysis of alkyl selenides[220] and the cleavage of some alkyl selenides with Br_2 or selenoxides with HBr,[221] when combined with selenium anion chemistry, provide synthetic pathways to alkanes and alkyl bromides.

Episelenides are only marginally stable at room temperature, decomposing to olefin and selenium. They are probably intermediates in the acid catalyzed deoxygenation of epoxides using $Ph_3P{=}Se$ discovered by Clive and Den-

[219] W. Dumont, P. Bayet, and A. Krief, *Angew. Chem., Int. Ed. Engl.* **13**, 274 (1974); D. V. Ende and A. Krief, *Tetrahedron Lett.* p. 457 (1976); D. V. Ende, W. Dumont and A. Krief, *Angew Chem. Int. Ed. Engl.* **14**, 709 (1975).

[220] M. Sevrin, D. V. Ende, and A. Krief, *Tetrahedron Lett.* p. 2647 (1976).

yer,[222] as well as several related reactions using KSeCN[223,224] or a selenoxobenzothiazole.[225] In fact, under appropriate conditions, solutions of episelenides can be prepared using $Bu_3P{=}Se$.[226] Terminal epoxides can also be deoxygenated by $(EtO)_2POTe^-$ via the epitelluride.[226a]

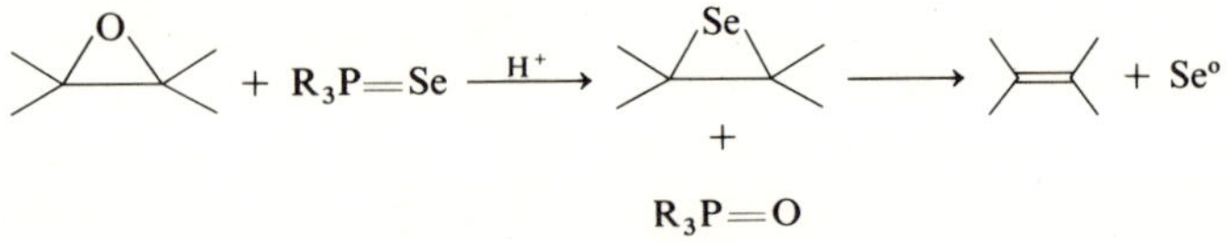

A stereochemical study has shown that the debromination of 2,3-dibromobutanes using Na_2Se does not proceed via the episelenide.[227]

A number of carbonylations using amorphous selenium, some of them catalytic have been discovered by Sonoda, Kondo, and co-workers,[228] and by Koch.[229] These procedures have allowed the synthesis of carbamates,[228a–d,229a] thiocarbamates,[228d,229b] ureas,[228e] formanilides,[228f] carbonates,[228g] and thiocarbonates[229a] by reaction of amines, alcohols, or thiols with carbon monoxide in the presence of selenium and oxygen.

Selenium has also been shown to be active as a catalyst for the reduction of aromatic nitro compounds,[230] and the oxidation of hydrazine to diimide.[231]

[222] D. L. J. Clive and C. V. Denyer, *Chem. Commun.* p. 253 (1973).
[223] J. M. Behan, R. A. W. Johnstone, and M. J. Wright, *J. Chem. Soc., Perkin Trans. 1* p. 1216 (1975).
[224] D. V. Ende and A. Krief, *Tetrahedron Lett.* p. 2709 (1975).
[225] V. Calo, L. Lopez, and A. Mincuzzi, *Synthesis* p. 200 (1976).
[226] T. H. Chan and J. R. Finkenbine, *Tetrahedron Lett.* p. 2091 (1974).
[226a] D. L. J. Clive and S. M. Menchen, *Chem. Commun.* p. 658 (1977).
[227] M. Prince and B. W. Bremer, *J. Org. Chem.* **32**, 1655 (1967).
[228a] N. Sonoda, G. Yamamoto, and S. Tsutsumi, *Bull. Chem. Soc. Jpn.* **45**, 2937 (1972).
[228b] K. Kondo, N. Sonoda, and S. Tsutsumi, *Chem. Lett.* p. 373 (1972).
[228c] K. Kondo, N. Sonoda, and H. Sakurai, *Chem. Commun.* p. 160 (1974).
[228d] N. Sonoda, G. Yamamoto, K. Natsukawa, K. Kondo, and S. Murai, *Tetrahedron Lett.* p. 1969 (1975).
[228e] N. Sonoda, T. Yasuhara, K. Kondo, T. Ikeda, and S. Tsutsumi, *J. Am. Chem. Soc.* **93**, 6344 (1971). K. Kondo, N. Sonoda, and S. Tsutsumi, *Chem. Commun.* p. 307 (1972).
[228f] K. Kondo, N. Sonoda, and H. Sakurai, *Chem. Commun.* p. 853 (1973).
[228g] K. Kondo, N. Sonoda, and S. Tsutsumi, *Tetrahedron Lett.* p. 4885 (1971).
[229a] P. Koch and E. Perrotti, *Tetrahedron Lett.* p. 2899 (1974).
[229b] P. Koch, *Tetrahedron Lett.* p. 2087 (1975).
[230] K. Kondo, N. Sonoda, and H. Sakurai, *Chem. Commun.* p. 42 (1975).
[231] K. Kondo, S. Murai, and N. Sonoda *Tetrahedron Lett.* p. 3727 (1977).

CHAPTER II

Oxygenations with Microorganisms

ROY A. JOHNSON*

I. Introduction

Ever since the hydroxylation of progesterone by *Rhizopus nigricans* provided a key step in the synthesis of cortisone,[1] microorganisms have presented the chemist with a series of dramatic chemical oxidations. The potential of this method is virtually unlimited. For example, what other reagent could one use to carry out the selective oxidation of a molecule such as lucanthone (**1**) to its hydroxymethyl derivative hycanthone (**2**),[2] or to carry out the selective oxidation of pentaerythritol (**3**) to the monocarboxylic acid (**4**),[3] or the

* It is with pleasure that I dedicate this chapter to Dr. Herbert C. Murray on the occasion of his 65th birthday.

[1] D. H. Peterson and H. C. Murray, *J. Am. Chem. Soc.* **74**, 1871 (1952); D. H. Peterson, H. C. Murray, S. H. Eppstein, L. M. Reineke, A. Weintraub, P. D. Meister, and H. M. Leigh, *ibid.* p. 5933.

[2] D. Rosi, G. P. Peruzzotti, E. W. Dennis, D. A. Berberian, H. Freele, B. F. Tullar, and S. Archer, *J. Med. Chem.* **10**, 867 (1967); S. Archer and A. Yarinsky, *Prog. Drug Res.* **16**, 12 (1972).

[3] C. T. Goodhue and J. R. Schaeffer, *Proc. Int. Ferment. Sym., 3rd, 1968* Abstract G2-7 (1968).

ISBN 0-12-697252-4

oxidation of toluene (**5**) to the optically active diene-diol (**6**)?[4] These are but a few of the hundreds of chemical oxidations that have been carried out with the use of microorganisms.

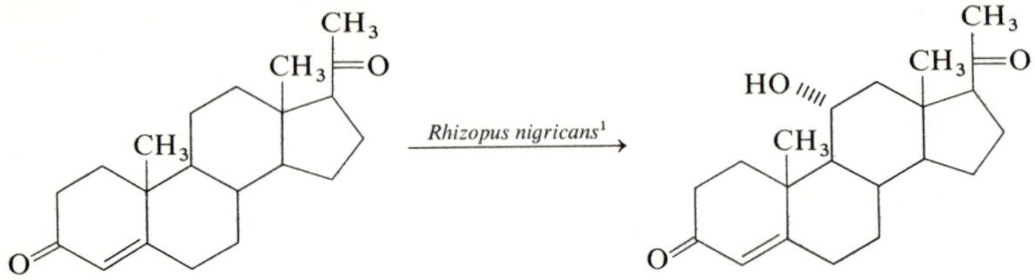

Progesterone 11α-Hydroxyprogesterone

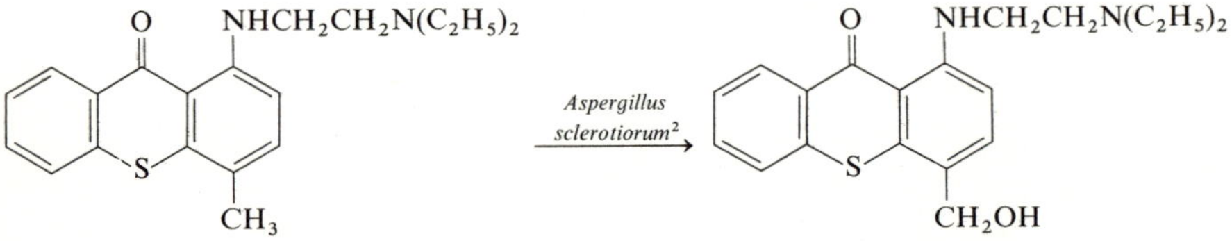

(**1**) (**2**)

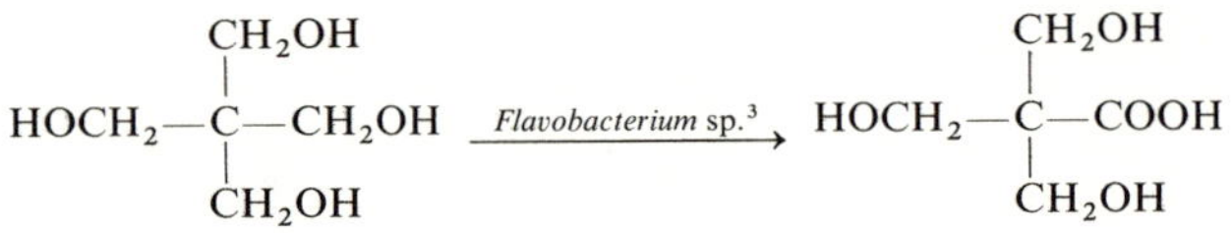

(**3**) (**4**)

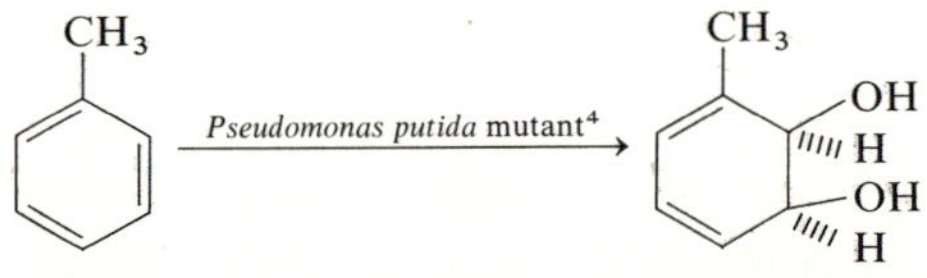

(**5**) (**6**)

Yet, despite such impressive reactions, the use of microbial oxidations has not become commonplace. There may be several reasons that account for this. The unfamiliarity on the part of the chemist with the techniques of fermentation may cause him to avoid this method. Faced with the inability to predict which microorganism to use to achieve a desired oxidation, the chemist may choose to devise an alternate synthesis rather than chance the use of a microorganism. The sometimes modest yields reported for these oxidations may discourage synthetic chemists from using them in their preparative schemes. This chapter will discuss various aspects of these particular problems. Thus, while a variety of fermentation techniques—some of them very sophisticated—have been used to attain the entire scope of

[4] D. T. Gibson, M. Hensley, H. Yoshioka, and T. J. Mabry, *Biochemistry* **9**, 1626 (1970).

results seen in this chapter, a surprisingly large number of them are the result of a relatively simple, general fermentation procedure. This procedure, which is not unlike the fermentation that can be carried out in one's own basement for the brewing of beer[5] is presented in some detail. Additional comments on fermentation technique can be found in various publications[6,7] including a previous monograph from these laboratories.[8] It should be possible to carry out such a fermentation in a chemical laboratory with no more difficulty than is required of such oxidative methods as ozonolysis, photooxygenation, or electrochemical oxidation.

Also discussed in detail are the oxygenations carried out by several microorganisms that have been used extensively. Particular attention is given to the patterns of oxygenation found in the products produced by these microorganisms.

Recent microbial oxidations of chemical interest are presented with examples taken largely from the period 1970 through 1975. Many such reactions before this period have been collected and discussed previously.[8–10] Reviews of microbial oxidations from other viewpoints have appeared frequently. A comprehensive review of steroid oxidations through 1966 has appeared in handbook form[11] and a recent review has brought this area up to date through 1973.[12] Other reviews of steroid oxidations include several monographs,[13] an excellent discussion of practical aspects,[14] and of bile acid transformations.[15] Chemical applications of microbial oxidations in the preparation of drugs have been extensively reviewed.[16] Reviews on hydrocarbon oxidations include alkane hydroxylations,[17,18] alkane hydroxylation

[5] L. P. Beadle, "Brew It Yourself" (revised ed.). Farrar, Straus & Giroux, New York, 1973.

[6] L. E. Casida, Jr., "Industrial Microbiology," p. 117. Wiley, New York, 1968.

[7] *The American Biology Teacher*, Vol. 30, No. 6 (1968).

[8] G. S. Fonken and R. A. Johnson, "Chemical Oxidations with Microorganisms." Dekker, New York, 1972.

[9] K. Kieslich, *Synthesis* 120, 147 (1969).

[10] C. J. Suckling and K. E. Suckling, *Chem. Soc. Rev.* **3**, 387 (1974).

[11] W. Charney and H. L. Herzog, "Microbial Transformations of Steroids." Academic Press, New York, 1967.

[12] L. L. Smith, *in* "Terpenoids and Steroids" (K. H. Overton, ed.), Spec. Period. Rep., Vol. 4, p. 394, Chem. Soc., London, 1974.

[13] A. Čapek, O. Hanč, and M. Tadra, "Microbial Transformations of Steroids." Academia, Prague, 1966; H. Siizuka and A. Naito, "Microbial Transformation of Steroids and Alkaloids." Univ. Park Press, State College, Pennsylvania, 1967.

[14] W. J. Marsheck, Jr., *Prog. Ind. Microbiol.* **10**, 49 (1971).

[15] S. Hayakawa, *Adv. Lipid Res.* **11**, 143 (1973).

[16] R. Beukers, A. F. Marx, and M. H. J. Zuidweg, *Drug Des.* **3**, 1 (1972).

[17] A. J. Markovetz, *Crit. Rev. Microbiol.* **1**, 225 (1971); P. Jurtshuk and G. E. Cardini, *ibid.* p. 239.

[18] B. J. Abbott and W. E. Gledhill, *Adv. Appl. Microbiol.* **14**, 249 (1971).

and further oxidation to fatty acids,[19,20] and oxidation of aromatic hydrocarbons.[21,22] Reviews on the transformations of pesticides[23] and lignin[24] by microorganisms include some oxidations. More specific topics are a review of the activities of the genus *Nocardia*,[25] modifications of antibiotics,[26,27] and monoterpene oxidation and metabolism.[28] A recent comprehensive review of the oxidation of molecules other than steroids has appeared in handbook form.[29]

II. Methods

There are two convenient methods for carrying out a microbial oxidation. The first of these is the use of a whole culture of the microorganism, i.e., the classical fermentation. A second method, that in principle has considerable appeal, is the use of microbial spores to carry out the oxidation. Each of these is discussed further below. Other methods of some promise include the use of dry cell preparations, the use of gel-entrapped cells of the microorganism,[30] or the use of cell-free enzyme preparations. These latter methods are considerably less well developed and they require additional work before the chemist can easily use them.

A. OXIDATIONS WITH WHOLE CULTURES

By far the most common approach to a microbial oxidation is to add the substrate to an organism actively growing in a medium (a fermentation). Both the differences and the similarities of fermentation techniques may be emphasized. It is clear that an anaerobic microorganism will require a fermentation technique different from an aerobic organism. However, numerous oxidations have been carried out under very similar conditions and for the chemist it is this aspect that we wish to emphasize here. To illustrate a typical aerobic fermentation procedure that can be used with many fungi we have reproduced below a description that has been reported from our own labo-

[19] C. Ratledge, *Chem. Ind. (London)* p. 843 (1970).
[20] C. W. Bird and P. M. Molton, *Top. Lipid Chem.* **3**, 125 (1972).
[21] R. L. Raymond, V. W. Jamison, and J. O. Hudson, *Lipids* **6**, 453 (1971).
[22] D. T. Gibson, *Crit. Rev. Microbiol.* **1**, 199 (1971).
[23] J. M. Bollag, *Crit. Rev. Microbiol.* **2**, 35 (1972).
[24] T. K. Kirk, *Annu. Rev. Phytopathol.* **9**, 185 (1971).
[25] R. L. Raymond and V. W. Jamison, *Adv. Appl. Microbiol.* **14**, 93 (1971).
[26] O. K. Sebek and D. Perlman, *Adv. Appl. Microbiol.* **14**, 123 (1971).
[27] B. J. Magerlein, *Adv. Appl. Microbiol.* **14**, 185 (1971).
[28] I. C. Gunsalus and V. P. Marshall, *Crit. Rev. Microbiol.* **1**, 291 (1971).
[29] K. Kieslich, "Microbial Transformations of Non-Steroid Cyclic Compounds." Georg Thieme, Stuttgart, 1976 (in English).
[30] Cf. K. Mosbach and P.-O. Larson, *Biotechnol. Bioeng.* **12**, 19 (1970).

ratories.[31] Interspersed between the lines of this description are additional comments about the procedure.

The culture used in these experiments was Sporotrichum sulfurescens V. Beyma (ATCC 7159). [*Sporotrichum sulfurescens* has recently been reclassified to the genus *Beauveria* and classified as *Beauveria sulfurescens* (V. Beyma) J. Taylor.[32]] V. Beyma originally isolated and identified the organism. The designation ATCC 7159 is the strain number given to this microorganism by the American Type Culture Collection. The ATCC is the primary source of cultures in the United States.* The ATCC catalog contains listings of microorganisms available to scientists as well as other useful information about handling and growing cultures. *Sporotrichum sulfurescens* will be found as *B. sulfurescens* in the most recent edition (1976) of this catalog. A similar agency in Europe is the Centraalbureaus voor Schimmelcultures, Baarn, Netherlands. Classifications (taxonomy) of bacteria and fungi indicate where a culture belongs in the world of microorganisms.[33] An abbreviated taxonomy containing only microorganisms that oxygenate or otherwise transform steroids is found in Charney and Herzog's handbook.[11] *A medium of commercial dextrose (10 gm/liter) and cornsteep liquor (20 gm/liter) was prepared with tap water (to 1 liter) and adjusted to pH 5.0 with sodium hydroxide.* The medium is an appropriate mixture of nutrients (commercially available) in which the microorganism can grow and multiply. The pH may be adjusted with either sodium hydroxide or hydrochloric acid. *Flasks of the sterilized medium were inoculated with spores of S. sulfurescens which were grown on malt (wort) agar slants.* The nutrient medium is parceled out among Erlenmeyer flasks so that each contains medium comprising one-fifth of its total volume. A 500 ml Erlenmeyer flask, containing 100 ml of the medium, is a convenient size to use and, as a first approximation, is suitable for the oxidation of about 20 mg of substrate. The desired number of flasks and their contents are each plugged with a wad of cotton and sterilized. An autoclave may be used but if not available, flasks may be sterilized in an ordinary pressure cooker using 15 psi steam for 20 minutes. The microorganism will have been received growing on an agar slant, that is, a test tube in which the agar has been allowed to solidify to give a slanted surface. Such slants can be purchased already prepared and consist of various compositions such as the malt agar indicated above. The technique of

* The address of ATCC is 12301 Parklawn Drive, Rockville, Maryland 20852.

31 R. A. Johnson, M. E. Herr, H. C. Murray, and G. S. Fonken, *J. Org. Chem.* **33**, 3187 (1968).

32 J. J. Taylor, *Mycologia* **62**, 797 (1970).

33 Bacteria: R. E. Buchanan and N. E. Gibbons, eds., "Bergey's Manual of Determinative Bacteriology," 8th ed. Williams & Wilkins, Baltimore, Marland, 1974; Fungi: G. C. Ainsworth, F. K. Sparrow, and A. S. Sussman, eds., "The Fungi," Vols. 4A and 4B. Academic Press, New York, 1973.

inoculating flasks (or new slants for the purpose of maintaining the culture) requires a Meeker burner, an inoculating loop, and an area relatively free of drafts. The loop is sterilized by heating it to redness in a flame and is then allowed to cool. The caps or plugs are removed from the slant containing the culture and from the sterilized flasks. The mouths of the containers are flamed briefly. The transfer is then made by scraping cells of the organism from the slant and introducing into the new container by immersing the loop into the liquid medium or by rubbing it gently across the surface of a new slant. The mouths of the containers are again flamed and immediately reclosed. The culture in the newly inoculated flask is allowed to grow at least 48 hours and preferably at a temperature of 30°–38°C. Newly inoculated slants may require considerably longer (1–2 weeks) to develop a heavy growth. *The flasks were shaken for 48 hours (until a heavy growth was apparent) and then used for seeding other flasks or tanks.* A rotary shaker is generally used to provide a vigorous swirling action. The purpose of the shaking is to aerate the culture in the flask. A heavy growth of a fungus such as *S. sulfurescens* will appear as a heavy mat or as small, soft balls of cells and mycelia. The seeding technique is similar to the transfer technique described above except that a sterile pipette is used to transfer a portion of the entire mixture to the new flasks. A pipette having a large mouth is useful here. *The tanks used in this work were inoculated with 5 parts of the vegetative culture to 100 parts of fresh medium.* "The tanks" refer to the use of fermentation tanks for larger scale fermentations. For the present discussion we will be concerned mainly with the use of flasks, which are sufficient in capacity for preliminary experimentation and initial reactions. The ratio of inoculum (the culture that has been growing in the flasks for for the past 48 hours) to fresh medium is useful for preparing new flasks as well as tanks. Note that we have now described two ways to inoculate flasks of fresh, sterile media. The first is inoculation by transfer from an agar slant. This will necessarily result in transfer of fewer cells of the microorganism and so may require a longer growth period. The second way is the transfer of a portion of heavy growth. This will result in more "seed" being used and the new flasks will show heavy growth sooner. The later method also allows inoculation of a considerable number of new flasks. *On any scale, the culture is aerated and agitated for the growth period.* For flasks this means use of a rotary shaker as previously mentioned. *In stirred vessels, with good agitation, aeration with 5–10 volumes of air per minute per 100 volumes of culture was satisfactory.* This again refers to fermentations in tanks, which must be stirred mechanically and must be aerated by passing a stream of sterile air through the culture. *The substance to be oxygenated was dissolved in dimethylformamide or acetone and was added after 24 hours of growth.* Our favorite solvent is dimethylformamide. Acetone, ethanol, and dimethyl sulfoxide all are satisfactory solvents. One need not be concerned when common sense dictates that the substance is likely to be precipitated from solution when added to the aqueous culture. This matter somehow resolves itself in most cases. *The amount of solvent used for the substance was kept to a minimum (or less than 1 part solvent to 100 parts culture). The level of substrate was 0.2 gm to 0.3 gm per liter.* A conversion time of 72 hours was used. Considerable variation may be made in the time a fermentation is allowed to proceed after the substrate has been added. It is often helpful to remove a sample (5–10 ml) from the fermentation from time to time and extract it with a solvent such as dichloromethane. Thin layer chromatography (TLC) examination of the extract will give a good indication of the progress of the transformation of the substrate. The time it takes to oxidize a substrate may vary from a few hours to 5 or 6 days. *The products and residual substrate were removed from the "beer" by extraction with dichloromethane, using a volume of solvent one-fourth that of the "beer."* The "beer" is the filtrate obtained by filtering the entire fermentation mixture following completion of the fermentation. Filtration will be aided by the use of a thick layer of Celite in the Buchner funnel. Filtration will likely be slow and may require frequent loosening of the Celite cake. We find dichloromethane to be a particularily satisfactory extraction solvent and four extractions, each with a one-fourth volume, is usually ample. Other solvents that may be used include ethyl acetate, chloroform, and 1-butanol. One should be aware of the possibility that the product may not be found in the "beer" in some cases. Then an extraction of the mycelium, which are the solids that have been filtered off, must be devised (see Ref. 34 for a procedure).

The procedure described above has been used for the oxygenation of over one hundred substrates with *S. sulfurescens* and has been used with several other microorganisms as well. Another procedure from a different laboratory is given in part below with the intent that the reader may note the similarities in the basic fermentation techniques. This procedure, from the laboratory of Jones, Meakins, and co-workers at Oxford University,[34] has been used with several microorganisms in their extensive studies of steroid oxygenations.

Cultures of the microorganisms were obtained from the Centraalbureau voor Schimmelcultures, Baarn, Netherlands. The specific microorganisms used are given elsewhere in the Experimental Section of the original reference. Slopes were prepared from 3% malt extract containing 2% agar in ion-free distilled water. The term slopes is synonomous with slants, as used in the preceding section. These were inoculated with the organism from the master

[34] J. W. Blunt, I. M. Clark, J. M. Evans, E. R. H. Jones, G. D. Meakins, and J. T. Pinhey, *J. Chem. Soc. C* p. 1136 (1971).

slope and developed for 10–15 days at 25°C. A continuous supply of slopes was maintained by preparing new ones from the existing slopes 2 weeks after the appearance of a healthy development. The slopes were used as a source of spore suspension for inoculating the flasks of medium described later.

Two nutrient media were used. Medium A was prepared from malt extract (1 gm), beef extract (1 gm), yeast (Difco) (1 gm), cornsteep liquor (1 ml), glucose (5 gm), and ion-free distilled water (1 liter); for medium B the amounts of the first 4 ingredients were doubled. The pH was brought to 5.5 with 2 *N* HCl and sucrose (2 gm) was added. The medium was introduced into 500 ml conical flasks (200 ml per flask for A, 80 ml for B) which were then sterilized in steam.

The flasks were inoculated with spore suspensions (3 or 4 flasks per slope), under sterile conditions, plugged with cotton wool, and swirled (200 revolutions per minute) on a New Brunswick gyrorotatory shaker at 25°C for 2 days. A solution of the steroid substrate in an organic solvent was added under sterile conditions, and the flasks were swirled for a specified time. Different extraction procedures (I, II, and III) were used according to the type of substrate and microorganism involved. The reader is referred to the original reference for details of the extraction procedure,[34] which include a method for extraction of the mycelium.

A description of the epoxidation of alkenes by the microorganism, *Pseudomonas oleovorans*, that is suitable for organic chemistry laboratory students has appeared recently.[35]

B. OXIDATIONS WITH SPORES

While the procedures of the preceding section are not formidable, the chemist undoubtedly would favor a situation where he could take a container off the shelf (or out of the freezer), add the contents to a flask containing his substrate in the necessary solvent and then sit back and wait for the microbial oxidation to proceed. The possibility for doing this does, in fact, exist. Spores, which may be loosely defined as the fruit or seed by which microorganisms reproduce, can be separated from the cells of the microorganism and, in some cases, can be stored indefinitely as a paste at −20°C. Placement of these spores in suspension in water or in an aqueous buffer, together with a nutrient such as glucose and an organic substrate, can result in oxygenation of the substrate. At the same time, the spores remain relatively dormant, allowing for an easy recovery of product from the mixture.

The use of spores for oxygenation reactions has been studied extensively by Vézina, Sehgal, and Singh at the Ayerst Laboratories in Montreal.[36]

[35] P. L. Kumler and P. J. DeJong, *J. Chem. Educ.* **52**, 475 (1975).

[36] C. Vézina, S. N. Sehgal, and K. Singh, *Adv. Appl. Microbiol.* **10**, 221 (1968).

Their primary interests have been in both the 11α-hydroxylation and the C-1 dehydrogenation of progesterone and its analogs. Spores of *Aspergillus ochraceus* are used for the former reaction,[37,38] while spores of *Septomyxa affinis* are used for the latter.[39] For example, 6,19-oxido-4-pregnene-3,20-dione (**7**) is reported to be converted completely to the 11α-hydroxy derivative (**8**).[38] The spores of several other organisms have been used for similar transformations of steroids.[36] Oxygenations with spores produce the same products that are obtained from fermentations using the same microorganism.

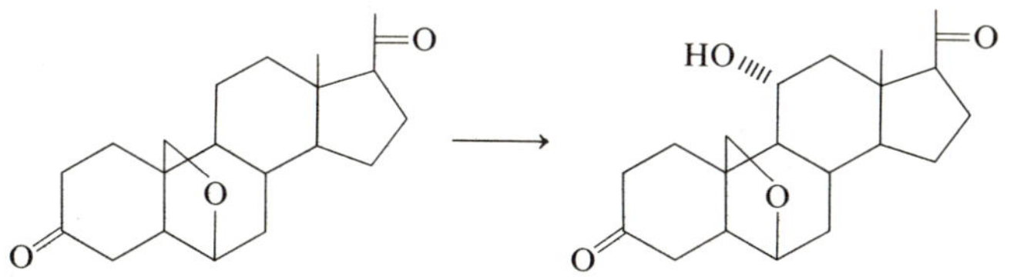

(**7**) (**8**)

An interesting synthesis of methyl ketones is carried out by the spores of *Penicillium roqueforti*.[40] The substrates for this reaction are the homologous fatty acids. The reaction apparently proceeds via β-oxidation to the stage of the β-hydroxy acid, at which point it is postulated that decarboxylation occurs.[41]

$$RCH_2CH_2COOH \longrightarrow R{-}\overset{\overset{\large O}{\|}}{C}{-}CH_3$$

The use of spores for oxidation requires two main processes. The spores must first be produced and harvested. Spores for several experiments may thus be obtained at once. A recently described procedure for propagation of spores (conidia) on bakery bread appears to be very straightforward.[42] The second step is the oxidation reaction and it is carried out independently of the first procedure. As mentioned above, this is done in an aqueous system, usually at a pH of 5–6. Since the reactions are not run under aseptic conditions, this pH aids in the control of bacterial contamination as does the addition of a small amount of an antibiotic to the reaction.[36]

To summarize, we may note that the technique of using spores for oxidations has been worked out in considerable detail. The method has been applied with much success, but to only a very limited number of substrates.

[37] C. Vézina, S. N. Sehgal, and K. Singh, *Appl. Microbiol.* **11**, 50 (1963).
[38] S. N. Sehgal, K. Singh, and C. Vézina, *Can. J. Microbiol.* **14**, 529 (1968).
[39] K. Singh, S. N. Sehgal, and C. Vézina, *Steroids* **2**, 513 (1963).
[40] R. F. Gehrig and S. G. Knight, *Appl. Microbiol.* **11**, 166 (1963).
[41] S. G. Knight, *Ann. N. Y. Acad. Sci.* **139**, 8 (1966).
[42] G. A. Sansing and A. Ciegler, *Appl. Microbiol.* **26**, 830 (1973).

Its potential with other organic molecules awaits exploration. The fact that not all microorganisms produce spores is, however, a limitation to the method. An excellent review by the Ayerst group[36] describes the essence of this technique in much greater detail than does the discussion presented here. While their review[36] appeared in 1968, it remains quite up to date as a result of very little activity in the intervening years.

III. Oxygenation at Nonactivated Carbon

A. GENERAL ASPECTS

Oxygenation at nonactivated carbon is the most powerful chemical use for microbial oxidations. When taken with its often high specificity (both positional and stereochemical), the reaction exceeds in potential most classical chemical methods for the functionalization of the hydrocarbon portion of a molecule. The inability to predict the site of this oxygenation, however, remains a serious drawback to the application of this reaction. Classical chemical theory can offer little help. Even the likely preference for attack at tertiary over secondary and at secondary over primary sites often is overshadowed by the influences of stereochemistry. In turning to a consideration of stereochemical factors, one is again faced with a dearth of information. No hydroxylating enzymes from microorganisms have been purified (with exception of the camphor-specific hydroxylating enzyme from a strain of *Pseudomonas putida*[43]) and thus knowledge of enzyme stereochemistry is scarce. Until recently, the use of a single organism to oxygenate a wide variety of substrates has not been studied. Rather, most applications of microbial oxidations to chemical problems have followed the approach of trying different organisms until the one is found that will achieve the desired chemical transformation.

In recent years several microorganisms have been widely used to oxygenate a variety of substrates. At the same time there have been several proposals put forth with the intent of rationalizing why the oxygenating enzyme(s) of a particular organism attacks a substrate at certain positions.[44–48] A recurrent theme is found in these proposals for enzyme–substrate inter-

[43] D. W. Cushman, R. L. Tsai, and I. C. Gunsalus, *Biochem. Biophys. Res. Commun.* **26**, 577 (1967).

[44] M. G. Rogoff, *J. Bacteriol.* **83**, 998 (1962).

[45] R. D. Swisher, *Dev. Ind. Microbiol.* **6**, 39 (1964).

[46] G. S. Fonken, M. E. Herr, H. C. Murray, and L. M. Reineke, *J. Am. Chem. Soc.* **89**, 672 (1967).

[47] D. R. Brannon, F. W. Parrish, B. J. Wiley, and L. Long, Jr., *J. Org. Chem.* **32**, 1521 (1967).

[48] A. M. Bell, P. C. Cherry, I. M. Clark, W. A. Denny, E. R. H. Jones, G. D. Meakins, and P. D. Woodgate, *J. Chem.* Soc., *Perkin Trans. 1* p. 2081 (1972).

actions. This is the suggestion that there is an initial interaction between an electron-rich center of the substrate and a site on the enzyme (which consequently may be presumed to be electrophilic). This interaction may then serve to orient the substrate molecule with respect to the oxygenation site. From this general hypothesis it follows that a specific microorganism should impart a repetitive pattern of hydroxylation to its products. Examination of these patterns may then suggest where a new substrate may be expected to be oxygenated. Underlying all of these considerations is an assumption that all products from one organism are produced by the same enzyme system. Discussed below are the results observed with four different microorganisms that have been used widely for hydroxylation reactions.

Before discussing these microorganisms, several other generalizations about microbial oxidations may be made. An early observation in the steroid work was that hydroxylation proceeds with retention of the configuration held by the displaced hydrogen.[49] For example, in the hydroxylation of pregnanedione with *Rhizopus nigricans*, substrates labeled with deuterium at either the 11β-position (**9**) or at the 9α-, 11α-, and 12α-positions (**11**) have been hydroxylated.

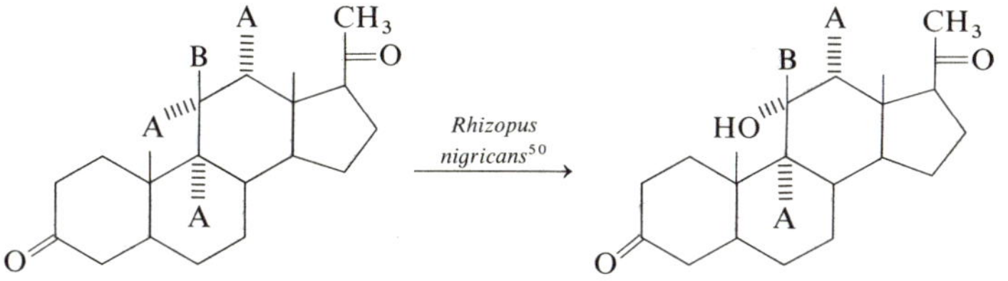

(**9**) A = H, B = D
(**11**) A = D, B = H

(**10**) A = H, B = D
(**12**) A = D, B = H

The ensuing 11α-hydroxyprogesterone retains the deuterium at C-11 in the case of **9** and loses it in the case of **11**.[50] Similar experiments with other substrates have subsequently been carried out and a similar retention of configuration has been observed.[51,52] "Retention" of configuration also occurs when a tertiary hydrogen located on a structurally rigid substrate is replaced by oxygen.

Another general feature of these oxygenations is that the oxygen is derived from molecular oxygen as opposed to the aqueous oxygen (H_2O) of the

[49] M. Hayano, M. Gut, R. I. Dorfman, O. K. Sebek, and D. H. Peterson, *J. Am. Chem. Soc.* **80**, 2336 (1958).

[50] E. J. Corey, G. Gregoriou, and D. H. Peterson, *J. Am. Chem. Soc.* **80**, 2338 (1958).

[51] M. Hayano, M. Gut, R. I. Dorfman, A. Schubert, and R. Siebert, *Biochim. Biophys. Acta* **32**, 269 (1959).

[52] D. F. Jones, *J. Chem. Soc.* C p. 2827 (1968); E. Heinz, A. P. Tulloch, and J. F. T. Spencer, *J. Biol. Chem.* **244**, 882 (1969).

medium. Isotopic labeling of both O_2 and H_2O in separate experiments has led to this conclusion.[53,54]

Both of the above generalizations have been rigorously demonstrated in only a few cases. Many hydroxylating organisms have not been tested for either property. While the generalizations seem likely to be found valid in most cases, further examples would be highly desirable. Any exceptions would, of course, be most interesting.

B. OXYGENATIONS WITH SPECIFIC MICROORGANISMS

1. *Sporotrichum sulfurescens* (ATCC-7159)

This fungus is a member of the class Fungi Imperfecti. Several species of the genus *Sporotrichum* are pathogenic (that is, the cause of a human disease). *Sporotrichum sulfurescens*, however, has been reclassified to the genus *Beauveria* and is quite safe for ordinary laboratory use.

Sporotrichum sulfurescens was first used for the hydroxylation of steroids and was found to be an efficient 11α-hydroxylator with certain steroids.[55] 17-Methyltestosterone (**13**) is, for example, converted to 11α-hydroxy-17-methyltestosterone (**14**) in 70% yield.[55] However, other microorganisms are even more proficient in the highly desirable 11α-hydroxylation reaction and so *S. sulfurescens* has not been used widely in the steroid field.

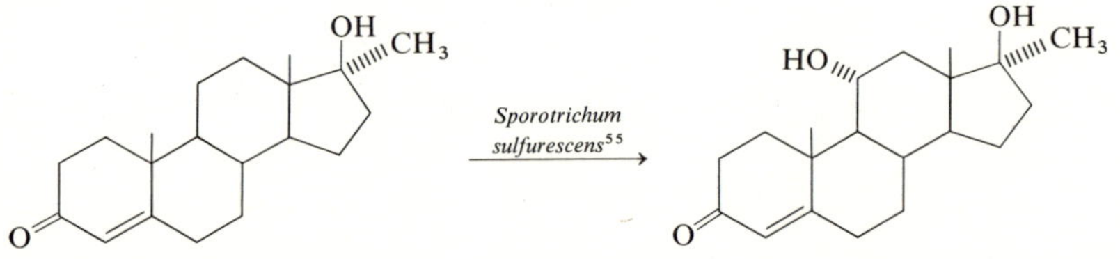

(**13**) (**14**)

In a departure from the use of steroids as substrates for microbial transformations, Fonken, Herr, and Murray found *S. sulfurescens* to be particularly useful in the oxygenation of various nonsteroidal substrates. Many of their early results are found in a mammoth patent that was issued in 1966.[56] Following this initial work, a series of reports has appeared on other results obtained with *S. sulfurescens*. Since many of these results have been summarized previously,[8] our aim here is to present mainly those examples that

[53] M. Hayano, A. Saito, D. Stone, and R. I. Dorfman, *Biochim. Biophys. Acta* **21**, 381 (1956).

[54] Y. Kurosawa, M. Hayano, M. Gut, R. I. Dorfman, A. Schubert, and C. A. Bunton, *Agric. Biol. Chem.* **25**, 424 (1961).

[55] P. D. Meister and A. Weintraub, U.S. Patent, 2,877,162 (1959).

[56] G S. Fonken, M. E. Herr, and H. C. Murray, U.S. Patent 3,281,330 (1966).

illustrate various principles that appear important in these oxygenation reactions.

It was observed early in the use of *S. sulfurescens* that hydroxylation of cyclic hydrocarbons was facilitated by the presence of a functional group in the molecule. Thus, the yield of *trans,trans*-4,4′-dihydroxybicyclohexyl (**17**) is significantly greater when the substrate is *trans*-4-hydroxybicyclohexyl (**16**) rather than bicyclohexyl (**15**).[56] Similarly, cyclododecane is not oxygenated under the same conditions used for the successful oxygenation of cyclododecanol. Cyclododecane can be oxygenated, however, if a wetting agent is added to the fermentation.[57] This is a technique that often is helpful in improving poor (or no) yield reactions.

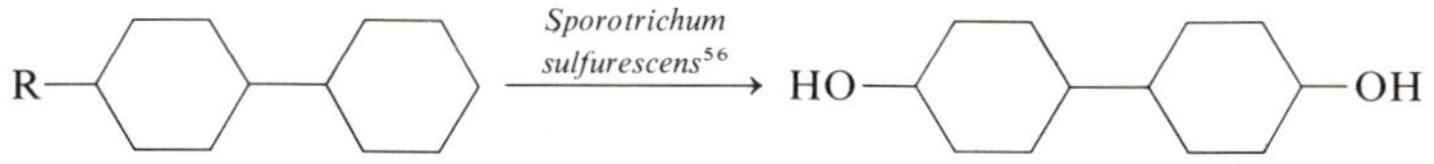

(**15**) R = H (**17**)
(**16**) R = OH

Another problem is encountered in the oxygenation of lower molecular weight substrates such as cyclohexane. The aerobic conditions used for *S. sulfurescens*, and most of the other microorganisms discussed in this chapter, may result in loss of the substrate due to volatility. Also problematic is the potential water solubility of di- or polyfunctional molecules of lower molecular weight. For example, cyclohexanol yielded no detectable products when fermented with *S. sulfurescens*, but when converted to the higher molecular weight *N*-phenylcarbamate (**18**), the cyclohexane ring was oxygenated to give a 35% yield of the 4-hydroxy product (**19**).[46]

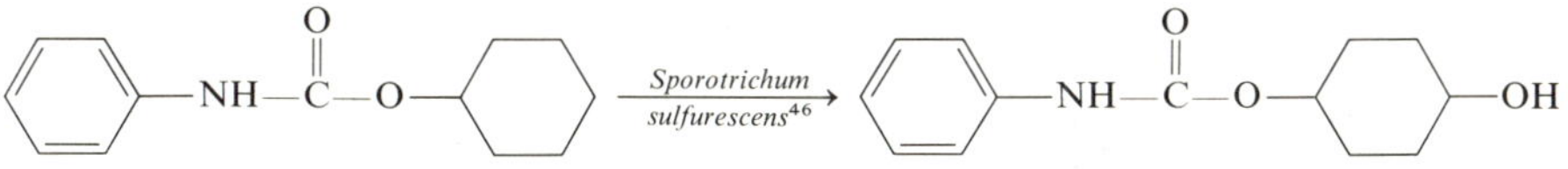

(**18**) (**19**)

Cyclododecanol (**20**) is oxygenated by *S. sulfurescens* primarily at the C-6 and C-7 positions, with only a trace of the C-5 oxygenation product being formed. In this early work, the complex mixture of hydroxylic and ketonic products was simplified by chemical oxidation to the diketones.[46]

Being "struck by the apparent lack of a rational explanation for the selection by a given microorganism of the particular carbon atom to be oxygenated," Fonken *et al.* formulated the following hypothesis based on the results obtained with *S. sulfurescens*, especially with cyclododecanol as the substrate.[46] They proposed that there is an initial binding of an electron-rich functional group of the substrate with the oxygenating enzyme (see

[57] H. C. Murray and R. A. Johnson, unpublished results.

Fig. 1). This is then followed by oxygenation of a saturated carbon of the substrate by the enzyme. The location of this carbon was characterized as being approximately 5.5 Å away from the initial site of binding.[46] With this hypothesis in mind, the oxygenation of other substrates can be considered.

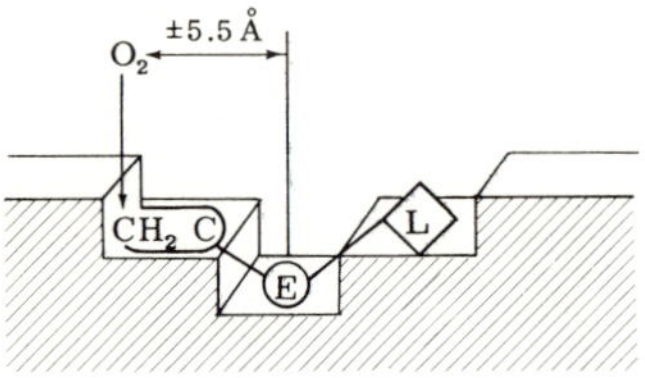

FIG. 1. Hypothesis to explain selection of carbon atom to be oxygenated by a microorganism.

Cyclododecylamine (**21**) is analogous to cyclododecanol (**20**) in that it has an electron-rich functional group that could act as a site for binding to the oxygenating enzyme. Oxygenation of cyclododecylamine with *S. sulfurescens* was successful, but only after the amine was first acylated by the microorganism.[58] The substrate actually oxygenated was cyclododecylacetamide (**22**). The main products, following chemical oxidation of alcohols to ketones, were the 6- and 7-ketones as well as a trace of the 5-ketone. These and other results have led to the generalization that substrates having an amino function are much more successfully oxygenated when the amine has been derivatized. In addition to carboxamides of various types, sulfonamides, urethans, ureas, and sulfonylureas have been used successfully as substrates.

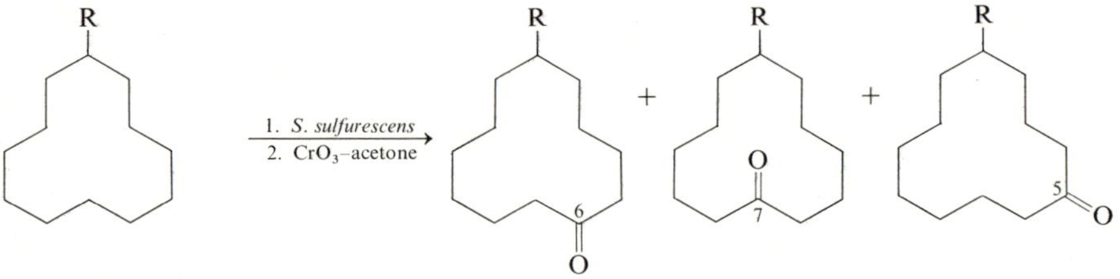

(**20**) R = —OH
(**21**) R = —NH_2
(**22**) R = —$NHCOCH_3$

R = =O
R = —$NHCOCH_3$

Many amides have since been used as substrates with *S. sulfurescens.* The results of these experiments provide a variety of examples that can be used to examine the enzyme–substrate model proposed by Fonken. The carbonyl oxygen has the highest electron density of the amide functionality and may be considered as the site at which the substrate interacts with the enzyme. In the following paragraphs several series of observations are discussed that

[58] G. S. Fonken, M. E. Herr, H. C. Murray, and L. M. Reineke, *J. Org. Chem.* **33**, 3182 (1968).

are consistent with the proposed enzyme–substrate model. Each is based on an analysis of the stereochemistry of the oxygenation products.

First, a general stereochemical relationship between the amide group and the hydroxyl group may be observed. This is seen as a recurring trans orientation between the two groups in the many rigid amide molecules that have been hydroxylated. Two examples are given to illustrate this point. Oxygenation of cyclohexylbenzamide (**23**) results in formation of *trans*-4-hydroxycyclohexylbenzamide (**24**) in 30% yield.[58,59] Oxygenation of *N*-benzoyl-*N*-methyl-1-adamantanamine (**25**) gave monohydroxylated product **26** in 14% yield and dihydroxylated product **27** in 55% yield.[60] Product **26** was shown to be an intermediate in the formation of **27**. The products of

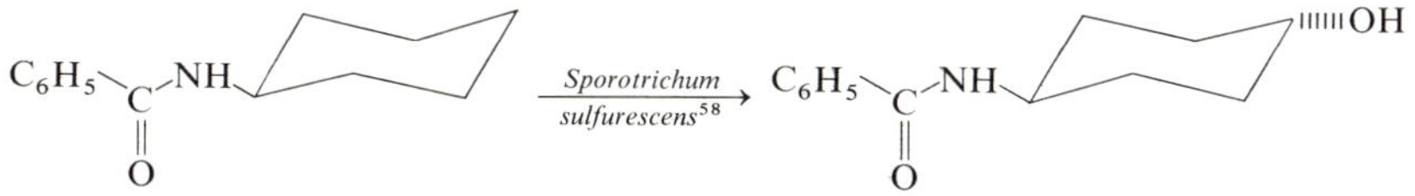

(**23**) (**24**)

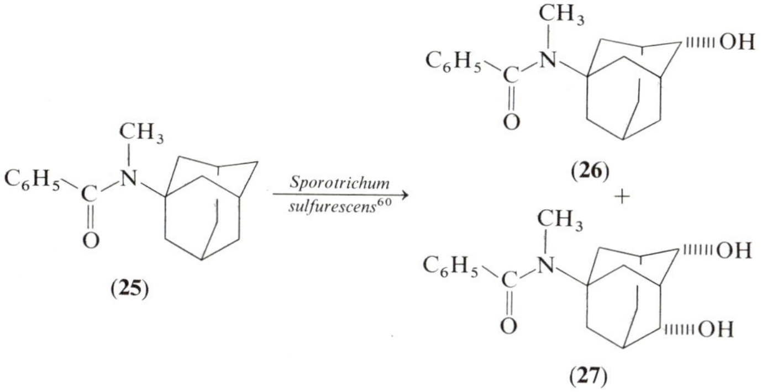

(**25**) (**26**) (**27**)

these two reactions clearly show the trans relationship between amide and hydroxyl groups. Another example illustrates that the above stereochemistry is not the result of simply formation of an equatorial hydroxyl group. Oxygenation of 3-benzoyl-3-azabicyclo[3.3.1]nonane (**28**) with *S. sulfurescens* gave 6-hydroxylated product **29** in 71% yield in which the hydroxyl group is axial with respect to the bicyclic ring structure, but is trans to the amide group.[61] This stereochemical relationship may be seen in many other products formed by *S. sulfurescens.*

[59] R. A. Johnson, M. E. Herr, H. C. Murray, and G. S. Fonken, *J. Org. Chem.* **35**, 622 (1970).

[60] M. E. Herr, R. A. Johnson, H. C. Murray, L. M. Reineke, and G. S. Fonken, *J. Org. Chem.* **33**, 3201 (1968).

[61] R. A. Johnson, M. E. Herr, H. C. Murray, L. M. Reineke, and G. S. Fonken, *J. Org. Chem.* **33**, 3195 (1968).

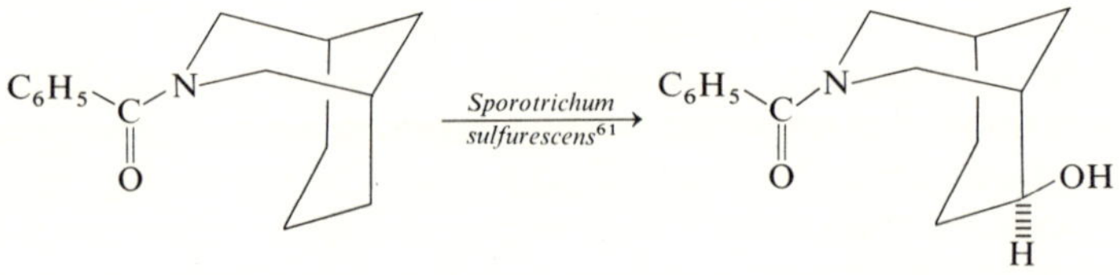

(28) (29)

The stereochemical results seen in the oxygenation of *N*-benzoyl-2-amino-4-methylpentane (**30**)[62] also bear on the enzyme–substrate model. Approximately equal quantities of tertiary alcohol **31** and primary alcohol **32** are obtained from hydroxylation of racemic (*R*,*S*) **30**. When either the (*R*) or

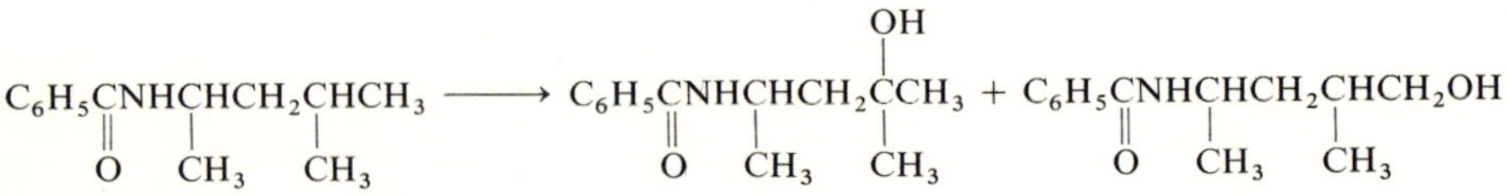

(30)	(31)	(32)
(*R*,*S*)	(22%)	(29%)
(*R*)	(41%)	(14%)
(*S*)	(12%)	(36%)

the (*S*) enantiomer of the substrate is used, the quantities of the two products significantly shifted in favor of the tertiary alcohol or the primary alcohol, respectively. These results are explained by assuming interaction of the amide group with a fixed electrophilic site on the enzyme. Since the asymmetric carbon is attached to the amide, this then results in either the tertiary carbon or the primary carbon being in more favorable proximity to the oxygen transfer site, depending on which enantiomer is being hydroxylated (see Fig. 4e and f for further illustration of this difference).

Finally we note the stereochemical difference caused by the presence or absence of an *N*-methyl group in the oxygenation of cycloheptylbenzamide. Both cycloheptylbenzamide (**33**) and *N*-methyl-cycloheptylbenzamide (**35**) are oxygenated at the 4-position (giving in each case a mixture of alcohol and

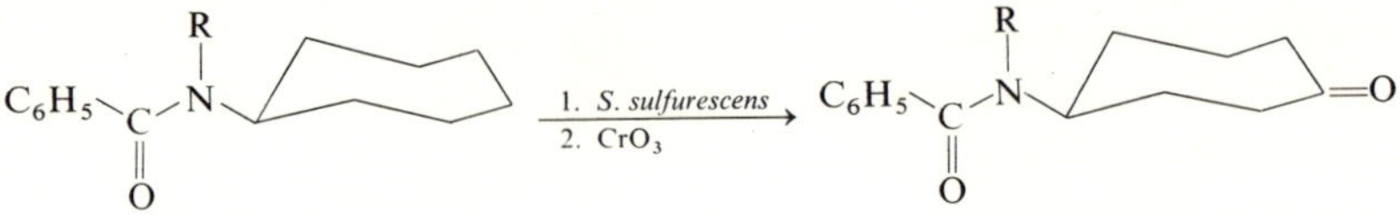

(**33**) R = H (**34**) R = H
(**35**) R = CH_3 (**36**) R = CH_3

[62] R. A. Johnson, H. C. Murray, and L. M. Reineke, *J. Am. Chem. Soc.* **93**, 4872 (1971).

ketone). However, **33** is oxygenated predominantly at the pro-*S* 4-position while **35** is oxygenated predominantly at the pro-*R* 4-position.[63] The powerful influence exerted by the small difference between amide substituents upon the outcome of this reaction seems reasonable only if the amide group is involved in an interaction with the enzyme.

If the presence of a functional group on a substrate simply serves to block oxygenation in its vicinity, these stereochemical relationships would not be observed. However, if they are the result of an enzyme–substrate interaction as proposed, it should be possible to develop this model further to include these stereochemical details. A previous attempt[64] to achieve this has not proven to be useful. Below another attempt is described in this direction that results in a successful stereochemical mapping of the enzyme.

Kinetic rate data are traditionally used in measuring the "fit" of a substrate to an enzyme active site. This approach requires at least a reproducible enzyme preparation and ideally a purified enzyme. Few hydroxylating enzymes have been obtained in cell-free preparations. Consequently, we use the absolute stereochemistry of the oxygenation products as the experimental base for mapping the active site of the *S. sulfurescens* enzyme. It is assumed that the same enzyme is responsible for all the oxygenations observed with *S. sulfurescens*. A second assumption is that conformationally flexible molecules assume the conformation of lowest energy when interacting with the enzyme.

FIG. 2. Amide group in a defined position in the diamond lattice.

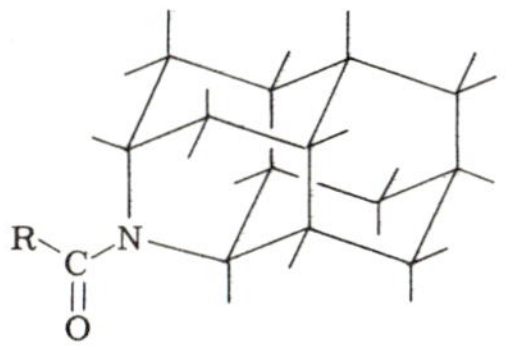

As a framework on which to develop the model, it is convenient to use the diamond lattice of carbon. The amide function

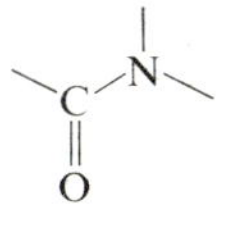

is given a fixed position in the lattice as shown in Fig. 2. As already noted, the carbonyl oxygen is assumed to be the site of initial interaction between the substrate and the enzyme.

[63] M. E. Herr, R. A. Johnson, and H. C. Murray, the Upjohn Company, 1971 (unpublished results).

[64] R. A. Johnson, M. E. Herr, H. C. Murray, and G. S. Fonken, *J. Org. Chem.* **33**, 3217 (1968).

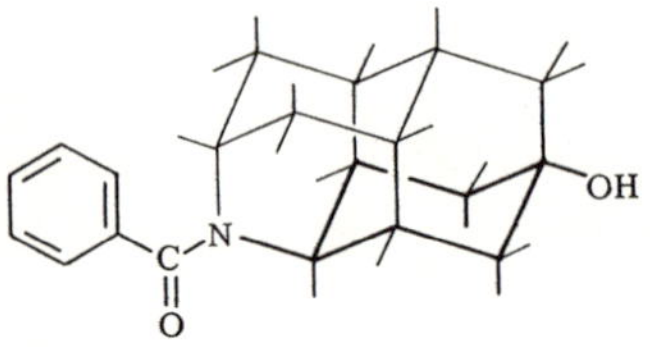

FIG. 3. *trans*-4-Hydroxy-*N*-cyclohexylbenzamide in diamond lattice.

Each hydroxy-amide obtained from *S. sulfurescens* is now placed into this lattice so that the amide group is in the defined position. The position of the hydroxyl group is then noted. For example, the product of hydroxylation of *N*-cyclohexylbenzamide[58] fits into the lattice as shown in Fig. 3. Other hydroxyamides whose structures coincide with the diamond lattice (5- and 7-membered rings, for example, cannot coincide with the diamond lattice) and whose absolute configurations are known, are shown in Fig. 4.[31,62,63,65,66] A composite of these structures, shown in Fig. 5, reveals a close grouping of hydroxyl groups having a definite stereochemical relationship to the amide group. The closeness of this grouping is somewhat distorted by the two-dimensional drawing. However, the precise positioning of the amide group into the diamond lattice introduces an artificial rigidity which, if released, will result in an increase in the closeness of the hydroxyl groups. The chirality of the enzyme is reflected by this composite model. Thus, if one were standing on the carbonyl carbon, as it is drawn in Fig. 5, and were looking down the plane of the amide group, he would see the hydroxyl groups lying either in that plane or to the right of it. It is interesting that when *N*-cycloheptylbenzamide is placed so that the amide coincides with that of the model, the pro-*S* C-4 position (with respect to absolute configuration at C-1) lies to the right of this plane while the pro-*R* C-4 position lies to the left. It is the pro-*S* C-4 that is predominantly oxygenated

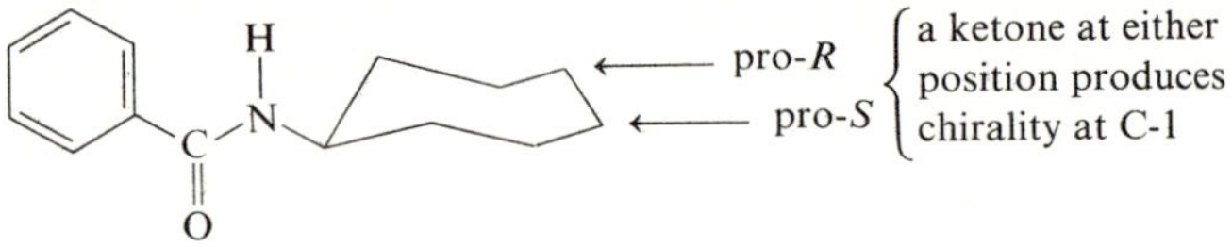

in this substrate by *S. sulfurescens*.[63] Less clear is why the *N*-methyl analog is oxygenated predominantly at the pro-*R* C-4 carbon. Perhaps the *N*-methyl group sufficiently lowers the energy difference between the above and the next favored conformation (rotation by 180° about the C-1–N bond) so that the latter is preferred by the enzyme.

[65] R. A. Johnson, H. C. Murray, L. M. Reineke, and G. S. Fonken, *J. Org. Chem.* **34**, 2279 (1969).

[66] R. A. Johnson, H. C. Murray, L. M. Reineke, and G. S. Fonken, *J. Org. Chem.* **33**, 3207 (1968.)

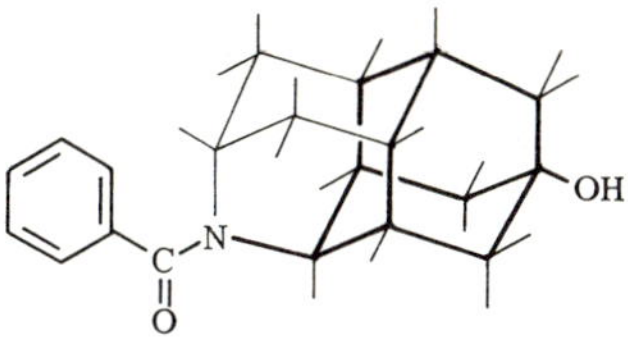

(a) *trans*-4-Benzamido-1-adamantanol (Ref. 63)

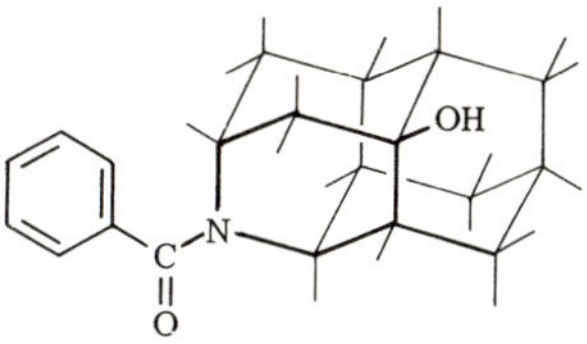

(b) 1-Benzoyl-4-piperidinol (Ref. 31)

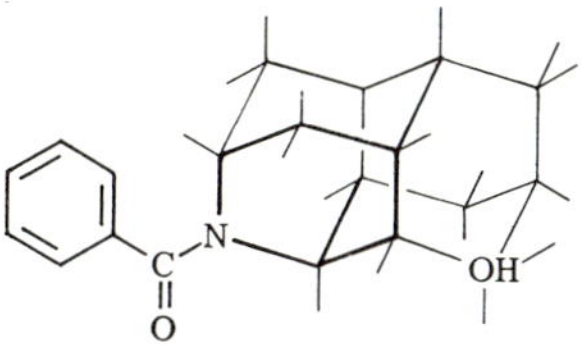

(c) (2*S*,3*S*)-1-Benzoyl-2-methyl-3-piperidinol (Ref. 65)

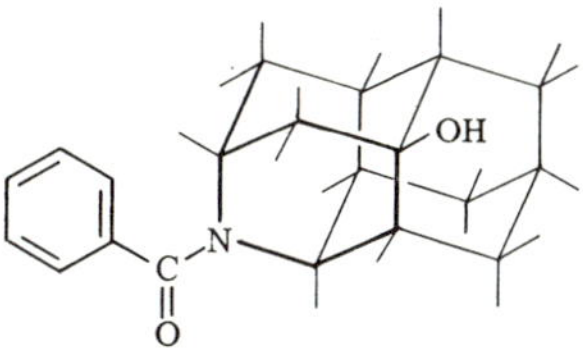

(d) (2*R*,4*S*)-1-Benzoyl-2-methyl-4-piperidinol (Ref. 65)

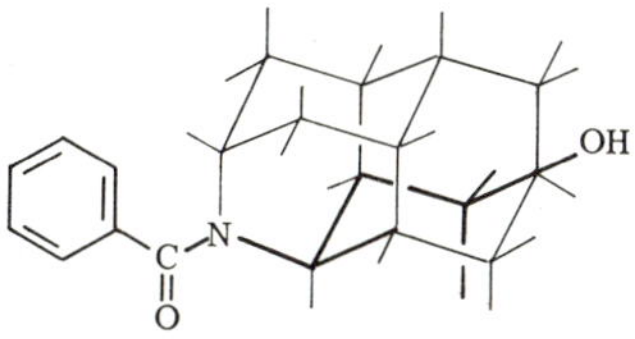

(e) (1*R*)-*N*-(1,3-dimethyl-3-hydroxybutyl)benzamide (Ref. 62)

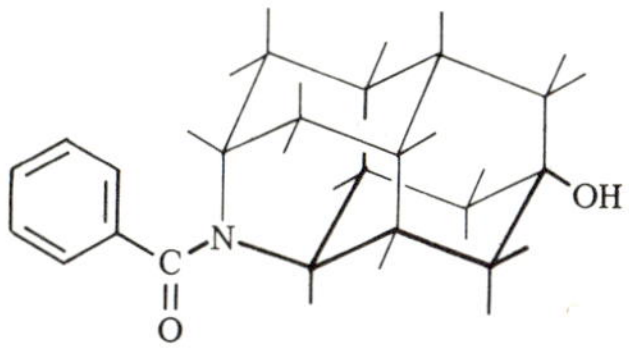

(f) (1*S*)-*N*-(1,3 dimethyl-4-hydroxybutyl)benzamide (Ref. 62)

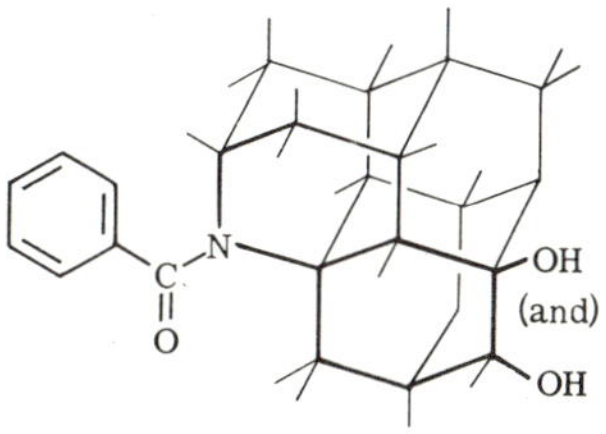

(g) (4*aS*,5*S*,8*cR*)-1-benzoyl-*trans*-decahydroquinolin-5-ol and (4*aR*,6*R*,8*aR*)-1-benzoyl-*trans*-decahydroquinolin-6-ol (Ref. 66)

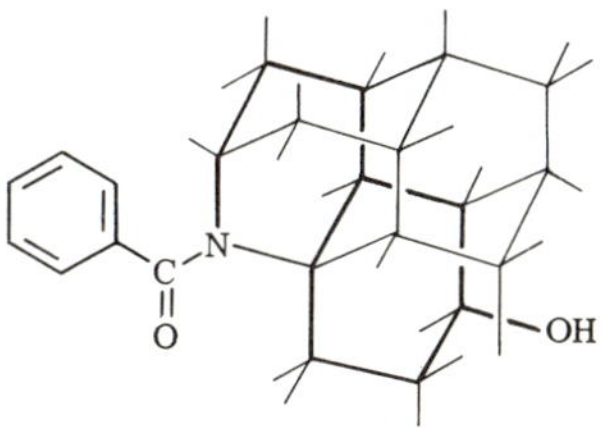

(h) (4*aS*,6*S*,8*aS*)-1-benzoyl-*trans*-decahydroquinolin-6-ol (Ref. 66)

FIG. 4. Additional examples of hydroxyamides that fit into the diamond lattice.

FIG. 5. Composite drawing of hydroxyamides in the diamond lattice. When hydroxyl groups from more than one hydroxyamide fall at the same position, the number doing so is shown in parentheses.

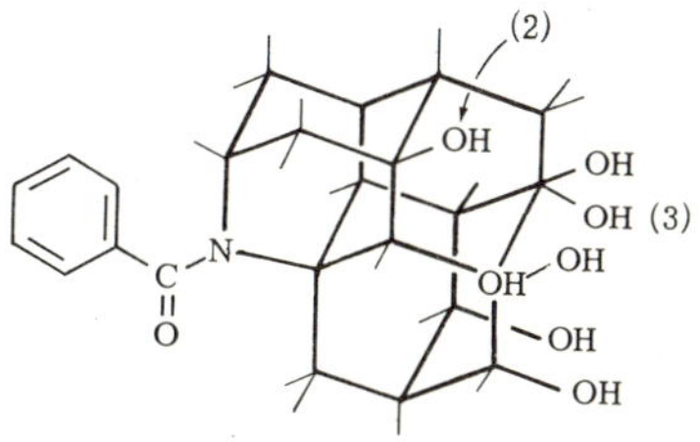

Chemists quite logically ask in these cases whether there is any preference for oxidation at primary, secondary, or tertiary carbon. In oxidations with *S. sulfurescens*, there appears to be a preference for attack at the more highly substituted carbon. This preference is subtle, however, and is easily overwhelmed by the steric requirements of the preceding model.

The question of preference has been examined by preparing several substrates having a methyl substituent on a position known to be oxygenated

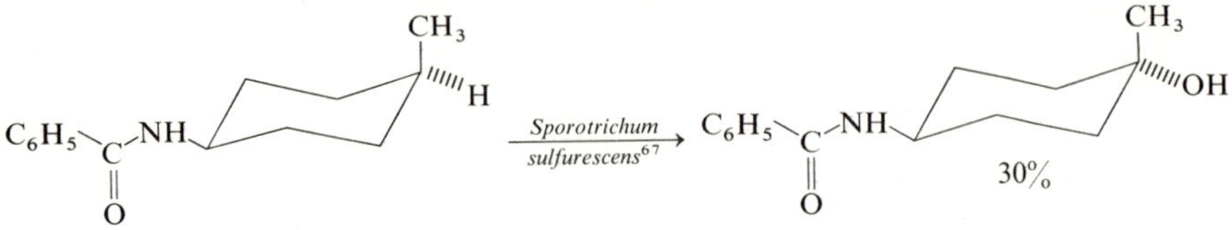

(37)

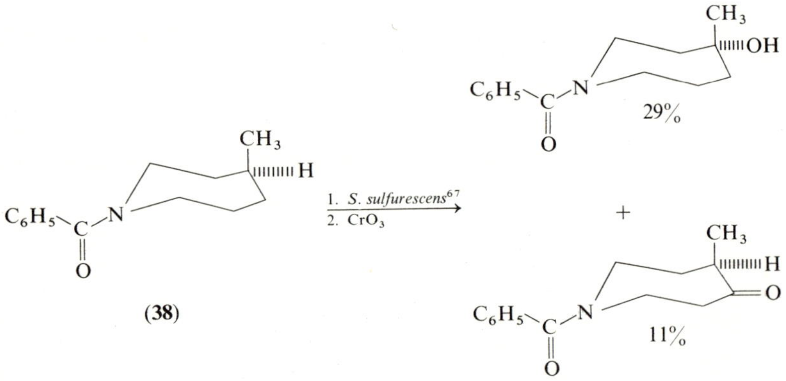

(38)

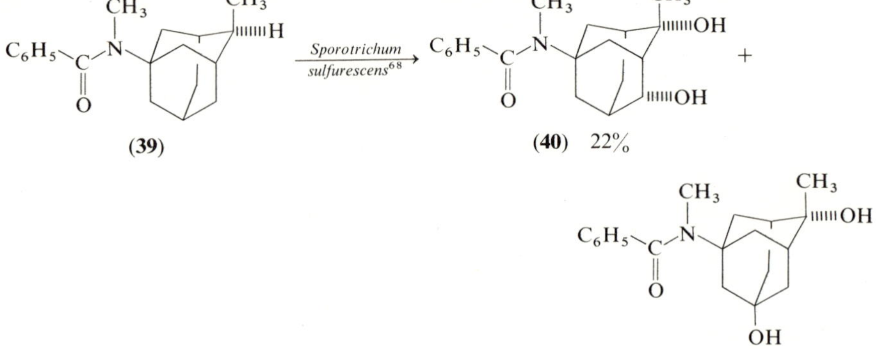

(39) **(40)** 22%

(tentative)

18%

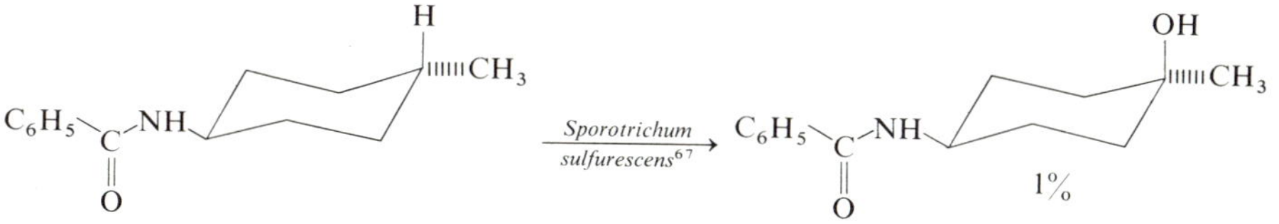

(41)

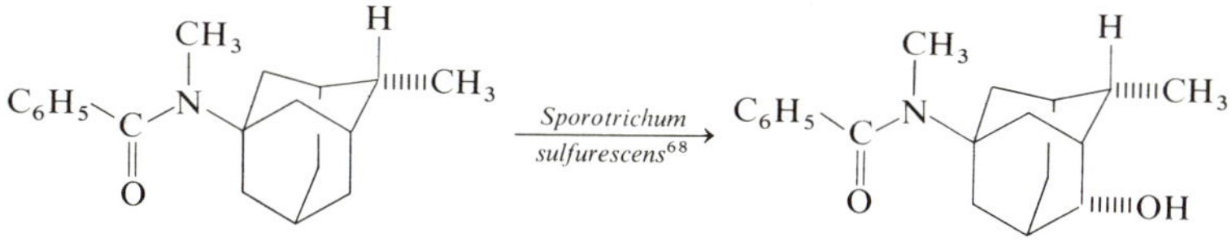

(42) 3%

when the methyl group is absent. *cis*-4-Methyl-*N*-cyclohexylbenzamide (**37**),[67] 1-benzoyl-4-methylhexahydro-4*H*-azepine (**38**),[67] and *cis*-4,*N*-dimethyl-*N*-benzoyl-1-adamantylamine (**39**)[68] are all oxygenated preferentially at the carbon bearing the methyl group. On the other hand, when the methyl group is in a trans configuration with respect to the amide, as in **41** and **42**, it appears to greatly hinder the oxygenation. Failure to oxygenate the methyl group itself may result from the fact that the optimum distance (~5.5 Å) between carbonyl oxygen (initial interaction site) and the site of oxidation is greatly exceeded.[67] Oxygenation of the methyl group is possible in other molecules as shown by formation of hydroxymethyl derivative **45** in the reaction of 1-benzoyl-4-methylpiperidine (**43**) with *S. sulfurescens* for here the distance is more favorable.[65]

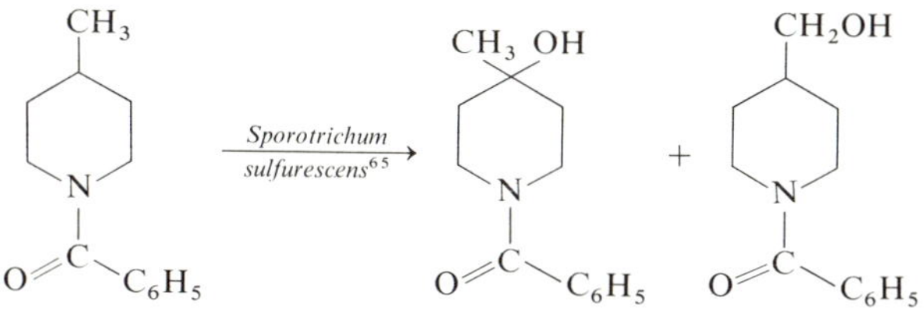

(43) **(44)** 13% **(45)** 23%

In the oxygenation of a series of acylic benzamides, it was observed that with longer chain lengths (five or six carbons from the amide group) oxygenation occurred almost exclusively at the subterminal carbon atom.[62] In substrates of four carbons in length from the amide group, oxygenation occurred in significant amounts at both the terminal and subterminal positions. Thus, it again appears that a preference for oxidative attack at

[67] R. A. Johnson, M. E. Herr, H. C. Murray, and L. M. Reineke, *J. Am. Chem. Soc.* **93**, 4880 (1971).

the more highly substituted carbon is subjugated to a need for optimum distance between amide group and the oxidation site.

This section on *S. sulfurescens* will be concluded with a number of examples illustrating synthetic applications and other recent uses of this microorganism. The oxygenation of molecules containing medium sized rings such as **46** provides an alternate route to aminoketones having a transannular disposition.[31]

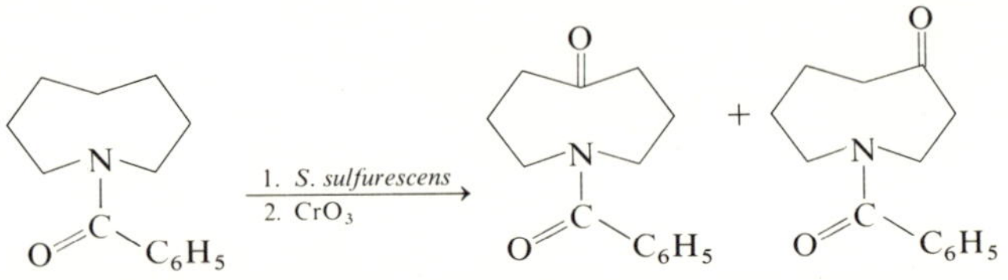

(46) **(47)** 59% **(48)** 19%

The production of optically active molecules by these enzymatic reactions can provide a convenient source of spectroscopic models. The optically active adamantane (**40**), obtained from oxygenation of **39**, was further modified to provide an example of an optically active cyclohexanone containing an axial methyl group in the 3-position.[68] The latter provided a model for consideration of the octant rule and "antioctant" effects. Similarly, the spectroscopic models, 5-oxo-, 6-oxo-, and 7-oxo-*trans*-decahydroquinoline, were prepared in optically active form by modification of the hydroxylation products from 1-benzoyl-*trans*-decahydroquinoline.[66]

The oxygenation of 1,4-diisopropylbenzene (**49**), giving (*R*,*R*)-1,4-di-(2-hydroxy-methylpropyl)benzene (**50**), and related dialkylbenzenes, led to the rational synthesis of a metabolite of the antiflammatory substance 2-(4-isobutylphenyl)propionic acid.[69] The substrate (**51**) was prepared chemically and oxygenated with *S. sulfurescens* to give the desired (2*R*)-2-[4-(2-hydroxy-2-methylpropyl)phenyl]propan-1-o1 (**52**) and the by-product (**53**). Chemical oxidation of **52** gave the desired metabolite, (2*R*)-2-[4-(2-hydroxy-2-methylpropyl)phenyl]propionic acid.[69]

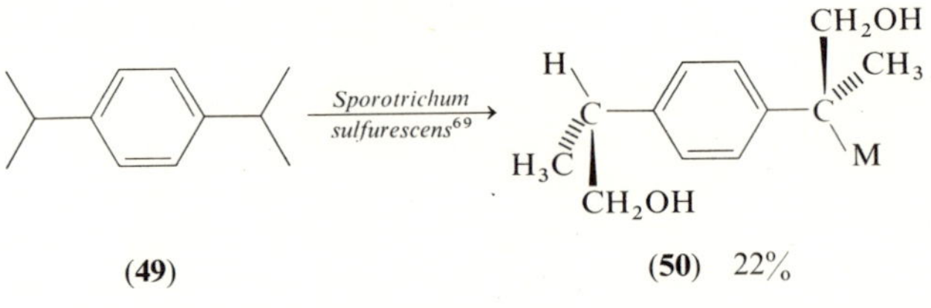

(49) **(50)** 22%

[68] M. E. Herr, R. A. Johnson, W. C. Krueger, H. C. Murray, and L. M. Pschigoda, *J. Org. Chem.* **35**, 3607 (1970).

[69] R. A. Johnson, C. M. Hall, W. C. Krueger, and H. C. Murray, *Bioorg. Chem.* **2**, 99 (1973).

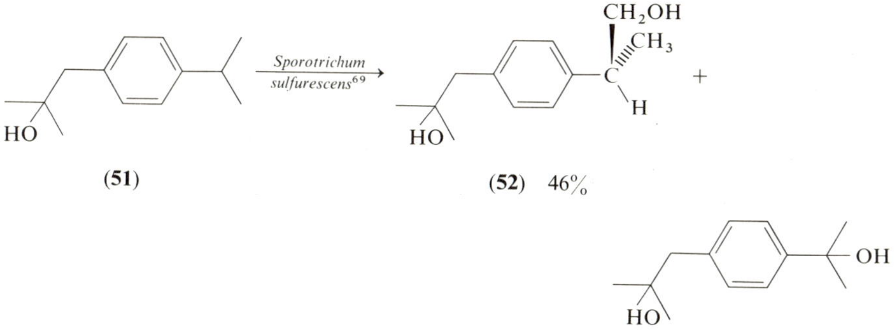

(51) **(52)** 46% **(53)** 10%

A convenient functionalization of one of the geminal methyl groups in the pinane skeleton has been achieved in the oxygenation (+)-*N*-benzoylisopinocamphenylamine (**54**).[59] The (−)-isomer is similarly oxygenated, but in lower (28%) yield.

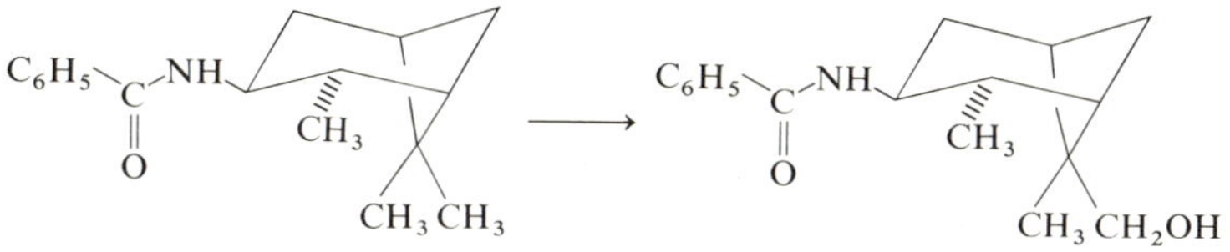

(54) **(55)** 62%

Microbial oxidations can be used to prepare compounds that could arise from the metabolism of drugs in animal systems. For example, oxygenation of the *N*-benzoyl derivative (**56**) of the hypotensive agent mecamylamine gives products oxygenated at either the 6- or the 7-positions.[70]

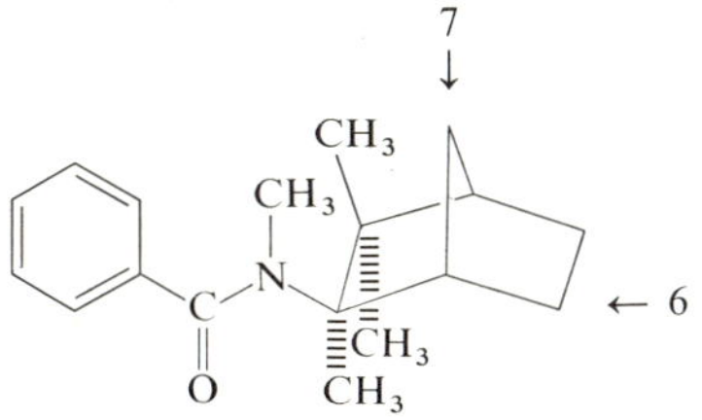

(56)

The hydroxylating enzyme of *S. sulfurescens* is able to tolerate a ferrocene moiety as evidenced by the successful oxygenation of 1,2-(α-oxotetramethylene)ferrocene.[71]

[70] M. E. Herr, H. C. Murray, and G. S. Fonken, *J. Med. Chem.* **14**, 842 (1971).
[71] R. A. Johnson and H. C. Murray, *Chem. Commun.* p. 989 (1971).

2. *Calonectria Decora* (ATCC 14767)

This fungus is a member of the class *Ascomycetes* and is characterized by having a distinctive pink coloration.

Early transformations using *C. decora* were done with steroids, on which it effects both dehydrogenation and hydroxylation reactions. It has come to be known as an efficient 12β, 15α-dihydroxylator of steroids. Progesterone (**57**), for example, is converted to 12β,15α-dihydroxyprogesterone (**58**) in 77% yield.[72,73]

(**57**) (**58**)

It was recognized by Jones, Meakins, and co-workers at Oxford University that the natural steroids used for most microbial oxygenations were commonly substituted at both "ends" of the molecule with an oxygen function.[48] They were interested in determining whether these functional groups had any influence on which positions in the steroids were oxygenated. To examine this question they prepared a series of androstanes having either one or two oxygen substituents at various positions. The functional group most oftenly used was the ketone, but other substrates containing hydroxyl, methoxyl, ethylene ketal, or halogen groups were also used. These compounds were then oxygenated with *C. decora*.[48,74–76] The results are summarized in Table I. Examination of these results leads to a number of interesting observations.

It soon becomes apparent that the position of a substituent on the androstane nucleus has a strong influence on the site of hydroxylation. Consider first the following entries from Table I. 3,17-Dioxoandrostane (**59**), 3β-hydroxy-17-oxoandrostane (**60**), and 3β-methoxy-17-oxoandrostane (**61**) all undergo 12β, 15α-dihydroxylation.[74] However, a change of the 3-substituent to the α-configuration, as in 3α-hydroxy-17-oxoandrostane (**62**)[74]

[72] A. Schubert, G. Langbein, and R. Siebert, *Chem. Ber.* **90**, 2576 (1957).

[73] A. Gubler and C. Tamm, *Helv. Chim. Acta* **41**, 301 (1958).

[74] A. M. Bell, W. A. Denny, E. R. H. Jones, G. D. Meakins, and W. E. Müller, *J. Chem.* Soc., *Perkin Trans. 1* p. 2759 (1972).

[75] J. M. Evans, E. R. H. Jones, G. D. Meakins, J. O. Miners, A. Pendlebury, and A. L. Wilkins, *J. Chem.* Soc., *Perkin Trans. 1* p. 1356 (1975).

[76] E. R. H. Jones, G. D. Meakins, J. O. Miners, and A. L. Wilkins, *J. Chem. Soc., Perkin Trans. 1* p. 2308 (1975).

TABLE I

OXYGENATION OF ANDROSTANES BY *Calonectria decora*[48,74–76]

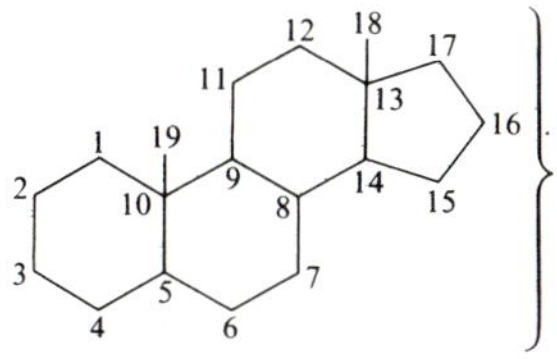

$[=O]_n$ where n = 1,2

Substrate position of carbonyl and other substituents	Substrate recovered (%)	Product position of OH groups	Yield (%)[a,b]
None	45	None	0
1	75	None	0
2	12	6α,12β (**68**)	26
		6α,11α (**69**)	11
3 (**65**)	23	12β,15α (**70**)	52
		(3β),[c]12β,15α	13
4 (**66**)	0	11α,15α (**71**)	36
		12β,15α (**72**)	36
6	90	None	0
7	72	diOH[d]	14
11	38	1β, 6α	11
		6α	
12	8	6α,15α (**73**)	15
		1β,6α,15α	12
15	4	(14β-H)[e], 6α,12β	21
		2α,12β	8
16 (**67**)	31	6α,11α (**74**)	33
		1β, 6α (**75**)	9
17 (**64**)	40	1β, 6α (**76**)	47
2,16	0	6α,12β	12
		6-C=O,12β	27
2,17	15	6α	13
3,11	27	15α	18
		(3β)[c],15α	23
3,12	10	15α	11
		(3β)[c], 15α	26
3,17,Δ^4	4	15α	36
3,7	26	(3β)[c], 12β	19
6,17	27	1β	64
7,17	25	1β	55
11,17	4	6α	32
3,17 (**59**)	23	12β,15α	5
3,17β-OH,Δ^4	25	15α	27
		6-C=O	11
3β-OH,17 (**60**)	6	(3-C=O)[f], 12β,15α	32
		12β,15α	21

(*cont.*)

TABLE I (*cont.*)

Substrate position of carbonyl and other substituents	Substrate recovered (%)	Product position of OH groups	Yield (%)[a,b]
3β-OCH_3,17 (**61**)	12	12β,15α	30
3α-OH,17 (**62**)	0	1β, 6α	37
		1β,15α	25
3α-OCH_3,17 (**63**)	3	1β, 6α	60
$3,3\substack{-O\\-O}]$	37	6α,12β	27
		12β,15α	14
$16,16\substack{-O\\-O}]$	29	6α,12β	71
$17,17\substack{-O\\-O}]$	53	7β,12β,15α	47
		1β,7β,12β	21
		7β,12β	6
$3,17,17\substack{-O\\-O}]$	12	12β,15α	25
$17,3,3\substack{-O\\-O}]$	2	1β,6α	22
		12β,15α	11 }
		(3-C=O)[f], 12β,15α	5 }
		7β,12β,15α	6
$3,3;17,17(\substack{-O\\-O}])_2$	10	12β,15α	25
		7β,12β	13
3,11α-OH	4	12β,15α	24
$3\alpha\text{-OH},17,17\substack{-O\\-O}]$	8	12β,15α	37 }
		(17-C=O),12β,15α	7 }
		(17-C=O), 1β,15α	16
$3\beta\text{-OH},17,17\substack{-O\\-O}]$	0	12β,15α	29 }
		(3-C=O),12β,15α	10 }
		(17 C=O),12β,15α	6 }
3α-F,17	12	1β,6α	41
3α-Cl,17	28	1β,6α	70
3α-Br,17	33	1β,6α	65
3β-Cl,17	22	Δ^2, 1-C=O,6α	20
		Δ^2, 1-C=O,6α,19	12
		1β,6α	4
3β-Br,17	26	Δ^2, 1-C=O,6α	25
		Δ^2, 1-C=O,6α,19	10
17α-F,3	42	12β,15α	17

[a] Yields based on recovered starting material.

[b] Yields enclosed by braces can be combined to give total yield for a particular hydroxylation. Modification of another portion of the molecule has resulted in formation of two products.

[c] 3β refers to the fact that the 3-ketone of the substrate has been partially reduced to the 3β-OH.

[d] Position of hydroxyl groups not determined.

[e] Epimerization of 14-hydrogen has occurred.

[f] The 3-OH group has, in part, been oxidized to the C=O group in the 3-position.

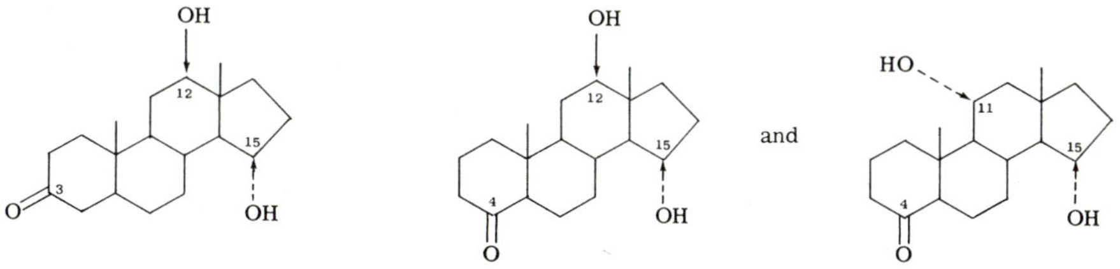

65 *66*

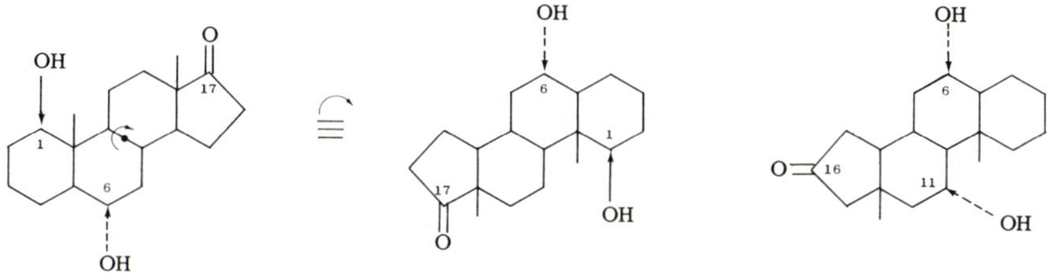

64 *67*

FIG. 6. Dihydroxylation of several androstane monoketones illustrating similar pattern of entering oxygens. Solid arrows indicate formation of a β-hydroxy group and dashed arrows indicate formation of an α-hydroxy group.

and 3α-methoxy-17-oxoandrostane (**63**),[74] or removal of the 3-substituent, as in 17-oxoandrostane (**64**),[48] results in a change to a $1\beta,6\alpha$-dihydroxylation pattern. Removal of the 17-substituent, as in 3-oxoandrostane (**65**), does not effect the $12\beta,15\alpha$-dihydroxylation pattern.[48] These results suggest direction of hydroxylation by the 3β-oxy or 3-oxo substituents to the $12\beta,15\alpha$-positions and by the 17-oxo substituent to the $1\beta,6\alpha$-positions. They also suggest the directing strengths of these substituents to be 3-keto, 3β-OH, and 3β-OCH_3 > 17-keto > 3α-OH and 3α-OCH_3.[74] The preceding rationale has been developed further as illustrated in Fig. 6.[48,77] Here hydroxylation patterns are shown for two A-ring ketones, 3-oxoandrostane (**65**) and 4-oxoandrostane (**66**) (see Table I), and two D-ring ketones, 17-oxoandrostane (**64**) and 16-oxoandrostane (**67**).[48] By rotating the D-ring ketones 180 degrees in the plane of the paper, as illustrated, both the positions of the D-ring ketones and the positions of the entering hydroxyl groups are seen to coincide closely to the A-ring examples. The variation from α-substitution to β-substitution in these products is not as significant as is the fact that the hydroxyl groups are all equatorial except for the 15-OH which is on a five membered ring and therefore is neither strictly axial nor equatorial.

[77] E. R. H. Jones, *Pure Appl. Chem.* **33**, 39 (1973).

Careful fitting of scale drawings of products **68–76** (Table I) so that carbonyl groups and carbons that are hydroxylated coincide results in the pattern shown in Fig. 7.[48] The close coincidence of the hydroxylated carbons and the reasonable grouping of the carbonyl groups has led to the suggestion that the dominant polar group of the substrate interacts with a hydrophilic portion of the hydroxylating enzyme, thereby determining the orientation of the steroid as presented to the hydroxylation site.[77] The distances between the two carbon atoms that are hydroxylated are on the order of 3.8–4.5 Å. The distances between the carbonyl group and each of the groupings of hydroxylated carbons are approximately 4–6 A.

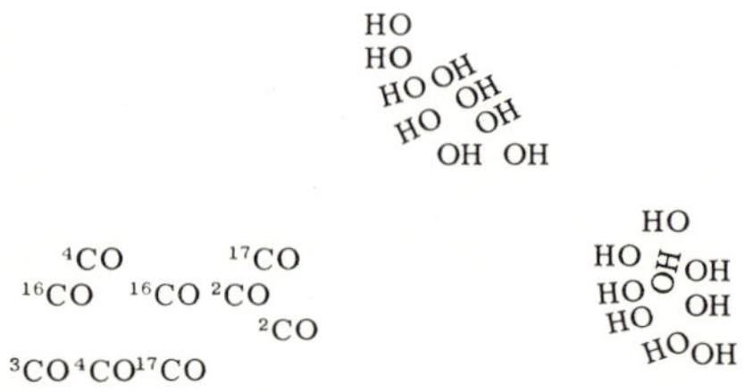

FIG. 7. Coincidence of hydroxylated carbons in the oxygenation of monoketoandrostanes by *Calonectria decora*.[48]

Several ketoandrostanes have been used as substrates for *C. decora* with their ketone groups masked as an ethylene ketal.[75] A pattern of 12β, 15α-dihydroxylation together with other dihydroxylations are generally seen in the products of the reactions. One exception of interest is the formation of the 6α,12β-dihydroxylated product when the ethylene ketal of 16-keto-5α-androstane is the substrate, whereas in the unprotected form the latter is dihydroxylated mainly at the 6α,11α-positions.

A representation of halogen substituted, mainly at the 3-position (see Table I), 17-ketoandrostanes have been used as substrates for *C. decora*.[76] A pattern of 1β,6α-dihydroxylation is seen for the 3-halo substrates. In the cases of β-halogen substituents, the 1β-hydroxylation is followed by dehydrogenation (to the 1-ketone) and apparently further dehydrohalogenation to the α,β-unsaturated ketone system. Failure to observe this sequence of events in the 3α-halo substrates suggests that the α-substituent (axial) may interfere in the enzymatic removal of the 1α-hydrogen atom (also axial).

It would be of considerable interest to use nonsteroidal substrates with *C. decora* and to determine whether the above patterns of hydroxylation would hold. As a first step in this direction, several perhydrochrysene (*dl*-D-homogonane) ketones have been prepared and oxygenated with *C. decora*.[78] Oxygenation of monoketone **77** gave the keto diol **78** in 27% yield. Oxygenation of the difunctional substrate **79**, which presumably is an

[78] M. J. Ashton, A. S. Bailey, and E. R. H. Jones, *J. Chem. Soc., Perkin Trans. 1* p. 1658 (1974).

intermediate in the formation of **78**, resulted in an increased yield of **78** (78%) and a trihydroxylated product **80** (17%). Several other related substrates also were oxygenated with *C. decora*. The 4,12-diketone **81** was oxygenated in the 15-position giving **82** in 19% yield. It can be seen that the patterns of

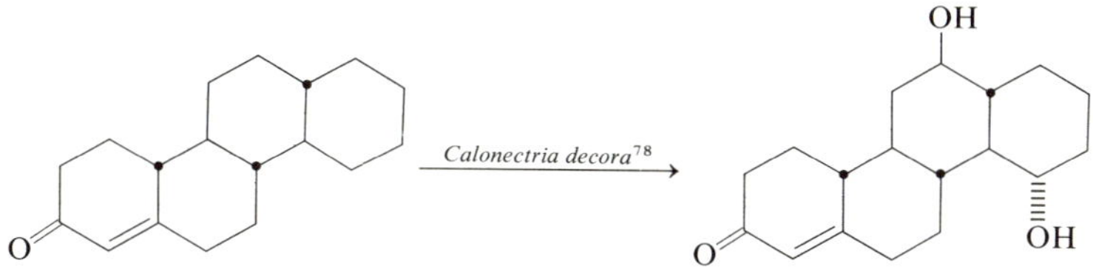

dl-(**77**) *dl*-(**78**)

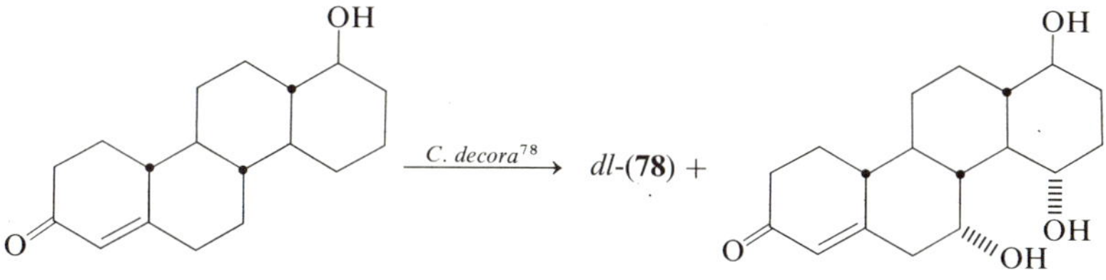

dl-(**79**) *dl*-(**80**)

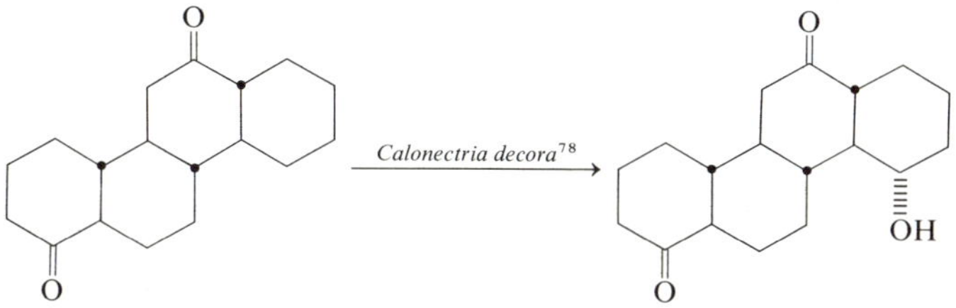

dl-(**81**) *dl*-(**82**)

hydroxylation in these products are very similar to those outlined for the androstanes above. None of the products were optically active nor were any of the recovered substrates optically active. This is surprising at first but from models it can be seen that both enantiomers can adopt the same orientation relative to the pattern shown in Fig. 7.

It has been reported that the oxygenations of steroids with *C. decora* may be more vigorous resulting in more highly oxygenated products when the substrate is added to the fermentation with dimethyl sulfoxide.[74,77] An interesting example is the oxygenation of 17-oxoandrostane (**83**) where hydroxylation at C-19 has been observed in addition to the normal oxygenation at C-1 and C-6. The product is shown by formula **84**. Such oxygenations of angular methyl groups in steroids are unusual. This reaction has been put to clever use in the case of the isoandrostane **85**. The latter is hydroxylated at both C-11 and C-19 as shown (**86**).[77,79] Isosteroids are easily converted back to the 3-hydroxy-5-ene structure from which they are generated.

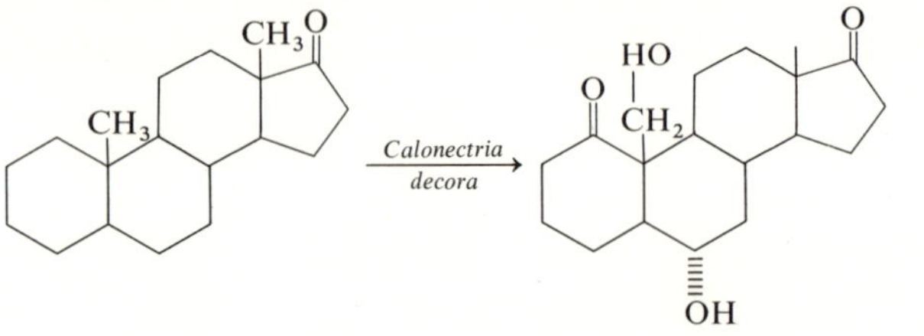

(**83**) (**84**)

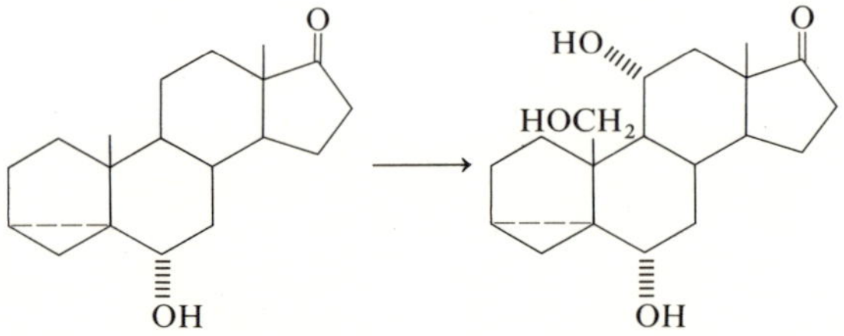

(**85**) (**86**)

Oxygenation of cyclododecanone with *C. decora* gives results very similar to those observed in the oxygenation of cyclododecanol (**20**) with *Sporotrichum sulfurescens*. The two predominant products result from oxygenation at the C-6 (18%) and C-7 (17%) positions of the cyclododecanone ring.[80] A smaller amount of oxygenation (2%) occurs at C-5 and polar products of an unidentified nature are also formed (5–30%).[80]

The oxygenation of a series of tetracyclic diterpenes having the *ent*-kaurane skeleton (structures **87–91** shown in Table II) with *C. decora* has been reported in preliminary form.[81] Positions C-1 or C-7 are the preferential sites of hydroxylation in these substrates (Table II).[82,83] It is an interesting exercise with models to see if there is any similarity between hydroxylation patterns for these substrates and the androstanes discussed earlier. Also included in this study were two diterpenes of the *ent*-beyerene series **92** and **94**. Both were oxygenated in 50% yield at the 6α-position, giving **93** and **95**.[81]

[79] V. E. M. Chambers, W. A. Denny, E. R. H. Jones, G. D. Meakins, J. O. Miners, J. T. Pinhey, and A. L. Wilkins, *J. Chem. Soc., Perkin Trans. 1* p. 1359 (1975).

[80] M. J. Ashton, A. S. Bailey, and E. R. H. Jones, *J. Chem. Soc., Perkin Trans.* p. 1665 (1974).

[81] J. P. Beilby, E. L. Ghisalberti, P. R. Jefferies, M. A. Sefton, and P. H. Sheppard, *Tetrahedron Lett.* p. 2589 (1973).

[82] K. D. Croft, E. L. Ghisalberti, P. R. Jefferies, J. R. Knox, T. J. Mahoney, and P. N. Sheppard, *Tetrahedron* **30**, 3663 (1974).

[83] R. McCrindle, J. K. Turnbull, and A. B. Anderson, *J. Chem. Soc., Perkin Trans. 1* p. 1202 (1975).

TABLE II

OXYGENATION OF SOME TETRACYCLIC DITERPENES

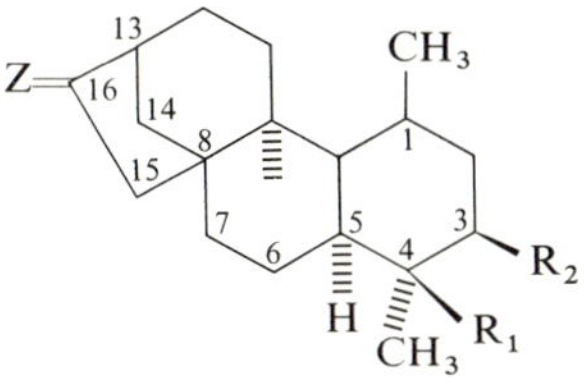

Substrate	Products from oxygenation with *Calonectria decora*	*Rhizopus nigricans*	*Aspergillus ochraceous*
87: $R_1 = CH_2OH$, $R_2 = H$, $Z = O$	1β-OH (10%) 7β-OH (10%)	1β-OH (20%) 7β-OH (20%)	16α-OH (10%)
88: $R_1 = COOH$, $R_2 = H$, $Z = O$	1β-OH (5%) 7β-OH (15%) 7α-OH (40%)	1β-OH (30%) 7β-OH (30%) 7α-OH (5%)	13-OH (5%) 13,16α-diOH (5%)
89: $R_1 = COOH$, $R_2 = H$, $Z = CH_2$	15β,7β-diOH (30%) 15β-OH (5%) 7β-OH (5%)	7α-OH (25%)	16β,17-diOH (20%)
90: $R_1 = CH_3$, $R_2 = OH$, $Z = O$	7β-OH (40%)	1β-OH (25%) 7β-OH (35%)	6α-OH (30%) 7β-OH (25%)
91[a]: $R_1 = COOH$, $R_2 = H$, Z = <(H)(CH$_2$OH)	7α-OH (23%) 7β-OH (14%)	7α-OH (20%)	
17-Norkauran-16-one[b]: $R_1 = CH_3$, $R_2 = H$, $Z = O$		3α,7α-diOH (21%) 3α,7β-diOH (8%) 3α,9β-diOH (15%)	

[a] Croft *et al.*[82]
[b] McCrindle *et al.*[83]

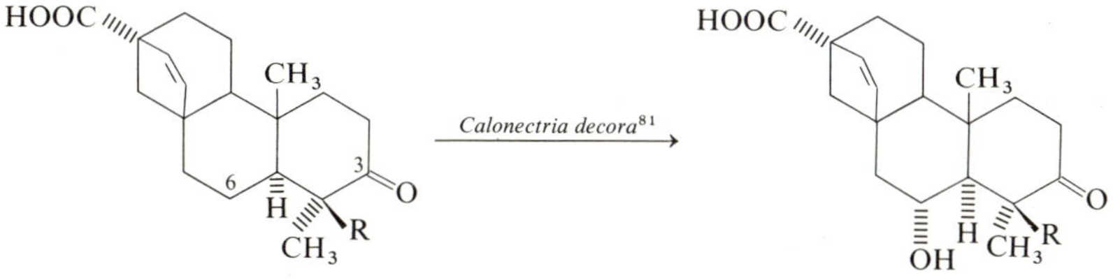

(92) R = H
(94) $R = CH_3$

(93) R = H
(95) $R = CH_3$

Finally, we note the oxygenation of a heterocyclic substrate with *C. decora*. 1-Acetyl-*cis*-2,3-cyclooctanoindoline **(96)** is converted to a mixture of isomeric monoketones. The 9-ketone (**97**) was obtained in 37% yield and the 8-ketone (**98**) in 6%.[84]

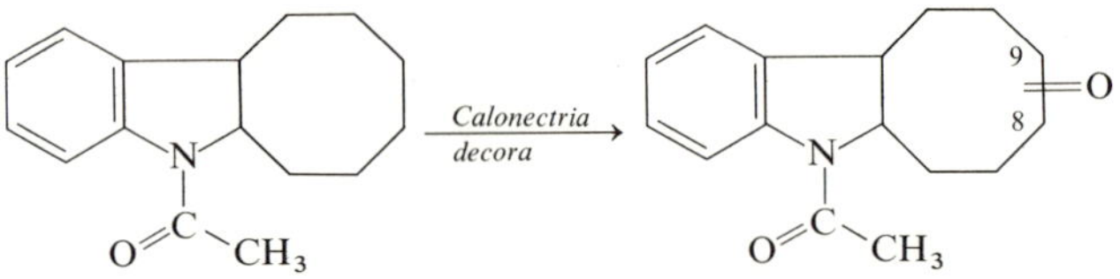

(**96**)

(**97**) 9-keto
(**98**) 8-keto

3. *Rhizopus nigricans*

Rhizopus nigricans is of the class Phycomycetes and the order Mucorales. This organism now will be found in the ATCC catalog as *Rhizopus stolonifer* for which nine different strains have been deposited.

The capability of this organism to hydroxylate progesterone and other steroids at the 11α-position is, to a large degree, responsible for the widespread use of microorganisms to oxygenate steroids that exists today. This genus was the first discovered that could achieve the highly desired hydroxylation of a steroid at the 11-position.[1] *Rhizopus nigricans* retains its central position in this field due to its use on a large scale for the industrial production of steroid intermediates. Numerous other steroids have been used as substrates with this organism and hydroxylation generally occurs at the 11α-position and occasionally at the 6β- or 7β-positions.

Just as with *Calonectria decora*, most past steroid work with *R. nigricans* has involved substrates with oxygen functionality at both ends of the molecule. Consequently, the Oxford group has carried out a detailed study of the oxygenation of mono-[85,86] and difunctional androstanes[87] by *R. nigricans* similar to that done with *C. decora*. Many of their results are summarized in Table III. A general observation was that monohydroxy androstanes were extremely poor substrates for *R. nigricans*.[86] Monoketones (Table III) generally underwent dihydroxylation but unchanged substrate was always recovered in substantial quantities.[86] Only when difunctional substrates were used, was oxygenation, usually monohydroxylation, achieved in good yield (Table III).[87]

[84] T. L. Lemke, R. A. Johnson, H. C. Murray, D. J. Duchamp, C. G. Chidester, J. B. Hester, Jr., and R. V. Heinzelman, *J. Org. Chem.* **36**, 2823 (1971).

[85] J. W. Browne, W. A. Denny, E. R. H. Jones, G. D. Meakins, Y. Morisawa, A. Pendlebury, and J. Pragnell, *J. Chem. Soc., Perkin Trans. 1* p. 1493 (1973).

[86] J. M. Evans, E. R. H. Jones, A. Kasal, V. Kumar, G. D. Meakins, and J. Wicha, *Chem. Commun.* p. 1491 (1969).

[87] V. E. M. Chambers, W. A. Denny, J. M. Evans, E. R. H. Jones, A. Kasal, G. D. Meakins, and J. Pragnell, *Chem. Commun.* p. 1500 (1973).

TABLE III

HYDROXYLATION OF 5α-ANDROSTANES BY RHIZOPUS NIGRICANS

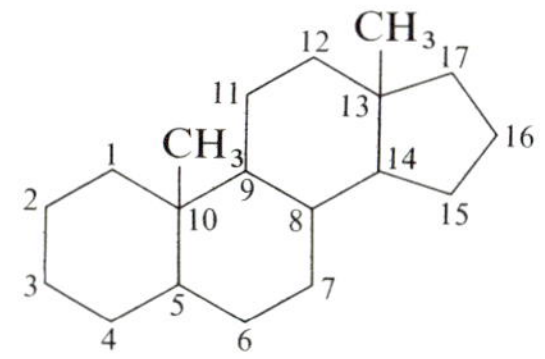

Position of substrate functional groups	Substrate recovered (%)	Position of product OH(groups)[a]	Yield (%)[b,c]
	Monoketo substrates		
2	51	6α,16α	10
		(2α),6α,16α	2
3 (**65**)	46	11α,16β	33
		6β,11α,	4
		11α	1
4	40	11α,16β	9
		(4β),11α,16β	7
6	90	None	
7	90	3α,16β	34
		4α,16β	33
		4β,16β	16
11	62	4α,16β	7
		3α,16β	4
		3β,16β	3
15	59	None	
16 (**67**)	34	3β,7α	46
17	59	3α,11α	10
		3β,7β	9
		6α,11α	9
		7β,11α	2
	Diketo substrates		
2,16	0	(2α),7α	55
		7α	7
2,17	0	(2α),7α	20
		(2α),7β	6
		(2α),6α	2
3,6	6	(3β),16β	26
		16β	9
		(3β),16α	23
		16α	23
3,7	4	16β	32
		(3β),16β	9

(cont.)

TABLE III (*cont.*)

Position of substrate functional groups	Substrate recovered (%)	Position of product OH(groups)[a]	Yield (%)[b,c]
3,11 (**99**)	4	16β	55 }
		(3β),16β	20 }
		9α,16β	6
		9α,16α	5
		(3β),16α	4 }
		16α	2 }
3,16 (**105**)	0	(3β),7α	50
3,17	32	11α	23
		6α	15 }
		(3β),6α	3 }
		(3β),7β	4
7,17	0	3α	37 }
		(17β),3α	6 }
		3β	7 }
		(17β),3β	4 }
		11α,3α	5
11,16	23	3β,7β	24
		3α	10
		3β	5
11,17	21	(17β),3α	36 }
		3α	14 }
		(17β),4α	15
		Hydroxyketo and dihydroxy subtrates	
3,16β-OH (**103**)	0	11α	71 }
		(3β),11α	27 }
2,17β-OH	17	6α	15
3,17β-OH	8	11α	34 }
		(3β),11α	3 }
		6α	30 }
		(3β),6α	5 }
		(3β),7β	7
		5α	6
3β-OH,16 (**104**)	0	7α	90
3α-OH,17	7	11α	23
		7β	24
		6α	5
3β-OH,17	0	7β	48
		7α	15
		6α	12
3,6α-OH	0	16α	30
		16β	27
		17α	7
		11α,16β	4

TABLE III (*cont.*)

Position of substrate functional groups	Substrate recovered (%)	Position of product OH(groups)[a]	Yield (%)[b,c]
3,6β-OH	0	16β	39 }
		(3β),16β	2 }
		11α,16β	30
3β-OH,6	0	16α	43
		16β	18
3β,6α-diOH	6	16α	37
		16β	17
		17α	10
3α,6β-diOH	17	16β	38
		15α	17
3,6,Δ^4	20	(3β),16α	32
		(3β),16β	7
3,6β-OH,Δ^4	4	16β	61
		11α,16β	22
3,7α-OH (	0	16β	33 }
		(3α),16β	24 }
3,7α-OH	31	16β	52
3β-OH,7	0	16α(or β)	55
		16β(or α)	22
3β-OH,11	0	16β	34
		16α	11
		6α	9
		7β,16β	8
		7β,16-CO	7
		16-CO	6
3,11α-OH	5	16β	45 }
		(3β,)16β	11 }
		16α	14
		16-CO	4
3,11β-OH	27	(11-CO),9α,16β	13
		16β	11 }
		(11-CO),16β	9 }
		(3β),(11-CO),16β	9 }
		(3β),16β	4 }
		(3β),(11-CO),6α	2
		(3β),(11-CO),16α	1
6,17β-OH	0	3β	23
		3α	21
		11α	11
		5α,11α	7
		3α,11α	7
6α-OH,17	16	11α	58
		3β	10

(*cont.*)

TABLE III (*cont.*)

Position of substrate functional groups	Substrate recovered (%)	Position of product OH(groups)[a]	Yield (%)[b,c]
7,17β-OH	0	3α	52
		3α	12
		4α	3
7α-OH,17 (**100**)	10	3β	77
7β-OH,17	25	3α	47
11,17β-OH (**102**)	26	3α	47
		4α	25
	Haloketo substrates		
3α-F,17	32	3CO,11α	12
		3β,11α	11
3α-Cl,17	30	3CO,11α	13
		3β,7β	12
		3β,11α	11
3α-Br,17	38	3CO,11α	13
		3β,7β	12
		3β,11α	11
3β-F,17	28	6α,11α	24
		3CO,11α	12
3β-Cl,17	29	3CO,11α	11
		3β,11α	10
3β-Br,17	34	3CO,11α	14
		3β,11α	10

[a] Group positions enclosed by parentheses are groups that were present in the substrate, but which have undergone either reduction or oxidation.

[b] Yields were calculated after adjustment for recovered substrate.

[c] Yields grouped together represent total yield of a specific oxygenation. Products differ due to oxidation or reduction of a functional group present in the substrate.

Examination of the results summarized in Table III reveals a pattern of steroid hydroxylation reminiscent of that seen with *C. decora*. There are, however, distinct differences between the two microorganisms. The basic pattern of hydroxylation with *R. nigricans* can be exemplified with several examples from Table III. Consider first the two monoketones: 3-oxoandrostane (**65**) and 16-oxoandrostane (**67**). The first (**65**) is hydroxylated at the 11α-and the 16β-positions, while the second (**67**), is hydroxylated at the 7α- and 3β-positions.[86] By rotating **67** in the plane of the paper, the similar patterns of hydroxylation for these two compounds can be seen.

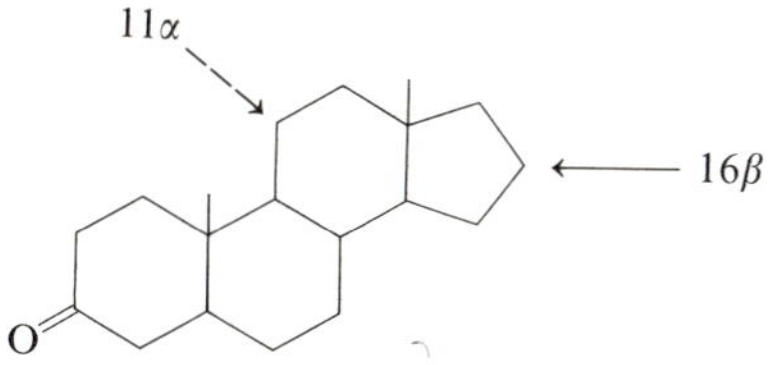

(65) 33%

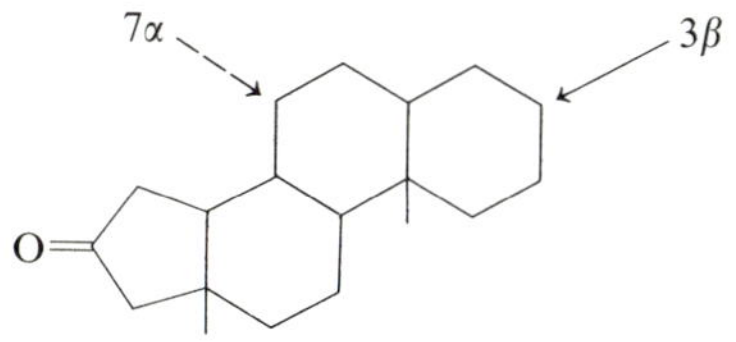

(67) 46%

Looking next at difunctional substrates such as 3,11-dioxoandrostane (**99**) and 7α-hydroxy-17-oxoandrostane (**100**), a similar pattern may be seen.[87] The position of the entering hydroxyl group as indicated by the arrows is

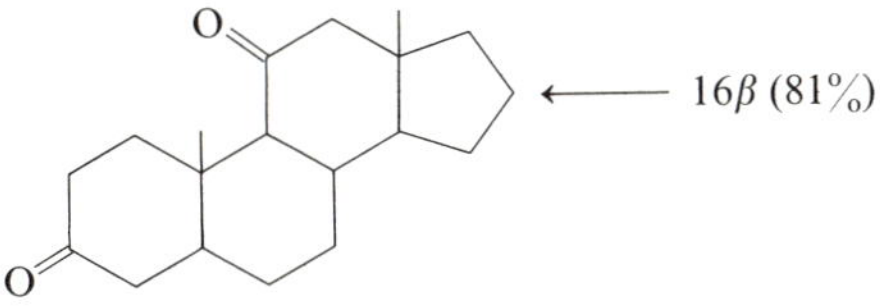

(99)

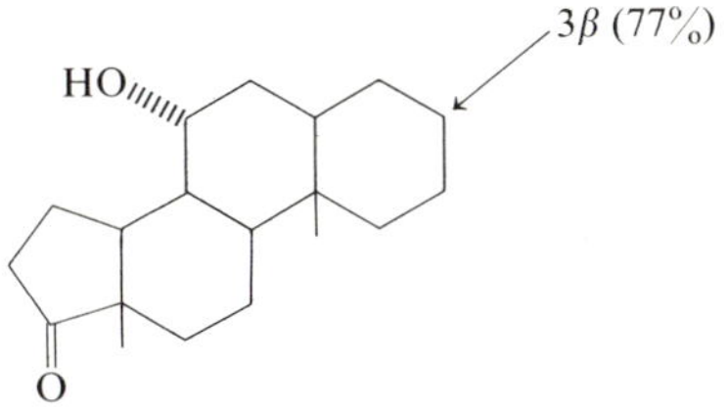

(100)

the same as above. Now, however, the second functional group present in the molecule occupies a position that was previously hydroxylated. This coincidence of functional group with hydroxylation site is not necessary as shown by the hydroxylation of 7α-hydroxy-3-oxoandrostane (**101**) and 17β-hydroxyl-11-oxoandrostane (**102**). Here the second functional group is not occupying a position of hydroxylation.

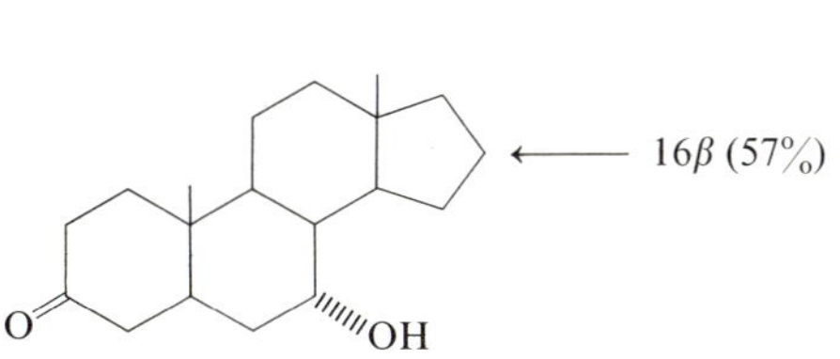

(101)

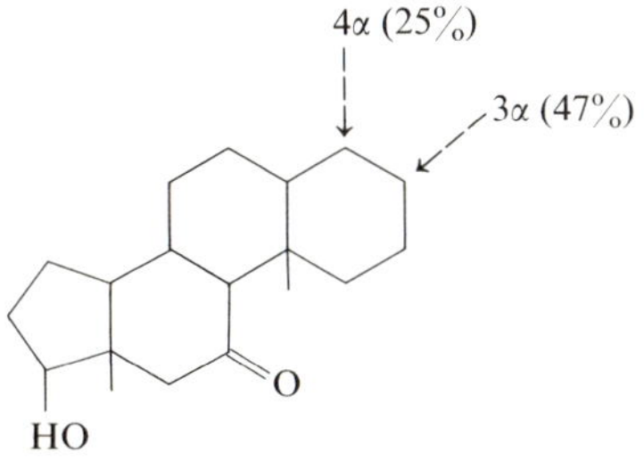

(102)

As noted previously, *R. nigricans* does dihydroxylate monoketo androstanes to a certain extent but does not hydroxylate monohydroxy androstanes.[86] This selectivity is also seen in the hydroxylation of difunctional molecules.[87] Substrate **103**, having a ketone at C-3 and a hydroxyl at C-16, is hydroxylated exclusively at the 11α-position. Substrate **104**, however, in which the ketone is at C-16 and the hydroxyl is at C-3, is hydroxylated exclusively at the 7α-position. When both C-3 and C-16 oxygen functions are ketones (**105**), hydroxylation is observed only at the 7α-position.

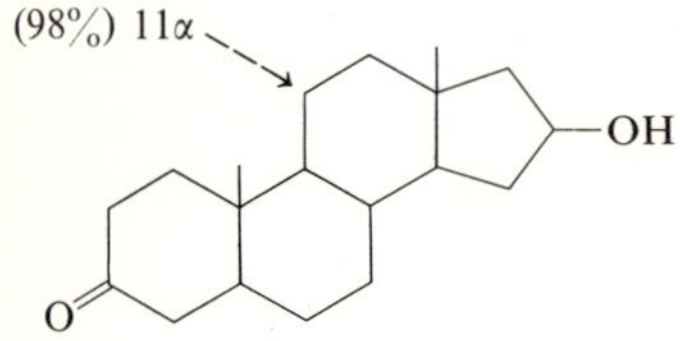

(**103**)

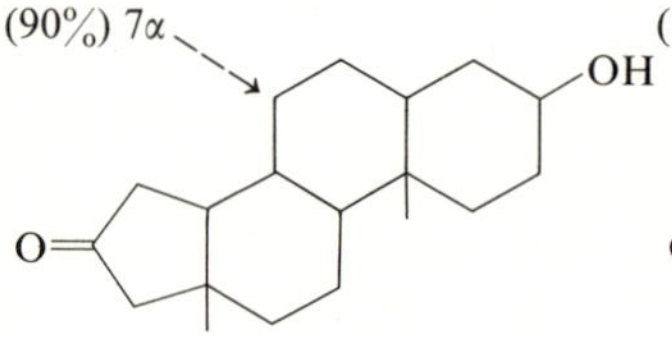

(**104**)

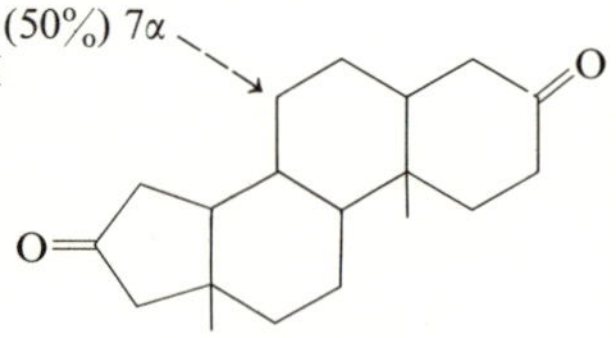

(**105**)

Nearly all of these results have been rationalized by the model shown in Fig. 8.[77] In this model, E represents an electrophilic (or hydrophilic) site on the enzyme that interacts with an oxygen function of the steroid substrate. Hydroxylation can then occur at sites A and B. Alternatively, sites A or B can accommodate a functional group already present on the substrate with hydroxylation occurring at the remaining site. Site E generally is dominant in orientation of the steroid molecule but some results can be explained only if the initial interaction is at site A or B.[77]

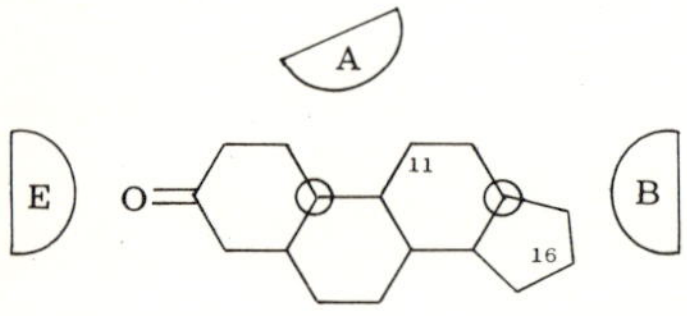

FIG. 8. Model representing sites of interaction between substrate and hydroxylating enzyme of *Rhizopus nigricans*. E = Electrophilic site that interacts with oxygen function of substrate. A and B = sites that can either oxygenate substrate or accommodate a functional group of the substrate. 3-Oxoandrostane is shown with the ketone interacting with site E. ⌽– indicates an angular methyl group extending above the plane of the paper.

A comparison of results from hydroxylation of various 5α-androstanes with three species of *Rhizopus* has been reported.[88] *Rhizopus arrhizus* and *R. circinnans* hydroxylate diketoandrostanes and hydroxyketoandrostanes in a manner very similar to that described for *R. nigricans*.

One of the more interesting chemical modifications of the steroid nucleus is the formation of the so-called isosteroids. Both isoprogestrone (**106**) and 17-oxo-isoandrostane (**108**) are readily available and both have been oxy-

[88] A. M. Bell, I. M. Clark, W. A. Denny, E. R. H. Jones, G. D. Meakins, W. E. Müller, and E. E. Richards, *Chem. Commun.* p. 2131 (1973).

genated at the 11α-position with *R. nigricans*, giving **107**[89] and **109**[90] respectively.

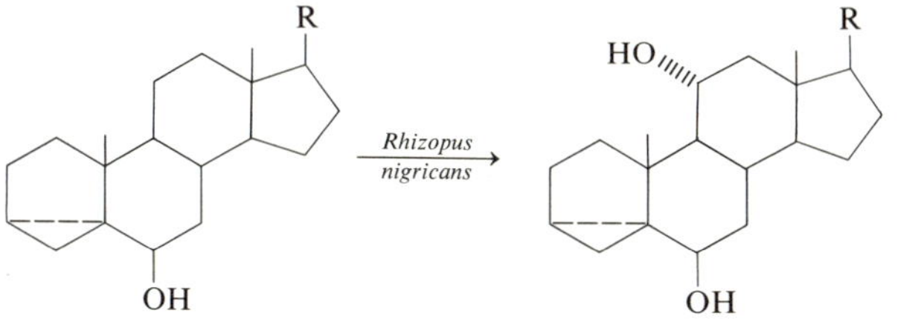

(**106**) R = —$COCH_3$
(**108**) R = =O

(**107**) R = —$COCH_3$ (85%)
(**109**) R = =O (30%)

The oxygenation of several 3-halo-17-ketoandrostanes with *R. nigricans* has been studied (see Table III).[76] Several products are isolated from each fermentation and the yield of any one is quite low and synthetically unpromising in this sense. What is interesting chemically, however, is the isolation of halgoen-free products having either ketone or hydroxyl groups at the 3-position. It is postulated[76] that hydroxylation occurs at the 3-position, giving the unstable halohydrin. The latter convert to ketones and microbial reduction of the ketone leads to the 3β-hydroxy compounds observed in the product mixtures. The substrates are also hydroxylated at a second position, generally of the 11α-configuration.

As a departure from steroid substrates, the closely related *dl*-perhydrochrysene derivative **79** was oxidized with *R. nigricans*.[78] The main product (**110**), 50% yield based on recovered starting material) was the result of hydroxylation at the 17α-position similar to what has been observed with the androstanes. The product (**110**) was found to be optically inactive.[78]

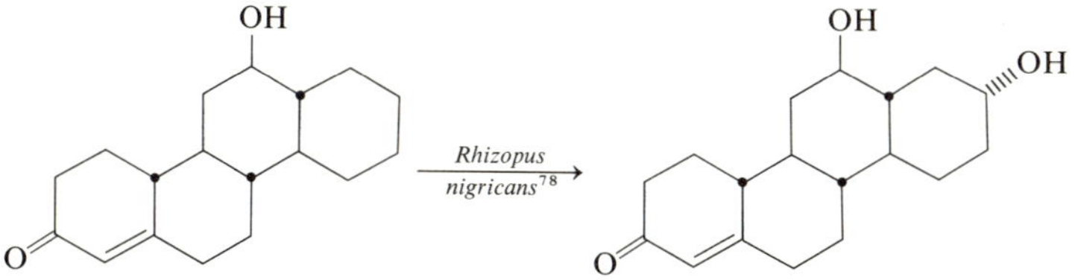

(**79**) (**110**)

The oxygenation of several tetracyclic diterpenes with *R. nigricans* has been studied. The results of the oxygenation of various 17-norkauranes are

[89] W. J. Wechter and H. C. Murray, *J. Org. Chem.* **28**, 755 (1963).

[90] Z. Procházka, M. Buděšinský, and P. Prekajski, *Collect. Czech. Chem. Commun.* **39**, 982 (1974).

summarized in Table II.[81–83] All substrates undergo hydroxylation at the 7-position (either α or β). Hydroxylation at the 1-position is also observed, and when neither the 3-position nor one of the geminal methyl groups at C-4 are functionalized in the substrate, hydroxylation also occurs at the 3-position. The selection of oxygenation sites in 17-norkauran-16-one and *ent*-17-norkauran-16-one has been rationalized in the same manner as just described for the oxygenation of the ketoandrostanes by *R. nigricans.*[83]

Several macrocyclic ketones have been oxygenated with *R. nigricans.*[80] The isolation of 5-hydroxycyclododecanone (8%), 6-hydroxycyclododecane (18%), and 7-hydroxycyclododecanone (8%) was typical of the results observed.

Diamantan-1-ol (**111**) and diamantan-4-ol (**112**), molecules that are difficult to functionalize at specific positions have been hydroxylated with *R. nigricans.*[91] The former, (**111**), is hydroxylated at the 7-position in 86% yield. The latter, (**112**), is hydroxylated at two different positions, giving diamantane-1,9-diol (**114**) in 58% yield, and diamantane-4,9-diol (**115**) in 11% yield. It is of interest to note that all three products are the result of hydroxylation at a tertiary carbon.

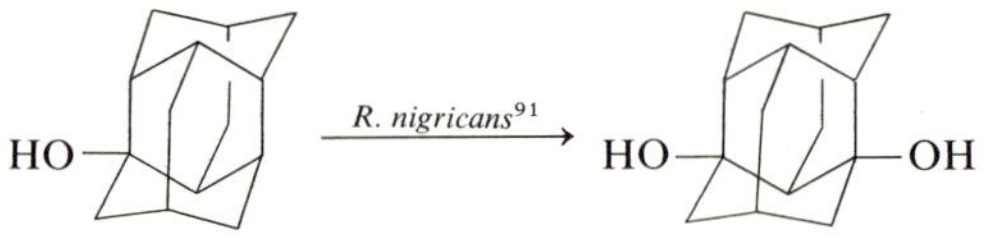

(**111**) (**113**)

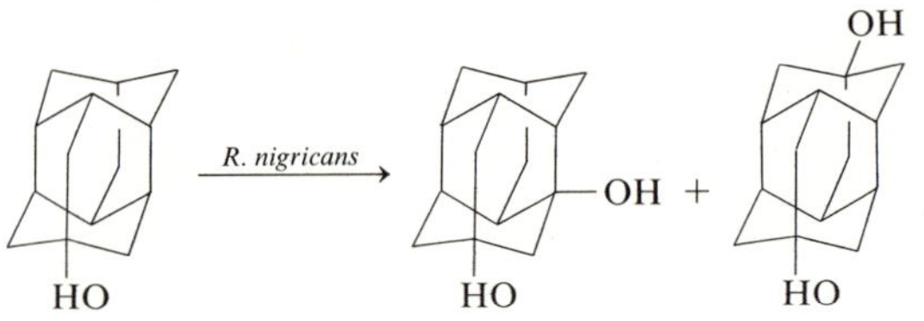

(**112**) (**114**) (**115**)

4. *Curvularia lunata* (ATCC 12017)

This fungus is from the class Fungi Imperfecti. This microorganism carries out the highly desirable 11β-hydroxylation of many steroids (11β being the the configuration of the hydroxyl in the steroid cortical hormones). Unfortunately, the yield of desired product is lower than that obtained with micro-

[91] F. Blaney, D. E. Johnston, M. A. McKervey, E. R. H. Jones, and J. Pragnell, *Chem. Commun.* p. 297 (1974).

organisms giving 11α-hydroxylation. Minor products are also formed by this organism as a result of hydroxylation at the 7α-, 9α-, or 14α-positions.

Examination of the indicated positions of hydroxylation reveals that all are axial positions. By contrast, *Calonectria decora* introduces hydroxyl groups into equatorial positions and *Rhizopus nigricans* introduces hydroxyl groups into an equatorial position to a much greater extent than it does to an axial position. This characteristic of *C. lunata* (it is not, however, limited to this microorganism) was recognized in early work with steroids and, in addition, a correlation between axial hydroxylation and epoxidation was observed. Bloom and Shull found that *C. lunata* can either hydroxylate steroids at the 11β- and 14α-positions or can epoxide steroids having a double bond between C-9 and C-11 or between C-14 and C-15.[92] The epoxides so formed also have an "axial" configuration. It has been suggested that the oxygenating enzyme does not distinguish between the π electrons of the double bonds and the electron pair of the axial C–H bonds.[93] Implied by this rationale is that the same enzyme is responsible for both hydroxylation and epoxidation.

A related aspect of hydroxylation with *C. lunata* is the reduction in the quantity of 7α-, 9α-, or 14α-hydroxy by-products formed when the substrate contains an α-substituent. Kieslich and co-workers have used several such substituents to reduce the amount of undesired α-hydroxylation. For example, addition of first a 16α-methyl group and then a 6α-fluoro group to 21-hydroxy-3,20-dioxo-4-pregnene (**116**) shifted the ratio of hydroxylation in favor of the 11β-position as shown in Table IV.[94] Total yields were not

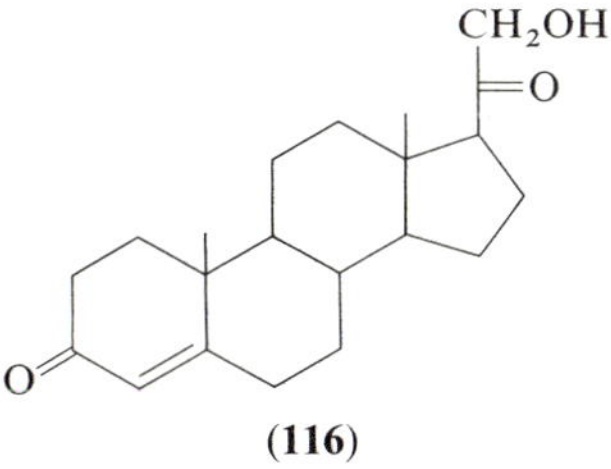

(**116**)

given. When the compound 6β-fluoro-5α-bromo-3β-hydroxy-21-acetoxy-16α-methyl-20-oxopregnane (**117**) was used as a substrate, only 11β-hydroxylation was observed (45% yield), giving **118** as the product.[94] In addition to hydroxylation, it is clear that other modifications have occurred. These

[92] B. M. Bloom and G. M. Shull, *J. Am. Chem. Soc.* **77**, 5756 (1955).

[93] Y. Kurosawa, M. Hayano, and B. M. Bloom, *Agric. Biol. Chem.* **25**, 838 (1961).

[94] K. Kieslich, K. Petzoldt, H. Kosmol, and W. Koch, *Justus Liebigs Ann. Chem.* **726**, 168 (1969).

TABLE IV

SHIFT IN THE RATIO OF 14α-, 7α-, AND 9α- TO 11β-HYDROXYLATION BY C. *lunata* DUE TO α-SUBSTITUENTS

Substrate	11β-	14α-	7α-	9α-
116	1	: 1	: 0.5	: 0
16α-methyl-**116**	4	: 1	: 1	: 0
6α-fluoro-16α-methyl-**116**	6	: 1	: 0	: 1

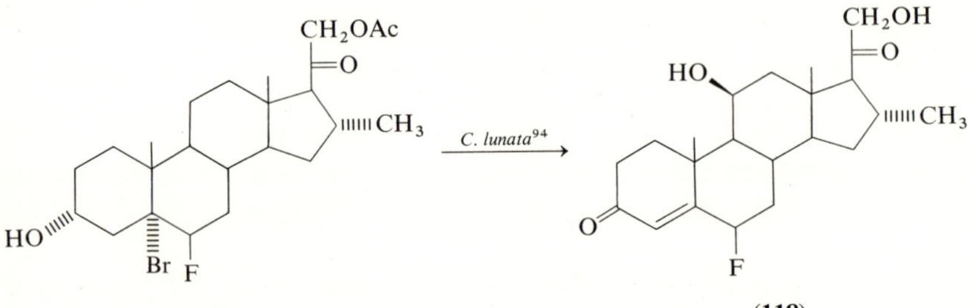

(**117**) (**118**)

may be attributed to the conditions of workup to which the fermentation product was exposed.

The influence of α-epoxides, particularly at the 5- and 6-positions, on hydroxylation with *C. lunata* has also been studied by Kieslich and co-workers.[95,96] The results are summarized in Table V.

The hydroxylation of a series of 19-norsteroids by *C. lunata* has been reported by Smith, Lin, and co-workers.[97] Particularly interesting in their work was the fact that the substrates used were synthetic steroids and were racemates. Two substrates were studied in detail: *dl*-19-nortestosterone (**119**) and the homolog *dl*-17β-hydroxy-13β-ethylgon-4-en-3-one (**120**).[97] A third

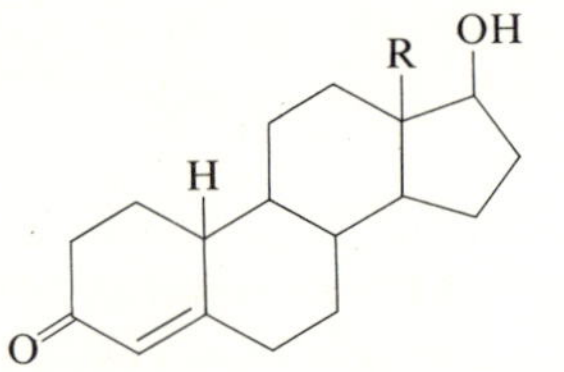

(**119**) R = —CH_3
(**120**) R = —CH_2CH_3

[95] K. Kieslich and H. Wieglepp, *Chem. Ber*. **104**, 205 (1971).
[96] G. Cleve, G. A. Hoyer, K. Kieslich, and H. Wieglepp, *Chem. Ber*. **105**, 658 (1972).
[97] Y. Y. Lin, M. Shibahara, and L. L. Smith, *J. Org. Chem*. **34**, 3530 (1969).

TABLE V

11β-HYDROXYLATION OF 5α,6α-EPOXYSTEROIDS WITH *C. lunata*[a]

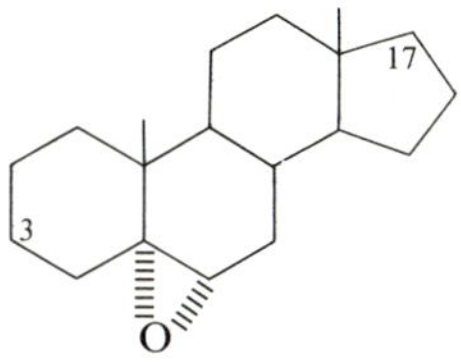

Androstane Series

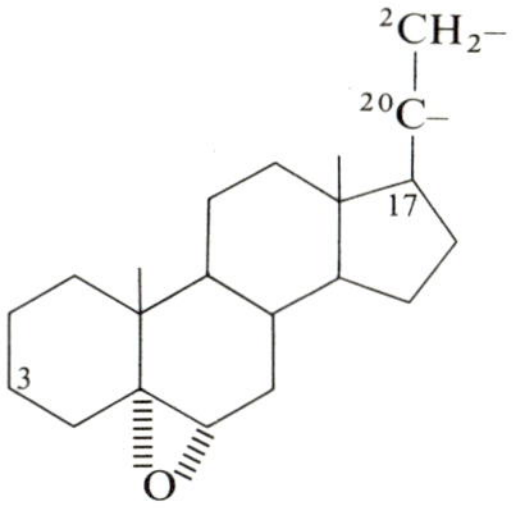

Pregnane Series

Position of substrate substituents	Yield (%) of 11β-hydroxylated product
Androstane series	
3β-OH; 17 =O	24
3β,17β-diOH	26
3β,17α-diOH	43
3β,17β-diOH; 17α-CH_3	21
3β,17β-diOH; 17α-$COCH_3$	31
3β-OH; 16α,17α-epoxide; 20 =O	45[b]
Pregnane series	
3β-OH; 20 =O	24
3β,20β-diOH	52
3β-OH; 20 =O; 21-$OCOCH_3$	55
3β,17β-diOH; 20 =O	20
3β,17,20α-triOH	33
3β,17-diOH; 20 =O; 21-$OCOCH_3$	28
17,20α-epoxide; 3β,21-diOH	28
3β-OH; 16α-CH_3; 20 =O; 21-$OCOCH_3$	55
3β,21-di-$OCOCH_3$; 16α-CH_3; 20 =O	36

[a] Except where noted, all results are from Kieslich and Wieglepp.[95]
[b] Clevc *et al.*[96]

substrate, *dl*-17β-hydroxyestra-4,8(14)-dien-3-one,[98] is also included in a summary of their work shown in Table VI. The results listed there show that the *d*-antipodes (the natural configuration of steroids) are hydroxylated primarily at the 6-, 10-, 11-, or 14-positions, giving axial hydroxyl groups. The *l*-antipodes are similarly hydroxylated at the 6-, 10-, 11-, or 12-positions,

[98] Y. Y. Lin and L. L. Smith, *Biochim. Biophys. Acta* **210**, 319 (1970).

TABLE VI

HYDROXYLATION OF 19-NORSTEROIDS BY *Curvularia lunata*

Products from hydroxylation of:			
d-19-Nor-testosterone	*dl*-19-Nor-testosterone	*dl*-17*β*-Hydroxy-13*β*-ethylgon-4-en-3-one	*dl*-17*β*-Hydroxy-estra-4-8(14)-dien-3-one
d-6*β*-OH (1%)	*dl*-6*β*-OH (4%)	*dl*-6*β*-OH (0.8%)	
d-10*β*-OH (52%)	*d*-10*β*-OH (19%)	*dl*-10*β*-OH (16%)	*dl*-10*β*-OH (25%)
d-11*β*-OH (2%)	*dl*-11*β*-OH (6%)		*dl*-11*β*-OH (15%)
	l-12α-OH (1%)	*l*-12α-OH (14%)	
d-14α-OH (9%)	*d*-14α-OH (0.5%)	*d*-14α-OH (4%)	
			dl-7α-OH (5%)

again giving axial hydroxyl groups. The similarity of the oxidative attack on the two antipodes can be seen from molecular models. Representations of such models are shown in Fig. 9.[99] The positions (11*β*-, 7α-, 9α-, 14α-) at which steroids of the progesterone class (natural configuration) are hydroxylated can be similarly illustrated.

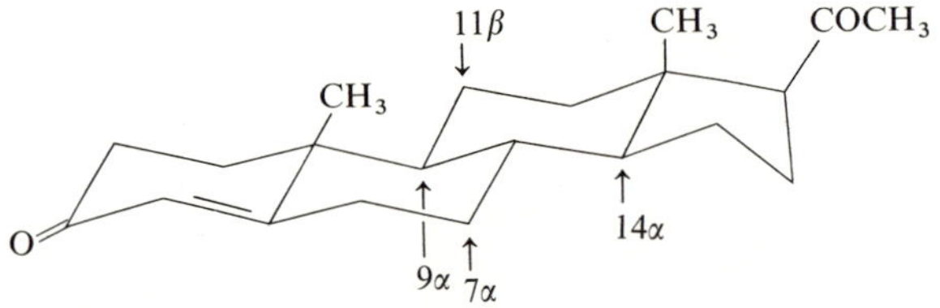

From these illustrations it is clear that both faces of the steroid are subject to hydroxylation in the middle portions of the molecules. The fact that all the hydroxyl groups are axial, and therefore perpendicular to the plane of the steroid nucleus, suggests that the enzymatic oxygenation site is situated above and/or below the steroid during hydroxylation.

The question of whether the observed hydroxylations are the result of one or more hydroxylase enzymes has been considered. Smith and Lin have carried out a kinetic analysis of their hydroxylations with an enzyme preparation from *C. lunata*.[100] They also have examined induction of the hydroxylating enzyme(s) by various 19-nortestosterone analogs.[101] From the results of both experimental approaches, they have suggested that a

[99] This representation was first portrayed by Y. Y. Lin, personal communication to G. S. Fonken, September 28, 1973.

[100] Y. Y. Lin and L. L. Smith, *Biochim. Biophys. Acta* **218**, 515 (1970).

[101] Y. Y. Lin and L. L. Smith, *Biochim. Biophys. Acta* **218**, 526 (1970).

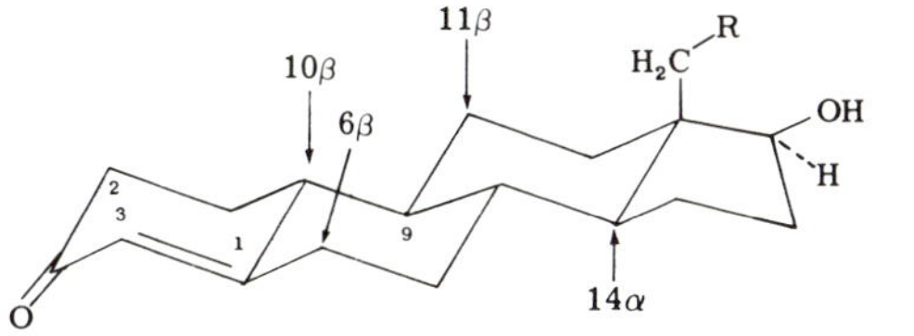

d-Antipode

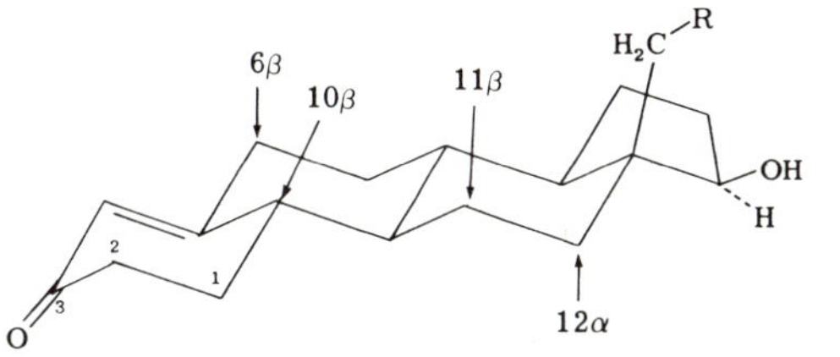

l-Antipode

FIG. 9. Positions at which *dl*-19-norsteroids are hydroxylated by *Curvularia lunata*. Substrates are **119** (R = H) and **120** (R = CH_3).

single hydroxylase is responsible for the introduction of all hydroxyl groups on both faces of the steroid substrates.

It is interesting to note that in the results with *C. lunata* shown in Table VI, homologation of the 18-methyl group to the 18-ethyl group completely blocks 11β-hydroxylation.[97] The same observation has been made for the hydroxylation of 18-methyltestosterone with *C. lunata*.[102] Small amounts of products resulting from monohydroxylation at the 14α-position were isolated from the latter case.

C. OTHER OXYGENATIONS OF SYNTHETIC INTEREST

1. CYCLIC SUBSTRATES

In this section a number of oxygenations of cyclic compounds from the recent literature which are not discussed in the preceding sections are summarized. The discussion is arranged approximately according to increasing complexity of ring systems.

Preparative oxygenations of simple monocyclic compounds such as cyclohexane are seldom encountered. This may be because such compounds are quite volatile and therefore are easily lost from the fermentor or because

[102] K. Kieslich, H. D. Berndt, R. Wiechert, U. Kerb, G. Schulz, and H. J. Koch, *Justus Liebigs Ann. Chem.* **726**, 161 (1969).

the expected products are already available from conventional chemical sources. On the other hand, many microbiologists prefer to grow microorganisms using a single, pure hydrocarbon as the only carbon source. In these endeavors, they are frequently successful with the result that the substrate must be oxygenated but is then further metabolized as a source of energy for the microorganism. As a consequence, little useful chemical product is found. These procedures represent oxidations of potential value. Several recent examples include the growth of an organism on cyclohexane,[103] methylcyclohexane,[104] and 1-menthene.[105]

Macrocycles have been successfully oxygenated by various microorganisms. The oxygenation of cyclododecane and cyclododecanol by *Sporotrichum sulfurescens* and of cyclododecanone by *Calonectria decora* and *Rhizopus nigricans* has already been mentioned in the respective sections on these microorganisms. An additional example of potential synthetic interest is the conversion of cyclododecanone to 6-hydroxycyclododecanone in 34% yield (50% based on recovery of starting material) by *Ophiobolus herpotrichus.*[80]

Although the subject has been reviewed extensively,[106,107] the hydroxylation and further dissimilation of camphor (**121**) by several *Pseudomonas putida* strains requires mention here. Extended investigations of this oxidation by Gunsalus and co-workers at the University of Illinois have led to the isolation and purification of the enzyme system responsible for the initial hydroxylation giving **122**. The key component in this system that is respon-

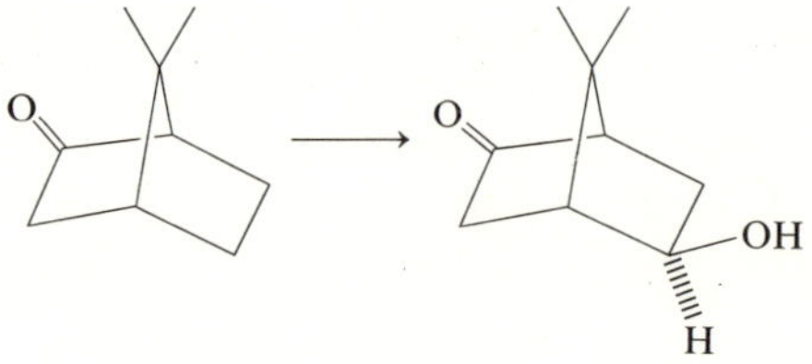

(**121**) (**122**)

sible for the activation of molecular oxygen is the protein, cytochrome P-450$_{cam}$. This protein has been isolated in crystalline form.[108] Elucidation

[103] H. deKlerk and A. C. van der Linden, *Antonie van Leeuwenhoek* **40**, 7 (1974).

[104] G. M. Tonge and I. J. Higgins, *J. Gen. Microbiol.* **81**, 521 (1974).

[105] B. B. Mukherjee, G. Kraidman, and I. D. Hill, *Appl. Microbiol.* **27**, 1070 (1974).

[106] I. C. Gunsalus, J. R. Meeks, J. D. Lipscomb, and P. Debrunner, *in* "Molecular Mechanisms of Oxygen Activation" (O. Hayaishi, ed.), p. 559. Academic Press, New York, 1974.

[107] I. C. Gunsalus, J. D. Lipscomb, V. Marshall, H. Trauenfelder, E. Greenbaum, and E. Munck, *in* "Biological Hydroxylation Mechanisms" (G. S. Boyd and R. M. S. Smellie, eds.), p. 135. Academic Press, New York, 1972.

[108] C.-A. Yu and I. C. Gunsalus, *Biochem. Biophys. Res. Commun.* **40**, 1431 (1970).

of the molecular details of this protein holds much promise for the further understanding of the intimate chemical details of the hydroxylation reaction. Transformations of camphor and related monoterpenes that are similar to the preceding have been reported for a strain of *P. pseudomallei*.[109]

The bicyclic substrate 3-benzoyl-3-azabicyclo[3.3.1]nonane (**28**) has been hydroxylated with *Rhizopus arrhizus* giving two monohydroxy products.[110] One of these, **29**, is identical to the only product isolated when this substrate is hydroxylated with *Sporotrichum sulfurescens*. The second product (**123**) results from hydroxylation at the tertiary bridgehead carbon and is optically active. The absolute configuration (1*R*,5*R*) is as shown in the drawing below.

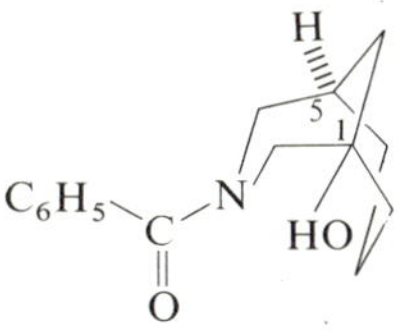

(**123**)

The sesquiterpenoids, guaioxide (**124**) and liguloxide (**125**), have been the substrates for a series of oxygenations reported by Ishii and co-workers. Two organisms, *Mucor parasiticus*[111–113] and *Streptomyces purpurescens*,[114] have been used in this work and each gave numerous products. None of the products was produced in synthetically useful quantities but were valuable to the structure determination of guaioxide itself and of several related

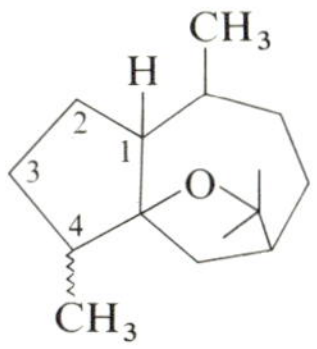

(**124**) 4β-CH_3
(**125**) 4α-CH_3

sesquiterpenes. The hydroxylation of several diterpenes with *C. decora*, *R. nigricans*, and *A. ochraceus* are summarized in Table II. To those results we can add the hydroxylation of 7α- (**126**) and 7β-hydroxykaurenolide (**127**)

[109] T. Hayashi, T. Kakimoto, H. Ueda, and C. Tatsumi, *Nippon Nogei Kagaku Kaishi* **44**, 401 (1970).

[110] R. A. Johnson, H. C. Murray, and L. M. Reineke, *J. Org. Chem.* **34**, 3834 (1969).

[111] H. Ishii, T. Tozyo, M. Nakamura, and H. Minato, *Tetrahedron* **26**, 2751 (1970).

[112] E. Funke, T. Tozyo, H. Ishii, and K. Takeda, *J. Chem. Soc. C* p. 2548 (1970).

[113] H. Ishii, T. Tozyo, and M. Nakamura, *Chem. Pharm. Bull.* **19**, 842 (1971).

[114] H. Ishii, T. Tozyo, and M. Nakamura, *Tetrahedron* **27**, 4263 (1971).

by *R. arrhizus.*[115] Two principal products were obtained from each substrate and were the result of hydroxylation at either the 11α-positions or the 13α-positions.

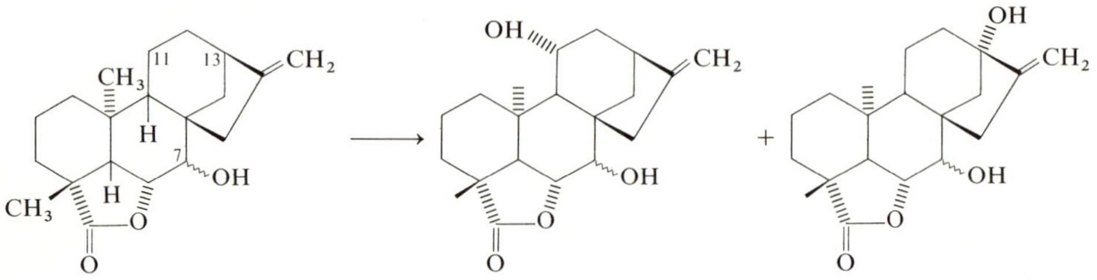

(**126**) 7α-OH
(**127**) 7β-OH

(**128**) 7α-OH (10%)
(**129**) 7β-OH (8%)

(**130**) 7α-OH (20%)
(**131**) 7β-OH (11%)

Various mutants and strains of *Gibberella fujikuroi* continue to be used for oxidative transformation of diterpenes. Steviol acetate (**132**), for example, is converted to the gibberellin GA_{20}-acetate (**133**) in 30% yield by a mutant of *G. fujikuroi.*[116] A strain of *G. fujikuroi* converts gibberellin A_{12}-aldehyde to four different products by oxidative reactions.[117]

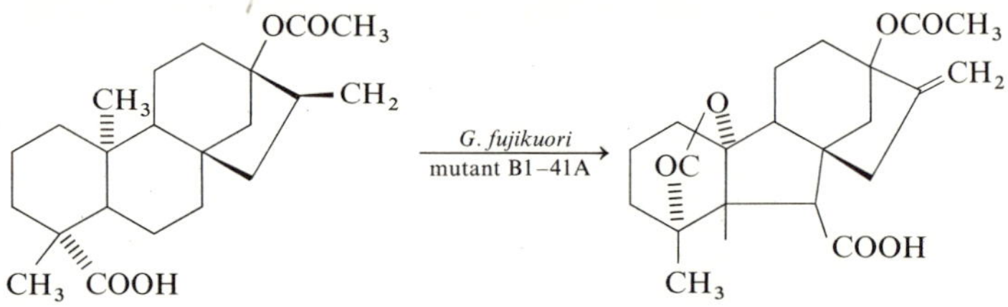

(**132**) (**133**)

The genus *Aspergillus* has found much use in the oxygenation of steroids as well as many aromatic compounds. Both *Aspergillus niger* and *A. ochraceus* commonly hydroxylate steroids at the 11α-position and, to a lesser extent, at the 6β-position. Recently, *A. niger* has been used to oxygenate the structurally similar, but sterochemically different, diterpenes: 17-norkauran-16-one (**134**) and 16-norphyllocladan-16-one (**135**).[118] Both are hydroxylated at the 3-position to give the equatorial hydroxyl groups as shown for **136** and **137**. Yields are not specified and in addition to **137**, the 3-ketone is obtained as a second product from oxygenation of **135**.

[115] J. R. Hanson, G. Savona, and M. Siverns, *J. Chem. Soc., Perkin Trans. 1* p. 2001 (1974).
[116] J. R. Bearder, J. MacMillan, C. M. Wells, and B. O. Phinney, *Chem. Commun.* p. 778 (1973).
[117] J. R. Bearder, J. MacMillan, and B. O. Phinney, *Phytochemistry* **12**, 2173 (1973).
[118] A. B. Anderson, R. McCrindle, and J. K. Turnbull, *Chem. Commun.* p. 143 (1973); for full details, see *Can. J. Chem.* **53**, 1181 (1975).

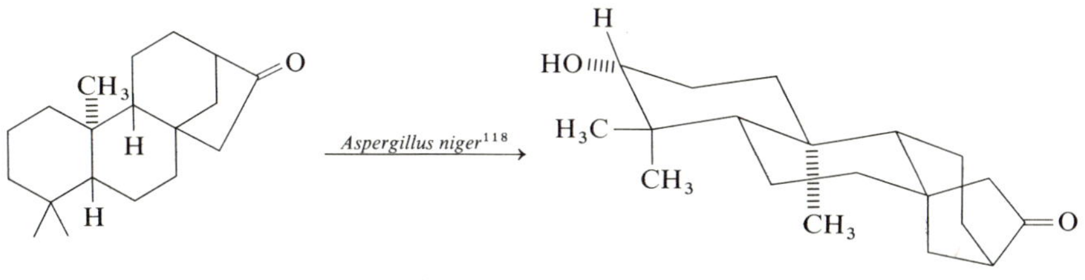

(134) **(136)**

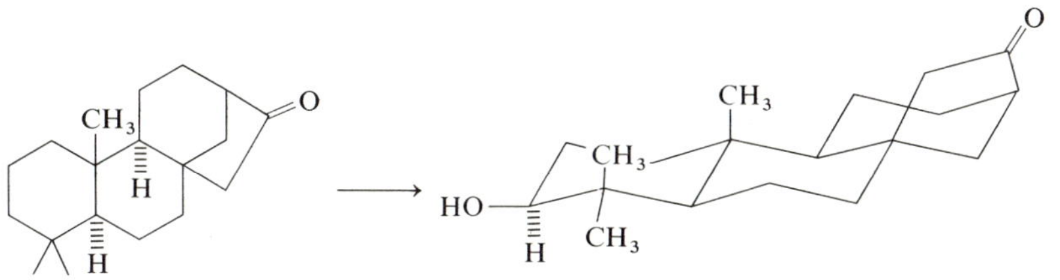

(135) **(137)**

The unusual hydroxylation of an angular methyl group in a steroid has been reported for *A. niger*. The 18-position of 3,17-dioxo-4-androstene (**138**) is hydroxylated in 45% yield, giving **139**.[119]

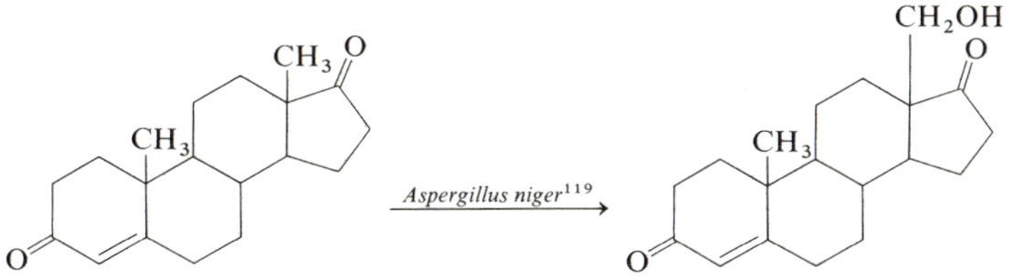

(138) **(139)**

An extension study of the hydroxylation of mono- and difunctional androstanes by *A. ochraceus* has been made much in the manner described in the sections on *Calonectria decora* and *Rhizopus nigricans*.[120,121] The results are summarized in Table VII. As can be seen from these results and as expressed by the authors, *A. ochraceus* "shows a monotonous predilection for 11α-hydroxylation" of the 5α-androstanes.[120]

In the hydroxylation of progresterone by *A. ochraceus*, 11α-hydroxylation may be followed by hydroxylation of the 6β-position. It has been shown

[119] B. J. Auret and H. L. Holland, *Chem. Commun.* p. 1157 (1971).

[120] A. M. Bell, J. W. Browne, W. A. Denny, E. R. H. Jones, A. Kasal, and G. D. Meakins, *J. Chem. Soc., Perkin Trans. 1* p. 2930 (1972).

[121] A. M. Bell, V. E. M. Chambers, E. R. H. Jones, G. D. Meakins, W. E. Müller, and J. Pragnell, *J. Chem. Soc., Perkin Trans. 1* p. 312 (1974).

TABLE VII

HYDROXYLATION OF 5α-ANDROSTANES WITH ASPERGILLUS OCHRACEUS[a]

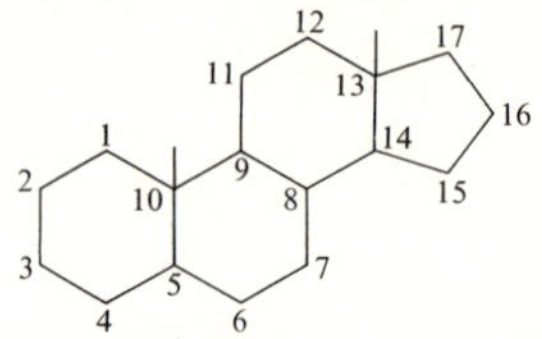

Position of substrate functional groups[b]	Substrate recovered (%)	Position of product OH groups	Yield (%)[c]
3-CO	56	6β,11α-diOH	51
3-CO,Δ^4	0	6β,11α-diOH	75
3-CO,Δ^1	67	6β,11α-diOH	39
17-CO	65	7β,11α-diOH	27
		11α-OH	5
2,16-diCO	4	11α-OH	61
2,17-diCO	9	11α-OH	57
2-CO,17β-OH	10	11α-OH	32
3,6-diCO	7	11α-OH	60
		(3β),11α-diOH	25
3-CO,6β-OH	0	11α-OH	76
3-CO,Δ^4,6β-OH	3	11α-OH	61
3,7-diCO	5	11α-OH	74
		(3β),11α-diOH	20
3β,7β-diOH	16	11α-OH	78[d]
3,11-diCO	85	none	
3-CO,11α-OH	79	6β-OH	34
3-CO,Δ^4,11α-OH	37	6β-OH	76
3,16-diCO	0	11α-OH	55
		(3β),11α-diOH	14
3-CO,16β-OH	0	11α-OH	41
3β-OH,16-CO	0	11α-OH	71
3,17-di-CO	0	11α-OH	75[d]
3-CO,17β-OH	0	11α-OH	79
3-CO,Δ^4,17β-OH	0	11α-OH	73
3-OH,17-CO	0	11α-OH	81
3β-OH,17-CO	2	11α-OH	89[d]
3β-OAc,17-CO	3	(3β),11α-diOH	61
3β-OH,Δ^5,17-CO	48	11α-OH	87
3β-OMe,Δ^5,17-CO	27	11α-OH	41

[a] Except where noted, results are from Bell *et al.*[120]

[b] CO refers to a ketone functional group.

[c] Yields are based on corrections for recovered starting material.

[d] Bell *et al.*[121]

using cell-free preparations[122,123] that the enzyme carrying out the 11α-hydroxylation is induced by progesterone but that this does not give 6β-hydroxylation.[122] The latter results only when the hydroxylating enzymes are induced with 11α- hydroxyprogesterone. The two hydroxylations therefore occur sequentially and require two different enzymes. It has been suggested that this sequence also occurs in the hydroxylations of the androstanes (Table VII).[120]

Hydroxylation of several 5α-pregnanes with *A. ochraceus* has led to some interesting results. [124] Most interesting is the hydroxylation of 3β-hydroxy-20-oxopregnane (**140**) which gave 1β-,3β-,11α-trihydroxy-20-oxopregnane (**141**) in 60% yield. The steric relationship of the 1β- and 11-hydroxyl groups

(**140**) (**141**)

is emphasized by the fact that the acetonide derived therefrom is remarkably resistant to hydrolysis. It is suggested that these two groups are introduced onto the substrate by the same enzyme.[124] By simply changing the 3-hydroxyl to a ketone or introducing a C-5–C-6 double bond, the amount of hydroxylation at the 1-position is greatly reduced (< 15%).[124]

Two other fungi have been used to hydroxylate a variety of 5α-androstanes.[125] These are *Wojnowicia graminis* and *Ophiobolus herpotrichus.* The results are summarized in Tables VIII and IX, respectively. Minor products and low yielding reactions have been omitted from these tables. The point has been made that both organisms are useful for hydroxylation of the terminal ring of the steroid nucleus provided oxygen functions are present on the other terminal ring and on one of the middle rings.[12] Hydroxylations of dioxygenated 5α-androstanes with *Absidia regnieri* or with *Syncephelastrum racemosum* generally produce complex mixtures of products.[126]

[122] M. Shibahara, J. A. Moody, and L. L. Smith, *Biochim. Biophys. Acta* **202**, 172 (1970).
[123] L. Tan and P. Falardeau, *J. Steroid Biochem.* **1**, 221 (1970).
[124] A. S. Clegg, W. A. Denny, E. R. H. Jones, G. D. Meakins, and J. T. Pinhey, *J. Chem. Soc., Perkin Trans. 1* p. 2137 (1973).
[125] V. E. M. Chambers, E. R. H. Jones, G. D. Meakins, J. O. Miners, and A. L. Wilkins, *J. Chem. Soc., Perkin Trans. 1* p. 55 (1975).
[126] A. M. Bell, E. R. H. Jones, G. D. Meakins, J. O. Miners, and A. L. Wilkins, *J. Chem. Soc., Perkin Trans. 1* p. 2040 (1975).

TABLE VIII

HYDROXYLATION OF 5α-ANDROSTANES WITH *Wojnowicia graminis*

Position of substrate functional groups	Substrate recovered (%)	Product positions hydroxylated	Yield (%)[a]
2,7-diCO	15	2β,17β-diOH	19
		2β,18-diOH	10
3,7-diCO	32	(3β)[b],17β-diOH	41
		(3β),17β-diOH	27
3α-OH,7-CO	6	17β-OH	55
3β-OH,7-CO	4	17β-OH	63
3,11-diCO	24	17β-OH	28
		(3β),17β-diOH	13
3β-OH,11-CO	48	16β-OH	48
		17β-OH	22
7,17-diCO	13	2α-OH	34
		(17β,2α)-diOH	30
7-CO,17β-OH	14	2α-OH	58
		(7α),2α-diOH	15
		3β-OH	11
11-CO,17β-OH	39	3β-OH	35
		3α-OH	11

[a] Yields are based on correction for recovered starting material.

[b] Hydroxyl positions enclosed by parentheses indicate that a ketone present on the substrate has been reduced enzymatically.

TABLE IX

HYDROXYLATION OF 5α-ANDROSTANES WITH *Ophiobolus herpotrichus*

Position of substrate functional groups	Substrate recovered (%)	Product position hydroxylated[a]	Yield (%)[b]
3,7-diCO	12	(3β),17β-diOH	62
3,11-diCO	46	16-CO[c]	38
		(3β)-OH,16-CO	21
		16β-OH	11
		17β-OH	11
7,17-diCO	38	(17β),3β-diOH	43
		3β-OH	32
11,17-diCO	63	3α-OH	45
		(17β),3α-diOH	44
11-CO-17β-OH	72	3α-OH	95

[a] Hydroxyl positions enclosed by parentheses indicate that a ketone present on the substrate has been reduced enzymatically.

[b] Yields are based on correction for recovered starting material.

[c] A 16-hydroxylation has been followed by enzymatic oxidation to the ketone.

Hydroxylation at C-1 of steroids with *A. ochraceus* has already been described.[124] Several other organisms can hydroxylate selected steroids at the 1-position. *Botrodiplodia malorum* hydroxylates certain 19-nor steroids at the 1β-position, but this reaction is very sensitive to structural modifications as illustrated by the following comparisons.[127–129]

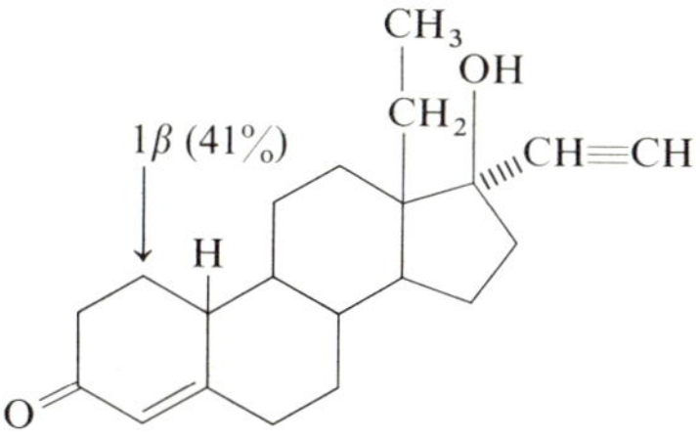

vs.[129]

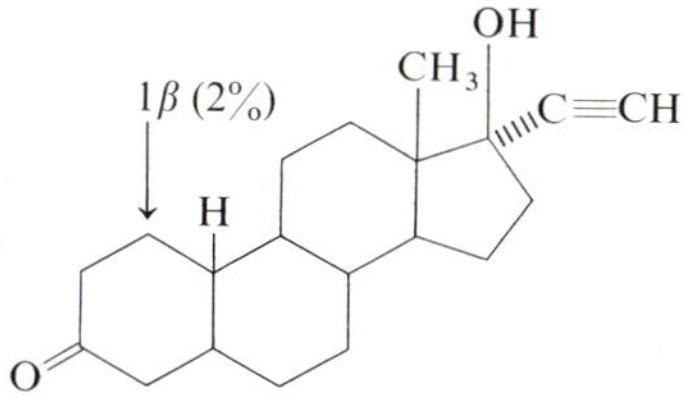

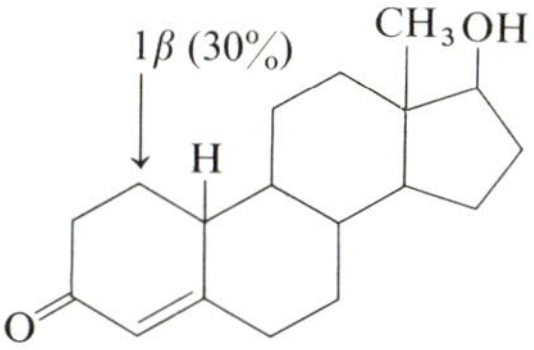

vs.[127]

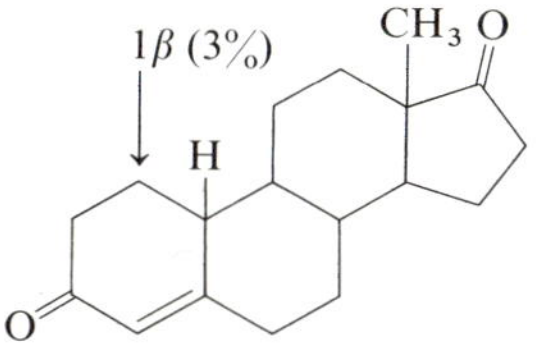

Another organism that oxygenates various pregnanes at the 1β-position in modest (9–16%) yields is *Coniosporium rhizophilum*.[130] *Wojnowicia graminis* hydroxylates 3,7-difunctional androstanes at the 1α-position in similar yields but several other positions may also be hydroxylated to give a mixture of products.[125]

The fungus *Diaporthe celastrina* has been found to hydroxylate certain disubstituted androstanes at single positions in good yield (see Table X).[131] Of additional interest is the fact that all the hydroxyl groups (with the exception of minor amounts of 16β-OH products) have been introduced from the α face of the steroid nucleus and have the axial configuration.

The influence of various methyl substituents on the hydroxylation of 21-hydroxy-3,20-dioxo-4-pregnene (**142**) by *Mucor griseocyanus* (+) has been

[127] H. J. Brodie, C. E. Hay, and J. D. Townsley, *Biochim. Biophys. Acta* **239**, 103 (1971).
[128] I. Kim, C. E. Hay, and H. J. Brodie, *J. Biol. Chem.* **248**, 2134 (1973).
[129] G. Greenspan, R. W. A. Rees, G. D. Link, C. P. Boyd, R. C. Jones, and H. E. Alburn, *Experientia* **30**, 328 (1974).
[130] V. Schwarz and J. Protiva, *Collect. Czech. Chem. Commun.* **37**, 1577 (1972).
[131] A. M. Bell, A. D. Boul, E. R. H. Jones, G. D. Meakins, J. O. Miners, and A. L. Wilkins, *J. Chem. Soc., Perkin Trans. 1* p. 1364 (1975).

TABLE X

HYDROXYLATION OF 5α-ANDROSTANES WITH *Diaporthe celastrina*

Position and type of substrate functional group	Substrate recovered (%)	Product position hydroxylated[a]	Yield (%)[b,c]
3,7-diCO	7	(3β),16α-diOH	48
		(3β),16β-diOH	11
3-CO,9α-OH	3	(3β),16α-diOH	47
		16α-OH	15
3,11-diCO	0	(3β),16α-diOH	55
		16α-OH	10
3-CO,12α-OH	0	(3β),16α-diOH	48
		16β-OH	
3β-OH,17-CO	0	9α-OH	40
		(3-CO),9α-OH	15
		7α-OH	12
7,17-diCO	0	3α-OH	52

[a] Positions enclosed by parentheses indicate modification of a functional group present in the substrate.

[b] Yields are based on correction for recovery of starting material.

[c] Yields grouped together indicate total yield of a specific oxygenation. Products differ due to oxidation or reduction of a functional group present in the substrate.

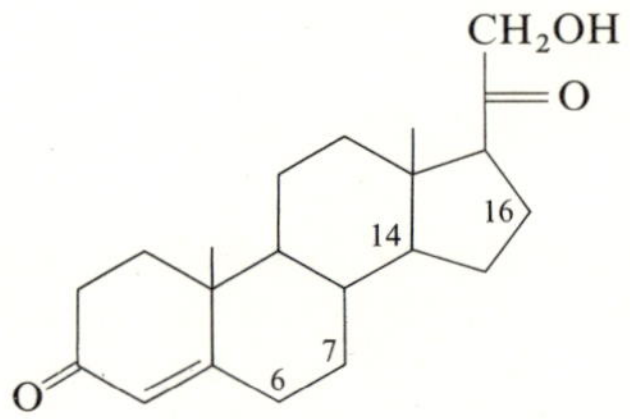

(142)

examined.[132] With a 16β-methyl substituent hydroxylation is limited to the 7α-position (13%) while a 16α- methyl substituent results in hydroxylation at the 7α- and 12β-positions. A 6α-methyl group causes direction of hydroxylation to the 14α- and 6β-positions. When no methyl groups are present, hydroxylation occurs first at the 14α-position and then at the 7α-position.

Finally, the fungus (incompletely classified as yet) that is cultivated by the ant species *Acromyrmex octospinosus*, may be added to the list of unusual sources of microorganisms. This fungus, referred to as "*Acromyrmex* fungus,"

[132] V. Valcavi, B. Corsi, and S. Tedeschi, *Farmaco, Ed. Sci.* **28**, 873 (1973).

oxygenates various androstanes and pregnanes at a variety of positions.[133] It also carries out Baeyer–Villiger type oxidation of A-ring ketones and generally produces mixtures of products. The reader is referred to the original reference for the details.

2. Acyclic Substrates

The oxygenation of acyclic hydrocarbons requires consideration of a new problem. Oxygenation of cyclic substrates usually stops at the hydroxyl level, or at most, may be followed by oxidation to a ketone. Only rarely does further oxidation occur, usually via a microbiological Baeyer–Villiger reaction.[8] In the case of acyclic hydrocarbons, however, the rate of initial hydroxylation is often exceeded by subsequent oxidation steps. Thus, one must consider the problem of how to stop the oxidation at an early stage if one wishes to retain the carbon skeleton of the substrate.

The two main pathways of oxidative attack on alkanes are either at the terminal carbon (α-carbon) or at the subterminal carbon (β-carbon). In recent years, attack at other interior carbons of an alkane chain have been observed but these are not yet recognized as major modes of oxidation.

Terminal hydroxylation may be followed by dehydrogenation reactions, converting the alcohol to an acid via the aldehyde. The carboxylic acid may then be degraded by β-oxidation, a process that is an important source of energy to the microorganism. Subterminal hydroxylation may be followed first by dehydrogenation to the ketone and then by a Baeyer–Villiger type oxidation to an acetate ester. Hydrolysis of the latter provides a primary alcohol that is two carbons shorter than the original hydrocarbon and that now is subject to the same route of metabolism as is the terminal hydroxylation product.

The problem of over-oxidation is most serious with simple *n*-alkanes. This is fortunate from a synthetic viewpoint however, since simple hydroxyalkanes generally are available. Many of the examples of alkane oxygenation given below are discussed with an emphasis on the experimental or structural feature that has prevented over-oxidation.

A method of hydrocarbon oxygenation termed cooxidation, has developed from the practice of growing microorganisms on a single hydrocarbon as the sole carbon source.[21,25] This method has developed as a means to oxygenate hydrocarbons that alone will not support the growth of a microorganism. A second hydrocarbon, upon which the microorganism can grow, is used as a growth substrate in the fermentation. This then results in a limited oxygenation of the first substrate. As the definition of a growth

[133] E. R. H. Jones, G. D. Meakins, J. O. Miners, J. H. Pragnell, and A. L. Wilkins, *J. Chem. Soc., Perkin Trans. 1* p. 1552 (1975).

substrate has expanded, it has become apparent that, in principle, the method of cooxidation is essentially a fermentation using a very simple medium. The concept, nevertheless, has been most useful in attempts to convert the simple aliphatic and aromatic petroleum hydrocarbons into useful chemical intermediates by microbial oxidation.[21]

A second principle by which over-oxidation of acyclic hydrocarbons may be limited is to use only those enzymes capable of an initial oxygenation reaction. One experimental approach to this is the preparation of mutants of the microorganism in which further oxidation steps have been blocked. Such an approach has been described for the organism *Candida cloacae*, which is able to convert *n*-alkanes of five to sixteen carbon atoms to dicarboxylic acids very efficiently.[134] A mutant, M-1, of this organism was found which could not further metabolize the dicarboxylic acids. Use of this mutant to oxidize the *n*-alkanes results in accumulation of the diacids. For example, 21.7 gm/liter of hexadecane-1,16-dicarboxylic acid was produced after 72 hours in a medium containing 10% (v/v) hexadecane.[134]

Fungi of the genus *Torulopsis* have developed a mechanism whereby they themselves protect the initial products of alkane oxygenation from further degradation.[135,136] Hexadecane, for example, is first converted by oxidation to 15-hydroxyhexadecanoic acid. This hydroxy acid is then incorporated into a glycolipid which, after the fermentation, is found as an extracellular oil. Chemical methanolysis of this oil releases the hydroxy acid as the methyl ester.

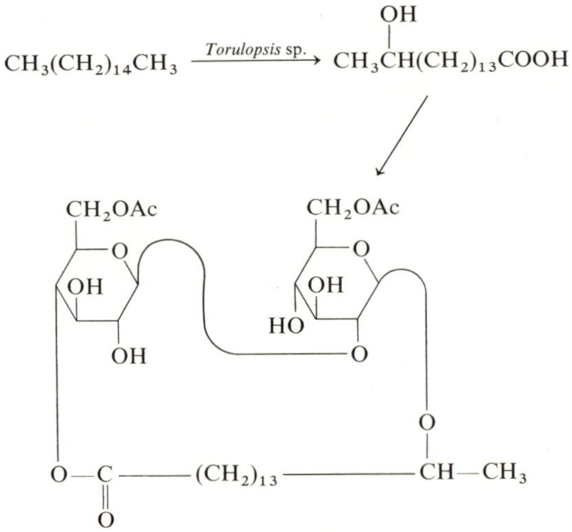

[134] R. Uchio and I. Shiio, *Agric. Biol. Chem.* **36**, 426 (1972).
[135] A. P. Tulloch, J. F. T. Spencer, and P. A. J. Gorin, *Can. J. Chem.* **40**, 1326 (1963).
[136] A. P. Tulloch, A. Hill, and J. F. T. Spencer, *Chem. Commun.* p. 584 (1967).

Alkanes longer than sixteen carbons are shortened by β-oxidation before being incorporated into the glycolipid.[135] Hydrocarbons,[135,137] fatty acids,[135,137] fatty alcohols,[137] alkyl halides,[138] alkoxyalkanes,[138] alkyl amides,[139] nitroalkanes,[139] and even cyanoalkanes[138,139] have all been converted into the glycolipid. The results of the extensive work done with these fungi has been reviewed previously.[8]

1,4-Diethylbenzene has been used as a substrate with a *Pseudomonas* sp. isolated from a fuel storage tank.[140] 4-Ethylphenylacetic acid is produced in this fermentation.

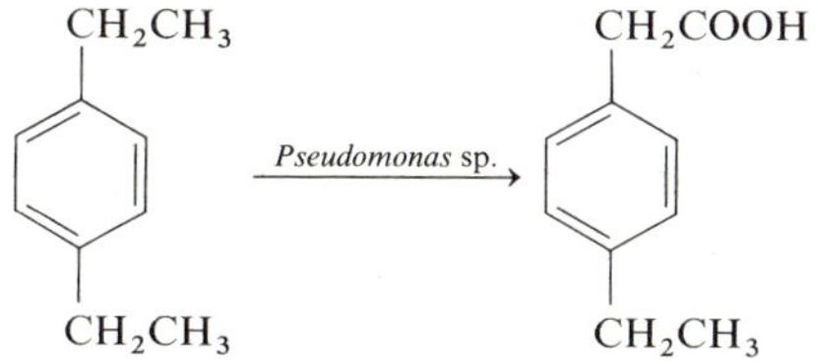

Deuterium labeling of one methyl group exhibited an isotope effect in this oxidation of $k_H/k_D = 1.7$. When deuterium was placed on a methylene carbon, it was retained in the product, indicating direct oxygenation of the methyl group rather than dehydrogenation to form an olefin in the initial step of the oxidation.

Pseudomonas aeruginosa is a bacterium that has frequently been used to oxygenate alkanes as well as other substrates.[141] A cell-free extract of *P. aeruginosa* (strain 473) has been used by van Ravenswaay Claasen and van der Linden to study the substrate specificity of the hydroxylase present in this organism.[142,143] Three classes of substrates were used: alkanes, alkylbenzenes, and alkylcycloalkanes. A large number of compounds were oxygenated by the enzyme preparation and the results are summarized in Table XI. In addition to these results, several compounds of a similar nature but containing a polar group were found not to be substrates. 4-Hydroxyisopropylbenzene and cyclohexylmethanol, for example, were not hydroxylated. From these results, the following conclusions were drawn.[142] The site of hydroxylase activity is a hydrophobic cavity in the enzyme surface. The width of this cavity is about 5 Å as shown by the selectivity between hydroxylation of isopropyl and isopentylbenzenes and the failure to hydroxylate

[137] D. F. Jones, *J. Chem. Soc. C* p. 2809 (1968).
[138] D. F. Jones and R. Howe, *J. Chem. Soc. C* p. 2816 (1968).
[139] D. F. Jones and R. Howe, *J. Chem. Soc. C* p. 2821 (1968).
[140] M. Tanabe, R. L. Dehn, and M. H. Kuo, *Biochemistry* **10**, 1087 (1971).
[141] J. E. Allen, F. W. Forney, and A. J. Markovetz, *Lipids* **6**, 448 (1971).
[142] J. C. van Ravenswaay Claasen and A. C. van der Linden, *Antonie van Leeuwenhoek* **37**, 339 (1971).
[143] A. C. van der Linden and J. C. van Ravenswaay Claasen, *Lipids* **6**, 437 (1971).

TABLE XI: Oxygenation of Alkanes, Alkylbenzenes, and Alkylcycloalkanes with a Cell-Free Extract of *Pseudomonas aeruqinosa* (strain 473)

Substrate	Product	Relative yield[a]
Alkanes		
Pentane	Pentanoic acid	45
Hexane	Hexanoic acid	80
Heptane	Heptanoic acid	80
Octane	Octanoic acid	100
Nonane	Nonanoic acid	100
Decane	Decanoic acid	80
2-Methylhexane	[b]	60
3-Methylhexane	[b]	40
2,2-Dimethylhexane	[b]	5
2,3-Dimethylhexane	[b]	15
2,4-Dimethylhexane	[b]	30
2,5-Dimethylhexane	[b]	30
3,4-Dimethylhexane	[b]	10
2,2,4-Trimethylpentane	—	0
2,2-Dimethylpentane	—	0
2,2-Dimethylhexane	[b]	5
2,2-Dimethylheptane	[b]	90
2,2-Dimethyloctane	[b]	100
2,2-Dimethylnonane	[b]	100
Alkylbenzenes		
Benzene	—	0
Toluene	Benzyl alcohol	20
Ethylbenzene	Phenylacetic acid	100
Propylbenzene	3-Phenylpropionic acid	25
Butylbenzene	4-Phenylbutyric acid	100
Isopropylbenzene	2-Phenylpropanol	55
Isobutylbenzene	—	0
Isopentylbenzene	2-(2-Phenylethyl)propanol	60
2-Methylisopropylbenzene	—	0
3-Methylisopropylbenzene	2-(3-Methylphenyl)propanol	35
4-Methylisopropylbenzene	2-(4-Methylphenyl)propanol	50
4-Methyl-*tert*-butylbenzene	—	0
Alkylcycloalkanes		
Cyclohexane	Cyclohexanol	30
Methylcyclohexane	*trans*-4-Methylcyclohexanol	100
Ethylcyclohexane	*trans*-4-Ethylcyclohexanol	100
Propylcyclohexane	*trans*-4-Propylcyclohexanol	100
Butylcyclohexane	*trans*-4-Butylcyclohexanol	100
Isopropylcyclohexane	*trans*-4-Isopropylcyclohexanol	100
tert-Butylcyclohexane	*trans*-4-*tert*-Butylcyclohexanol	5
trans-4-Methylisopropylcyclohexane	2-(*trans*-4-Methylcyclohexyl)-propionic acid	60
cis-4-Methylisopropylcyclohexane	[b]	3
trans-Decalin	2-Hydroxydecalin	20
cis-Decalin	[b]	1

[a] If the product is formed to an extent of 1 mg with 15 mg of enzyme protein in one hour, the relative yield is given as 100. The experiments used 3 mg of substrate with approximately 30 mg of protein.

[b] Product was detected by glpc but the structure was not determined.

isobutylbenzene. Molecular models show the former to be about 4 Å thick while the latter is more than 5 Å thick. Other molecules that can remain relatively flat can penetrate the cavity while those that contain a *tert*-butyl group or have an axial substituent are poor substrates. The depth of the cavity is estimated to be about 8 Å since the cutoff point for oxidation of a series of 2,2-dimethylalkanes falls at 2,2-dimethylhexane.[142] This molecule is about 8 Å in length from the dimethyl branch point to the end of the alkyl chain. 2,2-Dimethylpentane has a shorter length from branch point to the methyl terminus and is not oxygenated.

Cell-free extracts of several other microorganisms capable of oxygenating alkanes have been prepared. These will only be listed here for purposes of reference and include *Cladosporium resinae*,[144] *Candida tropicalis*,[145] *Corynebacterium* sp.,[146] *Pseudomonas oleovorans*,[147] *Nocardia salmonicolor*,[148] and *Bacillus megaterium*.[149,150] Such enzyme preparations have the potential for avoiding oxidative degradation of initially formed products as is demonstrated by the results cited above for *P. aeruginosa*.

The presence of branching in an alkane often has a strong influence on the course of an oxygenation. The type of branching most frequently encountered in studies of such substrates is the presence of a methyl group on the chain. Alkane oxidizing organisms such as *P. aeruginosa*,[151] *T. apicola*, and *T. gropengiesseri*[137] appear to prefer attack at the end of the alkane furthest removed from the methyl substituent. More examples are needed, however, before this may become a valid generalization. Certainly, attack at the branched end of a hydrocarbon is only somewhat slower in rate.

The presence of a branch in an alkane chain also serves to inhibit β-oxidation. An often used branched hydrocarbon is 2,6,10,14-tetramethylpentadecane (**143**) better known as pristane. Pristane is found distributed throughout nature. Although the resistance of pristane to degradation has enabled it to be used as a biological time marker, microorganisms have been found that can oxygenate this molecule. *Torulopsis gropengiesseri* incorporates pristane into its extracellular glycolipid fraction. From this, the 1,15-diol (**144**), for example, can be released by hydrolysis in 16% yield.[137] A variety of organisms have been found that can grow and multiply with pristane as the sole carbon source. These include a *Corynebacterium* sp.,[152]

[144] J. D. Walker and J. J. Cooney, *J. Bacteriol.* **115**, 635 (1973).
[145] J. M. Lebeault and E. Azoulay, *Lipids* **6**, 444 (1971).
[146] C. Cardini and P. Jurtshuk, *J. Biol. Chem.* **245**, 2789 (1970).
[147] E. J. McKenna and M. J. Coon, *J. Biol. Chem.* **245**, 3882 (1970).
[148] F. S. Sariaslani, D. B. Harper, and I. J. Higgins, *Biochem. J.* **140**, 31 (1974).
[149] Y. Miura and A. J. Fulco, *Biochim. Biophys. Acta* **388**, 305 (1975).
[150] R. S. Hare and A. J. Fulco, *Biochem. Biophys. Res. Commun.* **65**, 665 (1975).
[151] G. J. E. Thijsee and A. C. van der Linden, *Antonie van Leeuwenhoek* **27**, 171 (1961).
[152] E. J. McKenna and R. E. Kallio, *Proc. Natl. Acad. Sci. U.S.A.* **68**, 1552 (1971).

a *Nocardia* sp.,[153] and *Mycobacterium fortuitum*,[154] and *Brevibacterium erythrogenes*.[155] Products isolated from these experiments generally have been pristanol (**145**) and pristanoic acid (**146**) as well as carboxylic acids of shorter chain length.

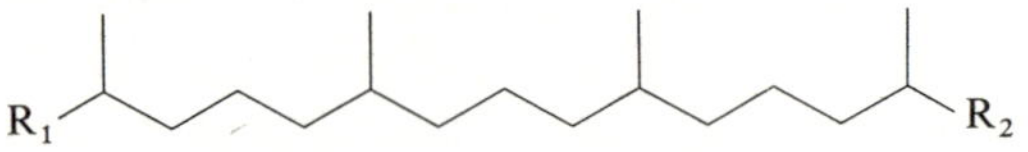

(**143**) $R_1 = R_2 = CH_3$
(**144**) $R_1 = R_2 = CH_2OH$
(**145**) $R_1 = H, R_2 = CH_2OH$
(**146**) $R_1 = H, R_2 = COOH$

A notable synthetic application of alkane hydroxylation is found in the oxygenation of 5,8,11,14-eicosatetraenoic acid by *Ophiobolus graminis*.[156] The unsaturated fatty acid is hydroxylated at 19-position (16%) and at the 18-position (19%). These products have been converted into prostaglandin molecules by further biochemical synthetic steps.

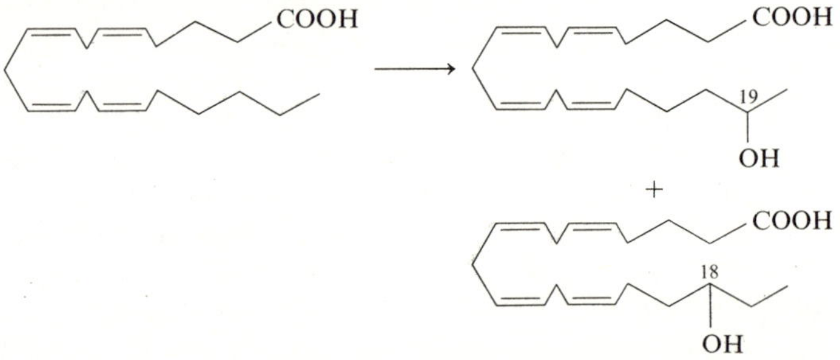

Similar use of the hydroxylation of a dialkylbenzene by *Sporotrichum sulfurescens* in the synthesis of a drug metabolite has been described in the section of this chapter dealing with *S. sulfurescens*.[69] This organism has also been used to oxygenate various acyclic alkylbenzamides.[62] In molecules such as these, further degradation of the alkyl group is probably retarded by the presence of the large nonalkyl substituent found on the hydrocarbon chain. Several additional examples of this type follow.

A series of alkyl substituted 2-(benzenesulfonylamino)pyrimidines has been hydroxylated with *Streptomces griseolus*.[157] The results are shown in

[153] K. Nakajima, A. Sato, T. Misono, T. Iida, and K. Nagayasu, *Agric. Biol. Chem.* **38**, 1859 (1974).

[154] R. E. Cox, J. R. Maxwell, R. G. Ackman, and S. N. Hooper, *Biochim. Biophys. Acta* **360**, 166 (1974).

[155] M. P. Pirnik, R. M. Atlas, and R. Bartha, *J. Bacteriol.* **119**, 868 (1974).

[156] C. J. Sih, G. Ambrus, P. Foss, and C. J. Lai, *J. Am. Chem. Soc.* **91**, 3685 (1969).

[157] G. Siewert, K. Kieslich, G. A. Hoyer, and D. Rosenberg, *Chem. Ber.* **106**, 1290 (1973).

TABLE XII

HYDROXYLATION OF 5-ALKYL-2-(BENZENESULFONYLAMINO)-PYRIMIDINES WITH *Streptomyces griseolus*

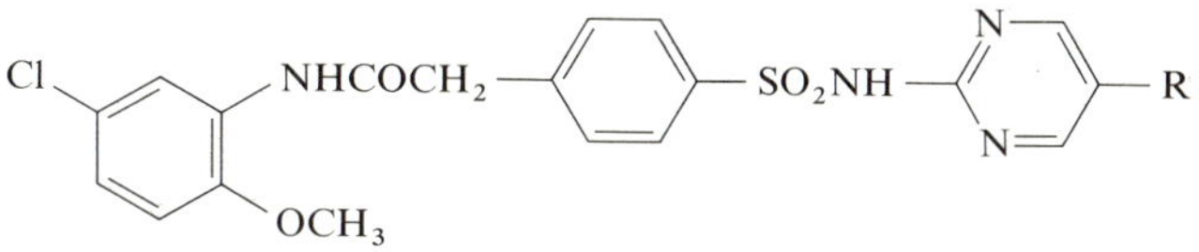

Substrate R =	Product R =	Yield (%)
$-CH_2CH(CH_3)_2$	$-CH_2C(OH)(CH_3)_2$	67
$-CH(CH_3)_2$	$-C(OH)(CH_3)_2$	5
	$-CH(CH_3)CH_2OH$	4
$-CH_2CH_3$	$-CH(OH)CH_3$	5
$-CH_2CH_2CH_3$	$-CH_2CH(OH)CH_3$	2
$-(CH_2)_3CH_3$	$-CH_2CH(OH)CH_2CH_3$	4
$-CH_2C(CH_3)_3$	$-CH_2C(CH_3)_2CH_2OH$	35
	$-CH_2C(CH_3)_2CH_2OH$[a]	40

[a] Methoxyl group of chloroaniline group was cleaved to the phenol.

Table XII. Hydroxylation at the more highly substituted carbon of the alkyl group is preferred in these reactions.

We can be grateful to the wood rotting fungi for keeping the forest floor free from the clutter of fallen trees. An important observation of alkyl hydroxylation by a wood rotting fungus, *Polyporus dichrous*, has been made by Krik and Lorenz.[158] Alkyl ethers of vanillic acid were prepared as lignin model compounds and used as substrates with this white rot fungus. The results are summarized in Table XIII. Although yields of products were not determined for these reactions, levels of the products in the fermentation were estimated to reach from 47 to 80% of the initial concentration of substrate. Competing and subsequent reactions are cleavages of the ether linkages.[158] The use of this culture is of special significance because it is able to accommodate the highly oxygenated, polar nature of the acidic substrates used in these experiments.

The isopentanoyl group of the macrolide antibiotics, maridomycin and josamycin, is hydroxylated at the tertiary position by *Streptomyces olivaceus*.[159]

[158] T. K. Kirk and L. F. Lorenz, *Appl. Microbiol.* **27**, 360 (1974).

[159] K. Nakahama, T. Kishi, and S. Igarasi, *J. Antibiot.* **27**, 433 (1974).

TABLE XIII

HYDROXYLATION OF 4-ALKOXYBENZOIC ACIDS WITH *Polyporus dichrous*

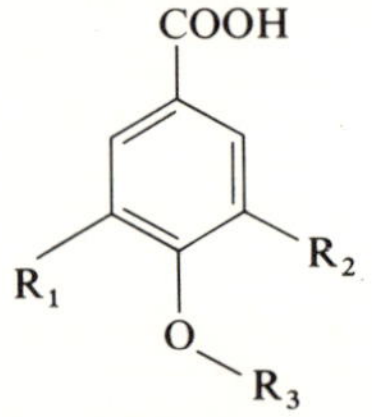

Substrate			Product
R_1	R_2	R_3	R_3
H	$—OCH_3$	$—CH(CH_3)_2$	$—CH(CH_3)CH_2OH$
H	$—OC_2H_5$	$—CH(CH_3)_2$	$—CH(CH_3)CH_2OH$
H	$—OC_2H_5$	$—CH_2CH_2CH_3$	$—CH_2CH_2CH_2OH$ + $—CH_2CH(OH)CH_3$
H	$—OC_2H_5$	$-CH(CH_3)CH_2CH_3$	[a]
$—OCH_3$	$—OCH_3$	$—CH(CH_3)_2$	$—CH(CH_3)CH_2OH$[b]

[a] Three products detected but structures were not determined.
[b] This product was optically active, $[\alpha]_{365} = -14.7°$.

The use of a microbial oxygenation to prepare metabolites of drug molecules is again demonstrated by oxygenation of methyl 5(6)-butyl-2-benzimidazolecarbamate, an anthelmintic agent, by *Cunninghamella bainieri.*[160] Hydroxylation of the terminal methyl group of the butyl side chain produces a hydroxymethyl compound. The latter undergoes further oxidation to a carboxyl group. Both of these oxidation products have been found as metabolites of the drug after administration to test animals.

Finally, we note that the herbicide, *N*-(3,4-dichlorophenyl)-2-methylpentanamide (Karsil), has been hydroxylated in the alkane chain by *Rhizopus japonicus.*[161] In this work, *R. japonicus* is considered typical of the microbial

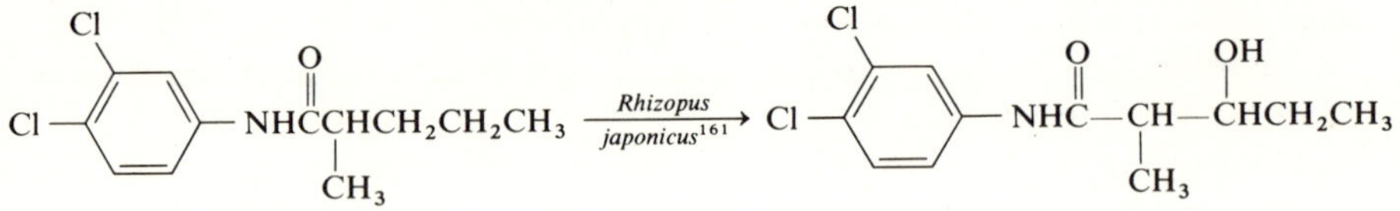

Karsil

[160] J. R. Valenta, C. J. DiCuollo, L. R. Fare, J. A. Miller, and J. F. Pagano, *Appl. Microbiol.* **28**, 995 (1974).

[161] P. R. Wallnöfer, S. Safe, and O. Hutzinger, *Chemosphere* **1**, 155 (1972).

flora found in soil and, consequently, the product is assumed to represent a microbial metabolite of the herbicide.

IV. Oxygenation at Allylic Carbon

Allylic oxygenations by microorganisms may be of three types: (i) oxygenation of a classical allylic carbon system, (ii) oxygenation of a benzylic carbon, and (iii) oxygenation at carbon next to a carbonyl group. Examples of all three types have been collected and reviewed previously.[8] Thus, only several recent examples are presented here to exemplify each of these oxygenations.

A. CLASSICAL ALLYLIC CARBON

Numerous oxygenations of steroids at allylic positions have been encountered, particularly at C-6 when it is allylic to the 3-keto-Δ^4 functionality. An additional recent example may be cited. Oxygenation of the steroidal compound **147** with *Chaetomium cochloides* gave three products, two of which (**148** and **150**) had been oxygenated at the allylic 6-position.[162] The formation of these products provided a convenient synthesis of three human metabolites of the antialdosterone agent, spironolactone.

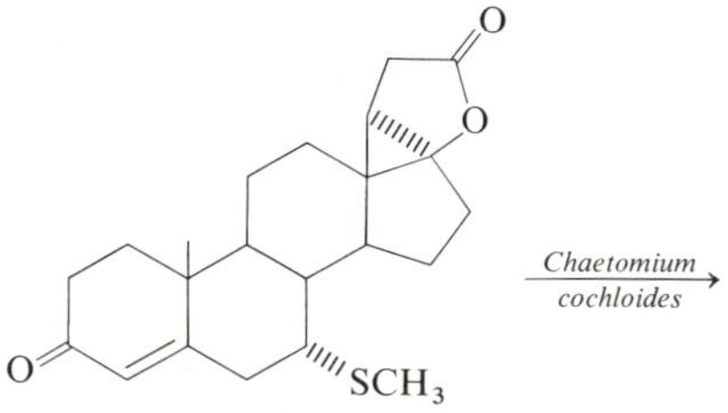

(**147**)

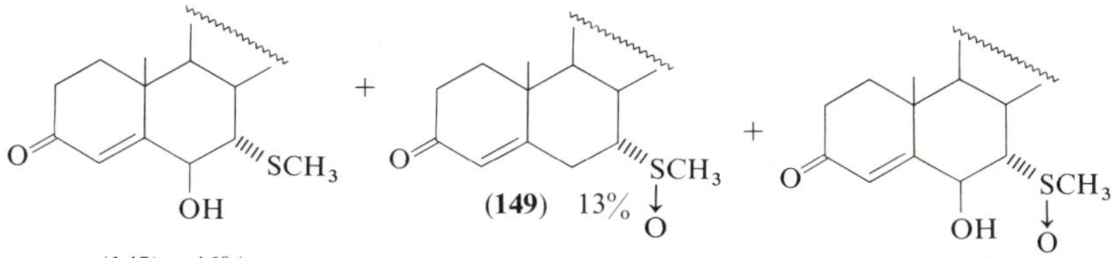

(**148**) 46% (**150**) 8%

It has been speculated that hydroxylation at the allylic 6-position of steroids involves enolic intermediates. The parallel between microbiological

[162] W. J. Marsheck and A. Karim, *Appl. Microbiol.* **25**, 647 (1973).

hydroxylation of this system, giving the axial 6β-hydroxyl group, and chemical oxidation of $\Delta^{3,5}$ enol ethers, also giving 6β-hydroxylation, led to this speculation. Recently, Holland and Auret have presented several additional lines of evidence supporting this hypothesis.[163,164] Hydroxylation of (±)-4a-methyl-4,4a,5,6,7,8-hexahydro-2(3*H*)-naphthalenone with *Rhizopus arrhizus* gave both α- and β-hydroxyl groups at the allylic carbon. Again,

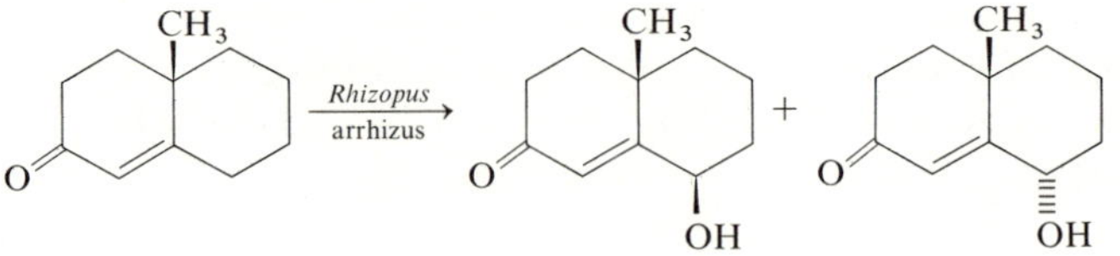

chemical oxidation of the analogous enol-ether paralleled the microbiological result.[163] More convincing are several experiments that compare hydroxylation of androst-4-ene-3,17-dione, 6α-methylandrost-4-ene-3,17-dione, and 6β-methylandrost-4-ene-3,17-dione with their corresponding enol-acetates.[164] The enol-acetates are readily hydrolyzed during the fermentation, allowing formation of the enone system and, in fact, the androst-4-ene-3,17-diones are major products isolated from the fermentation. Nevertheless, the authors show that hydroxylation of the enol-acetates proceeds more rapidly than does hydroxylation of the 3-keto-Δ^4 system. They also show that when there is a methyl substituent at the 6-position, the enol-acetate again undergoes 6β-hydroxylation. When either 6α-methyl or 6β-methyl androst-4-ene-3,17-dione are used as substrates under the same conditions, no hydroxylation at C-6 is observed. These results are interpreted to support a role for the $\Delta^{3,5}$-enol in the allylic hydroxylation of steroids.[164]

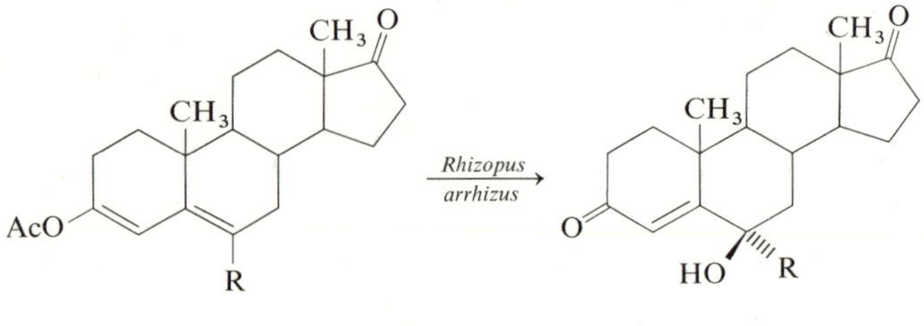

R = H or CH_3

The prostaglandin synthon **152** has been prepared in 67% yield by allylic hydroxylation of 7-(5-oxo-1-cyclopentenyl)heptanoic acid (**151**) with *Aspergillus niger*.[165] In addition, the product was optically active although the optical purity was not given.

[163] H. L. Holland and B. J. Auret, *Can. J. Chem.* **53**, 2041 (1975).
[164] H. L. Holland and B. J. Auret, *Tetrahedron Lett.* p. 3787 (1975).
[165] S. Kurozumi, T. Toru, and S. Ishimoto, *Tetrahedron Lett.* p. 4959 (1973).

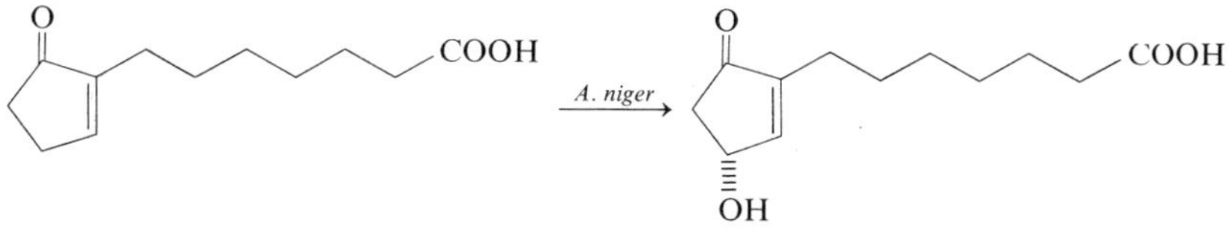

(151) (152)

Useful quantities of hydroxylated products have been obtained in a study of the microbial transformation of piperitone (**153**).[166] This monoterpene represents a major component of the steam volatile leaf oils of eucalyptus. The two major products, **154** and **155**, are the result of allylic hydroxylation.

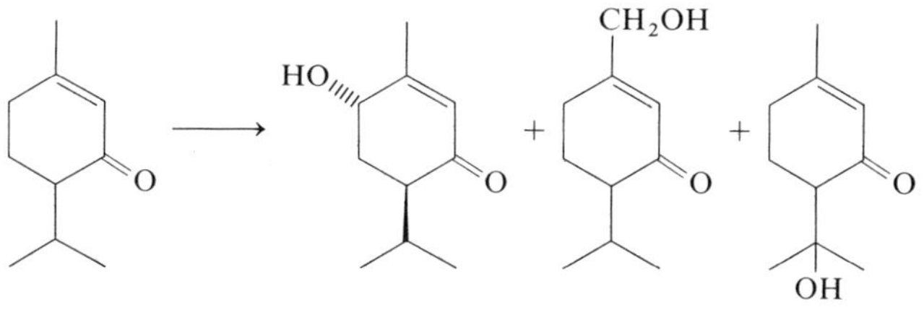

	(153)	(154)	(155)	(156)
Fusarium sp.		6.5%	45%	4.5%
Proactinomyces roseus		20%	~30%	
Aspergillus niger			20%	

An allylic oxygenation is also found in the conversion of linalool (**157**) to the two acids (**158** and **159**) by a *Pseudomonas* sp.[167]

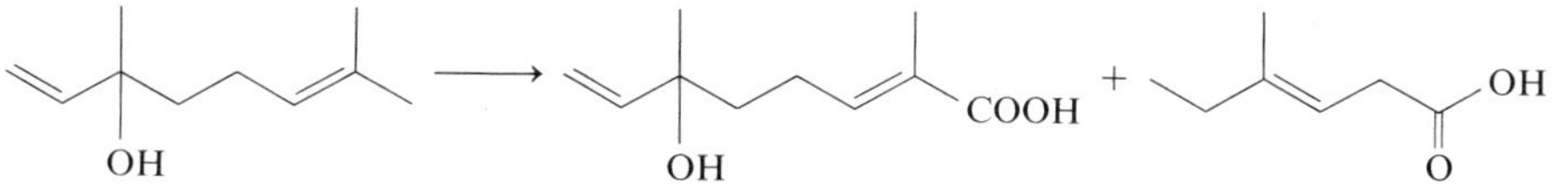

(157) (158) (159)

Morphine is hydroxylated at the allylic 14-position by either resting cells or cell-free systems of an *Arthrobacter* species.[168]

B. BENZYLIC CARBON

The selective hydroxylation of an aromatic methyl group of lucanthone (**1**) has been mentioned in the introduction to this chapter. The micro-organism performing this hydroxylation has also hydroxylated the benzylic carbon in a series of compounds of type **160**.[2] Similar benzylic hydroxylations

[166] E. V. Lassak, J. T. Pinhey, B. J. Ralph, T. Sheldon, and J. J. H. Simes, *Aust. J. Chem.* **26**, 845 (1973).

[167] T. Murakami, I. Ichimoto, and C. Tatsumi, *J. Agric. Chem. Soc.* **47**, 699 (1973).

[168] P. Liras and W. W. Umbreit, *Appl. Microbiol.* **30**, 262 (1975).

of nalidixic acid analogs (**161**) with *Penicillium adametzi* have been reported by the same group.[169]

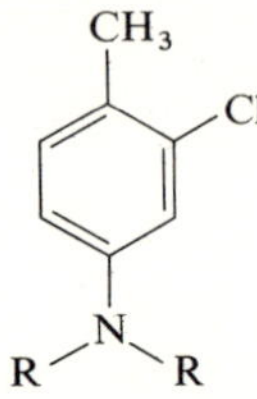

(**160**)

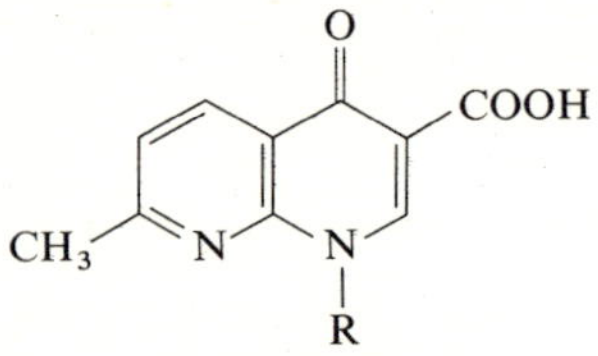

(**161**)

Recent examples further illustrate these benzylic hydroxylations. A strain of *Pseudomonas*, isolated from the edge of a polluted creek, has been found to grow using either *p*-xylene or *m*-xylene as the only source of carbon.[170] Mutation of this microorganism with *N*-methyl-*N'*-nitro-*N*-nitrosoguanidine gave two mutant strains in which complete metabolism of the xylenes was blocked. In one strain, benzylic oxygenation proceeded to the stage of the 3- or 4-methyl benzoic acids. In the second mutant strain, the methylbenzoic acids accumulated together with 3- or 4-methylcatechol. The latter was postulated to arise via a cis dihydroxylation of the methylbenzoic acids followed by decarboxylation.[170]

The quinoline derivative (**162**) is hydroxylated on the 7-methyl group (25% yield) by *Penicillium adametzi*, giving the hydroxymethyl product (**163**).[171] An allylic hydroxylation of a cyclic system by *Sporotrichum sulfurescens* has been reported for the related substrate (**164**). The product (**165**) is obtained in 39% yield.[171]

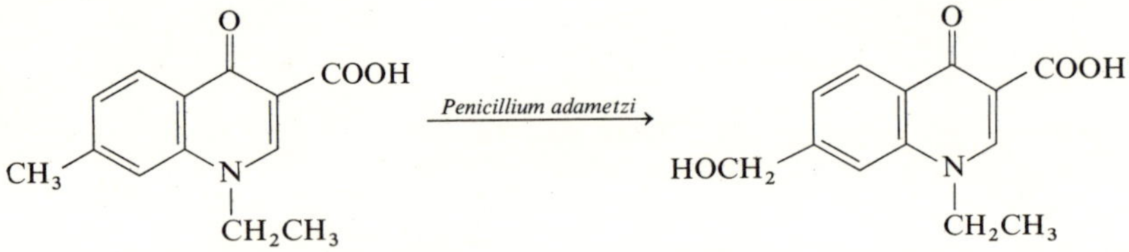

(**162**) (**163**)

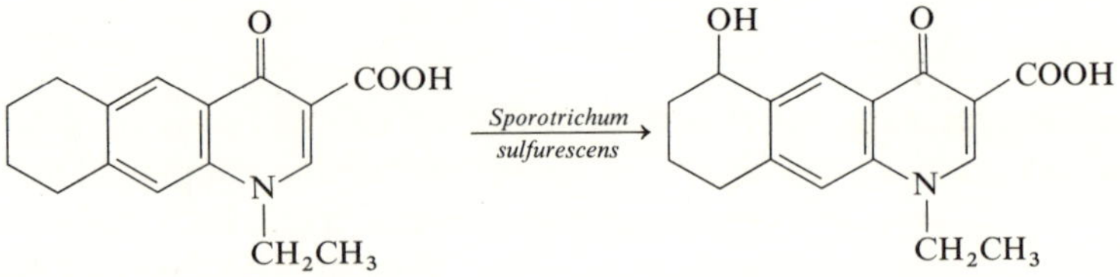

(**164**) (**165**)

[169] P. B. Hamilton, D. Rosi, G. P. Peruzzotti, and E. D. Nielson, *Appl. Microbiol.* **17**, 237 (1969).
[170] J. F. Davey and D. T. Gibson, *J. Bacteriol.* **119**, 923 (1974).
[171] K. Kieslich, H. Wieglepp, G. A. Hoyer, and D. Rosenberg, *Chem. Ber.* **106**, 2636 (1973).

Spontaneous decarboxylation of 2-(4-chlorophenyl)thiazolyl-4-acetic acid (**166**) in the fermentation medium can be apparently followed by hydroxylation of the methyl group, (**167**), to give **168**. A number of different microorganisms could effect this benzylic-type oxidation.[172]

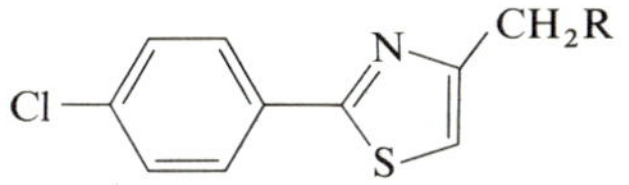

(**166**) R = COOH
(**167**) R = H
(**168**) R = OH

Among the numerous transformations of mycophenolic acid (**169**) by many microorganisms is the hydroxylation of the aromatic methyl group to give **170**.[173]

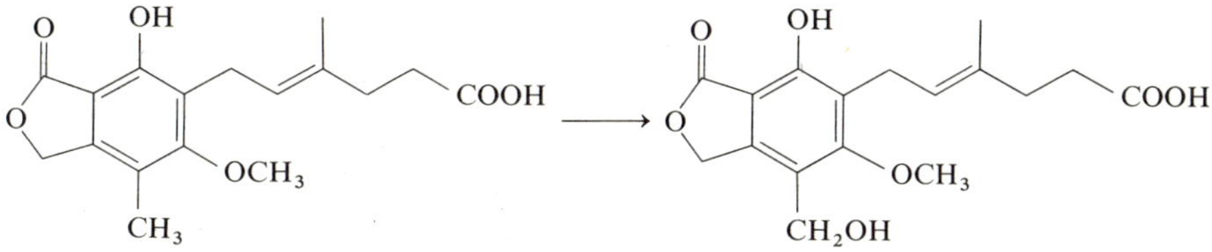

(**169**) (**170**)

The hydroxylation of either methyl group found on the furan unit of the fungicide (**171**) by *Rhizopus japonicus* also represents a benzylic-type hydroxylation.[174] The two products, **172** and **173**, are formed to the extent of 20% and 38%, respectively.

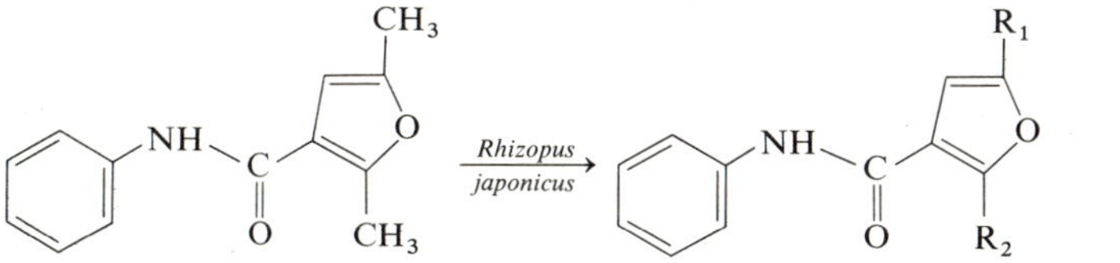

(**171**)

(**172**) $R_1 = CH_3$, $R_2 = CH_2OH$
(**173**) $R_1 = CH_2OH$, $R_2 = CH_3$

C. CARBONYL ACTIVATED CARBON

Such oxygenations are occasionally encountered in steroids; for example, next to the carbonyl group of the acetyl sidechain of the pregnane family.

Hydroxylation next to the 4′-carbonyl group of griseofulvin (**174**) has been reported for two microorganisms. One, *Streptomyces cinereocrocatus*,[175]

[172] R. Howe, R. H. Moore, B. S. Rao, and A. H. Wood, *J. Med. Chem.* **15**, 1040 (1972).
[173] D. F. Jones, R. H. Moore, and G. C. Crawley, *J. Chem. Soc. C* p. 1725 (1970).
[174] P. R. Wallnöfer, M. Koniger, S. Safe, and O. Hutzinger, *J. Agric. Food Chem.* **20**, 20 (1972).
[175] W. W. Andres, W. J. McGahren, and M. P. Kunstmann, *Tetrahedron Lett.* p. 3777 (1969).

gives the pseudoequatorial 5′-hydroxy derivative (**175**) while the other, *Cephalosporium curtipes*,[176] gives the pseudoaxial 5′-hydroxy product (**176**).

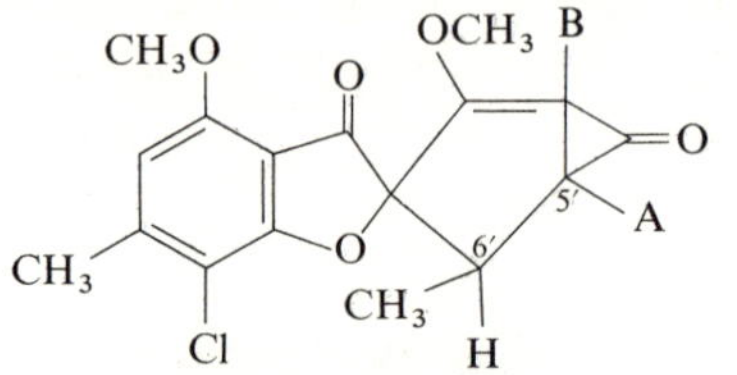

(**174**) A = B = H
(**175**) A = OH, B = H
(**176**) A = H, B = OH

A *Streptomyces* species has been found that hydroxylates several benzodiazepine analogs at the methylene carbon of the seven-membered ring.[177] Dealkylation of the amide nitrogen precedes the oxygenation reaction; the product of dealkylation alone is also isolated from these fermentations. Unfortunately, the interesting question of stereoselectivity of the hydroxylation was not discussed.

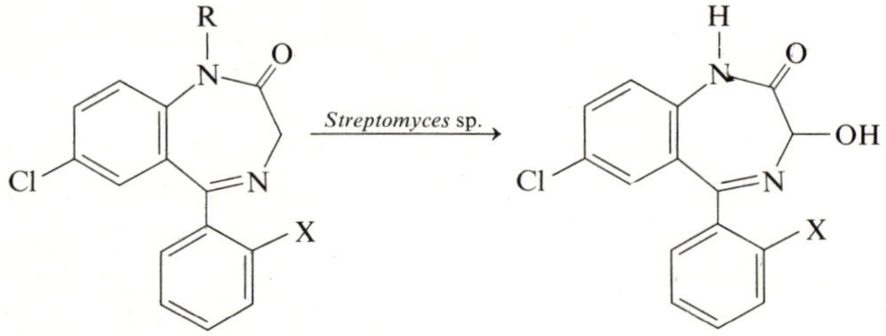

X = H, F
R = H, CH_3, CH_2–cyclopropyl

V. Oxygenation of Olefins

Olefins may be hydrated or epoxidized by microorganisms and the latter may be further hydrated. Thus, one may convert an olefin to an alcohol, to an epoxide, or to a diol. Although these reactions may be readily achieved chemically, there may be situations where the selectivity of the enzymatic method offers some advantages. Several examples from the recent literautre are offered. Fosfomycin (**178**) is an antibiotic produced by a species of *Streptomyces*. Chemical synthesis of **178** by epoxidation of the appropriate olefin (**177**) proceeds satisfactorily, but produces the (±)-mixture of the antibiotic.

[176] P. Bod, E. Szarka, J. Gyimesi, G. Horváth, Z. Vajna-Mehesfalvi, and I. Horváth, *J. Antibiot.* **26**, 101 (1973).

[177] F. Kishimoto, K. Suzuki, S. Ogino, and H. Yamamoto, *Agric. Biol. Chem.* **39**, 953 (1975).

However, epoxidation of **177** with *Penicillium spinulosum* gives the desired (−)-enantiomer of the antibiotic in 90% conversion.[178]

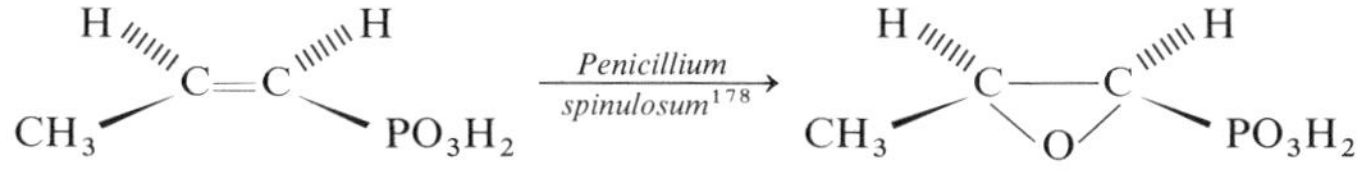

(177) **(178)**

The combined efforts of two microorganisms provide an interesting synthesis of *l*-malic acid.[179] One culture, *Candida hydrocarbofumarica*, efficiently produces fumaric acid when grown on a mixture of *n*-paraffins. A second microorganism, either *Pichia membranaefacians* or *Candida utilis*, is then added to the fermentation and converts the fumarate to *l*-malic acid. The first culture is allowed to grow 5 days before the second is added. Conversion of the *n*-paraffin to *l*-malic acid is stated to be 70%.[179]

Pseudomonas oleovorans is known to hydroxylate *n*-alkanes on the terminal methyl carbon.[147] Cell-free enzyme preparations obtained from this organism have been studied extensively for this activity.[147] It now has been found that this culture can also epoxidize various 1-alkenes[180–182] and 1,(ω-1)-alkadienes.[182–183a] Synthetically, the most interesting reactions reported to date are the stereospecific mono epoxidation of 1,7-octadiene (**179**) to form (*R*)-(+)-7,8-epoxy-1-octene (**180**) in about 25% yield[183] and a stereochemical study of the diepoxidation of either of these compounds.[183a] The experimental method is such that it has been adapted for use in instructional organic chemistry laboratory courses.[35]

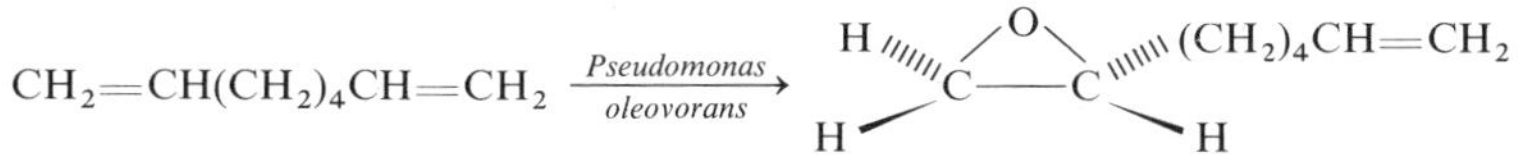

(179) **(180)**

The same cell-free enzyme system that hydroxylates *n*-alkanes is able to carry out this epoxidation reaction as well.[182] From comparison of the experimental details, it has been suggested that both reactions are carried out by the same active site on the enzyme.[182] A mutant strain of *Pseudomonas*

[178] R. F. White, J. Birnbaum, R. T. Meyer, J. ten Broeke, J. M. Chemerda, and A. L. Demain, *Appl. Microbiol.* **22**, 55 (1971).
[179] T. Furukawa, T. Nakahara, and K. Yamada, *Agric. Biol. Chem.* **34**, 1833 (1970).
[180] B. J. Abbott and C. T. Hou, *Appl. Microbiol.* **26**, 86 (1973).
[181] R. D. Schwartz, *Appl. Microbiol.* **25**, 574 (1973).
[182] S. W. May and B. J. Abbott, *J. Biol. Chem.* **248**, 1725 (1973).
[183] S. W. May and R. D. Schwartz, *J. Am. Chem. Soc.* **96**, 4031 (1974).
[183a] S. W. May, M. S. Steltonkamp, R. D. Schwartz, and C. J. McCoy, *J. Am. Chem. Soc.* **98**, 7856 (1976).

putida (*P. putida* 39/D) that has been used to effect the *cis*-1,2-dihydroxylation of various aromatic rings (see Table XIV)[4,184–195] has been found to hydroxylate 3-methylcyclohexene in a similar fashion.[196] Two products, (3*R*)-methyl-(1*S*,2*R*)-dihydroxycyclohexane and (3*S*)-methyl-(1*S*,2*R*)-dihy-

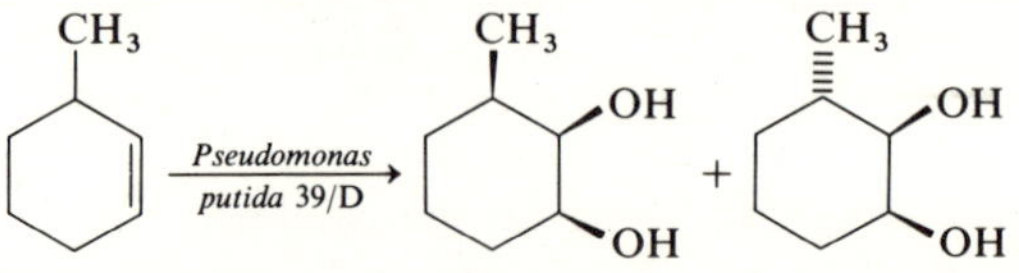

droxycyclohexane, were isolated and characterized, but yields were not stated in the preliminary communication.

The herbicide *N*-(3,4-dichlorophenyl)methylacrylamine (**181**) is converted to a diol (**182**) by *Rhizopus japonicus* as a model for fungal metabolism in the soil.[197] Whether an epoxide is formed as an intermediate in this reaction is not known. The various activities of *R. japonicus* also include hydroxylation

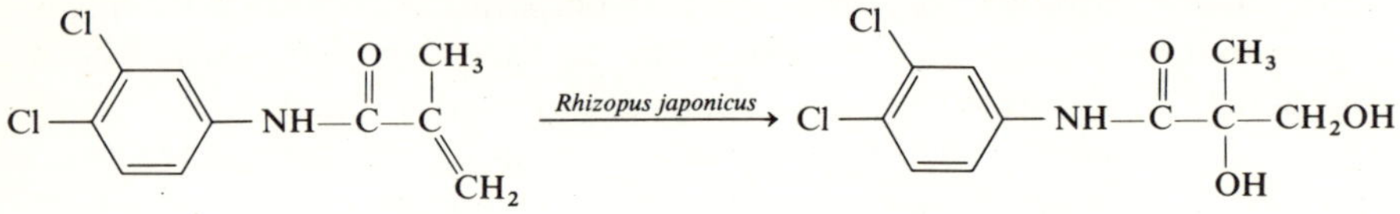

(**181**) (**182**)

at aliphatic[161] and benzylic[174] positions as well as the oxygenation of sulfur.[198] These results suggest that this organism may be useful for the oxygenation of many other organic substrates.

[184] D. T. Gibson, G. E. Cardini, F. C. Maseles, and R. E. Kallio, *Biochemistry* **9**, 6131 (1970).
[185] T. Hogn and L. Jaenicke, *Eur. J. Biochem.* **30**, 369 (1972).
[186] H. Ziffer, D. M. Jerina, D. T. Gibson, and V. M. Kobal, *J. Am. Chem. Soc.* **95**, 4048 (1973).
[187] D. T. Gibson, B. Gschwendt, W. D. Yeh, and V. M. Kobal, *Biochemistry* **12**, 1520 (1973).
[188] T. Omori, Y. Jigami, and Y. Minoda, *Agric. Biol. Chem.* **38**, 409 (1974).
[189] D. T. Gibson, V. Mahadevan, and J. F. Davey, *J. Bacteriol.* **119**, 930 (1974).
[190] D. T. Gibson, J. R. Koch, C. L. Schuld, and R. E. Kallio, *Biochemistry* **7**, 3795 (1968).
[191] A. M. Reiner and G. D. Hegeman, *Biochemistry* **10**, 2530 (1971).
[192] D. M. Jerina, J. W. Daly, A. M. Jeffery, and D. T. Gibson, *Arch. Biochem. Biophys.* **142**, 394 (1971); for more recent details, see A. M. Jeffrey, H. J. C. Yeh, D. M. Jerina, T. R. Patel, J. F. Davey, and D. T. Gibson, *Biochemistry* **14**, 575 (1975).
[193] N. Walker, N. F. Janes, J. R. Spokes, and P. Van Berkum, *J. Appl. Bacteriol.* **39**, 281 (1975).
[194] E. deFrenne, J. Eberspächer, and F. Lingens, *Eur. J. Biochem.* **33**, 357 (1973).
[195] D. T. Gibson, V. Mahadevan, D. M. Jerina, H. Yagi, and H. J. C. Yeh, *Science* **189**, 295 (1975).
[196] H. Ziffer and D. T. Gibson, *Tetrahedron Lett.* p. 2137 (1975).
[197] P. R. Wallnöfer, S. Safe, and O. Hutzinger, *J. Agric. Food Chem.* **21**, 502 (1973).
[198] P. R. Wallnöfer, M. Koeniger, S. Safe, and O. Hutzinger, *Int. J. Environ. Anal. Chem.* **2**, 37 (1972).

Several monocyclic terpenes have been oxidized to diols with different species of the genus *Cladosporium*.[199] The diols were produced in experiments where the pure terpene was used as the sole carbon source for the microorganism. As a result, the yields of products are low when based on the substrate levels used, but the quantities of products obtained are still significant. Diol **184** is formed in about 2% yield from 1-menthene (**183**), but the quantity produced in the fermentation is 200 mg/liter. Similarly, diols **186** and **187** are produced to the extent of 1.5 gm/liter and 0.2 gm/liter from *d*-limonene (**185**). Diol **189** (0.5 gm/liter) is formed from 3-menthene (**188**).

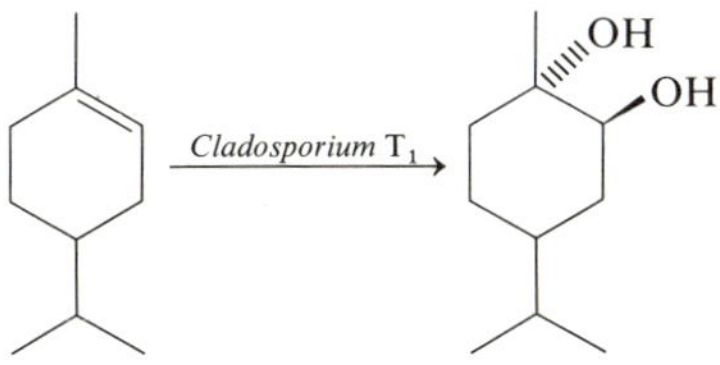

(183) **(184)**

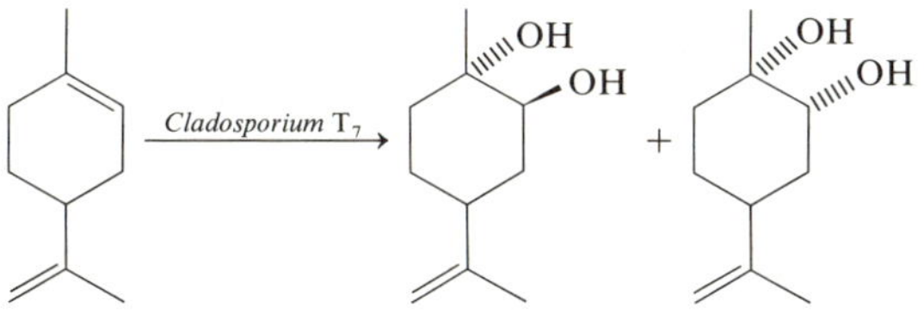

(185) **(186)** **(187)**

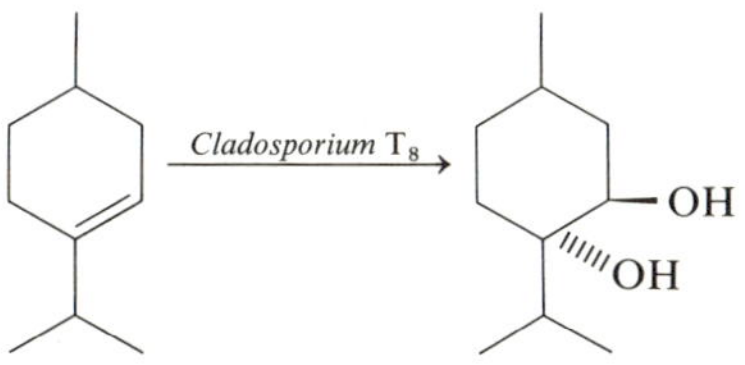

(188) **(189)**

Germacrone (**190**) is an unsaturated, macrocyclic compound thought to be a precursor of several bicycle sesquiterpenoids. It contains no asymmetric carbons. Incubation of **190** with *Cunninghamella blakesleeana* does not produce a bicyclic sesquiterpene but it does epoxidize several of the double bonds to produce three optically active products.[200] The structures and absolute configurations of these products are suggested to be as shown on the following page.[200]

[199] B. B. Mukherjee, G. Kraidman, and I. D. Hill, *Appl. Microbiol.* **25**, 447 (1973).
[200] H. Hikino, C. Konno, T. Nagashima, and T. Takemoto, *Tetrahedron Lett.*, p. 337 (1971).

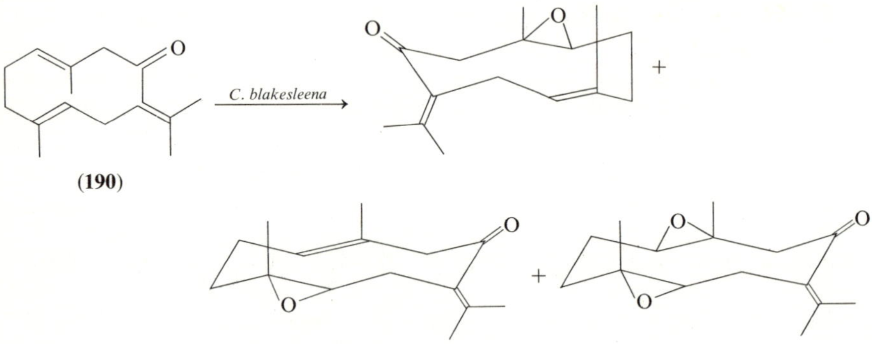

(190)

Gibberellin A_9 (GA_9), methyl ester, contains an exocyclic double bond that is oxygenated in good yield by *Rhizopus nigricans.*[201] The main product isolated, in yields up to 48% depending on conditions, was 16α,17-dihydroxy GA_9 methyl ester. A number of other products of hydroxylation were also isolated in yields of 2–9%

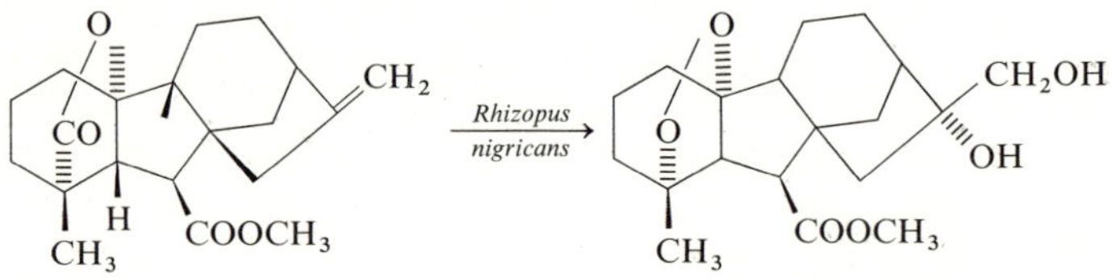

GA_9Me ester 16α, 17-Dihydroxy GA_9Me ester

VI. Oxygenation of Aromatic Rings

Microbiological oxygenation of aromatic rings has been widely studied, particularly with regard to the metabolism of such compounds. A limited number of synthetically useful reactions have come from this work and many of these have been reviewed previously.[8,21,22] In this section we will give examples of monohydroxylations from the recent literature that illustrate synthetic possibilities. Also included is a summary of the interesting *cis*-1,2-dihydroxylations of aromatic rings that have been reported in recent years.

Synthetic potential for monohydroxylations of aromatic rings is very nicely demonstrated with the substrate 5-anilino-1,2,3,4-thiatriazole (**191**).

[201] J. R. Bearder, V. M. Frydman, P. Gaskin, I. K. Hatton, W. E. Harvey, J. MacMillan, and B. O. Phinney, *J. Chem. Soc., Perkin Trans. 1* p. 178 (1976).

This substance is a potent experimental hypotensive agent, and as such,

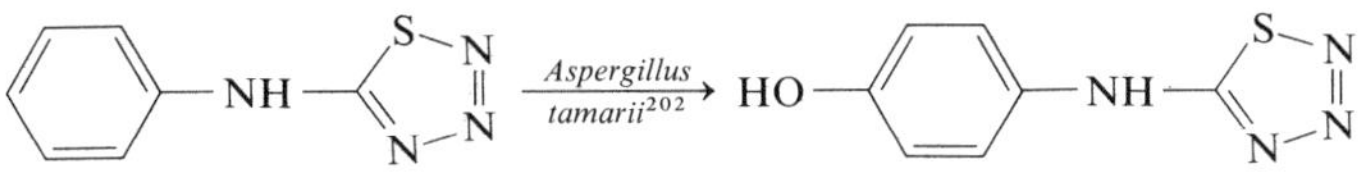

(191) (192)

synthesis of *p*-hydroxy derivative (**192**) as a potential metabolite was sought.[202] Initial chemical synthesis was unsuccessful so a number of microorganisms were screened for their ability to hydroxylate **191** in the para position. The fungus *Aspergillus tamarii* was found to carry out this reaction quite efficiently. A 79% conversion of **191** to **192** at a substrate level of 10 gm/liter is reported for shake flasks, while 69% conversion was achieved at a level of 6 gm/liter in 30 liter fermentors.[202]

Acronycine (**193**) is an alkaloid having antitumor activity in several experimental tumor systems. In order to obtain sufficient quantities of acronycine metabolites to determine if they too may have desirable activities, several organisms were screened to determine if they would give the desired compounds.[203,204] From this work, two organisms have been reported to hydroxylate **193** at the 9-position, giving the main mammalian metabolite **194**. *Aspergillus alleaceus* gave **194** in 77% yield, [203] while *Cunninghamella echinulata* produced **194** in 30% yield.[204] Another microorganism, *Streptomyces spectabilicus*, gave **195** resulting from hydroxylation of a methyl group of **193**.[203]

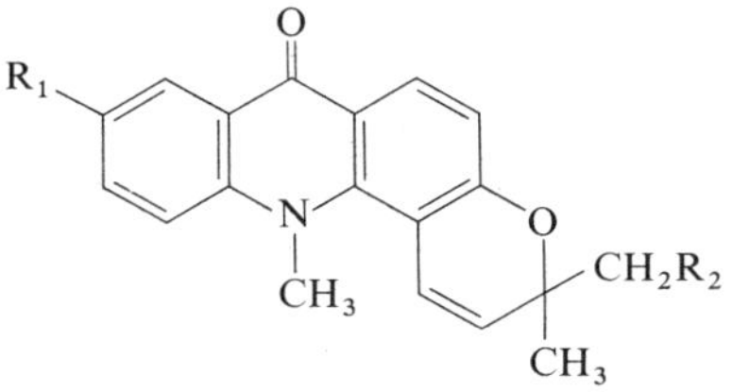

(**193**) $R_1 = R_2 = H$
(**194**) $R_1 = OH; R_2 = H$
(**195**) $R_1 = H; R_2 = OH$

The two preceding examples demonstrate synthetically useful hydroxylations as well as the preparation of drug metabolites. Smith and Rosazza have selected a battery of eleven microorganisms and have suggested that

[202] R. J. Theriault and T. H. Longfield, *Appl. Microbiol.* **25**, 606 (1973).
[203] D. R. Brannon, D. R. Horton, and G. H. Svoboda, *J. Med. Chem.* **17**, 653 (1974).
[204] R. E. Betts, D. E. Walters, and J. P. Rosazza, *J. Med. Chem.* **17**, 599 (1974).

they can be used as models of mammalian aromatic hydroxylation.[205] To demonstrate their approach they have used each organism to attempt the hydroxylation of thirteen selected aromatic compounds. Only one of the thirteen, nitrobenzene, was not hydroxylated by at least one of the microorganisms.

A microbiological synthesis of L-dopa (**197**) from L-tyrosine (**196**) required that the amino function of the amino acid be protected during the fermentation.[206] A second group now claims to have found conditions for the direct conversion of **196** to **197** in almost 90% yield with *Aspergillus oryzae*.[207] The use of washed mycelia of the microorganism in a medium of low pH (3–4) appears important to the success of this hydroxylation.

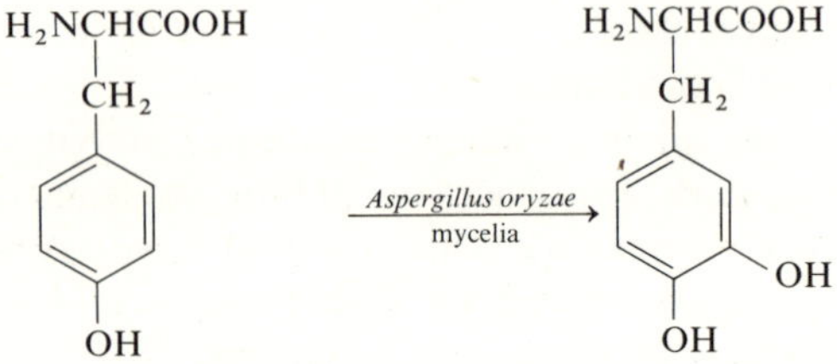

(**196**) (**197**)

Of mechanistic interest has been the elucidation of several early chemical steps in the enzymatic formation of monohydroxylated aromatic compounds. Now thoroughly documented and reviewed,[208,209] the formation of arene oxides followed by rearrangement to the phenols via the "NIH shift" was first discovered in the mammalian oxygenation of aromatic compounds. It was soon found that this sequence applied to several microbiological hydroxylations of aromatics as well.[210] In a preliminary report, the hydroxylation of anisole with nine different fungi was studied with regard to the migration of deuterium from the position of hydroxylation to the adjacent carbon.[211,211a] All nine fungi were found to effect the migration of deuterium to some extent during the hydroxylation, consistent with the mechanism of the NIH shift.

[205] R. V. Smith and J. P. Rosazza, *Arch. Biochem. Biophys.* **161**, 551 (1974).
[206] C. J. Sih, P. Foss, J. P. Rosazza, and M. Lemberger, *J. Am. Chem. Soc.* **91**, 6204 (1969).
[207] K. Haneda, S. Watanabe, and I. Takeda, *J. Ferment. Technol.* **51**, 398 (1973).
[208] D. M. Jerina and J. W. Daly, *Science* **185**, 573 (1974).
[209] J. W. Daly, D. M. Jerina, and B. Witkop, *Experientia* **28**, 1129 (1972).
[210] G. Guroff, J. W. Daly, D. M. Jerina, J. Renson, B. Witkop, and S. Udenfriend, *Science* **157**, 1524 (1967).
[211] B. J. Auret, D. R. Boyd, P. M. Robinson, C. G. Watson, J. W. Daly, and D. M. Jerina, *Chem. Commun. p.* 1585 (1971).
[211a] D. R. Boyd, R. M. Campbell, H. C. Craig, C. G. Watson, J. W. Daly, and D. M. Jerina, *J. Chem. Soc., Perkin Trans. 1* p. 2438 (1976).

TABLE XIV

cis-1, 2-Dihydroxylation of Aromatic Rings

Substrate	Microorganism	Product	$[\alpha]_D$	Yield (%)	Ref.
Benzene	*Pseudomonas putida* (mutant 39/D) *Moraxella* sp.	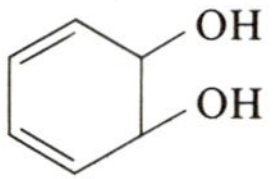			184, 185
Toluene	*P. putida* 39/D	(1*S*,2*R*) — 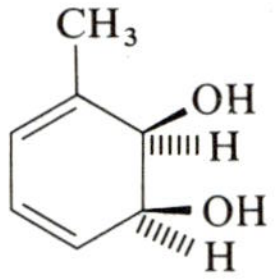	+25°	20	4, 186
Ethylbenzene	As above	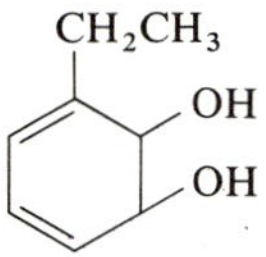	+42°	38	187
		+ 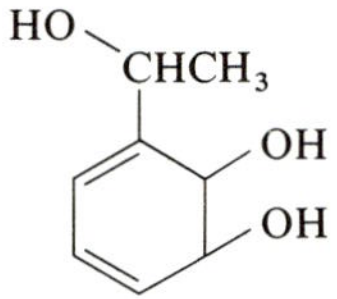	+54°	<2	
(±)-1-Phenylethanol	As above	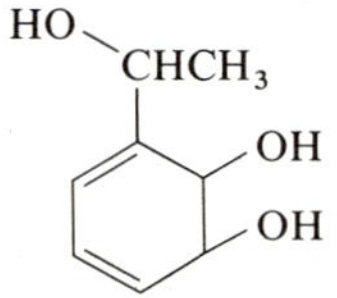	+29°	10	187

(*cont.*)

TABLE XIV (*cont.*)

Substrate	Microorganism	Product	$[\alpha]_D$	Yield (%)	Ref.
(+)-1-Phenylethanol	As above	As above	+55°	13	186
(−)-1-Phenylethanol	As above	As above	+6°	6	187
Acetophenone	As above	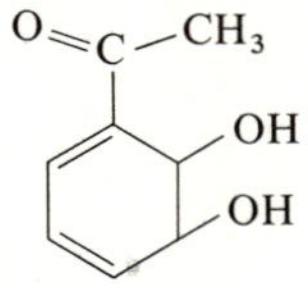	+104°	4	187
α-Methylstyrene	Bacterial strain S10781	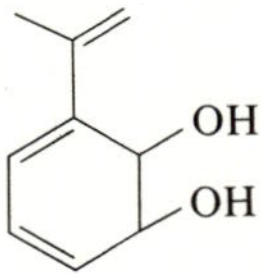	−105°	11	188
p-Xylene	*P. putida* 39/D	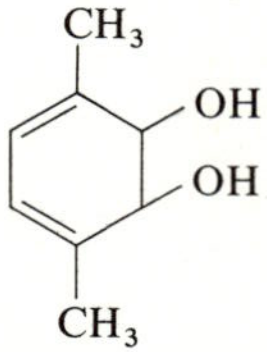	—	[a]	189
m-Xylene	As above	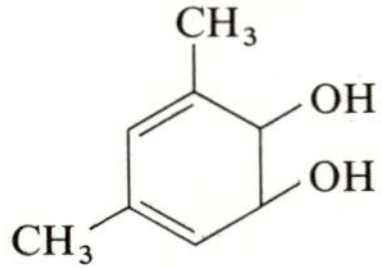	—	—	189

4-Chlorotoluene	*P. putida*	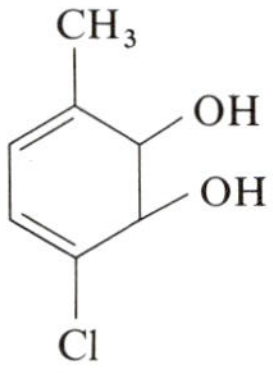	+46°	[a]	
Benzoic acid	*Alcaligenes eutrophus* (mutant strain)	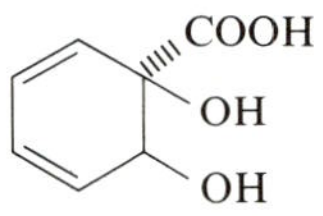	−123°[b]	50	191
3-Methylbenzoic acid	As above	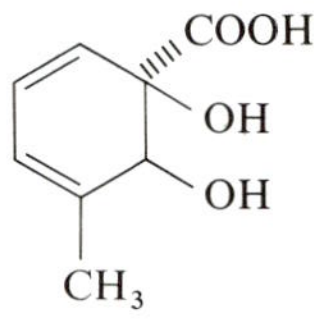	—	—	191
3-Chlorobenzoic acid	As above	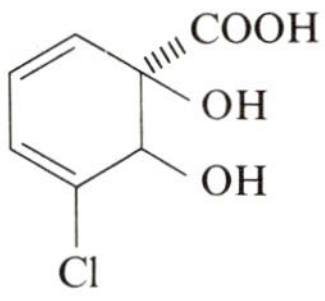	—	—	191
3-Fluorobenzoic acid	As above	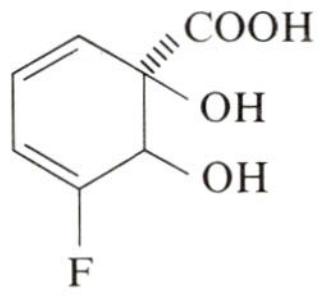	—	—	191

(*cont.*)

TABLE XIV (*cont.*)

Substrate	Microorganism	Product	$[\alpha]_D$	Yield (%)	Ref.
4-Fluorobenzoic acid	As above	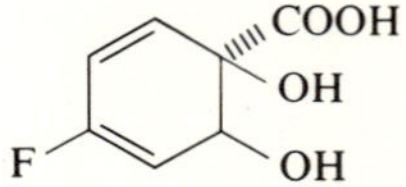	—	—	191
Naphthalene	*P. putida* 39/D	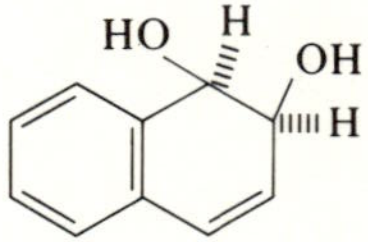 (1*S*,2*R*)	+211°	47	192
1-Naphthol	*Pseudomonas* sp.	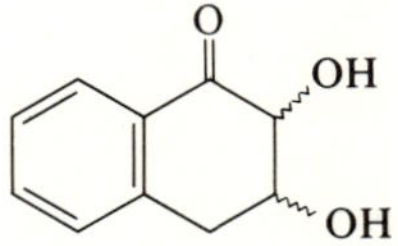	—	~10	193
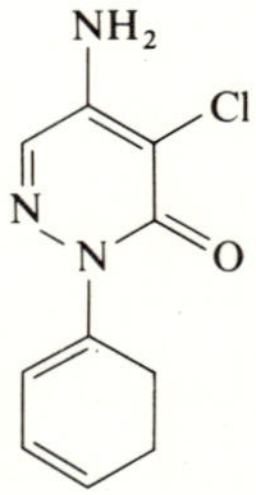	Bacterium	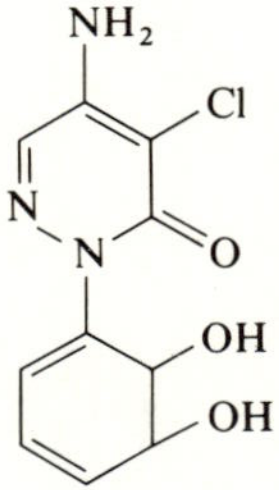	—	[a]	194

Biphenyl	*Beijerinckia* mutant strain B-836	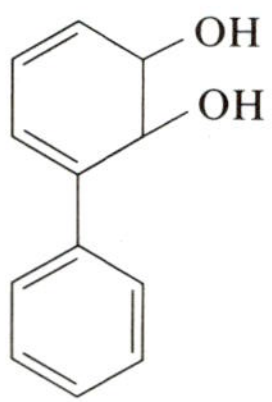		46	195
Phenanthrene	As above	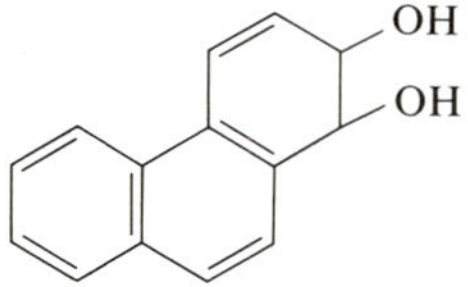		45	195
Anthracene	As above	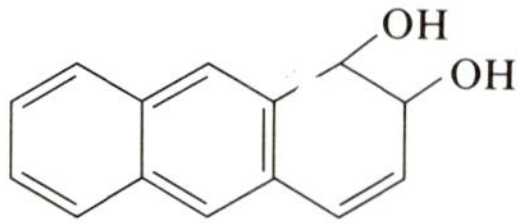	—	15	195
Benz[*a*]anthracene	As above	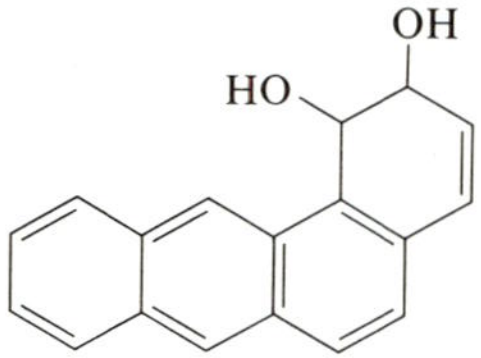	—	—	195
Benzo[*a*]pyrene	As above	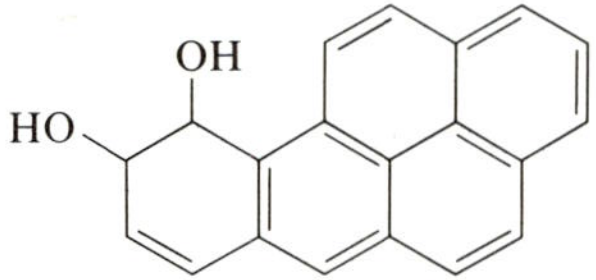	—	—	195

[a] Product isolated but yield not determined.
[b] Specific rotation observed on sodium salt of the acid.

A full report of these experiments has recently been published.[211a] Detailed reports of studies of the NIH shift with a *Pseudomonas* sp.[212] and with *Cunninghamella bainieri*[213] have also appeared.

Several of the early chemical steps in the *o*-dihydroxylation of aromatic rings by microorganisms have been elucidated. In these reactions both atoms of the oxygen molecule are added to the ring. The first product of this reaction that has been isolated is the *cis*-1,2-dihydroxy-1,2-dihydro aromatic ring structure. An example illustrating this reaction is the oxygenation of toluene given in the introduction to this chapter. A listing of this and other examples is given in Table XIV.

Following the enzymatic formation of the *cis*-1,2-dihydroxy-1,2-dihydro structure, a second enzyme catalyzes the dehydrogenation of this functional array to restore the aromaticity of the ring. This second reaction is facile, and consequently, it was only by using mutants in which this reaction is blocked, that the dihydro structures could be isolated on a large scale. The enzyme responsible for this dehydrogenation has been isolated and purified in at least two cases.[214,215]

o-Dihydroxy aromatic ring systems are quite susceptible to further enzymatic oxidations. These reactions often lead to cleavage of the aromatic ring between the two hydroxyl groups and can result either in complete metabolism of the molecule or in production of useful chemical intermediates. Notable in the latter regard is the conversion of *p*-xylene (**198**) to α,α-dimethyl-

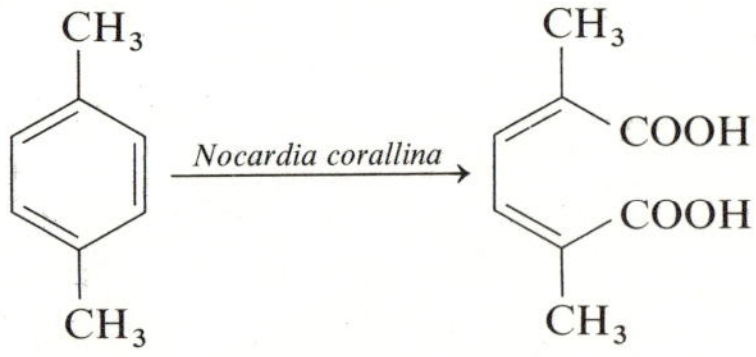

(**198**)

cis,cis-muconic acid in good yield by *Nocardia corallina*.[216,217] Additional examples of aromatic ring cleavage have been reviewed previously.[8]

212 W. R. Bowman, W. R. Gretton, and G. W. Kirby, *J. Chem. Soc., Perkin Trans. 1* p. 218 (1973).

213 J. P. Ferris, M. J. Fasco, F. L. Stylianopoulou, D. M. Jerina, J. W. Daly, and A. M. Jeffery, *Arch. Biochem. Biophys.* **156**, 97 (1973).

214 B. C. Axcell and P. J. Geary, *Biochem. J.* **136**, 927 (1973).

215 T. R. Patel and D. T. Gibson, *J. Bacteriol.* **119**, 879 (1974).

216 V. W. Jamison, R. L. Raymond, and J. O. Hudson, *Appl. Microbiol.* **17**, 853 (1969).

217 P. Hosler and R. W. Eltz, *in* "Fermentation Advances" (D. Perlman, ed.), p. 789. Academic Press, New York, 1969.

CHAPTER **III**

Oxidations with Peroxy Acids and Other Peroxides

Božo Plesničar

I. Introduction

Organic peroxides may be regarded as derivatives of hydrogen peroxide, HOOH, in which one or both hydrogen atoms are replaced by organic

ISBN 0-12-697252-4

radicals.[1–4] The reactivity of these compounds with organic substrates depends largely on whether hydrogen is one of the attached groups.

H—O—O—H	R—O—O—H	R—O—O—R
Hydrogen peroxide	Alkyl hydroperoxide	Dialkyl peroxide

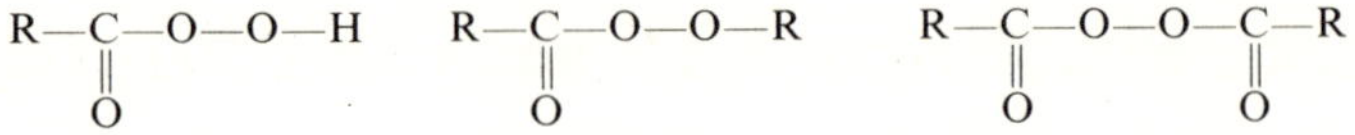

Peroxy acid	Peroxy ester	Diacyl peroxide

The reactions of diacyl peroxides,[5,6] peroxy esters,[7–9] and dialkyl peroxides[6,9,10] involve mainly the homolytic cleavage of the peroxidic bond, and require higher temperatures, ultraviolet irradiation, or the presence of a transition metal catalyst. Mixtures of products are usually obtained in these cases. Except for copper-catalyzed reactions of allylic and other activated carbon–hydrogen bonds with peroxy esters, these reactions are of limited synthetic interest and will not be discussed in this chapter.

On the other hand, the reactions of monosubstituted derivatives, i.e., peroxy acids,[11–15] alkyl hydroperoxides,[14–16] and hydrogen peroxide[14,15,17] with organic substrates are largely heterolytic and usually proceed under

[1] R. Criegee, *in* "Methoden der organischen Chemie" (E. Müller, ed.), Vol. 8, p. 1. Thieme, Stuttgart, 1952.

[2] A. G. Davies, "Organic Peroxides." Butterworth, London, 1961.

[3] E. G. E. Hawkins, "Organic Peroxides." Van Nostrand-Reinhold, Princeton, New Jersey, 1961.

[4] D. Swern, ed., "Organic Peroxides," Vol. 1, 2, and 3, Wiley (Interscience), New York, 1970, 1971, and 1972 resp.

[5] R. Hiatt, *in* "Organic Peroxides" (D. Swern, ed.), Vol. 2, Chapter 8, p. 799. Wiley (Interscience), New York, 1971.

[6] G. Sosnovsky and D. J. Rawlinson, *in* "Organic Peroxides" (D. Swern, ed.), Vol. 1, Chapter 9, p. 561. Wiley (Interscience), New York, 1970.

[7] L. A. Singer, *in* "Organic Peroxides" (D. Swern, ed.), Vol. 1, Chapter 5, p. 265. Wiley (Interscience), New York, 1970.

[8] G. Sosnovsky and D. J. Rawlinson, *in* "Organic Peroxides" (D. Swern, ed.), Chapter 10, Vol. 1, p. 585. Wiley (Interscience), New York, 1970.

[9] S. O. Lawesson and G. Schroll, *in* "The Chemistry of Carboxylic Acids and Esters" (S. Patai, ed.), Chapter 14, p. 669. Wiley (Interscience), New York, 1969.

[10] R. Hiatt, *in* "Organic Peroxides" (D. Swern, ed.), Vol. 3, Chapter 1, p. 1. Wiley (Interscience), New York, 1972.

[11] D. Swern, *Chem. Rev.* **45**, 1 (1949).

[12] D. Swern, *Org. React.* **7**, 378 (1953).

[13] D. Swern, *in* "Organic Peroxides" (D. Swern, ed.), Vol. 1, Chapter 6, p. 313. Wiley (Interscience), New York, 1970.

[14] S. N. Lewis, *in* "Oxidation, Techniques and Applications in Organic Synthesis" (R. L. Augustine, ed.), Vol. 1, Chapter 5, p. 213. Dekker, New York, 1969.

[15] H. O. House, "Modern Synthetic Reactions," 2nd ed. Chapter 6, p. 292. Benjamin, Menlo Park, California, 1972.

mild conditions. Transfer of the oxygen atom from the peroxide to the substrate in most cases involves a nucleophilic displacement at the electrophilic oxygen [Eq. (1)]. The oxidizing power of a particular peroxide depends on the acidity of the conjugated acid (ROH) of the leaving group RO^-. The following order of reactivity of compounds under consideration is observed: $RCO_3H > HOOH \geq ROOH$. The reactivity of the substrate will depend on the electron availability at the reaction center.[18,19]

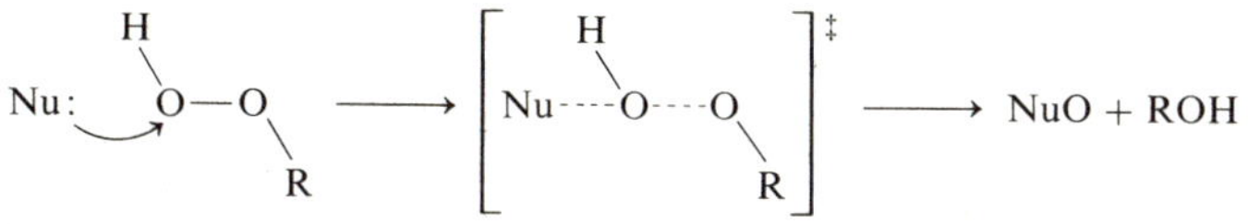
(1)

The present chapter discusses the oxidation of various organic compounds with peroxy acids of the type just described. Due to their great selectivity, simple reduction products, and relatively simple analysis, these reagents represent the major application of organic peroxy compounds in preparative chemistry. Organic peroxy acids are the reagents of choice for the preparation of epoxides and glycols from olefins, sulfoxides and sulfones from sulfides, *tert*-amine oxides from *tert*-amines, and esters or lactones from ketones. In all these oxidations, organic peroxy acids have many advantages over inorganic oxidants. The only reduction product, the corresponding carboxylic acid, can easily be removed. The oxidation product is thus obtained in high yield and purity. Some oxidations with alkyl hydroperoxides and hydrogen peroxide are also included. All these reagents complement each other in that alkyl hydroperoxides and hydrogen peroxide can also be used in neutral and basic solution.

Copper-catalyzed peroxy ester oxidations of activated carbon–hydrogen bonds are discussed in Section VIII. A brief outline of the preparation and properties of the reagents is also included.

The subject matter is organized according to organic functional groups and mechanisms. No attempts have been made to list all references in the field. Rather, most typical examples of synthetically useful reactions are presented, and whenever possible, considerable emphasis is given to the reaction mechanism.

---

[16] R. Hiatt, *in* "Organic Peroxides" (D. Swern, ed.), Vol. 2, Chapter 1, p. 1. Wiley (Interscience), New York, 1971.

[17] J. G. Wallace, "Hydrogen Peroxide in Organic Chemistry." Du Pont Publication, Wilmington, Delaware, 1962.

[18] J. O. Edwards, *in* "Peroxide Reaction Mechanisms" (J. O. Edwards, ed.), Chapter 4, p. 67. Wiley (Interscience), Publishers, New York, 1962.

[19] R. Curci and J. O. Edwards, *in* "Organic Peroxides" (D. Swern, ed.), Vol. 1, Chapter 4, p. 199. Wiley (Interscience), New York, 1970.

II. Reagents

A. Organic Peroxy Acids

1. Preparation

Two general methods for preparation of organic peroxy acids are used most frequently in the laboratory.[13] The first one involves introduction or

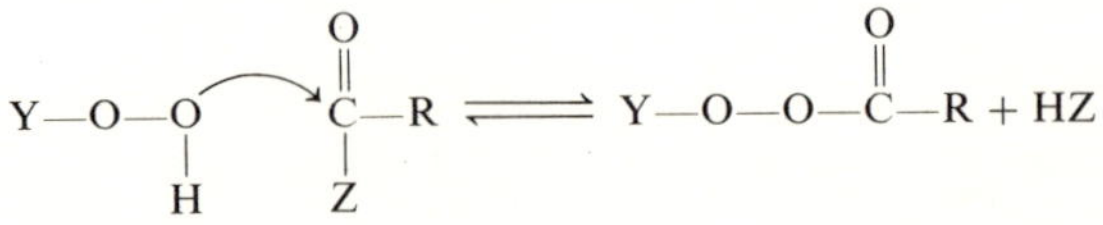

(2)

displacement of the peroxy group at the acyl carbon in a bimolecular nucleophilic reaction ($B_{AC}2$ or $A_{AC}2$)[2] [Eq. (2)]. The second method is the autoxidation of aldehydes [Eq. (3)]. The key reagent for the preparation of organic

$$R{-}C({=}O){-}H + O_2 \longrightarrow R{-}C({=}O){-}O{-}O{-}H \quad (3)$$

peroxy acids is hydrogen peroxide, which is available commercially in aqueous solutions having concentrations ranging from 3 to 98%.[17,20,21]

The active oxygen of hydrogen peroxide, which is not readily available for oxidation of organic compounds, can be transformed to the active peroxy acid form by direct oxidation of carboxylic acids with 30 to 98% hydrogen peroxide in the presence of mineral acid as a catalyst [Eq. (4)]. Strong aliphatic acids, i.e., formic or trifluoroacetic acid readily form peroxy acids

$$RCO_2H + H_2O_2 \overset{H^+}{\rightleftharpoons} RCO_3H + H_2O \quad (4)$$

even in the absence of the acid catalyst.[22] Water-insoluble aliphatic peroxy acids can be prepared in concentrated sulfuric acid as a solvent by using an excess of 50–60% hydrogen peroxide.[23]

Concentrated solutions of peroxyacetic acid (**1**) can be made by the reaction of acetic anhydride with 90% hydrogen peroxide in 1,2-dichloroethane [Eq. (5)].[24] Pure peroxyacetic acid is obtained by fractional distillation under

$$(CH_3CO)_2O + H_2O_2 \xrightarrow[ClCH_2CH_2Cl]{CF_3CO_2H\ (cat.)} \underset{(\mathbf{1})}{CH_3CO_3H} \quad (5)$$

vacuum or by fractional freezing of concentrated solutions.

[20] W. C. Schumb, C. N. Satterfield, and R. L. Wentworth, "Hydrogen Peroxide." Van Nostrand-Reinhold, Princeton, New Jersey, 1955.

[21] E. S. Shanley, *in* "Peroxide Reaction Mechanisms" (J. O. Edwards, ed.), Chapter 7, p. 129. Wiley (Interscience), Publishers, New York, 1962.

[22] W. D. Emmons and A. F. Ferris, *J. Am. Chem. Soc.* **75**, 4623 (1953).

[23] W. E. Parker, C. Ricciuti, C. L. Ogg, and D. Swern, *J. Am. Chem. Soc.* **77**, 4037 (1955).

[24] M. F. Hawthorne, W. D. Emmons, and K. S. McCullum, *J. Am. Chem. Soc.* **80**, 6393 (1958).

Peroxyacetic acid is now available commerically as a 40% solution in acetic acid with a trace of sulfuric acid.[25]

Anhydrous peroxytrifluoroacetic acid (**2**)[26] and peroxymaleic acid (**3**),[27] both the strongest peroxy acids known, can be prepared analogously in the reaction of the corresponding anhydride with 90% hydrogen peroxide in dichloromethane. Substituted peroxymaleic acids can also be prepared by this procedure.[28,29] For example, the isolation and characterization of diperoxyphenylmaleic acid has recently been reported.[29]

$$(CF_3CO)_2O + H_2O_2 \xrightarrow{CH_2Cl_2} \underset{\textbf{(2)}}{CF_3CO_3H} + CF_3CO_2H \qquad (6)$$

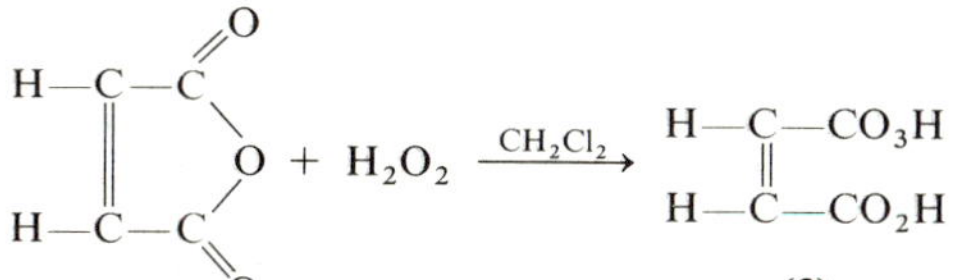

(**3**) (7)

Peroxybenzoic acid (**4**) has been prepared by the reaction of benzoyl peroxide with methanolic sodium methoxide in dichloromethane [Eq. (8a)].[30,31] *m*-Chloroperoxybenzoic acid (**5**), which is also available com-

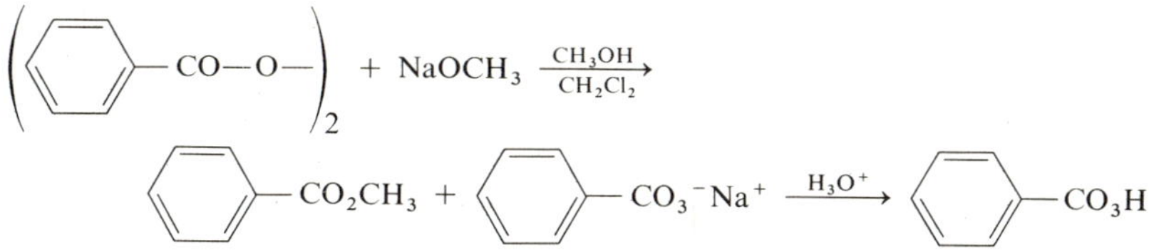

(8a)

(**4**)

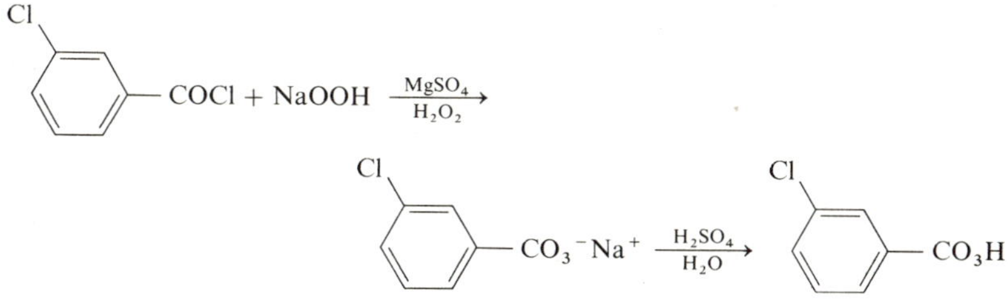

(8b)

(**5**)

[25] The FMC Corp., Inorganic Chemical Div., New York.

[26] W. D. Emmons, *J. Am. Chem. Soc.* **76**, 3468 (1954).

[27] R. H. White and W. D. Emmons, *Tetrahedron* **17**, 31 (1962).

[28] A. Pollak, M. Zupan, and B. Šket, *Synthesis* p. 495 (1973).

[29] M. M. Martin and J. M. King, *J. Org. Chem.* **38**, 1588 (1973).

[30] G. Braun, *in* "Organic Syntheses" (H. Gilman and A. H. Blatt, eds.), Collect. Vol. I, 2nd ed., p. 431. Wiley, New York, 1941.

[31] R. F. Cleinschmidt and A. C. Cope, *J. Am. Chem. Soc.* **66**, 1929 (1944).

mercially as a white crystalline solid containing ca. 85% of the peroxy acid,[32] can be prepared by the reaction of the corresponding benzoyl chloride with 30% hydrogen peroxide in the presence of excess sodium hydroxide and a a trace of magnesium sulfate.[32a]

Acyl chlorides can be converted to peroxy acids by the reaction with pyridine and 85% hydrogen peroxide in what is believed to be a two-step reaction. The reaction is usually carried out in tetrahydrofuran to provide a homogeneous medium.[32b] The use of methanesulfonic acid as a solvent and

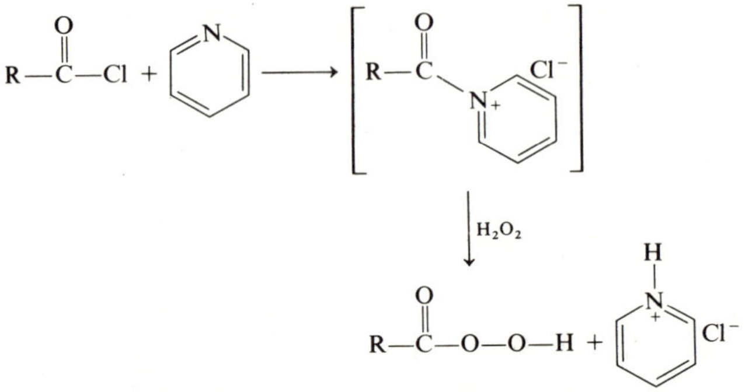

(9)

catalyst enables an elegant direct conversion of benzoic and high molecular weight aliphatic acids to peroxy acids with 70–95% hydrogen peroxide [Eq. (10)].[33] The reaction mixture is diluted with ice and cold, saturated

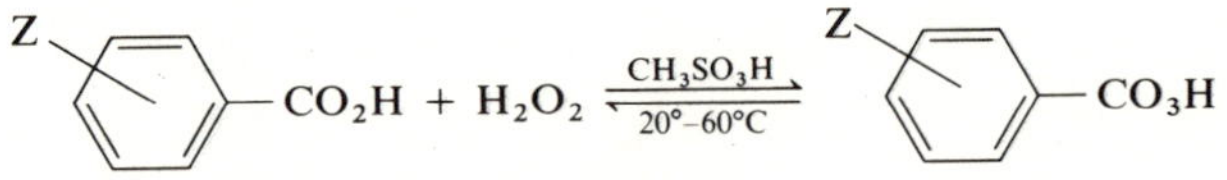

(10)

$Z = NO_2$, CN, F, Cl, Br, CH_3, *t*-Bu

ammonium sulfate, and the peroxy acid is then extracted with benzene or filtered (solid peroxy acids). This procedure is now the most convenient method for the preparation of peroxybenzoic acids.

Polymer-supported aromatic peroxy acids (**6**), which seem to be especially suitable for large scale oxidations, have been prepared by the treatment of carboxy-substituted polystyrene resins with hydrogen peroxide in methanesulfonic acid [Eq. (11)].[34] Polymer-supported higher aliphatic peroxy acids

[32] L. M. Fieser and M. Fieser, "Reagents for Organic Synthesis," Vols. I–V. Wiley, New York, 1968–1976.

[32a] R. N. McDonald, R. N. Steppel, and J. E. Dorsey, *Org. Synth.* **50**, 15 (1970).

[32b] J. Y. Nedelec, J. Sorba, and D. Lefort, *Synthesis* p. 821 (1976).

[33] L. S. Silbert, E. Siegel, and D. Swern, *J. Org. Chem.* **27**, 1336 (1962); *Org. Synth.* **43**, 93 (1963).

[34] C. R. Harrison and P. Hodge, *J. Chem. Soc., Perkin Trans. 1* pp. 605 and 2252 (1976).

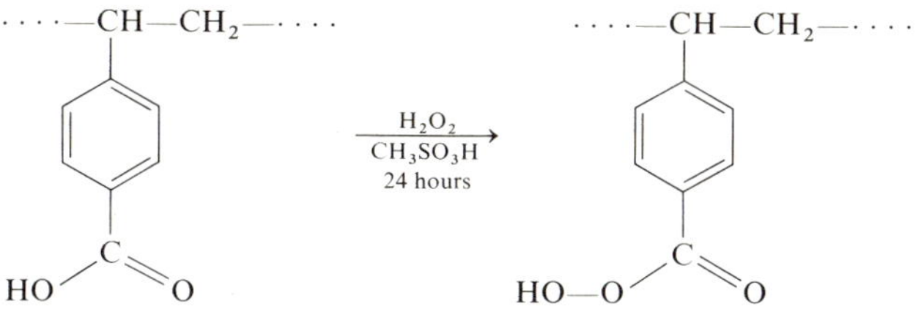

(11)

(6)

75–85% conversion of the carboxy groups

would also be very convenient for oxidation of acid sensitive compounds owing to a relatively lower acidity of the reduction product as compared to aromatic carboxylic acids.

Monoperoxyphthalic acid (**7**) can conveniently be made by treating phthalic anhydride with 30% hydrogen peroxide in diethyl ether or with 90% hydrogen peroxide in chloroform [Eq. (12)].[35,36] In cases where the

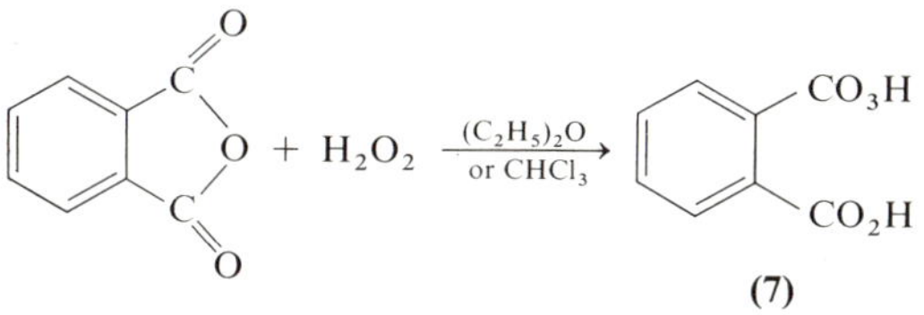

(12)

(7)

substrate or the oxidized product is sensitive to acidic conditions, peroxybenzimidic acid (**8**), prepared *in situ* from benzonitrile and alkaline hydrogen peroxide, can be used[37–40] [Eq. (13)]. The acidic alkyl hydroperoxide (**9**),

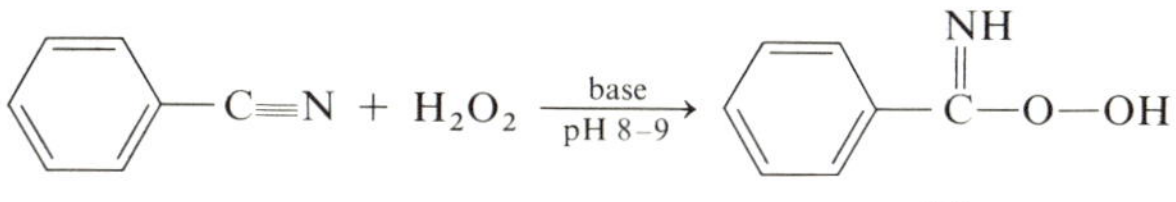

(13)

(8)

presumably formed *in situ* by the reaction of hexafluoroacetone and hydrogen peroxide, appears also to be of some synthetic interest [Eq. (14)].[41]

[35] H. Böhme, *in* "Organic Syntheses" (E. C. Horning, ed.), Collect. Vol. III, p. 619. Wiley, New York, 1955.

[36] E. E. Royals and L. L. Harrel, Jr., *J. Am. Chem. Soc.* **77**, 3405 (1955).

[37] G. B. Payne, *J. Org. Chem.* **24**, 1354 (1959).

[38] K. B. Wiberg, *J. Am. Chem. Soc.* **75**, 3961 (1953).

[39] G. B. Payne, P. H. Deming, and P. H. Williams, *J. Org. Chem.* **26**, 659 (1961).

[40] G. B. Payne, *Tetrahedron* **18**, 763 (1962).

[41] R. D. Chambers and M. Clark, *Tetrahedron Lett.* p. 2741 (1970).

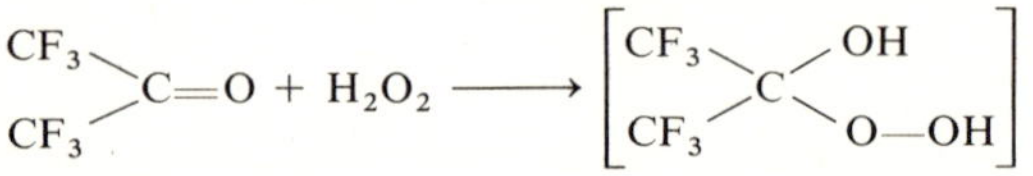 (14)

(9)

As mentioned above, an alternative method for the preparation of peroxy acids involves air oxidation of aldehydes. In this way, peroxyacetic acid free of acetic acid has been prepared [Eq. (15)].[42,43] Air oxidation of benzaldehyde to peroxy acid has also been used successfully for *in situ* oxidations of various organic substrates.[44,45]

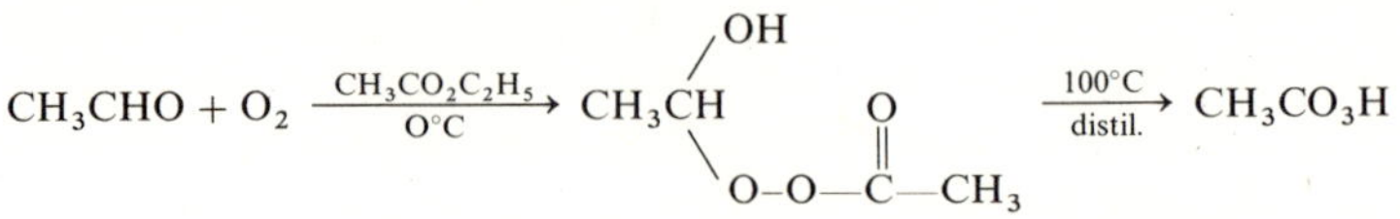 (15)

2. Physical Properties

Infrared,[46–50] nuclear magnetic resonance,[51,52] dipole moment,[53] and molecular weight[53] studies indicate that organic peroxy acids exist in "inert" solvents in the form of intramolecularly hydrogen-bonded chelates.[54]

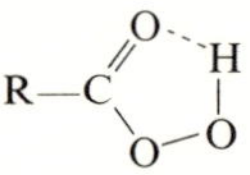

Consistent with the chelated form of peroxy acids in solution are lower dissociation constants as well as higher volatility compared to parent carboxylic acids. The pK values of peroxy acids lie between those of carboxylic acids and alkyl hydroperoxides (peroxyacetic acid, pK = 8.2;

[42] B. Phillips, F. C. Frostick, and P. S. Starcher, *J. Am. Chem. Soc.* **79**, 5982 (1957).
[43] P. S. Starcher, B. Phillips, and F. C. Frostick, *J. Org. Chem.* **26**, 3568 (1961).
[44] D. Swern, T. W. Findley, and T. Scanlan, *J. Am. Chem. Soc.* **66**, 1925 (1944).
[45] C. R. Dick and R. F. Hanna, *J. Org. Chem.* **29**, 1218 (1964).
[46] Z. K. Maizus, G. Ya. Timofeeva, and N. M. Emanuel, *Dokl. Akad. Nauk SSSR* **70**, 655 (1950).
[47] P. A. Giguère and A. W. Olmos, *Can. J. Chem.* **30**, 821 (1952).
[48] D. Swern, L. P. Witnaeuer, C. R. Eddy, and W. E. Parker, *J. Am. Chem. Soc.* **77**, 5537 (1955).
[49] R. Kavčič, B. Plesničar, and D. Hadži, *Spectrochim. Acta, Part A* **23**, 2483 (1967).
[50] D. Hadži, R. Kavčič, and B. Plesničar, *Spectrochim. Acta, Part A* **27**, 179 (1971).
[51] D. Swern, A. H. Clements, and T. M. Luong, *Anal. Chem.* **41**, 412 (1969).
[52] B. Plesničar, unpublished results.
[53] J. R. Rittenhouse, N. Lobunez, D. Swern, and J. G. Miller, *J. Am. Chem. Soc.* **80**, 4850 (1958).
[54] L. S. Silbert, *in* "Organic Peroxides" (D. Swern, ed.) Vol. 2, Chapter 7, p. 637. Wiley (Interscience), New York, 1971.

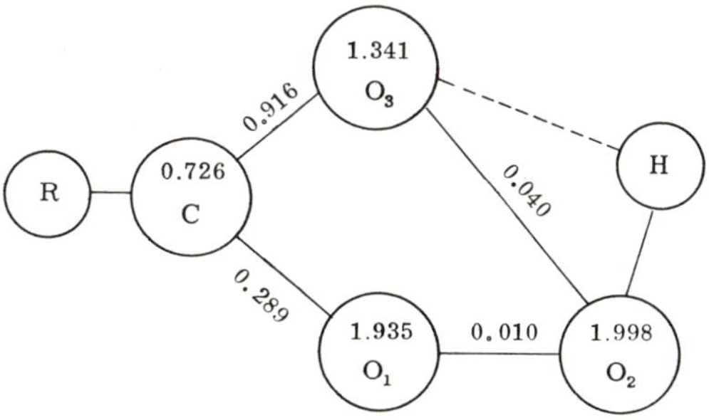

FIG. 1. Charges and bond orders in the intramolecularly hydrogen-bonded form of the peroxycarboxyl group.

peroxybenzoic acid, pK = 7.8).[55,56] The reasons for lower acidity of peroxy acids are evident. The anion derived from the peroxy acid lacks resonance stabilization of the carboxylate anion. At the same time, the polar and

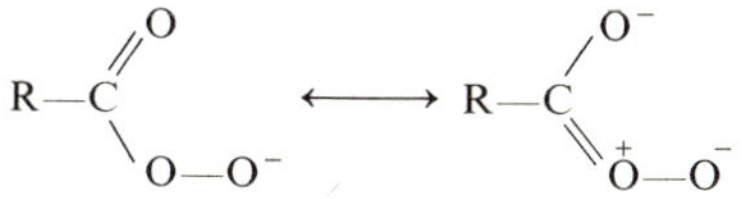

electronic effects of the carbonyl group and substituents in aromatic peroxy acids are weakened owing to an additional oxygen atom.

Lower acidity of peroxy acids compared to carboxylic acids enables the purification of the former by washing with various inorganic buffers which react only with carboxylic acids. The purification of commercially available *m*-chloroperoxybenzoic acid, which contains ca. 15% of *m*-chlorobenzoic acid with a phosphate buffer (pH 7.5) is an example.

Normal coordinate analysis has been used to study the intramolecular hydrogen bonding and potential functions of carboxylic and peroxy acids. The results indicate weak O ··· H bonds in carboxylic acids and considerably stronger O ··· H bonds in peroxy acids.[57] Electron deficiency of the oxygen atom bonded to hydrogen (O_2, electrophilic center) has been confirmed by the molecular orbital Pariser–Parr–Pople SCF calculations.[58] Charges and bond order in the intramolecularly hydrogen-bonded form of the peroxycarboxyl group are shown in Fig. 1. The localization energies on atoms O_1 and O_2 are 0.5 and 0.1 eV, respectively, and the frontier electron densities are 0.16 and 0.02, respectively.

[55] E. Koubek, M. L. Hagget, C. J. Battaglia, K. M. Ibne-Rasa, H. -Y. Pyun, and J. O. Edwards, *J. Am. Chem. Soc.* **85**, 2263 (1963).

[56] J. F. Goodman, P. Robson, and E. R. Wilson, *Trans. Faraday Soc.* **58**, 1846 (1962).

[57] W. V. Brooks and C. M. Haas, *J. Phys. Chem.* **71**, 650 (1967).

[58] A. Ažman, B. Borštnik, and B. Plesničar, *J. Org. Chem.* **34**, 971 (1969).

The question of the stereochemistry of the five-membered ring of the peroxy acid is still not settled. The conformation with a dihedral angle of 72° has been proposed for higher aliphatic peroxy acids on the basis of the dipole moment study.[53] A recent *ab initio* molecular orbital study indicates a totally planar ring structure for peroxyacetic acid.[59] Both possibilities are shown in Fig. 2.

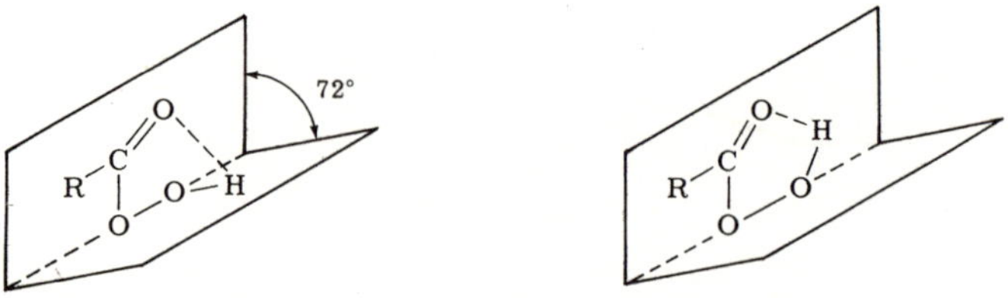

FIG. 2. Stereochemistry of the five-membered ring of peroxy acid.

A dipole moment study (graphical method of analyzing the data of the vectorial additive scheme) of meta- and para-substituted peroxybenzoic acids shows that experimental values agree with the planar cis conformation of the peroxycarboxyl group, provided that a significant contribution of the hydrogen bond to the moment is accounted for.[60] The energy of the intramolecular hydrogen bond has been roughly estimated to be in the range of ca 2–5 kcal/mole.[47,59] Solvents and systems with basic oxygen are able to disrupt the intramolecularly hydrogen-bonded chelate [Eq. (16)].[61–63] This type of association has, as will be shown below, a considerable influence on the rate of oxidation with peroxy acids when alcohols, esters, ethers, and amides are used as solvents.

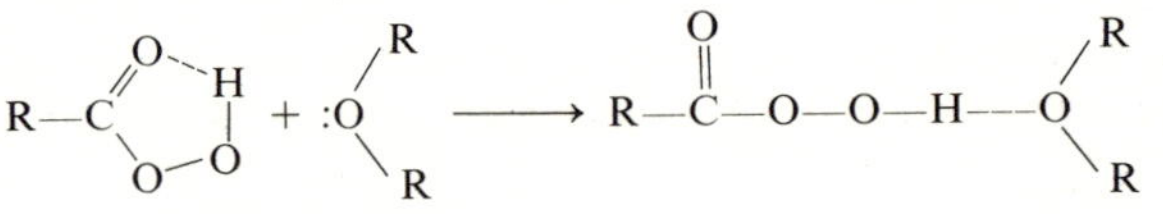

(16)

The heats of reaction between a dilute solution of *m*-chloroperoxybenzoic acid in carbon tetrachloride and various bases such as tetrahydrofuran THF, *N,N*-dimethylformamide, *N,N*-dimethylacetamide have been measured. The calculated enthalpy change per mole of hydrogen-bonded adduct formed is in the range of 5.5 to 8 kcal/mole, and is sensitive to basicity.[64] This is consistent with the shift of OH stretching frequencies of peroxy acids

[59] L. M. Hjelmeland and G. H. Loew, *Chem. Phys. Lett.* **32**, 309 (1975).

[60] O. Exner and B. Plesničar, unpublished results.

[61] R. F. Vasiljev, A. N. Terenin, and N. M. Emanuel, *Izv. Akad. Nauk SSSR* p. 403 (1956).

[62] N. N. Schwartz and J. Blumbergs, *J. Org. Chem.* **29**, 1967 (1964).

[63] R. Kavčič and B. Plesničar, *J. Org. Chem.* **35**, 2033 (1970).

[64] J. Škerjanc and B. Plesničar, unpublished results.

to lower values in infrared spectra,[63] and considerable deshielding of OH protons in NMR spectra[52] when these bases are used as solvents.

Supporting all the above is the isolation of crystalline 1:1 adducts of peroxybenzoic acids with triphenyl phosphine oxide, triphenyl arsine oxide, and pyridine oxides. The infrared spectra indicate relatively strong intermolecular hydrogen bonds in these adducts.[65] X-Ray crystallography has shown that peroxy acids exist in the solid state in the form of polymeric intermolecularly hydrogen-bonded associates.[66,67]

B. ALKYL HYDROPEROXIDES

tert-Alkyl hydroperoxides, which are most frequently used in the laboratory because of their great stability, are prepared by the reaction of an alcohol with hydrogen peroxide in the presence of a strong mineral acid (S_N1 nucleophilic displacement).[16,68,69]

$$\ce{>C-OH + H_2O_2 \longrightarrow >C-O-O-H + H_2O} \qquad (17a)$$

Alkyl hydroperoxides can also be prepared from the appropriate alkyl bromide or iodide and hydrogen peroxide in the presence of silver trifluoroacetate.[69a]

$$RX + H_2O_2 + AgOCOCF_3 \xrightarrow[0^\circ C]{(C_2H_5)_2O} ROOH + AgX + CF_3COOH \qquad (17b)$$

Oxidation of *N*-alkyl-*N*-tosyl hydrazines with hydrogen peroxide also provides a convenient method for the preparation of alkyl hydroperoxides.[69b]

$$R{-}NH{-}NH{-}Ts \xrightarrow[THF]{H_2O_2,\ Na_2O_2} \left[R{-}\overset{+}{N}{\equiv}N{-}Ts\right] \longrightarrow ROOH \qquad (17c)$$

Two *tert*-alkyl hydroperoxides, *tert*-butyl hydroperoxide (**10**) and cumyl hydroperoxide (**11**), are available commercially from several sources. The oxidizing power of these compounds is comparable to that of hydrogen

$(CH_3)_3C{-}O{-}O{-}H$

(10)

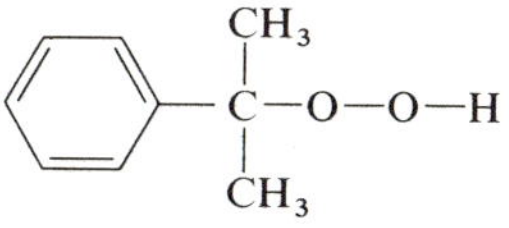

(11)

[65] B. Plesničar, R. Kavčič, and D. Hadži, *J. Mol. Struct.* **20**, 457 (1974).
[66] M. Sax, P. Beurskens, and S. Chu, *Acta Crystallogr.* **18**, 252 (1965).
[67] D. Belitskus and G. A. Jeffrey, *Acta Crystallogr.* **18**, 458 (1965).
[68] N. A. Milas and D. M. Surgenor, *J. Am. Chem. Soc.* **68**, 205 and 643 (1946).
[69] N. A. Milas and L. H. Perry, *J. Am. Chem. Soc.* **68**, 1938 (1946).
[69a] P. G. Cookson, A. G. Davies, and B. P. Roberts, *J. Chem. Soc., Chem. Commun.* p. 1022 (1976).
[69b] L. Caglioti, F. Gasparrini, and G. Palmieri, *Tetrahedron Lett.* p. 3987 (1976).

peroxide. Primary and secondary alkyl hydroperoxides tend to rearrange and are susceptible to decomposition. They are thus not used for oxidizing purposes.

Alkyl hydroperoxides are stronger acids than the corresponding alcohols ($pK_a = 11.5–12.8$).[70] They form salts with aqueous sodium hydroxide which enables their easy separation from the reaction mixture and purification.[71] Some commercial samples of *tert*-butyl hydroperoxide contain up to 30% water. This can conveniently be removed by dissolving the crude product in petroleum ether and extracting with a saturated sodium chloride solution. Subsequent passage of the organic layer through anhydrous magnesium sulfate, removal of the solvent under reduced pressure, and fractional distillation yield pure *tert*-butyl hydroperoxide.[72] Alkyl hydroperoxides tend to dimerize in nonpolar solutions but form intermolecularly hydrogen-bonded associates with alcohols, ethers, amines, amides, and carboxylic acids.[73]

C. PEROXY ESTERS

Peroxy esters can be prepared by acylation of an alkyl hydroperoxide with an acid chloride or anhydride.[7] *tert*-Butyl peroxy esters, which are normally not sensitive to base, are usually prepared by a Shotten–Baumen-type of procedure either in aqueous alkali or pyridine as a solvent and base.[74–76]

$$R{-}\overset{\overset{\displaystyle O}{\|}}{C}{-}Cl + (CH_3)_3COOH \xrightarrow{\text{pyridine}} R{-}\overset{\overset{\displaystyle O}{\|}}{C}{-}O{-}O{-}C(CH_3)_3 \qquad (18)$$

Acylation of alkyl hydroperoxides can also be accomplished with ketenes. A series of peroxyacetates has been prepared by this method.[77]

$$CH_2{=}C{=}O + (CH_3)_3COOH \longrightarrow CH_3{-}\overset{\overset{\displaystyle O}{\|}}{C}{-}O{-}O{-}C(CH_3)_3 \qquad (19)$$

An elegant method for the preparation of peroxy esters, which involves imidazolides as intermediates, has also frequently been used [Eq. (20)].[78]

[70] A. J. Everett and G. J. Minkoff, *Trans. Faraday Soc.* **49**, 410 (1953).
[71] P. D. Bartlett and J. M. McBride, *J. Am. Chem. Soc.* **87**, 1727 (1965).
[72] W. S. Trahanovsky and M. P. Doyle, *J. Org. Chem.* **32**, 146 (1967).
[73] C. Walling and L. Heaton, *J. Am. Chem. Soc.* **87**, 48 (1965).
[74] N. A. Milas and D. M. Surgenor, *J. Am. Chem. Soc.* **80**, 1398 (1958).
[75] P. D. Bartlett and R. Hiatt, *J. Am. Chem. Soc.* **80**, 1398 (1958).
[76] L. S. Silbert and D. Swern, *J. Am Chem. Soc.* **81**, 2464 (1959).
[77] R. Naylor, *J. Chem. Soc.* p. 244 (1945).
[78] R. Hecht and C. Rüchardt, *Chem. Ber.* **96**, 1281 (1963).

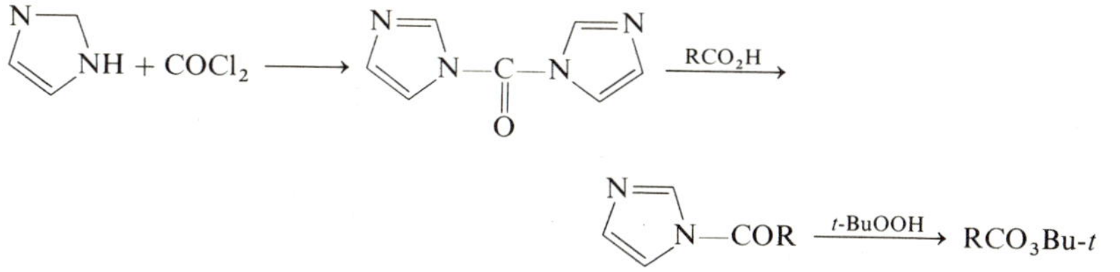

(20)

The carboxylic acid is thus converted directly to the peroxy ester without the preparation of the acid chloride.

Direct conversion of carboxylic acids to peroxy esters is also possible by treating the former with *p*-toluenesulfonyl chloride and pyridine in the presence of an alkyl hydroperoxide [Eq. (21)].[79] Two peroxy esters, *tert*-butyl

$$RCO_2H + TsCl \xrightarrow{\text{pyridine}} R{-}\overset{\overset{\displaystyle O}{\|}}{C}{-}OSO_2Ar \xrightarrow{t\text{-BuOOH}} R{-}\overset{\overset{\displaystyle O}{\|}}{C}{-}O{-}OBu\text{-}t \qquad (21)$$

peroxyacetate (**12**) and *tert*-butyl peroxybenzoate (**13**), are available commercially from several sources. Primary and secondary alkyl peroxy esters

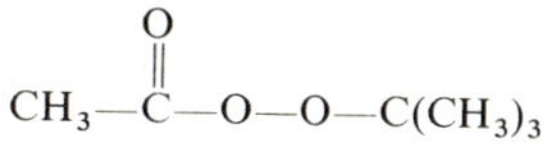

(12)

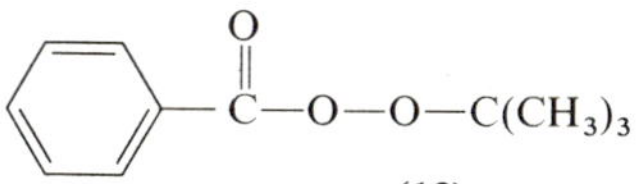

(13)

have found little application as oxidants owing to their tendency to decomposition and rearrangement. Relatively little is known about the structure of these compounds. A dipole moment study of *tert*-butyl peroxy esters indicates a nonplanar structure.[80]

D. EXPERIMENTAL PROCEDURE

Oxidations with organic peroxy acids can be run in a variety of organic solvents. The most frequently used solvents for aromatic peroxy acids are benzene, chlorobenzene, dichloromethane, 1,2-dichloroethane, and chloroform. Decomposition of peroxy acids in these solvents is rather slow at room and slightly elevated temperatures. If higher temperatures are required, decomposition may be subpressed by the addition of a radical inhibitor. For example, 4,4-thiobis(6-*tert*-butyl-3-methylphenol) completely prevents decomposition of *m*-chloroperoxybenzoic acid in 1,2-dichloroethane on heating at 90°C for 3 hours.[81] Alcohols and ethers are not recommended as reaction media, especially in cases where higher temperatures and longer reaction

[79] N. A. Milas and A. Golubović, *J. Am. Chem. Soc.* **81**, 3361, 5824, and 6461 (1959).

[80] F. D. Verderame and J. G. Miller, *J. Phys. Chem.* **66**, 2185 (1962).

[81] Y. Kishi, M. Aretani, H. Tanino, T. Fukuyama, T. Goto, S. Inoue, S. Sugiura, and H. Kakoi, *J. Chem. Soc., Chem. Commun.* p. 64 (1972).

times are required. A relatively fast free radical decomposition of peroxy acids in these solvents has been reported.[82,83]

The workup procedure depends on the nature of the oxidized products. Unreacted peroxy acid and its reduction product, the corresponding carboxylic acid, are frequently removed by shaking the reaction mixture with sodium sulfite and sodium bicarbonate, respectively. Since peroxy acids are more soluble than carboxylic acids in the above mentioned solvents, the latter may, in some cases, be removed by filtration. Lower aliphatic acids are soluble in water as well, thus allowing convenient isolation of the oxidized products by pouring the reaction mixture into water with subsequent isolation of the product. Peroxyformic, peroxyacetic, and peroxytrifluoroacetic acids can also be removed by distillation under reduced pressure due to their high volatility.

One of the mechanistic advantages of the study of peroxy acid oxidations is that the unconsummed peroxy acid can easily be determined by titrating the liberated iodine from the acidified aqueous potassium iodide with standard thiosulfate solutions.[13,84,84a]

Although the use of pure peroxy acids is preferred for kinetic and other quantitative studies, these reagents need not be isolated. The so-called *in situ* oxidation technique can be employed. The substrate is dissolved or dispersed in the carboxylic acid or anhydride and the hydrogen peroxide is added. A strong acid catalyst is usually required. Since an equilibrium mixture is thus obtained, a differential titrimetric procedure has been developed. Hydrogen peroxide is first titrated with a standard ceric sulfate solution followed by the iodometric titration of the peroxy acid.[85] Chromatographic as well as spectral techniques are also very suitable for the detection or quantitative determination of organic peroxy acids alone or in the presence of other peroxides.[84,84a,86]

Since some reactants and products are sensitive to the acidic conditions that are most frequently encountered in oxidation with organic peroxy acids, neutral or alkaline solutions of hydrogen peroxide and alkyl hydroperoxides are employed. Analysis of alkyl hydroperoxides can be performed by iodome-

[82] H. Tokumaru and O. Simamura, *Bull. Chem. Soc. Jnp.* **35**, 1955 (1962).

[83] M. Gruselle, M. Tichy, and D. Lefort, *Tetrahedron* **28**, 3885 (1972).

[84] R. N. Johnson and I. W. Siddiqi, "The Determination of Organic Peroxides." Pergamon, Oxford, 1970.

[84a] R. D. Mair and R. T. Hall, *in* "Organic Peroxides" (D. Swern ed.), Vol 2, Chapter 6, p. 535. Wiley (Interscience), New York, 1971.

[84b] G. A. Olah, D. G. Parker, N. Yoneda, and F. Pelizza, *J. Am. Chem. Soc.* **98**, 2245 (1976).

[85] F. P. Greenspan and D. G. MacKellar, *Anal. Chem.* **20**, 1061 (1948).

[86] R. Kavčič, B. Plesničar, and A. Perdih, *J. Chromatogr.* **38** 515 (1968).

tric as well as chromatographic and spectroscopic techniques.[84,84a] Carbon-13 NMR spectroscopy is a particulary useful technique in this respect.[84b] In general, the choice of a preferred oxidizing agent will thus depend on the reactivity of the substrate as well as on the stability of reactants and products under reaction conditions.

A word of caution is appropriate at this time. All peroxides are potentially explosive and should be handled with extreme care. The use of safety shielding is strongly recommended, especially in work with pure lower aliphatic peroxy acids, and with mixtures containing highly concentrated hydrogen peroxide.

III. Oxidation of Carbon–Carbon Double and Triple Bonds

A. OXIDATION OF OLEFINS WITH PEROXY ACIDS (EPOXIDATION)

1. ALKENES

The largest volume of literature concerning oxidation with peroxy acids involves olefins. Several excellent and comprehensive reviews on the topic are available.[11,12,87–88a] Only an outline of the most characteristic features of this important reaction will be given here.

The reaction of olefins with peroxy acids to produce epoxides (oxiranes) has been known for almost 70 years (Prilezhaev, 1908).[89] It provides the most convenient method for the preparation of one of the most important intermediates in organic chemistry.

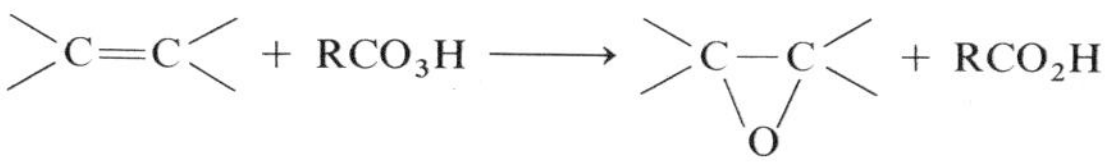

(22)

The choice of the preferred peroxy acid for epoxidation depends on the reactivity of the olefin and stability of the epoxide. Carboxylic acids, which are formed in the reaction (especially lower aliphatic acids), can react with the epoxide to form monoesters of 1,2-diols. This type of reaction will be discussed later in this chapter.

The use of peroxybenzoic acid, *m*-chloroperoxybenzoic acid, and mono-peroxyphthalic acid in a wide variety of inert solvents allows the isolation

[87] D. Swern, *in* "Organic Peroxides" (D. Swern ed.), Vol 2, Chapter 5, p. 355. Wiley (Interscience), New York 1971.

[88] G. Berti, *Top. Stereochem.* 7, p. 93 (1973).

[88a] E. N. Prilezhaeva, "Prilezhaev Reaction, Electrophilic Oxidation." Nauka, Moscow, 1974.

[89] N. Prilezhaev *Ber. Dtsch. Chem. Ges.* **42**, 4811 (1909).

of epoxides. Although the use of pure peroxy acids is preferable, the commercially available peroxyacetic acid is frequently employed. Some typical epoxidations with peroxy acids are shown below [Eqs. (23),[90] (24),[91] (25)[92]].

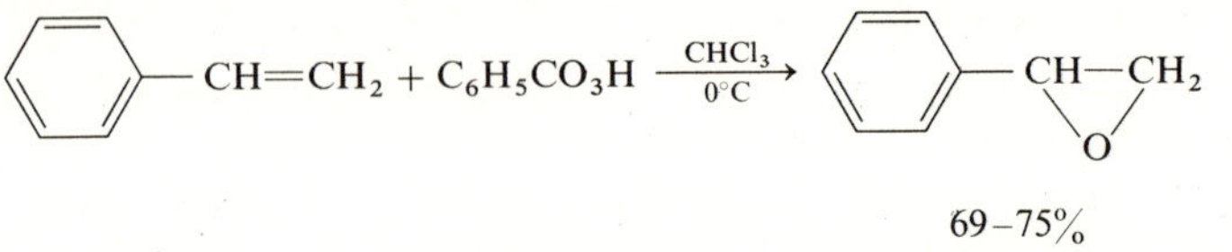

(23)

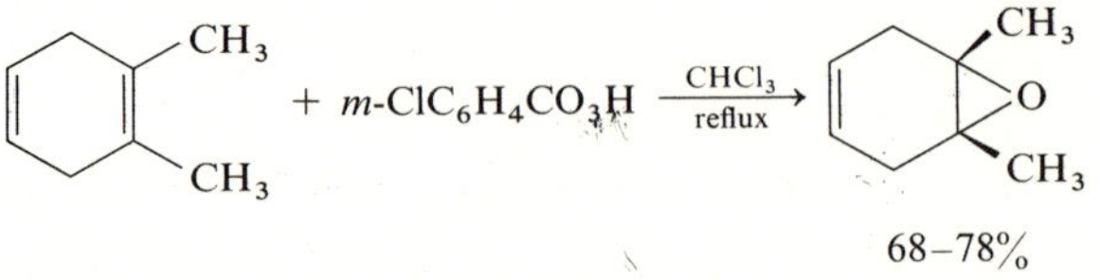

(24)

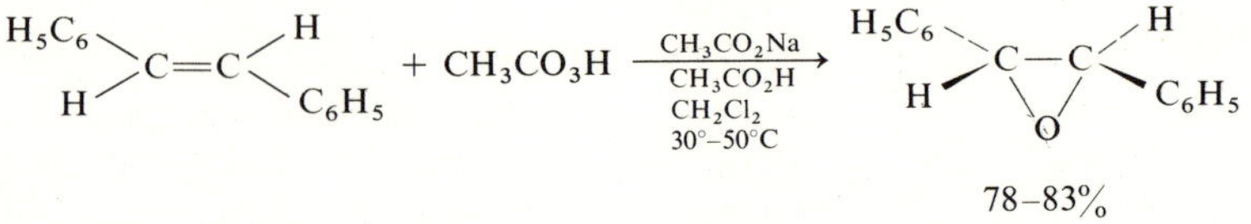

(25)

The reaction times are usually short depending largely on the number and nature of the group bonded to the unsaturated linkage. Tri- and tetra-substituted olefins are epoxidized rather quickly even with relatively weak peroxyacetic acid, while mono-substituted olefins require stronger peroxy acids.[93] This permits selective oxidation of the more substituted double bond in polyenes. For example, limonene (**14**) is epoxidized on the endocyclic double bond with peroxy acids.[94]

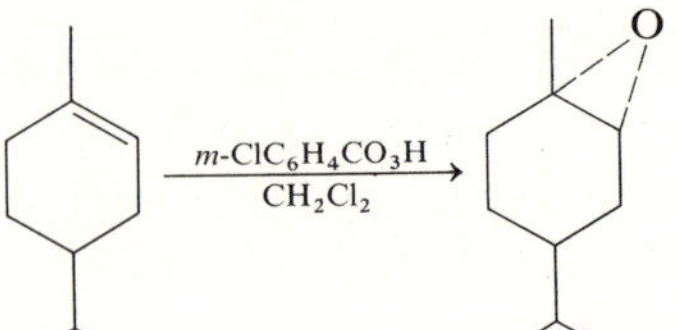

(26)

(14)

[90] H. Hibbert and P. Burt, *in* "Organic Syntheses" (H. Gilman and A. H. Blatt, eds.), Collect. Vol. I, 2nd ed., p. 494. Wiley, New York, 1941.

[91] L. A. Paquette and J. H. Barrett, *Org. Synth.* **49**, 62 (1960).

[92] D. J. Reif and H. O. House, *in* "Organic Syntheses" (N. Rabjohn, ed.), Collect. Vol. IV, p. 860. Wiley, New York, 1963.

[93] D. Swern, *J. Am. Chem. Soc.* **69**, 1692 (1947).

[94] E. E. Royals and J. C. Leffingwell, *J. Org. Chem.* **31**, 1937 (1966).

Relatively unreactive monosubstituted olefins can be converted to epoxides with peroxytrifluoroacetic acid. Anhydrous sodium carbonate or disodium hydrogen phosphate must be added to the reaction mixture to neutralize the trifluoroacetic acid present.[95]

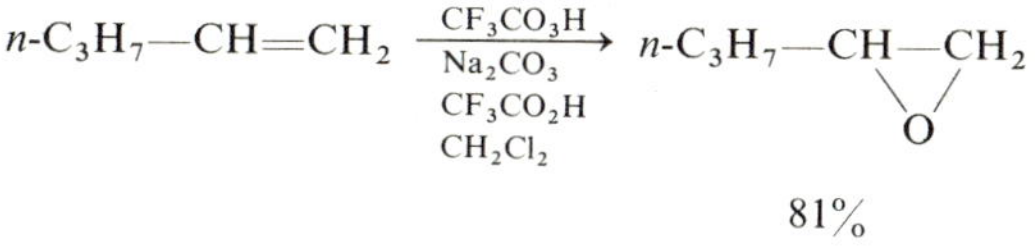

(27)

Optically active epoxides have been prepared by epoxidation with (+)-peroxycamphoric acid but the optical purity of these products is rather low (< 8%).[96–98]

Epoxidation with peroxy acids provides also a selective method for oxidation of olefins with hydroxyl and carbonyl functions [Eqs. (28),[99] (29)[100]].

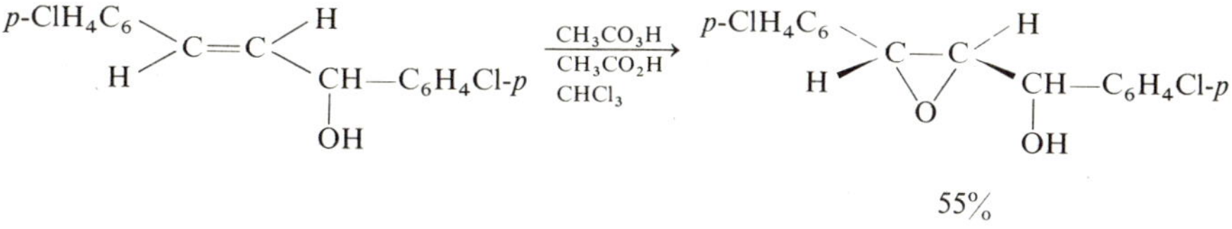

(28)

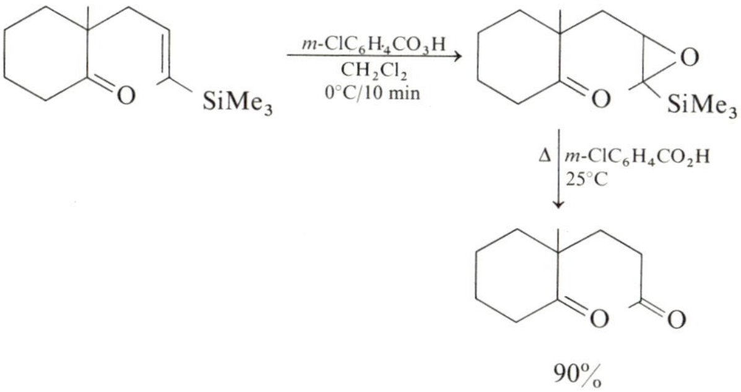

(29)

Oxidation of olefinic alcohols to epoxy ketones can be accomplished by using *m*-chloroperoxybenzoic acid and catalytic amounts of a nitroxide to

[95] W. D. Emmons and A. S. Pagano, *J. Am. Chem. Soc.* **77**, 89 (1955).
[96] R. C. Ewins, H. B. Henbest, and M. A. McKervey, *Chem. Commun.* p. 1085 (1967).
[97] F. Montanari, I. Morretti, and G. Torre, *Chem. Commun.* p. 135 (1969).
[98] D. R. Boyd, D. M. Jerina, and J. W. Daly, *J. Org. Chem.* **35**, 3170 (1970).
[99] H. O. House, *J. Am. Chem. Soc.* **78**, 2298 (1956).
[100] G. Stork and M. E. Jung, *J. Am. Chem. Soc.* **96**, 3682 (1974).

effect the oxidation of the hydroxyl function. The nitroxide catalyst is generated *in situ* by the reaction of 2,2,6,6-tetramethylpiperidine hydrochloride (TMP·HCl) with peroxy acid. For example, oxidation of 5-norbornen-2-ol (**15**) with this reagent affords *exo*-5,6-epoxy-2-norbornanone (**16**) in 86% yield.[101]

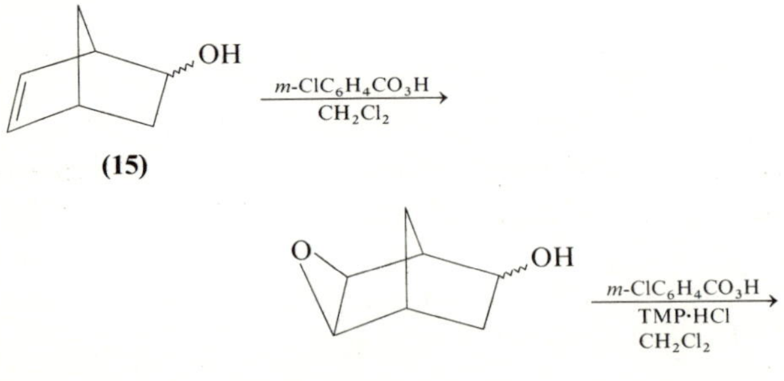

(30)

Acid-sensitive olefins, which usually form rearranged products with aliphatic and aromatic peroxy acids, can conveniently be epoxidized with *m*-chloroperoxybenzoic acid in a dichloromethane–aqueous sodium bicarbonate two-phase system.[102] For example, 6-methylhept-5-en-2-one (**17**) gives epoxide **18** when oxidized in this way, while only the rearranged product, i.e., 1,3,3-trimethyl-2,7-dioxabicyclo[2.2.1]heptane (**19**) is formed when the reaction is performed with peroxyacetic or *m*-chloroperoxybenzoic acid alone.

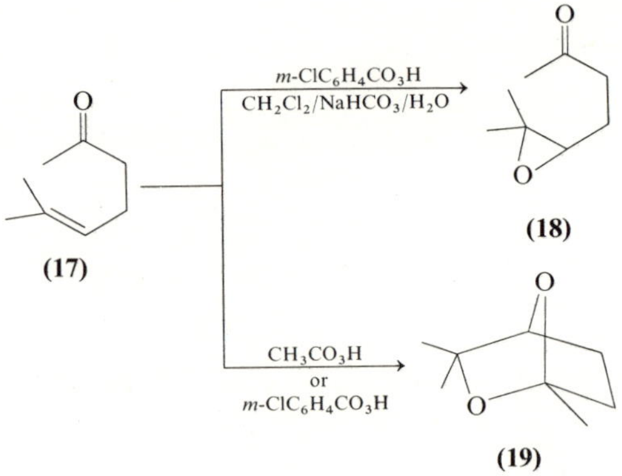

(31)

[101] J. A. Cella, J. A. Kelley, and E. F. Kenehan, *J. Org. Chem.* **40**, 1860 (1975).

[102] W. K. Anderson and T. Veysoglu, *J. Org. Chem.* **38**, 2267 (1973); J. Gaoni, *J. Chem. Soc. C* p. 2925 (1968).

Epoxides which are unstable under acidic conditions can also be prepared by oxidation with peroxybenzimidic acid, derived from benzonitrile and hydrogen peroxide in a buffered solution at pH 8 [Eq. (32)].[40] Selective

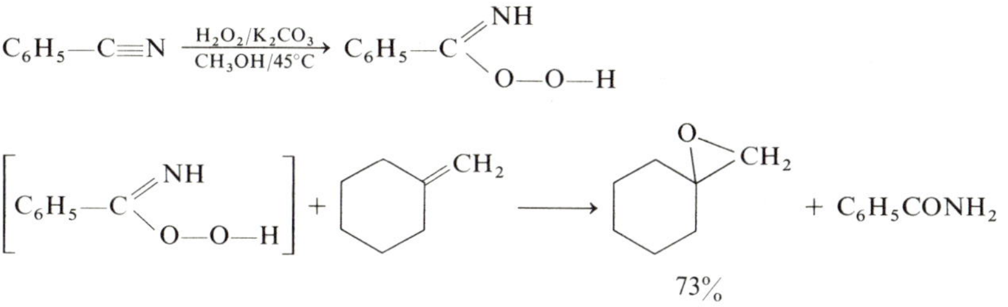

(32)

oxidation of carbon–carbon double bonds in the presence of carbonyl functions can also be performed with this peroxy acid. For example, **20** gives a good yield of the epoxide when oxidized with peroxybenzimidic acid, while the corresponding lactone (**21**) is obtained with peroxyacetic acid.[40]

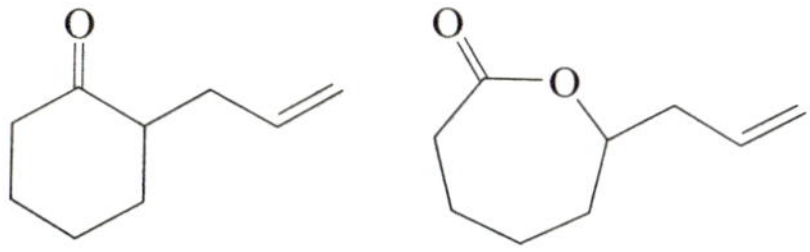

(20) **(21)**

Peroxybenzimidic acid appears to be less selective than other peroxy acids for the epoxidation of olefins. 4-Vinylcyclohexene (**22**) reacts with *m*-chloroperoxybenzoic acid to give predominantly 1,2-epoxide, while peroxybenzimidic acid epoxidizes both double bonds to almost the same degree.[103]

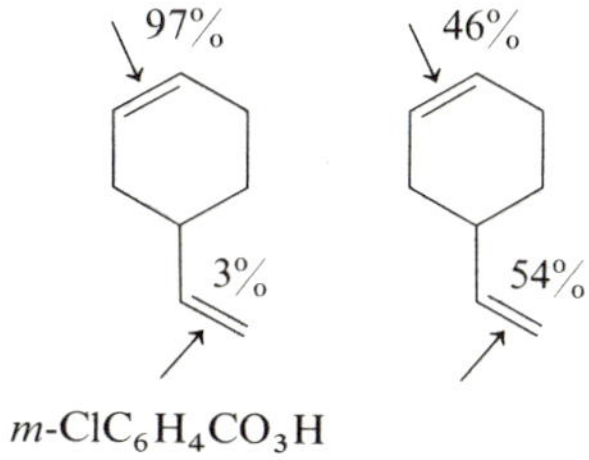

(22)

[103] R. G. Carlson, N. S. Behn, and C. Cowles, *J. Org. Chem.* **36**, 3832 (1971).

Examples mentioned previously have already shown that epoxidation is syn stereospecific. Thus, *cis*- and *trans*-stilbene always give *cis*- and *trans*-stilbene epoxide.[104] There are no reports in the literature of a cis olefin giving threo epoxides or trans olefin giving epoxides erythro epoxides.

Additional stereochemical features of epoxidation with peroxy acids have been obtained by studying the oxidation of cyclic systems.[88] It has clearly been demonstrated in some cases that neighboring functional groups may influence the direction of attack of the peroxy acid. For example, epoxidation of allylic compounds with a bulky group as a neighboring substituent proceeds with a preferential anti attack. 3-*tert*-Butylcyclohexene (**23**) approximately yields a 9:1 ratio of *trans*- to *cis*-epoxide (trans and cis with respect to the *tert*-butyl group) when oxidized with *m*-chloroperoxybenzoic acid in dichloromethane.[105]

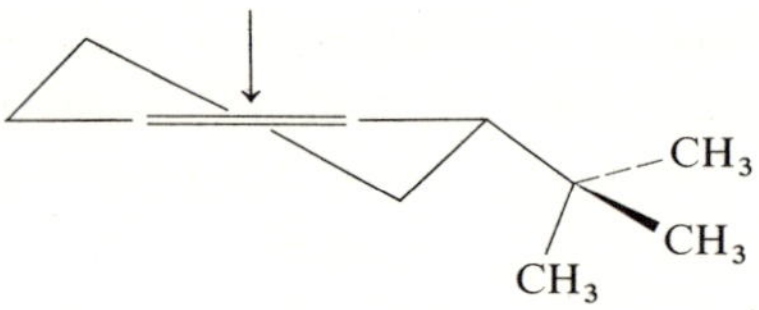

(23)

Remote groups can also influence the direction of the approach of the peroxy acid molecule. 4-Methylcyclopentene reacts with peroxylauric acid to give 73:27 ratio for *trans*- to *cis*-epoxide. This ratio does not change significantly in going from nonpolar to polar solvents thus indicating that the alkyl group exerts a steric rather than polar effect.[106] Steric influence is, in general, important only in cases where there are not more than two carbon atoms between the substituent and the reaction center.[107]

Preference for anti epoxidation has been observed also in systems in which a polar group is too far from the double bond to exhibit any steric influence. This preference for anti attack disappears in polar solvents indicating that a polar effect must be operative with such groups. For example, epoxidation of **24** with peroxylauric acid in cyclopentane gives a 72:28 ratio for *trans*- to *cis*-epoxide and a 55:45 ratio in acetonitrile.[106,108,109]

[104] B. M. Lynch and K. H. Pausacker, *J. Chem. Soc.* p. 1525 (1955).
[105] J. C. Richter and C. Preppel, *Can. J. Chem.* **46**, 3709 (1968).
[106] H. B. Henbest and J. J. McCullough, *Proc. Chem. Soc., London* p. 74 (1962).
[107] H. B. Henbest, *Chem. Soc., Spec Publ.* **19**, 23 (1964).
[108] H. B. Henbest, *Proc. Chem. Soc., London* p. 159 (1963).
[109] N. C. Crossley, A. C. Darby, H. B. Henbest, J. J. McCullough, B. Nicholls, and M. F. Stewart, *Tetrahedron Lett.* p. 398 (1961).

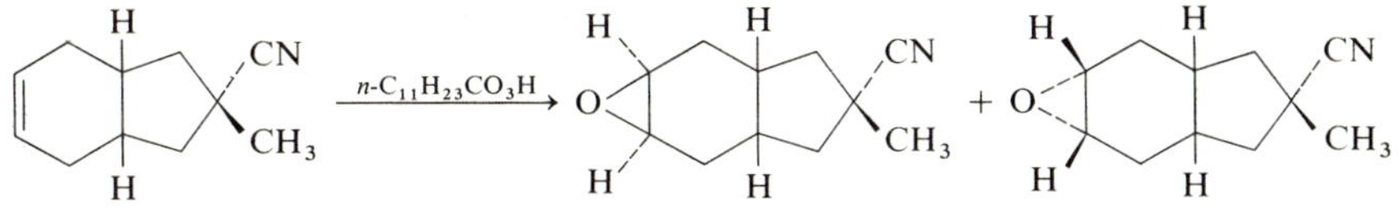

(24)

(33)

On the other hand, a 9:91 ratio for *trans*- to *cis*-epoxide has been reported for the epoxidation of the allylic alcohol 2-cyclohexene-1-ol (**25**) with little dependence on the solvent used.[108,110]

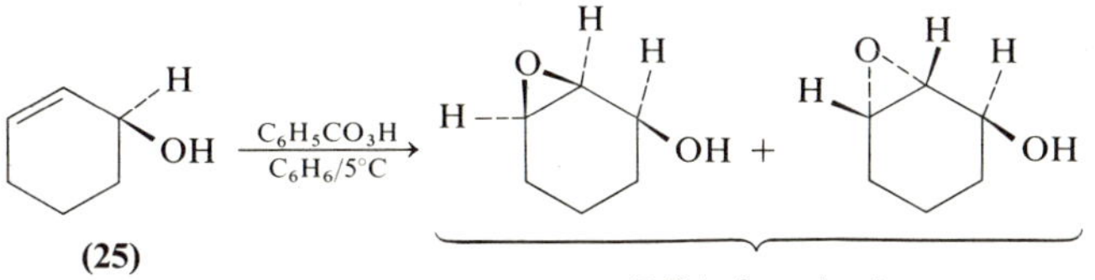 (34)

(25) (86% of product)

Similar results were reported for unsaturated steroids having a free hydroxyl substituent in the allylic position.[111–113] Allylic carboxylic acids and carboxylates are also effective syn-directing groups. For example, epoxidation of *trans*-1,2-dihydrophthalic acid (**26**) gives only the *cis* mono-epoxide [Eq. (35)][114] while epoxidation of dimethyl *trans*-1,2-dihydro-

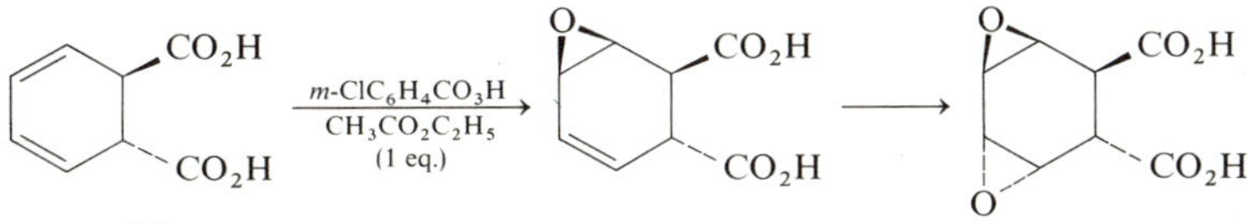 (35)

(26)

phthalate (**27**) yields a 9:1 mixture of *cis*- and *trans*-epoxide [Eq. (36)].[115]

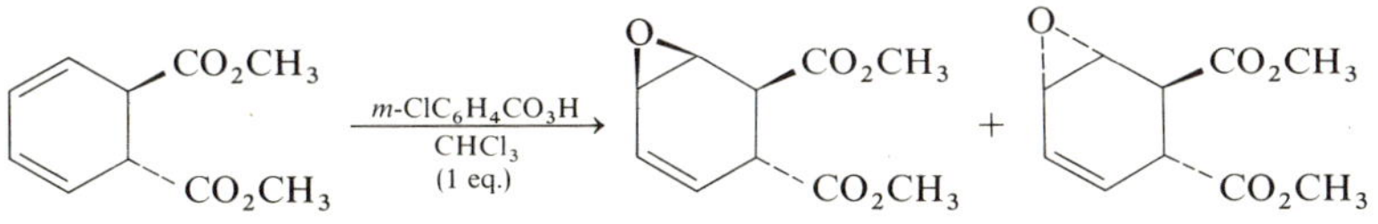 (36)

(27)

[110] P. Chamberlain, N. L. Roberts, and G. H. Whitham, *J. Chem. Soc. B* p. 1374 (1970).

[111] H. B. Henbest and R. A. L. Wilson, *J. Chem. Soc.* p. 1958 (1957).

[112] E. Glotter, S. Greenfield, and D. Lavie, *Tetrahedron Lett.* p. 5261 (1967).

[113] M. Lj. Mihailović, J. Foršek, L. Lorenc, Z. Maksimović, H. Fuhrer, and J. Kalvoda, *Helv. Chim. Acta* **52**, 459 (1969).

[114] B. A. Chiasson and G. A. Berchtold, *J. Am. Chem. Soc.* **96**, 2898 (1974).

[115] S. A. Cerefice and E. K. Fields, *J. Org. Chem.* **41**, 355 (1976).

Epoxidations of allylic acetates are considerably less stereospecific, but in some cases, more of the cis epoxide is formed than expected on steric grounds alone. For example, a 2:1 mixture of *cis*- and *trans*-epoxide has been obtained by oxidizing **28** with peroxybenzoic acid in benzene [Eq. (37)].

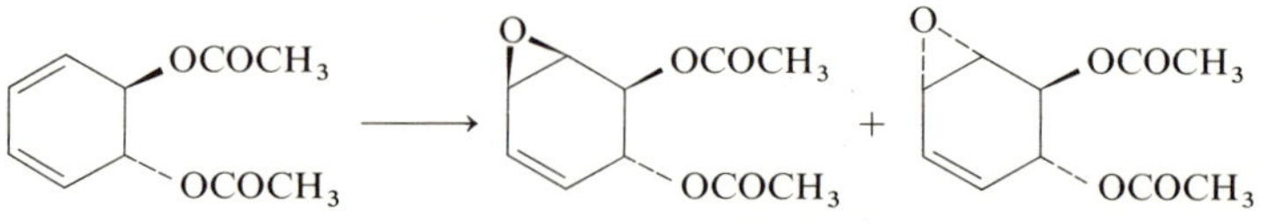

(28) (37)

Interactions of hydroxyl, carboxyl, and carboxylate functions with peroxy acid must be operative in all these reactions since epoxidation of allylic ethers do not undergo preferential syn attack. The relative order of syn-directing ability of substituents in allylic epoxidation is believed to be OH $\geq$ $CO_2H > CO_2R >$ OCOR.[115]

Steric hindrance most likely is operating in reactions of bridged cycloalkanes with peroxy acids.[116–119] Epoxidation of norbornene (**29**) with *m*-chloroperoxybenzoic acid affords the *exo*-epoxide in 99% yield.[116]

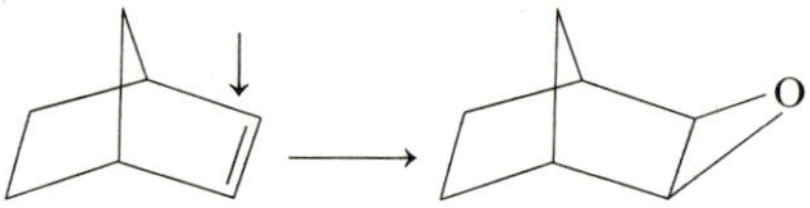

(29) (38)

7,7-Dimethylnorbornene (**30**) reacts with the same peroxy acid to give a 9:1 ratio of *endo*- to *exo*-epoxide.

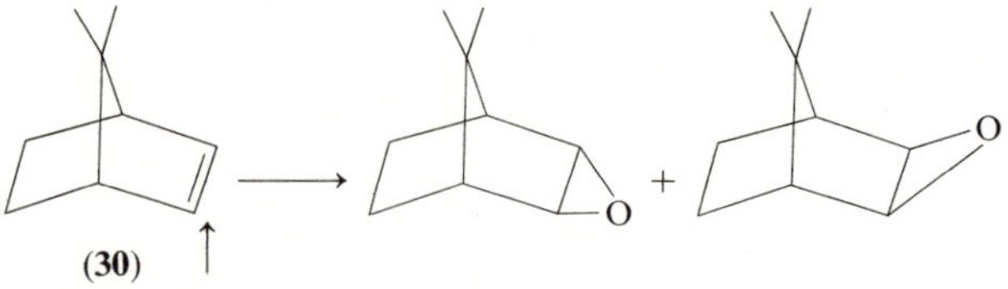

(30) (39)

Unfavorable endo attack becomes predominant most probably as a result of steric shielding to exo attack by the *syn*-7-methyl group.[116]

Unhindered unsymmetrically substituted alkyl cyclopropenes react with *m*-chloroperoxybenzoic acid in dichloromethane to afford two α,β-un-

[116] H. C. Brown, J. H. Kawakami, and S. Ikegami, *J. Am. Chem. Soc.* **92**, 6914 (1970).
[117] H. C. Brown, J. H. Kawakami, and K. -T. Liu, *J. Am. Chem. Soc.* **95**, 2209 (1973).
[118] P. G. Gassman and J. H. Dygos, *Tetrahedron Lett.* p. 4749 (1970).
[119] J. Haywood-Farmer, B. F. Friedlander, and L. M. Hammond, *J. Org. Chem.* **38**, 3145 (1973).

saturated carbonyl compounds.[120–123] For example, 1,3,3-trimethylcyclopropene (**31**) yields 69% of **32** and 31% of **33**.[123]

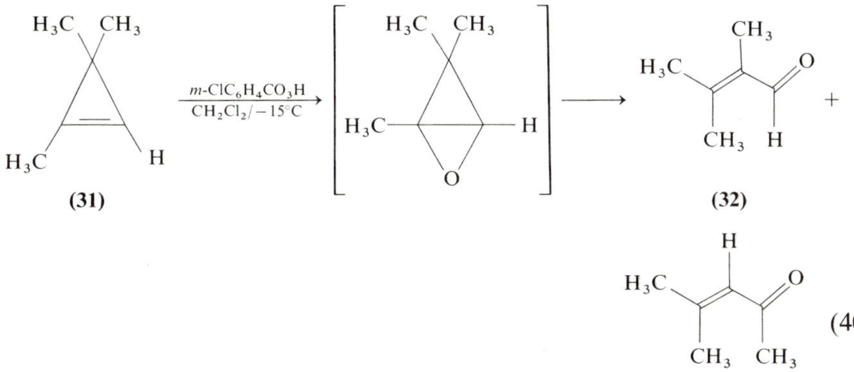

(31) (32) (33) (40)

If the double bond of the cyclopropene is sufficiently hindered, epoxidation does not take place. Rather, the peroxy acid abstracts an allylic C-3 cyclopropenyl hydrogen in **34** to form a cyclopropenyl cation which reacts further with the second molecule of peroxy acid to form an unstable peroxy ester (**35**). The latter decomposes to an acetylene and anhydride.[123]

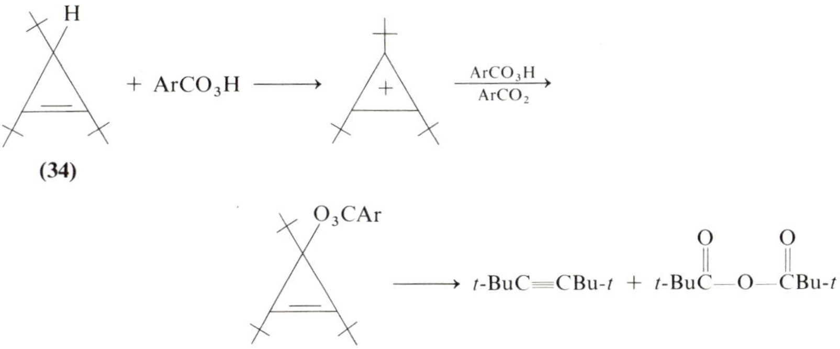

(34) (35) (41)

Oxidation of cyclobutene and various methyl-substituted derivatives with peroxybenzoic acid at 0° to −50°C gives the corresponding epoxides in yields ranging from 30–70%.[124]

[120] J. Ciabattoni and P. J. Kociensky, *J. Am. Chem. Soc.* **91**, 6534 (1969).
[121] L. E. Friedrich and R. A. Cormier, *J. Org. Chem.* **35**, 450 (1970).
[122] L. E. Friedrich and R. A. Fiatto, *J. Am. Chem. Soc.* **96**, 5783 (1974).
[123] P. J. Kociensky and J. Ciabattoni, *J. Org. Chem.* **39**, 388 (1974).
[124] J. L. Ripoll and J. M. Conia, *Bull. Soc. Chim. Fr.* p. 2755 (1965).

2. Allenes

Allenes are easily oxidized with peroxy acids but the initial products, mono- and diepoxides, usually undergo further changes in the reaction mixture.[125–131]

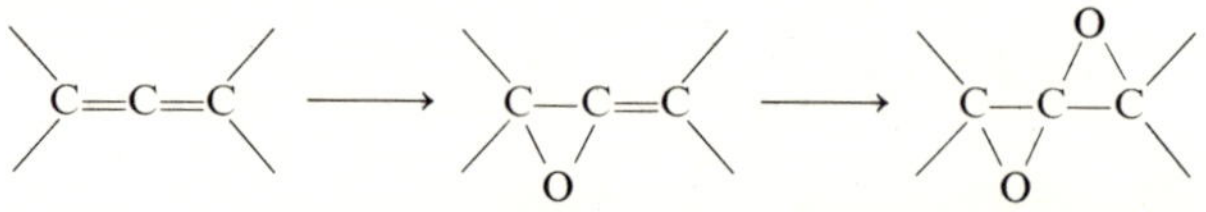

(42)

Oxidation of hindered allenes allows the isolation of mono- and diepoxides.[128,129,131] For example, 1,3-di-*tert*-butyl-1-methylallene (**36**) reacts with 1 equivalent of *p*-nitroperoxybenzoic acid to give allene oxide (**37**) and carbonyl compounds **38**, **39**, and **40** in the ratio 46:6:36:12, respectively.[131]

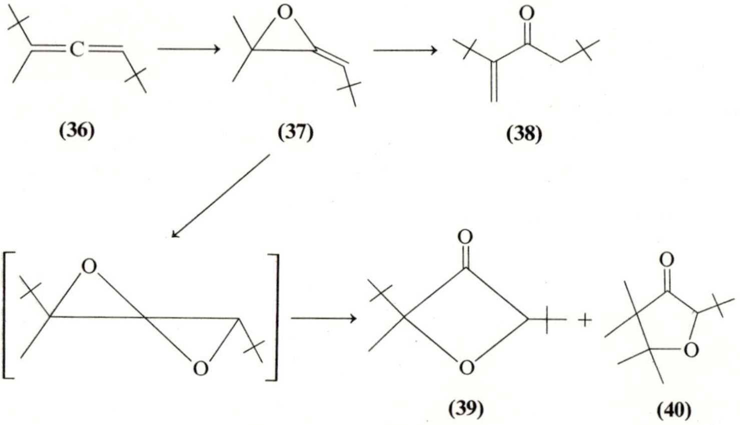

(43)

A high yield of diepoxide has been reported in the oxidation of 1,1-dimethyl-3-*tert*-butylallene (**41**) with two equivalents of peroxyacetic acid.[131]

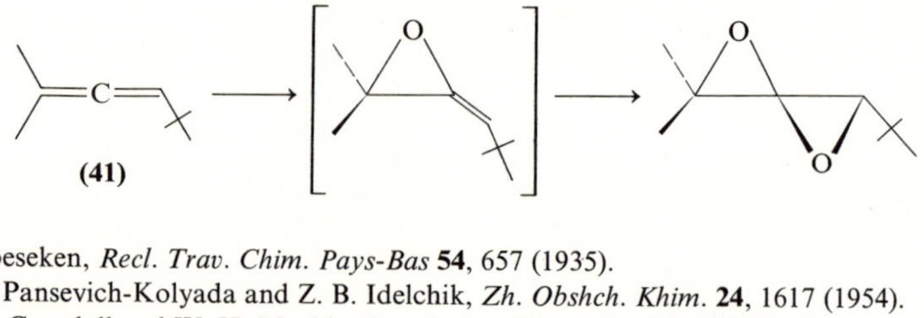

(44)

[125] J. Boeseken, *Recl. Trav. Chim. Pays-Bas* **54**, 657 (1935).
[126] V. I. Pansevich-Kolyada and Z. B. Idelchik, *Zh. Obshch. Khim.* **24**, 1617 (1954).
[127] J. K. Crandall and W. H. Machlender, *J. Am. Chem. Soc.* **90**, 7292 (1968).
[128] J. K. Crandall, W. H. Machlender, and M. J. Thomas, *J. Am. Chem. Soc.* **90**, 7346 (1968).
[129] R. L. Camp and F. D. Greene, *J. Am. Chem. Soc*, **90**, 7349 (1968).
[130] T. Greibrokk and L. Skattebol, *Acta Chem. Scand.* **27**, 1421 (1973).
[131] J. K. Crandall, W. W. Conover, J. B. Komin, and W. H. Machlender, *J. Org. Chem.* **39**, 1723 (1974).

The second step of this reaction appears to be much faster since only the diepoxide could be isolated in oxidation of **41** with 1 equivalent of peroxy acid. Oxidation of 1,2,5-trienes with *p*-nitroperoxybenzoic acid, involving allene oxide and cyclopropane valence tautomers as intermediates, has also been reported.[131a]

3. Vinyl Chlorides, Enol Ethers, and Enol Esters

As in the case of allenes, rearrangement of the initial epoxide is observed in oxidation of vinyl chlorides, enol ethers, and enol esters with peroxy acids according to the reaction scheme shown below.

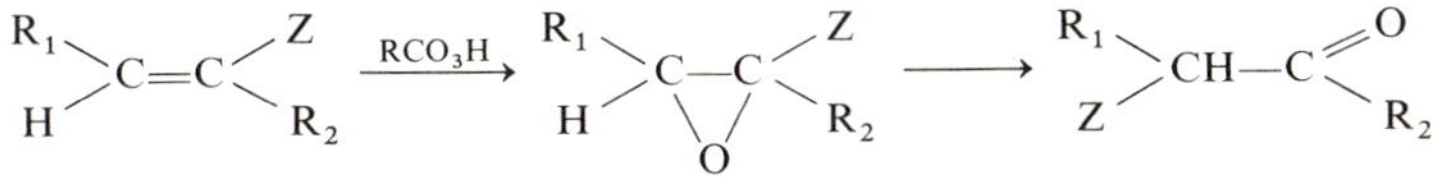

(45)

(Z = Cl, OCOR, OR)

When chlorine is attached directly to the double bond, the initially formed chloroepoxide rearranges to a chloroketone.[132–136] For example, (*Z*)-1-chloro-(*p*-tolyl)-2-phenylethene (**42**) reacts with peroxyacetic acid to produce the corresponding chloroketone (**43**) in 71% yield. Chloroepoxide, which has been proposed to be the initial product, rearranges with preferential chlorine migration.

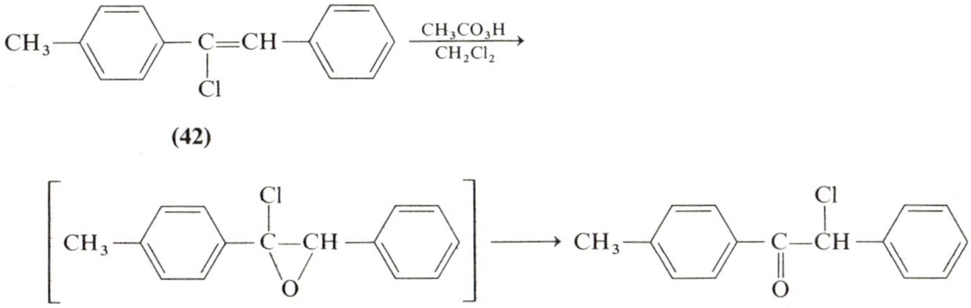

(42)

71%

(43)

(46)

[131a] J. Grimaldi, M. Malacria, and M. Bertrand, *Bull. Soc. Chim. Fr.* p. 1731 (1975).
[132] M. Mouseron and R. Jacquier, *Bull. Soc. Chim. Fr.* p. 698 (1950).
[133] R. N. McDonald and P. A. Schwab, *J. Am. Chem. Soc.* **85**, 820 and 4004 (1963).
[134] R. N. McDonald and P. A. Schwab, *J. Org. Chem.* **29**, 2458 (1964).
[135] R. N. McDonald and T. E. Tabor, *J. Org. Chem.* **33**, 2934 (1968).
[136] R. N. McDonald and R. N. Steppel, *J. Am. Chem. Soc.* **92**, 5664 (1970).

Low temperatures and weak peroxy acids are required for epoxidation of enol ethers[137–139] and enol esters[140–144] to prevent the initially formed epoxide from reacting with the carboxylic acids present or intramolecular isomerization.

Oxidation of 2-methyl-1-ethoxy-1-phenylpropene (**44**) with peroxybenzoic acid yields the corresponding epoxide in 90% yield under carefully controlled conditions [Eq. (47)].[137] 1-Ethoxycyclohexene (**45**) reacts similarly although the benzoate of α-hydroxycyclohexanone (**46**) has been isolated in 61% yield via the unstable epoxide as an intermediate.[137]

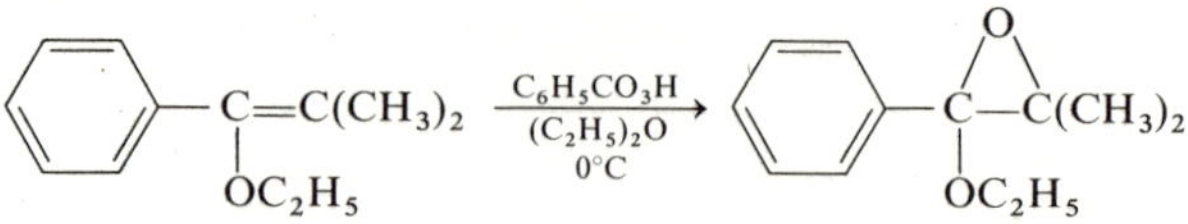

(44) 70% (47)

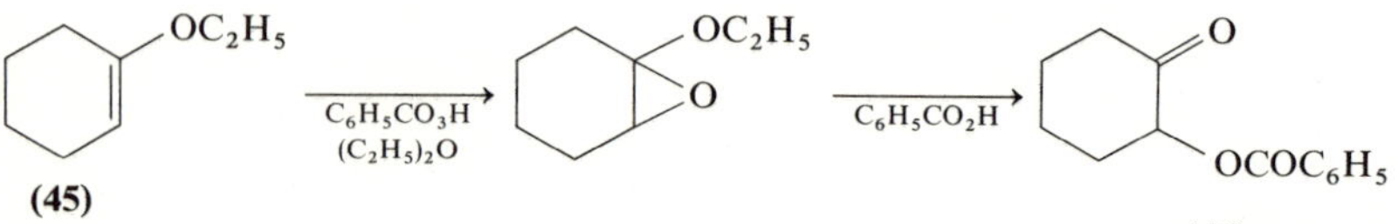

(45) **(46)** (48)

The epoxidation of vinyl acetate (**47**) with peroxybenzoic acid affords vinyl acetate epoxide in 25% yield. The epoxide is easily hydrolyzed by aqueous acid to glycolaldehyde presumably according to the following reaction scheme.[140]

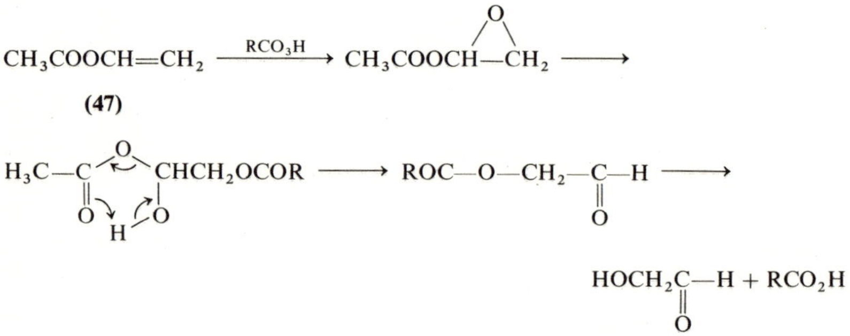

(49)

[137] C. L. Stevens and J. Tazuma, *J. Am. Chem. Soc.* **76**, 715 (1954).
[138] C. L. Stevens and S. J. Dykstra, *J. Am. Chem. Soc.* **75**, 5975 (1953).
[139] I. J. Borowitz, G. J. Williams, L. Gross, and R. Rapp, *J. Org. Chem.* **33**, 2013 (1968).
[140] L. S. Silbert, Z. B. Jacobs, W. E. Palm, L. P. Witnauer, W. S. Port, and D. Swern, *J. Polym. Sci.* **21**, 161 (1956); B. S. Gorton and J. A. Reeder, *J. Org. Chem.* **27**, 2920 (1962).
[141] K. L. Williams and W. S. Johnson, *J. Org. Chem.* **26**, 4563 (1961).
[142] P. D. Gardner, *J. Am. Chem. Soc.* **78**, 3421 (1956).
[143] N. S. Leeds, D. K. Fukushima, and T. F. Gallagher, *J. Am. Chem. Soc.* **76**, 2943 (1954).
[144] D. N. Kirk and J. M. Wiles, *Chem. Commun.* pp. 518 and 1015 (1970).

Aryl-substituted enol acetates also undergo rearrangement rather quickly. The corresponding acetoxyketone (**49**) can be isolated in 49% yield in the reaction of **48** with peroxybenzoic acid.

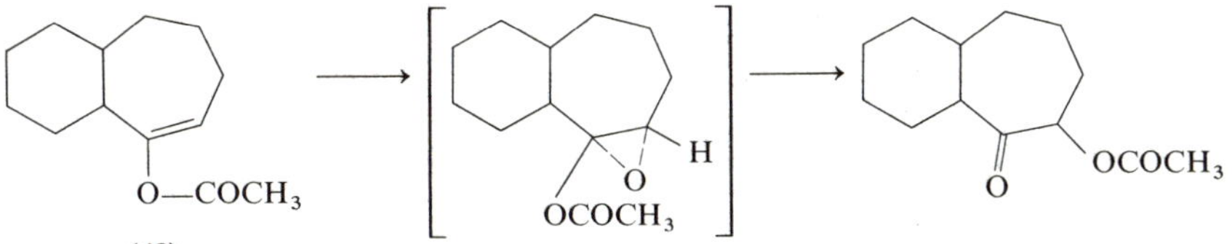

(**48**) (**49**)

(50)

Enol tautomers of dicarbonyl compounds react readily with peroxy acids.[145,146] Hydroxy epoxides, which are formed as intermediates, rearrange to 2-hydroxy-1,3-dicarbonyl compounds. The latter are further cleaved with an excess of peroxy acid [Eq. (51)].[145]

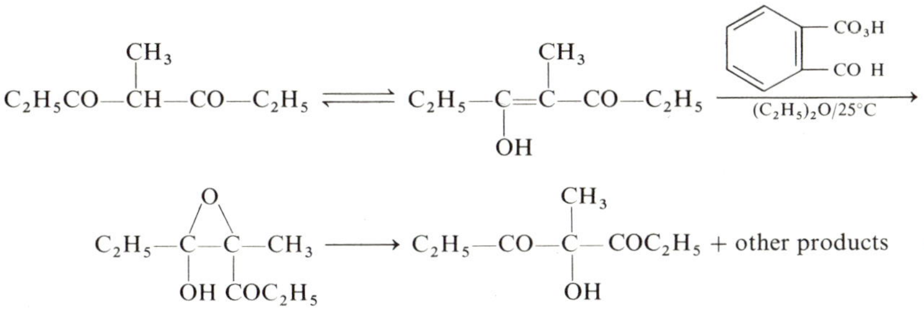

(51)

4. Kinetics and Mechanism

Epoxidation with peroxy acids is a second-order reaction; it is first-order in olefin and first-order in peroxy acid.

$$v = k[\mathrm{RCO_3H}][\mathrm{olefin}] \tag{52}$$

Electron-releasing groups in the olefin accelerate the rate and electron-withdrawing groups retard it thus indicating the nucleophilic nature of olefins in epoxidation.[93,104,147–153] Conjugation of the olefinic double bond

[145] H. O. House and W. F. Gannon, *J. Org. Chem.* **23**, 879 (1958).
[146] A. J. Hubert and P. S. Starcher, *J. Chem. Soc. C* p. 2500 (1968).
[147] J. B. Lee and B. C. Uff, *Q. Rev. Chem. Soc.* **21**, 431 (1967).
[148] D. I. Metelitsa, *Russ. Chem. Rev. (Engl. Transl.)* **41**, 807 (1972).
[149] S. Medvedev and O. Blokh, *J. Phys. Chem. (Moscow)* **4**, 721 (1933).
[150] J. Boeseken and J. Stuurman, *Proc. K. Ned. Akad. Wet.* **39**, 2 (1936).
[151] J. Boeseken and J Stuurman, *Recl. Trav. Chim. Pays-Bas* **56**, 1034 (1937).
[152] J. Boeseken and C. J. A. Nonegraaff, *Recl. Trav. Chim. Pays-Bas* **61**, 69 (1942).
[153] F. C. Frostick, Jr., B. Phillips, and P. S. Starcher, *J. Am. Chem. Soc.* **81**, 3350 (1959).

with phenyl rings reduces the rate of epoxidation. The trend of decreasing reactivity of some representative olefins with peroxybenzoic acid is styrene > *cis*-stilbene > *trans*-stilbene > triphenylethene > tetraphenylethene.[104] The planar structure of *trans*-stilbene enables the delocalization of electron density of the double bond. Such a delocalization is not possible in *cis*-stilbene owing to the nonplanarity of both rings. As a result, *cis*-stilbene is approximately twice as reactive as *trans*-stilbene. In general, cis double bonds in straight chain olefins are epoxidized faster than trans double bonds; the reverse holds for cis–trans isomers of medium-size cycloalkenes.[154–157]

Electron-withdrawing groups in the peroxy acid facilitate the rate of epoxidation.[158,159] The order of reactivity (electrophilicity) approximates the pK values of peroxy acids, i.e., peroxytrifluoroacetic > monoperoxymaleic > monoperoxyphthalic > *p*-nitroperoxybenzoic > *m*-chloroperoxybenzoic > peroxyformic > peroxybenzoic > peroxyacetic acid.

Kinetic parameters for epoxidation of several olefins with peroxy acids have been reported. For example, activation enthalpy, $\Delta H^{\ddagger}$, and activation entropy, $\Delta S^{\ddagger}$, for epoxidation of *trans*-stilbene with peroxybenzoic acid in benzene, are 15.2 kcal/mole and -22.8 eu, respectively.[104] Activation enthalpies are decreased by electron-donating groups and increased by withdrawing groups in *trans*-stilbene. On the other hand, electron-withdrawing groups in peroxybenzoic acids decrease this activation parameter and electron-donating groups increase it. Relatively large and negative activation entropies indicate a high degree of orientation of the activated complex. The ρ values of -0.9 (substituted stilbenes) and 0.9 (substituted peroxybenzoic acids) have been obtained by plotting the logarithms of rate constants vs. Hammett σ values.[104,160,161]

The study of the kinetics of epoxidation of meta- and para-substituted styrenes gave similar results; the Hammett ρ of -1.30 has been reported for the meta series.[162] The kinetics of epoxidation of strained olefins indicates that the rate is independent of the strain.[163–165] For example, 1,2-diphenyl-

[154] V. Prelog, K. Schenker, and H. H. Günthard, *Helv. Chim. Acta* **35**, 1598 (1952).
[155] V. Prelog, K. Schenker, and W. Küng, *Helv. Chim. Acta* **36**, 471 (1953).
[156] F. Asinger, B. Fell, G. Hadik, and G. Steffan, *Chem. Ber.* **97**, 1568 (1964).
[157] K. B. Sharpless, J. M. Townsend, and D. R. Williams, *J. Am. Chem. Soc.* **94**, 295 (1972).
[158] M. Vilkas, *Bull. Soc. Chim. Fr.* p. 1401 (1959).
[159] Y. Ogata and I. Tabushi, *J. Am. Chem. Soc.* **83**, 3440, 3444 (1961).
[160] D. R. Campbell, J. O. Edwards, J. Maclachlan, and R. Polgar, *J. Am. Chem. Soc.* **80**, 5308 (1958).
[161] V. S. Aksenov, *Sovrem. Probl. Org. Khim.* **58** (1971); *Chem. Abstr.* **78**, 3361 (1973).
[162] Y. Ishii and Y. Inamoto, *J. Chem. Soc. Jpn., Ind. Chem. Sect.* **63**, 765 (1960).
[163] K. D. Bingham, G. D. Meakins, and G. H. Whitham, *Chem. Commun.* p. 445 (1966).
[164] M. S. Newman, N. Gill, and D. W. Thompson, *J. Am. Chem. Soc.* **89**, 2059 (1967).
[165] L. E. Friedrich and R. A. Fiatto, *J. Org. Chem.* **39**, 416 (1974).

cyclopentene reacts 11.7 times faster than 1,2-diphenylcyclopropene with *m*-chloroperoxybenzoic acid in carbon tetrachloride at 0°C.[165] Norbornene reacts 2.4 times faster than cyclohexene with peroxylauric acid in chloroform at 25°C.[163] The kinetic evidence mentioned above, together with the syn stereospecificity of epoxidation, can be neatly accounted for with the mechanism first proposed by Bartlett.[166]

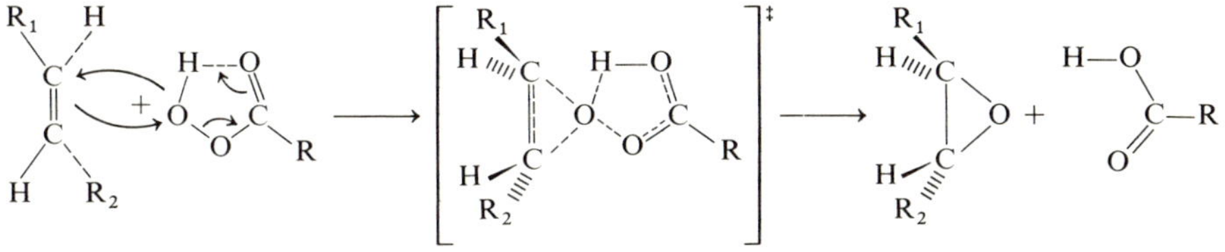

(50)

(53)

Transition state **50** is in accord with the infrared findings which support the chelated form of peroxy acids in inert solvents.

Considerable reduction in reaction rates of epoxidation in basic solvents, which are able to disrupt the chelated ring to form intermolecular hydrogen-bonded adducts, can also be regarded as supporting evidence for the proposed mechanism.[63,167,168] This type of peroxy acid–solvent interaction increases with increasing solvent basicity. In aprotic nonbasic solvents, the parallelism between the rate constants and E_T values has been reported.[169] The highest rates of epoxidation are usually observed in halogenated solvents ($CHCl_3$, CH_2Cl_2, C_6H_5Cl).[167] Steric requirements of solvents may also influence the rate.

Although there are only limited data available for the epoxidation with peroxybenzimidic acid, it seems reasonable to assume a nearly symmetrical three-membered cyclic activated complex (**51**) to be involved in these reactions.[170]

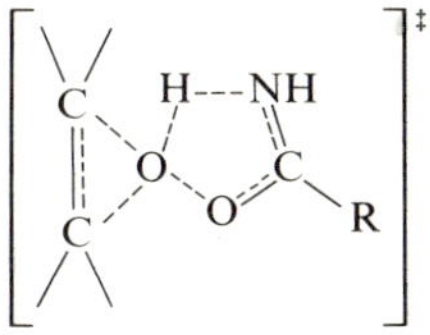

(51)

[166] P. D. Bartlett, *Rec. Chem. Progr.* **18**, 111 (1957).

[167] P. Renolen and J. Ugelstad, *J. Chim. Phys.* **57**, 1976 (1960).

[168] J. I. Itier, M. Tournaire, and A. Casadevall, *C. R. Hebd. Seances Acad. Sci., Ser. C* **271**, 878 (1970).

[169] H. Kropf and M. R. Yazdanbachsch, *Tetrahedron* **30**, 3455 (1974).

[170] R. G. Carlson and N. S. Behn, *J. Org, Chem.* **32**, 1363 (1967).

The directing effect of the hydroxyl, carboxyl, and carboxylate groups, discussed previously [Eqs. (34), (35), and (36)], has been suggested to arise as a result of the hydrogen bonding between these functional groups and the attacking peroxy acid molecule as depicted in the following transition states.[108,111,115]

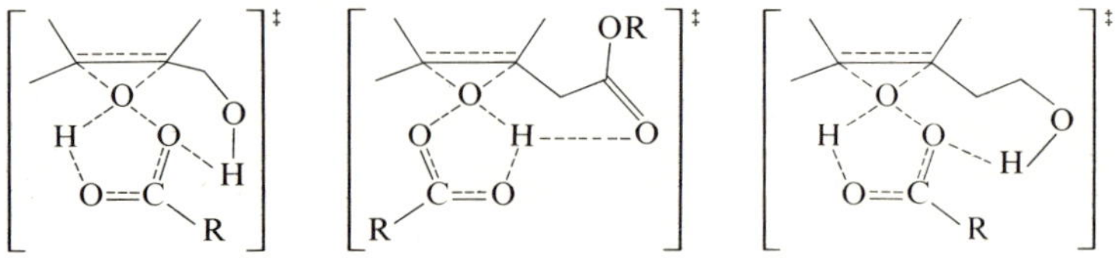

(52) **(53)** **(54)**

The highly negative entropy of activation ($\Delta S^{\ddagger} = -41$ eu) for the reaction of 3-hydroxycyclohexene with peroxybenzoic acid, as compared with the values obtained for epoxidation of cyclohexene and 3-methoxycyclohexene (-33 eu and -31 eu, respectively), is in accord with the highly oriented transition state proposed.[108,111]

There are some reports in the literature of strong acid catalysis in epoxidation.[171–173] This question is still not settled but it appears that only relatively strong acids, present in sufficiently large amounts, can increase the rate of reaction. Swern has proposed a mechanism of the catalysis in which the protonation of peroxy acid takes place first, thus facilitating the electrofilic attack of the protonated species on the double bond.

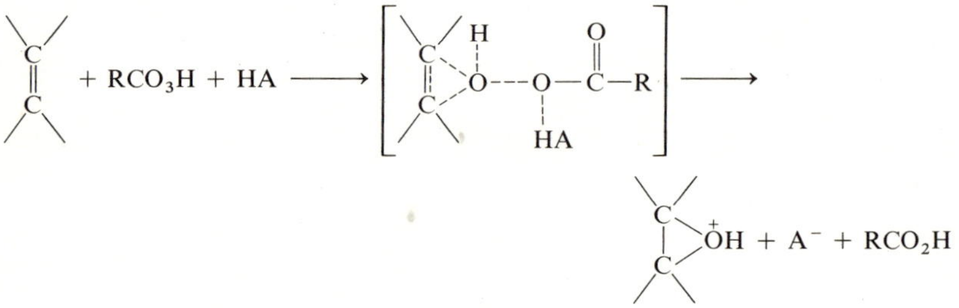

(54)

Recent studies of epoxidation of olefins of the type $ZCH_2CH{=}CH_2$ with various aliphatic and aromatic peroxy acids in acetic acid solution containing nitrogen base(urea, dimethylformamide), led Dryuk and co-workers to the conclusion that a charge transfer-type complex (**55**) between the peroxy acid and the olefin is formed first with subsequent transformation to the products

[171] G. Berti and F. Bottari, *J. Og. Chem.* **31**, 419 (1966).

[172] G. Berti and F. Bottari, *Gazz. Chim. Ital.* **89**, 2380 (1959).

[173] V. N. Sapunov and N. N. Lebedev, *Zh. Org. Khim.* **2**, 225 (1966).

via a transition state (**56**). The variations of the overall activation energies are believed to be mostly determined by the first step of the reaction, i.e., the formation of the complex.[173a]

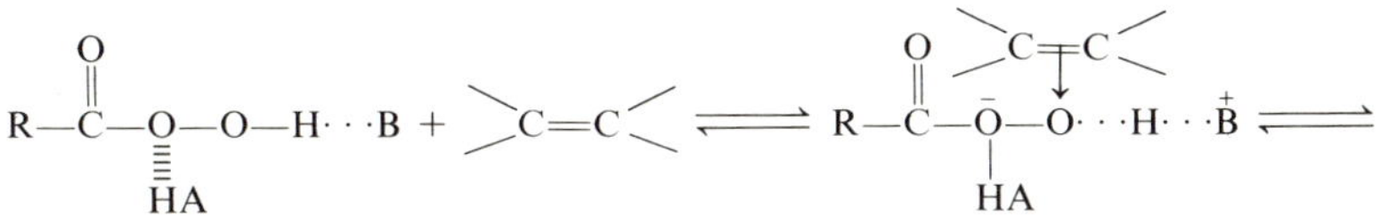

(55)

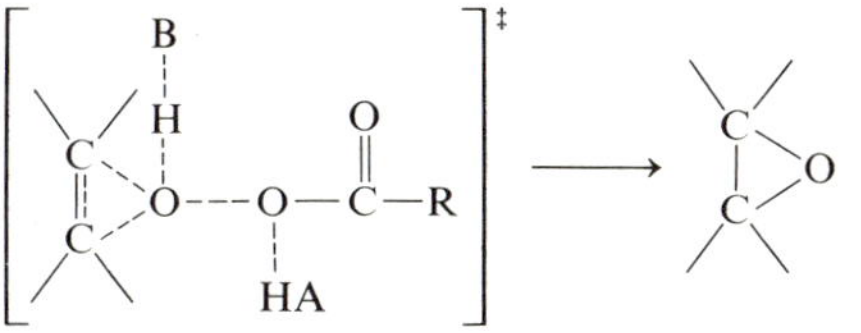

(55)

(56)

Waters proposed the initial attack of a hydroxyl cation, OH^+ (not necessarily as a free ion), on the olefin via the transition state (**57**).[174] A somewhat modified version of the originally proposed mechanism is shown below.

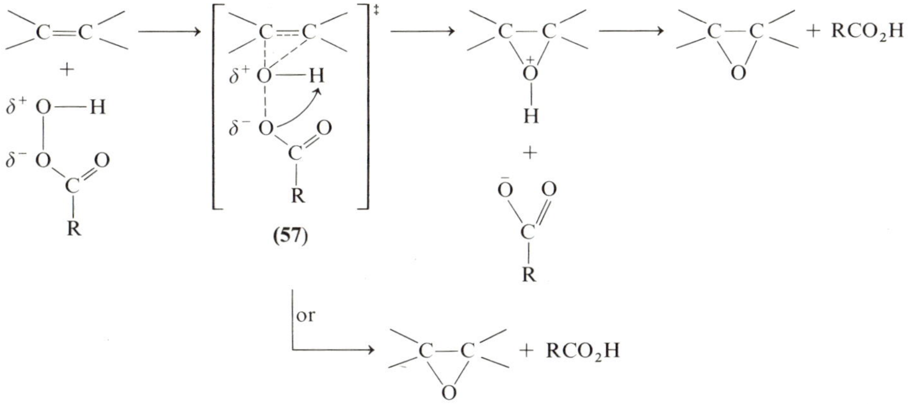

(56)

Kwart and Hoffman proposed as a possibility an epoxidation mechanism involving 1,3-dipolar addition of a hydroxycarbonyl oxide (**58**), derived from the intramolecularly hydrogen-bonded peroxy acid molecule to the olefinic

[173a] V. G. Dryuk, *Tetrahedron* **32**, 2855 (1976).

[174] W. A. Waters, "Mechanisms of Oxidation of Organic Compounds." Wiley, New York, 1965.

dipolarophile. Decomposition of the adduct (**59**) should produce the epoxide and carboxylic acid.[175]

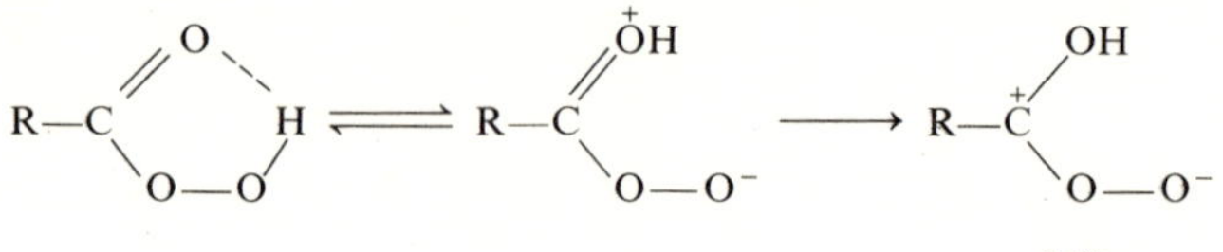

(**58**)

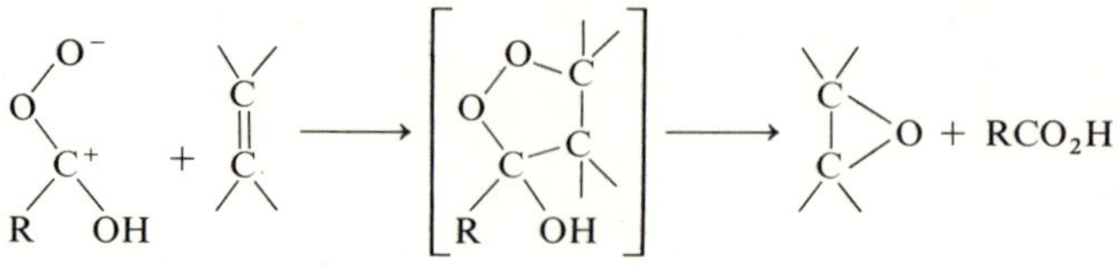

(**59**)

Strong parallelism between general kinetic characteristics and activation parameters of epoxidation and 1,3-dipolar addition (there are also some discrepancies[163]) has been taken as support for this mechanism. Since the proposed intermediates have actually been isolated and are stable under conditions of epoxidation, a modified transition state (**60**) has been suggested.[176]

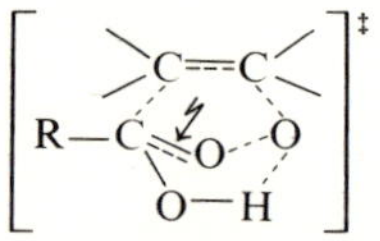

(**60**)

Quantum mechanical calculations of the electronic structure of the peroxy acid molecule in its intramolecularly hydrogen bonded and dipolar form, using the semiempirical PPP SCF-MO method, could not distinguish between both proposed mechanisms, i.e., 1,1- and 1,3-addition.[58]

A recent study of the secondary kinetic deuterium isotope effect for the epoxidation of *p*-phenylstyrene and three deuterated derivatives with *m*-chloroperoxybenzoic acid in 1,2-dichloroethane, revealed a clear distinction between the α and β carbon atoms of the olefin in the transition state. A symmetrical transition state of the "butterfly" mechanism requires that both carbon atoms become equally sp^3 hybridized. On the basis of these observations, an unsymmetrical transition state (**61**) has been proposed by Hanzlik and Shearer.[177]

[175] H. Kwart and D. H. Hoffman, *J. Org. Chem.* **31**, 419 (1966).
[176] H. Kwart, P. S. Starcher, and S. W. Tinsley, *Chem. Commun.* p. 335 (1967).
[177] R. P. Hanzlik and G. O. Shearer, *J. Am. Chem. Soc.* **97**, 5231 (1975).

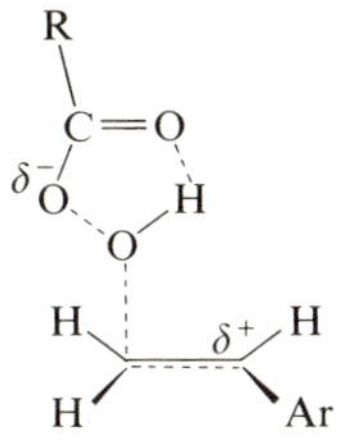

(61)

Stereospecificity of the reaction has been explained as being due to the considerably faster ring closure compared to rotation about the carbon–carbon bond. A small amount of π bond character would be sufficient to account for this phenomenon.

Owing to a great variety of plausible interpretations of available experimental results, it would be desirable to perform *ab initio* MO calculations on various symmetrical and asymmetrical transition states in order to get more insight into the mechanism of this reaction.

B. Epoxidation by Peroxides Other than Peroxy Acids

1. Hydrogen Peroxide and Alkyl Hydroperoxides

Selective epoxidation of α,β-unsaturated ketones and aldehydes may be accomplished by oxidation with the sodium salt of hydrogen peroxide[178] or the sodium salt of *tert*-butyl hydroperoxide[179] rather than peroxy acids [Eqs. (57)[180] and (58)[181]].

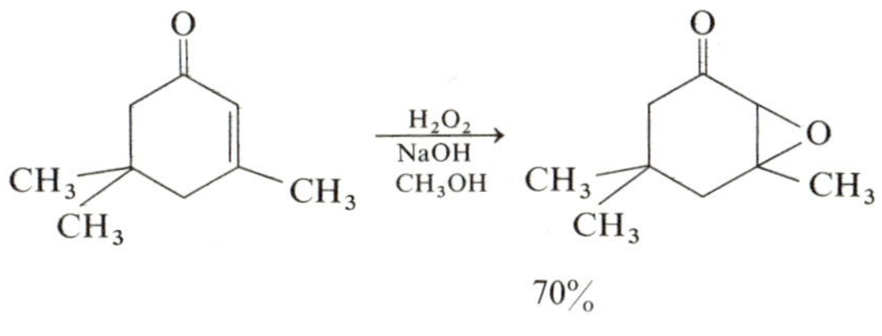

(57)

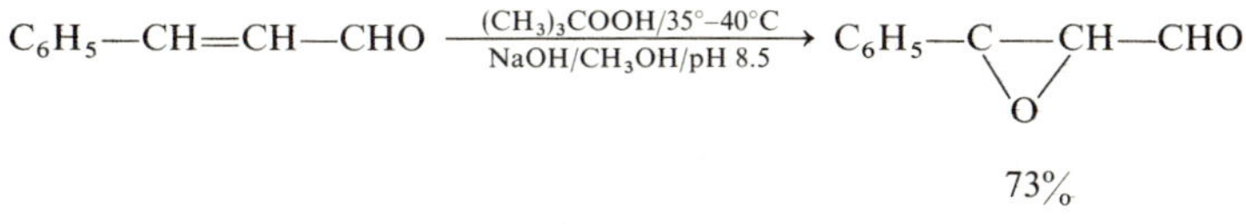

(58)

[178] C. H. Hassall, *Org. React.* **9**, 73 (1957).

[179] N. C. Yang and R. A. Finnegan, *J. Am. Chem. Soc.* **80**, 5845 (1958).

[180] R. L. Wasson and H. O. House, *in* "Organic Syntheses" (N. Rabjohn, ed.), Collect. Vol. IV, p. 552. Wiley, New York, 1963.

[181] G. B. Payne, *J. Org. Chem.* **25**, 275 (1960).

In contrast to epoxidation with peroxy acids, these reactions are stereoselective rather than stereospecific. For example, only one isomer is obtained by the epoxidation of either isomer of the unsaturated ketone (**62**).[182]

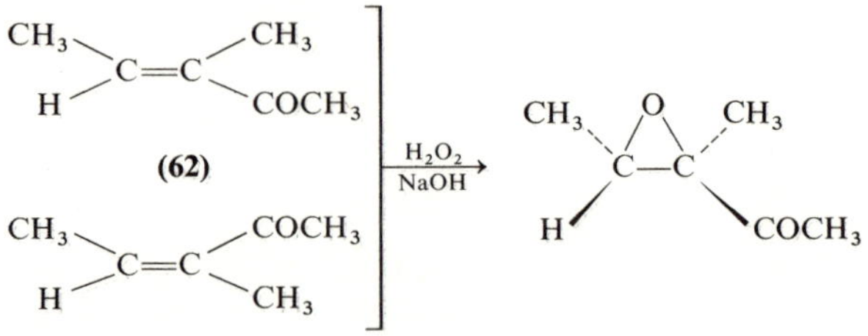

(59)

Nucleophilic addition of the hydroperoxide anion at the α-carbon of the α,β-unsaturated ketone or aldehyde is believed to be the mechanistic feature of this reaction.[183]

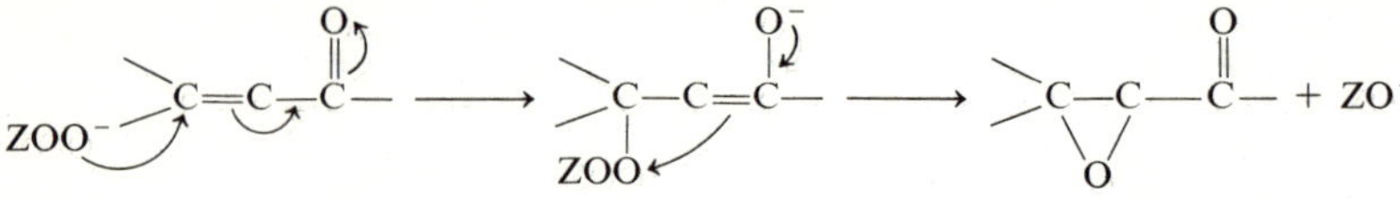

(60)

(Z = R,H)

It is evident that this type of epoxidation can only be employed for compounds undergoing Michael-type addition. Alkylidenemalonic esters [Eq. (61)],[184] unsaturated sulfones [Eq. (62)],[185] unsaturated nitro compounds [Eq. (63)],[186] and vinyl phosphonates [Eq. (64)][187] can also be epoxidized with these reagents.

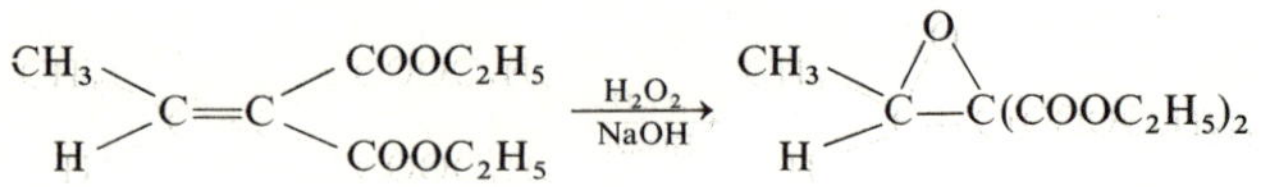

(61)

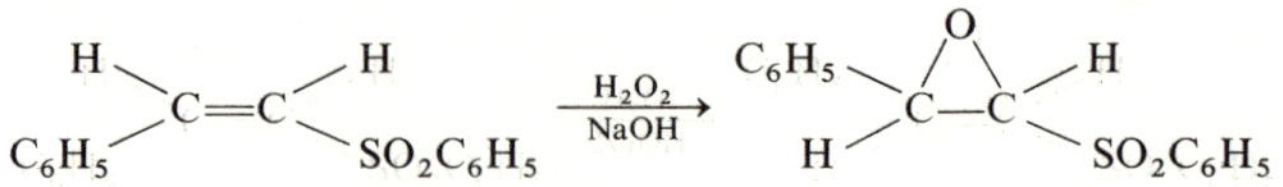

(62)

85%

[182] H. O. House and R. S. Ro, *J. Am. Chem. Soc.* **80**, 2428 (1958).
[183] C. A. Bunton and G. O. Minkoff, *J. Chem. Soc.* p. 665 (1949).
[184] G. B. Payne, *J. Org. Chem.* **24**, 2048 (1959).
[185] B. Zwanenburg and J. ter Wiel, *Tetrahedron Lett.* p. 935 (1970).
[186] H. Newman and R. B. Angier, *Tetrahedron* **26**, 825 (1970).
[187] C. E. Griffin and S. K. Kundu, *J. Org. Chem.* **34**, 1532 (1969).

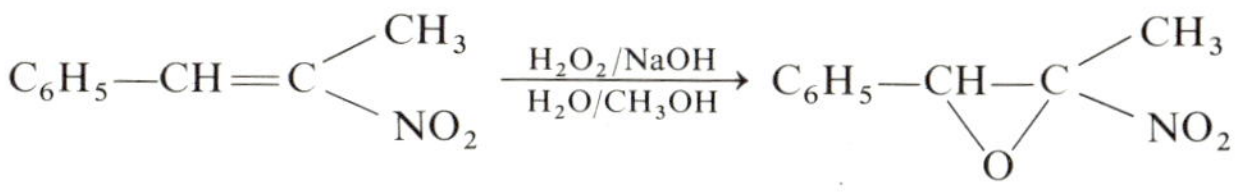

(63)

67%

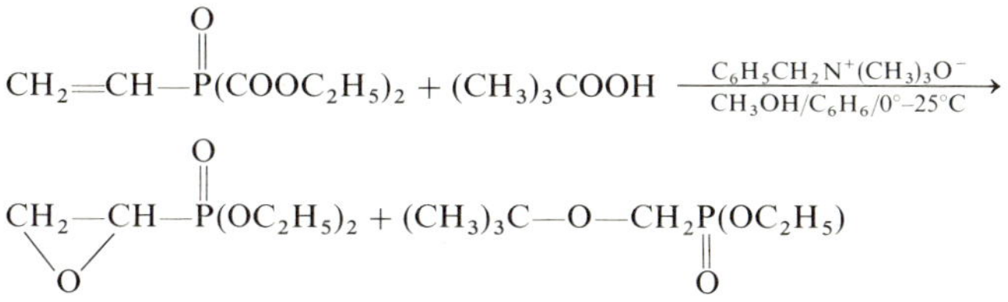

(64)

61% 23%

α,β-Unsaturated nitriles react with alkaline hydrogen peroxide to form epoxyamides [Eq. (65)].[188] This reaction involves a peroxyimidic acid as

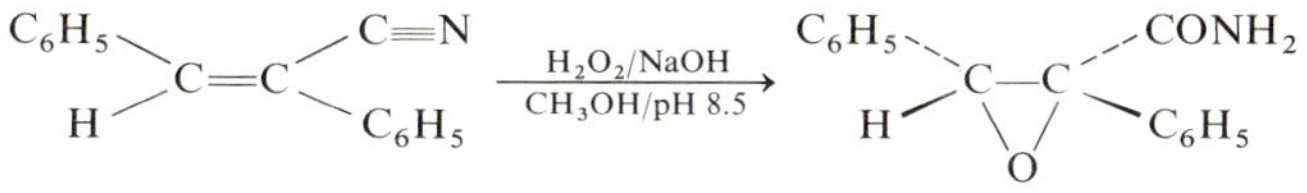

(65)

60%

an intermediate. Olefins with electron-withdrawing groups, which are not epoxidized readily with peroxy acids, can be converted to epoxides with hydrogen peroxide in the presence of vanadium pentoxide or tungsten trioxide.[189] For example, maleic acid is oxidized with such a reagent to form the epoxide (**63**) in 77% yield [Eq. (66)].[187] Peroxytunstic (**64**) and peroxyvanadic (**65**) acid are believed to be the epoxidizing agents in these cases.[149,190,191]

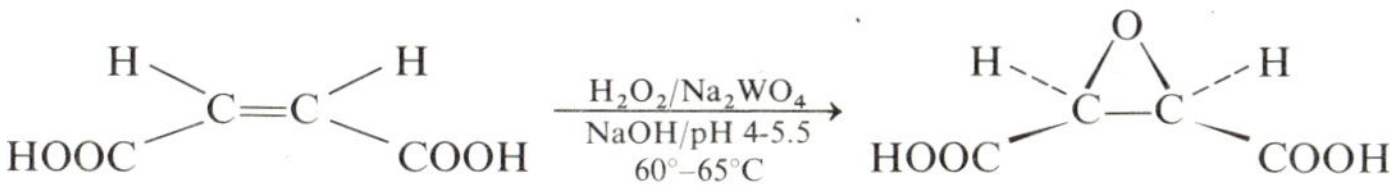

(66)

77%

(63)

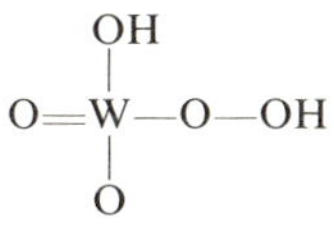

(64)

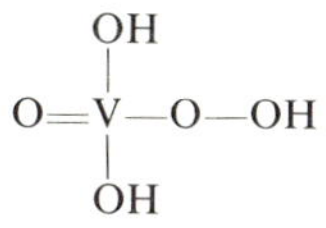

(65)

[188] G. B. Payne and P. H. Williams, *J. Org. Chem.* **24**, 54 (1959).
[189] N. A. Milas and S. Sussman, *J. Am. Chem. Soc.* **58**, 1302 (1936).
[190] Z. Raciszewski, *J. Am. Chem. Soc.* **82**, 1267 (1960).
[191] H. C. Stevens and A. J. Kaman, *J. Am. Chem. Soc.* **87** 734 (1965).

Molybdenum(VI) peroxo compounds have been found to epoxidize olefins stoichiometrically under anhydrous conditions in organic solvents. Originally a five-membered cyclic transition state (**66**) was proposed to be involved in these reactions.[192]

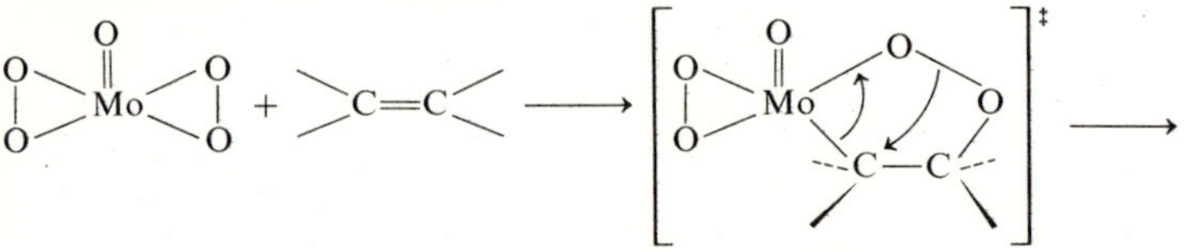

(66)

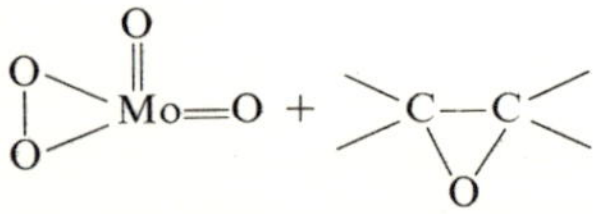

(67)

Recent evidence suggests an alternative mechanism involving a three-membered cyclic transition state, i.e., **67** or **68**.[157]

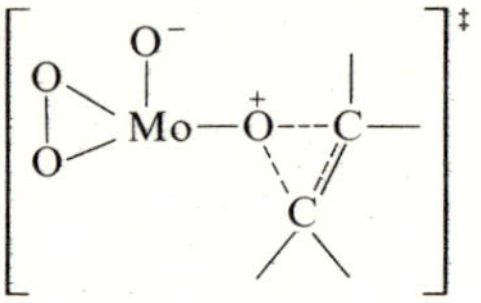

(67)

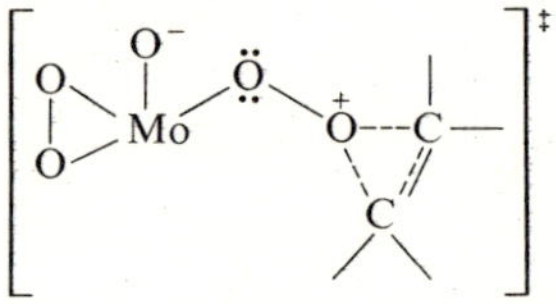

(68)

Alkyl hydroperoxides in the presence of molybdenum and vanadium compounds can also be used for epoxidation of olefins.[193–197] For example, homoallylic alcohol (**69**) affords only the *cis*-epoxy alcohol when oxidized with a molybdenum catalyst and *tert*-butyl hydroperoxide. Only the *trans*-epoxy isomer is formed in the peroxy acid oxidation of this system.[195]

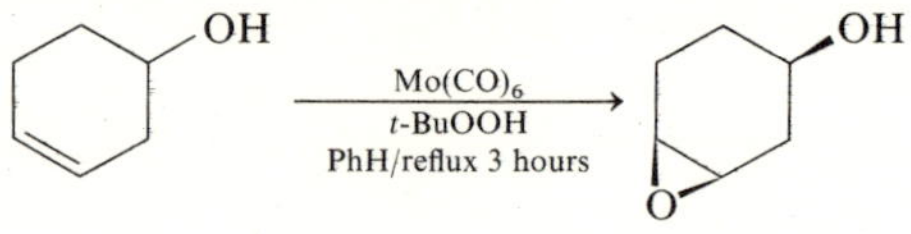

(68)

(69) 90%

[192] H. Mimoun, I. Seree de Roche, and L. Sajus, *Tetrahedron* **26**, 37 (1970); *Bull. Soc. Chim. Fr.* p. 1481 (1969).

[193] R. Hiatt, *in* "Oxidation, Techniques and Applications in Organic Synthesis" (R. L. Augustine and D. J. Trecker, eds.), Vol. 2, Chapter 3, p. 113. Dekker, New York, 1971.

[194] G. R. Howe and R. Hiatt, *J. Org. Chem.* **36**, 2493 (1971).

[195] K. B. Sharpless and R. C. Michaelson, *J. Am. Chem. Soc.* **95**, 6136 (1973).

[196] G. Descotes and P. Legrand, *Bull. Soc. Chim. Fr.* p. 2942 (1972).

[197] T. N. Baker, G. J. Mains, M. N. Sheng, and J. G. Zajacek, *J. Org. Chem.* **38**, 1145 (1973).

A heterolytic mechanism has been proposed for this type of reaction.[194,196,197]

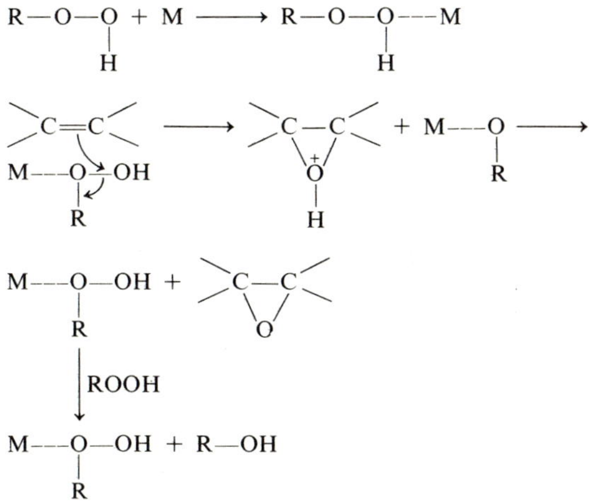

(69)

C. OXIDATION OF ACETYLENES

Theoretically, monoepoxidation of acetylenes should first produce oxirenes, which could be converted to dioxabicyclo derivatives [Eq. (70)].

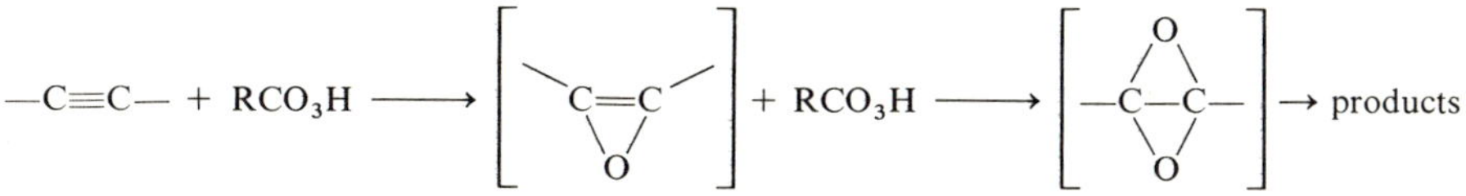

(70)

Neither of these products have yet been detected. Usually, a mixture of products is obtained in the oxidation of acetylenes with peroxy acids. The reaction products, however, can be explained on the basis of these intermediates.[198–202] The oxidation of di-*tert*-butylacetylene (**70**), for example, with *m*-chloroperoxybenzoic acid, gives 21% of 2,2,4,5-tetramethyl-4-hexen-3-one (**71**), 72% of the epoxide of that olefin (**72**), and 7% of cyclopropene derivative **73**.[201,202]

[198] V. Franzen, *Chem. Ber.* **88**, 717 (1955).

[199] R. N. McDonald and P. A. Schwab, *J. Am. Chem. Soc.* **86**, 4866 (1964).

[200] J. K. Stille and D. D. Whitehurst, *J. Am. Chem. Soc.* **86**, 4871 (1964).

[201] J Ciabattoni, R. A. Campbell, C. A. Renner, and P. W. Concannon, *J. Am. Chem. Soc.* **92**, 3826 (1970).

[202] P. W. Concannon and J. Ciabattoni, *J. Am. Chem. Soc.* **95**, 3824 (1973).

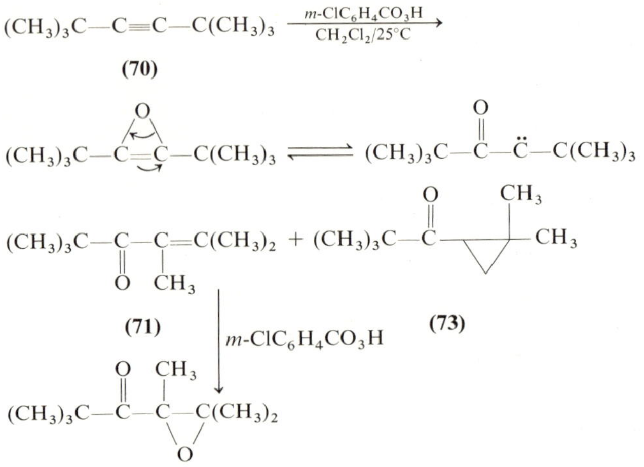

(70) **(71)** **(73)** **(72)**

(71)

An oxirene intermediate has been proposed in this reaction although the possibility of a ketocarbene as an alternative intermediate can not be excluded. In principle, oxirenes could exist in equilibrium with oxocarbenes. The decomposition of the corresponding diazoketone, 2,2,5,5-tetramethyl-4-diazohexan-3-one, which proceeds through a ketocarbene gives almost identical yields of **71** and **73** after taking into account that in this decomposition there is no peroxy acid to epoxidize the olefinic product.

Stronger evidence for an oxirene intermediate being involved in oxidation of acetylenes and a ketocarbene in diazoketone decomposition comes from

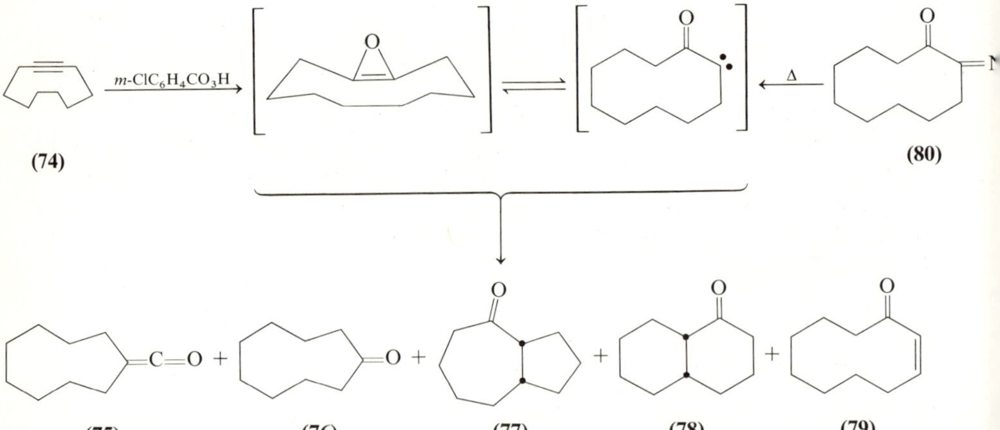

(74) **(80)** **(75)** **(76)** **(77)** **(78)** **(79)**

(72)

a study of oxidation of cyclic systems. Cyclodecyne (**74**) reacts with *m*-chloroperoxybenzoic acid in dichloromethane to yield less than 1% of cyclononanone (**76**), 67% of *cis*-bicyclo[5.3.0]decen-2-one (**77**), 12% of *cis*-bicyclo[4.4.0]decan-2-one (**78**), and 21% of *cis*-2-cyclodecen-1-one (**79**). Thermal decomposition of 2-diazocyclodecanone (**80**) in benzene similarly yields products **77**, **78**, and **79**, but in quite different amounts (4%:2%:49%). Octamethylketene (**75**) has been obtained in this case in 45% yield [Eq. (72)].[202]

Acetylenes react considerably slower with peroxy acids compared to structurally similar olefins.[203,204] The reaction is second-order; it is first-order with respect to acetylene and first-order with respect to peroxy acid.

$$v = k[\mathrm{RCO_3H}][\text{acetylene}] \qquad (73)$$

The rate of the reaction of substituted phenylacetylenes with peroxybenzoic acid has been measured in benzene at 25°C. The Hammett plot with σ^+ gives a ρ value of -1.40, thus indicating an electrophilic attack of peroxy acid. A transition state (**81**), resembling the one already proposed for olefin epoxidation, is consistent with these data.

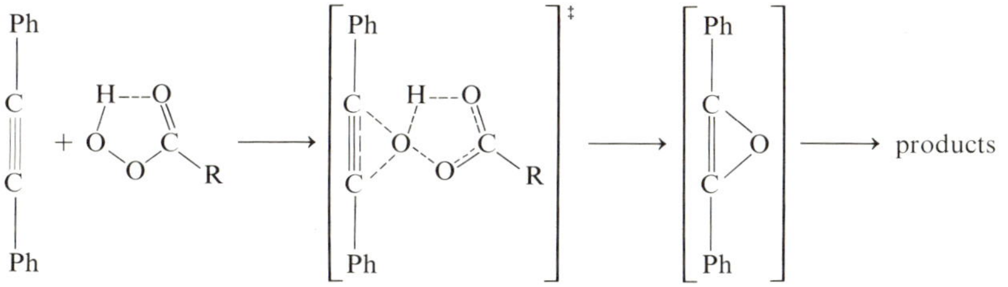

(74)

(81)

The rates of oxidation of 4-octyne with *m*-chloroperoxybenzoic acid in various solvents have been recently measured.[204] The results indicate that solvents play a similar role in epoxidation of 4-octyne as well as in epoxidation of cyclohexene. Peroxy acid–solvent interactions are indicated only. The trend of activation parameters in both cases is also the same. All these results strongly support oxirene intermediates in the peroxidation of acetylenes.

D. HYDROXYLATION OF OLEFINS

In epoxidation with peroxy acids, the corresponding carboxylic acids are formed as reduction products. The epoxides formed initially are reactive

[203] Y. Ogata, Y. Sawaki, and H. Inone, *J. Org. Chem.* **38**, 1044 (1973).

[204] K. M. Ibne-Rasa, R. H. Pater, J. Ciabattoni, and J. O. Edwards, *J. Am. Chem. Soc.* **95**, 7894 (1973).

and susceptible to attack by carboxylic acids to form monoesters of 1,2-diols, especially in cases when relatively strong aliphatic peroxy acids are prepared and used *in situ*. The monoesters can be hydrolized further to 1,2-diols [Eq. (75)].[11,12] Peroxyformic acid [Eq. (76)],[11] monoperoxymaleic acid,[27]

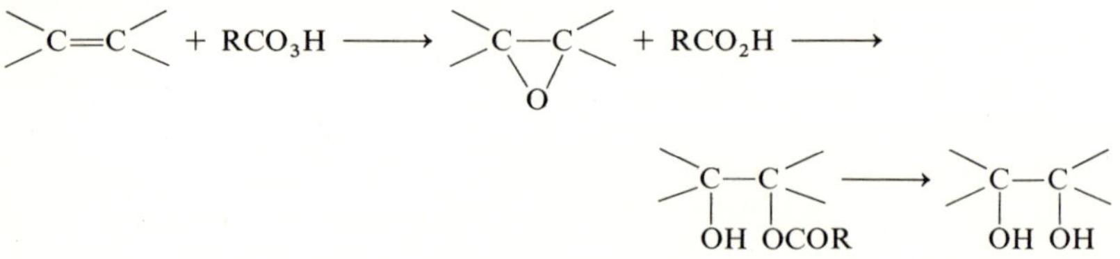

(75)

and peroxytrifluoroacetic acid [Eq. (77)][205] are most recommended for this purpose. The latter is the reagent of choice with terminal olefins which react rather slowly with peroxyformic acid.

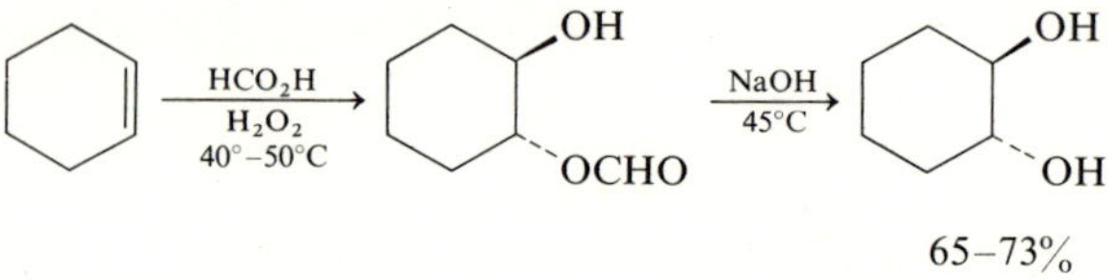

(76)

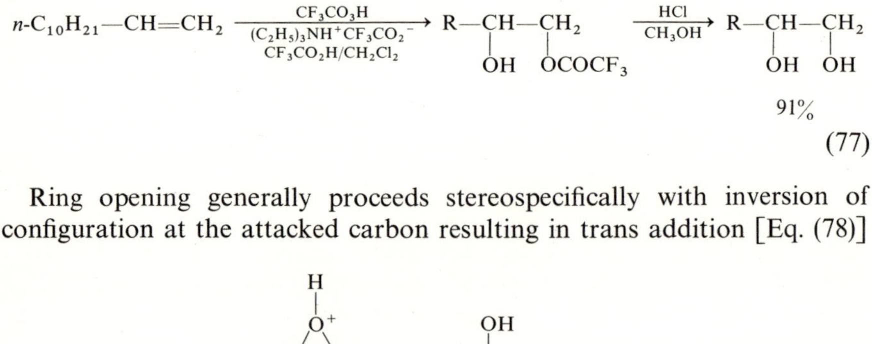

(77)

Ring opening generally proceeds stereospecifically with inversion of configuration at the attacked carbon resulting in trans addition [Eq. (78)]

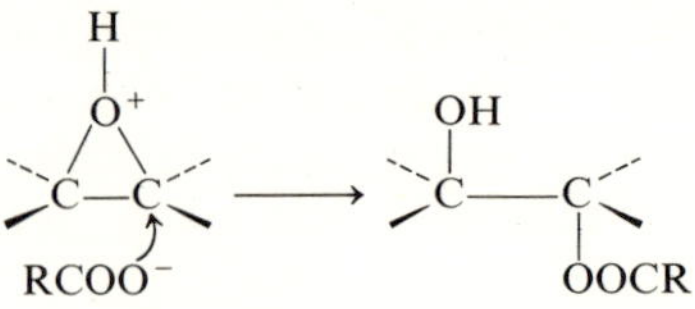

(78)

but cis hydroxylation has been observed in some cases.[206–210] Ring opening

[205] W. D. Emmons, A. S. Pagano, and J. P. Freeman, *J. Am. Chem. Soc.* **76**, 3472 (1954).
[206] D. Y. Curtin, A. Bradley, and Y. G. Hendrickson, *J. Am. Chem. Soc.* **78**, 4064 (1956).
[207] H. H. Wasserman and N. E. Aubrey, *J. Am. Chem. Soc.* **78**, 1726 (1956).
[208] G. Berti, F. Bottari, and B. Macchia, *Tetrahedron* **20**, 545 (1964).
[209] W. E. Rosen, L. Dorfman, and M. P. Linfield, *J. Org. Chem.* **29**, 1723 (1964).
[210] H. Riviere, *Bull. Soc. Chim. Fr.* p. 97 (1964).

with retention of configuration proceeds in some olefins which have at least one aryl substituent bonded to the epoxide ring [Eqs. (79)[206] and (80)[210]].

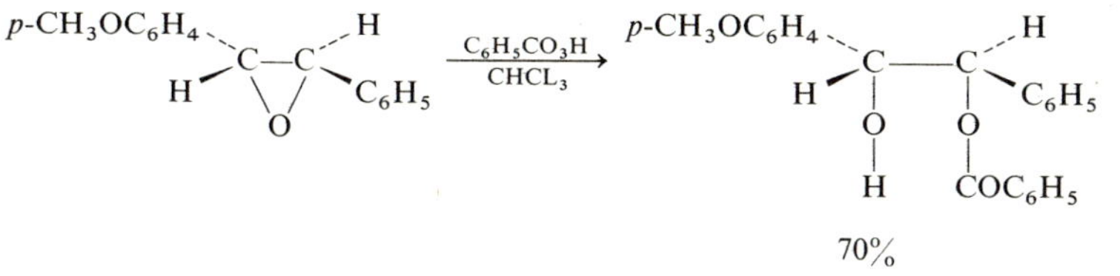

(79)

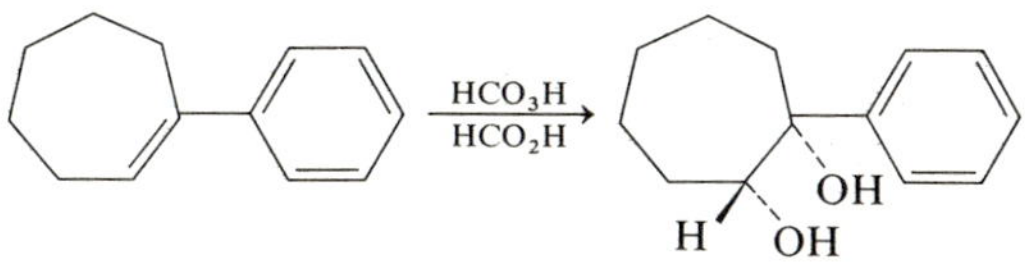

(80)

Ring opening with either retention or inversion of configuration by nucleophiles depends largely on the reaction conditions used. A full treatment of epoxide ring opening is beyond the scope of this chapter; the interested reader is referred to the available reviews.[211–214]

Solutions of *tert*-butyl hydroperoxide in the presence of base and OsO_4 have recently been used successfully for cis dihydroxylation of olefins. For example, (*E*)-4-octene (**82**) gives the corresponding diol and ketol in the

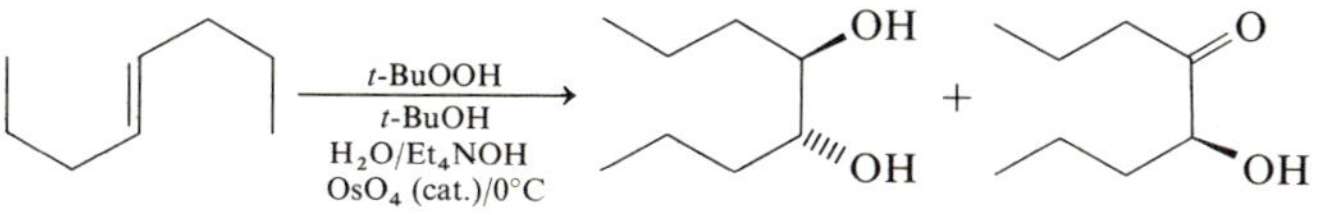

(82)

(81)

ratio 10:1 when oxidized with this reagent [Eq. (81)].[215] Hydroxylation of saturated carbon centers by iron-hydrogen peroxide,[215a] iron-peroxy acid[215b] acidic hydrogen peroxide (FSO_3H–SbF_5, FSO_3H, H_2SO_4, or

[211] R. E. Parker and N. S. Isaacs, *Chem. Rev.* **59**, 737 (1959).

[212] A Rosowsky, *Chem. Heterocycl. Compd.* **19**, 1 (1964).

[213] M. S. Malinovsky, "Epoxides and their Derivatives." Israel Program Sci. Transl., Jerusalem, 1965.

[214] J. Buchanan and H. Z. Sable, *in* "Selective Organic Transformations" (B. S. Thyagarajan, ed.), Vol. 2. Wiley (Interscience), New York, 1972.

[215] K. B. Sharpless and K. A. Kashi, *J. Am. Chem. Soc.* **98**, 1986 (1976).

[215a] C. Walling, *Acc. Chem. Res.* **8**, 125 (1975); J. T. Groves and M. Van der Puy, *J. Am. Chem. Soc.* **98**, 5290 (1976).

[215b] J. T. Groves and G. A. McClusky, *J. Am. Chem. Soc.* **98**, 859 (1976).

HF),[215c] and peroxytrifluoroacetic acid[215d] has also been reported. An ionic reaction path has been suggested recently for the hydroxylation of alkanes with the latter reagent.[215c]

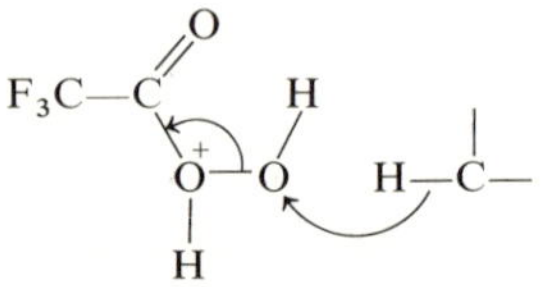

E. HYDROXYLATION OF AROMATIC COMPOUNDS

Aromatic hydrocarbons are oxidized to phenols by peroxy acids. Electron-donating groups on the aromatic ring facilitate the reaction. Good yields of mesitol (**84**) were obtained in the reaction of mesitylene (**83**) with hydrogen peroxide in a mixture of acetic and sulfuric acids.[216] Peroxytrifluoroacetic acid alone[217,218] or in combination with boron trifluoride[219] is now the reagent of choice for hydroxylation of aromatic systems. Quantitative yields of **84** were reported with this reagent even at $-40°C$.

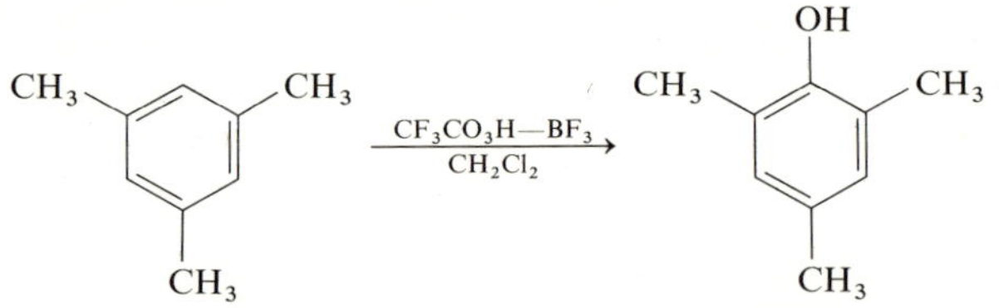

(82)

(83) **(84)**

One of the difficulties in controlled oxidation with peroxy acids is the activating effect of the hydroxyl group which is introduced. Further substitution in the ortho and para positions can take place to give *ortho*- and *para*-quinones. The latter are stable and can be isolated; the former, however, undergo ring fission to give diene acids. Naphthalene, for example, is converted to naphthoquinone with equimolar amount of peroxyacetic acid; the

[215c] G. A. Olah, N. Yoneda, and D. G. Parker, *J. Am. Chem. Soc.* **99**, 483 (1977).

[215d] N. C. Deno and L. A. Messer, *J. Chem. Soc., Chem. Commun.* p. 1051 (1976), and references cited therein.

[216] D. H. Derbyshire and W. A. Waters, *Nature* (*London*) **165**, 401 (1950).

[217] R. D. Chambers, P. Goggin, and W. K. R. Musgrave, *J. Chem. Soc.* p. 1804 (1959).

[218] R. O. C. Norman and J. R. L. Smith, *Oxidases Relat. Redox Syst., Proc. Symp., 1964*, Vol. 1, p. 131 (1965).

[219] H. Hart, *Acc. Chem. Res.* **4**. 337 (1971).

excess of the reagent, however, gives the fission product, *o*-carboxyallocinnamic acid (**85**) [Eq. (83)];[220] phenanthrenes react similarly.[221]

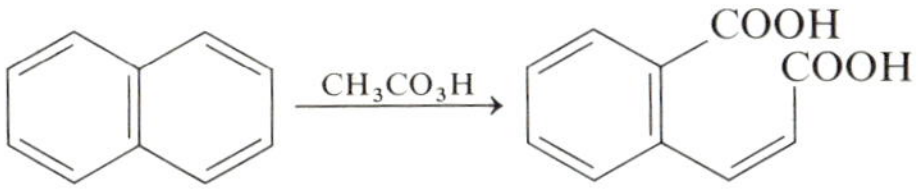

(85) (83)

Peroxytrifluoroacetic acid–boron trifluoride reagent attacks also at substituted positions. Hexamethylbenzene (**86**) gives the corresponding 2,4-cyclohexadienone (**87**) almost quantitatively.[222,223]

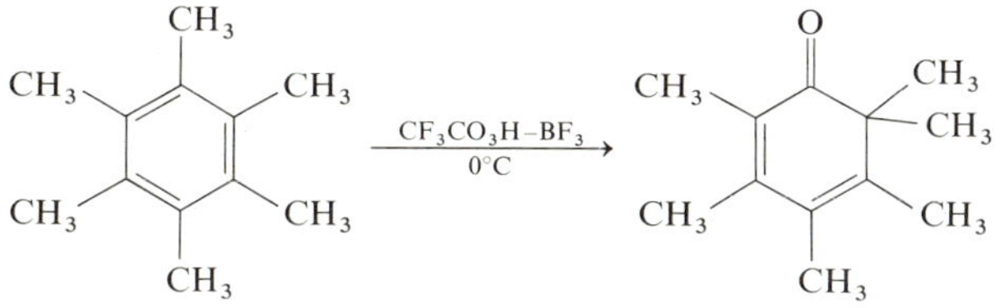

(86) **(87)** (84)

This reaction is general and provides the best method for preparation of previously inaccessible highly substituted cyclohexadienones.

The hydroxylation reaction is regarded as an example of electrophilic aromatic substitution although the nature of the attacking agent still remains unclear. Hydroxyl cation, OH^+, or its equivalent has been proposed to be the electrophilic agent.[223]

$$RCO_3H \longrightarrow RCO_2^-(OH^+)$$

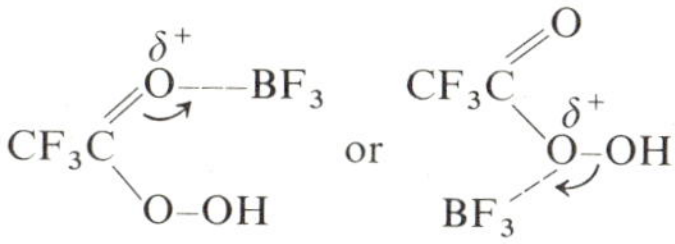

Another mechanistic possibility involves nucleophilic displacement by the aromatic compound on the oxygen atom of the peroxy acid.[224]

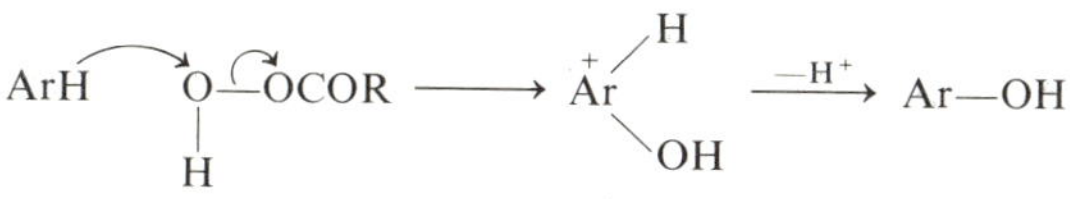

(85)

[220] J. Boeseken and G. Sloaf, *Recl. Trav. Chim. Pays-Bas* **49**, 100 (1930).

[221] W. F. O Connor and E. J. Moriconi, *J. Am. Chem. Soc.* **73**, 4044 (1951).

[222] A. J. Waring and H. Hart, *J. Am. Chem. Soc.* **86**, 1454 (1964).

[223] H. Hart, R. M. Lange, and P. M. Collins, *Org. Synth.* **48**, 87 (1968).

[224] A. J. Davidson and R. O. C. Norman, *J. Chem. Soc.* p. 5404 (1964).

The mechanism of electrophilic attack on the substituted position in hexamethylbenzene, as mentioned above, is presumed to proceed through a Wagner–Meerwein type rearrangement.[222]

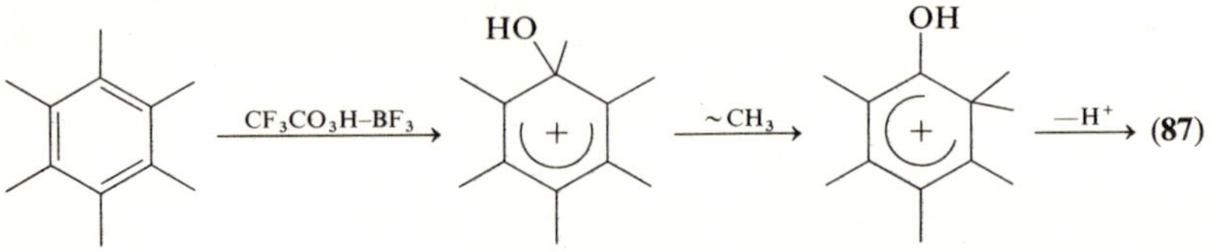

(86)

(87)

IV. Oxidation of Organic Carbonyl Compounds

A. KETONES

The oxidation of acyclic ketones to esters and cyclic ketones to lactones with peroxy acids was first demonstrated in 1899 by Baeyer and Villiger.[225] They used peroxysulfuric acid; since that time, several other organic peroxy acids have been used successfully.

$$R_1—\overset{\overset{\displaystyle O}{\|}}{C}—R_2 + R_3CO_3H \longrightarrow R_1—\overset{\overset{\displaystyle O}{\|}}{C}—OR_2 \quad \text{or} \quad R_1O—\overset{\overset{\displaystyle O}{\|}}{C}—R_2 + R_3CO_2H \qquad (87)$$

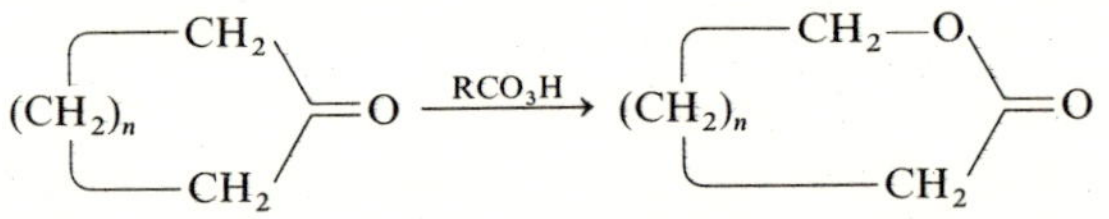

(88)

Since earlier work has been comprehensively reviewed,[147,178,226,226a] only the most characteristic synthetic and mechanistic features of this reaction will be given here.

Peroxy acid oxidation of ketones is slower than epoxidation and requires strong acids as catalysts. Rapid conversion of more reactive aryl ketones can be accomplished with peroxyacetic acid in the presence of sulfuric or *p*-toluenesulfonic acid [Eq. (89)].[227] Peroxytrifluoroacetic acid is generally the reagent of choice for the oxidation of simple ketones, which react slowly with peroxyacetic acid. Transesterification of the initially formed esters must

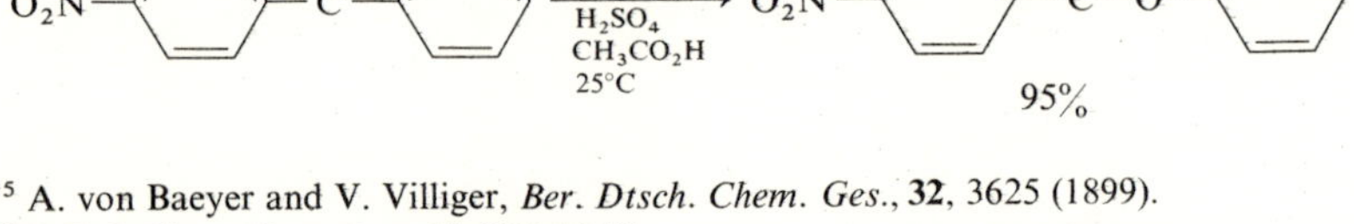

(89)

[225] A. von Baeyer and V. Villiger, *Ber. Dtsch. Chem. Ges.*, **32**, 3625 (1899).

[226] J. E. Leffler, *Chem. Rev.* **45**, 385 (1949).

[226a] P. A. S. Smith, *in* "Molecular Rearrangements" (P. deMayo, ed.), Vol. 1, p. 568. Wiley (Interscience), New York, 1963.

[227] W. von E. Doering and L. Speers, *J. Am. Chem. Soc.* **72**, 5515 (1950).

be prevented by the addition of an appropriate buffer, usually disodium hydrogen phosphate [Eq. (90)].[228] *m*-Chloroperoxybenzoic acid [Eq.

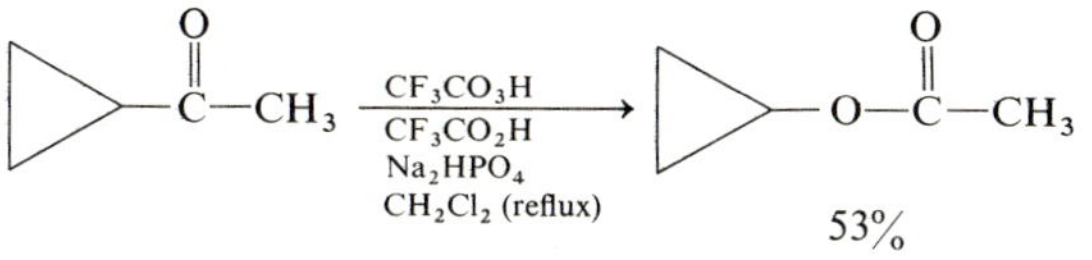

(90)

(91)][229] and monoperoxymaleic acid [Eq. (92)][27] have also been used successfully.

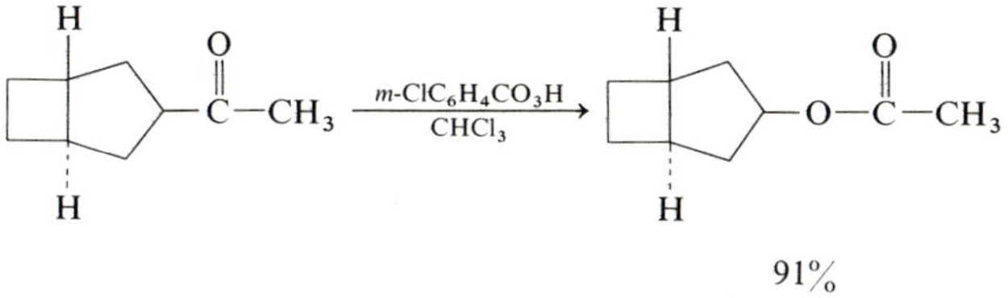

(91)

$$(CH_3)_2CH{-}CH_2COCH_3 \xrightarrow[\substack{CH_2Cl_2 \\ \text{reflux}}]{\substack{CH{-}CO_3H \\ \| \\ CH{-}CO_2H}} (CH_3)_2CH{-}CH_2{-}O{-}COCH_3 \quad (92)$$

72%

Unsymmetrical ketones afford two isomeric esters. Cyclohexyl phenyl ketone, for example, gives 71% of **88** and 14% of **89** when oxidized with peroxybenzoic acid[230]

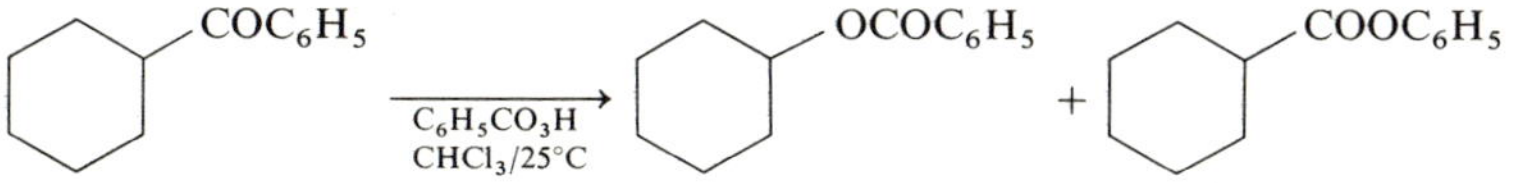

(93)

(88) **(89)**

Optically active esters have been prepared by the oxidation of optically active ketones, thus providing an elegant method for the preparation of optically active alcohols [Eq. (94)].[231]

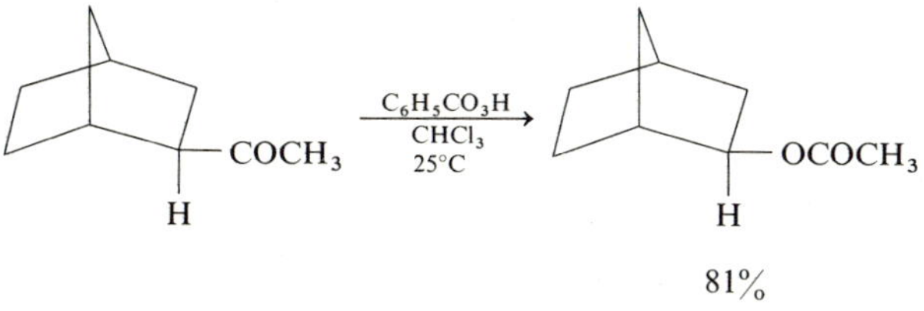

(94)

(optically active) (optically active)

[228] W. D. Emmons and G. B. Lucas, *J. Am. Chem. Soc.* **77**, 2287 (1965).
[229] J. Meinwald, J. Tufariello, and J. J. Hurst, *J. Org. Chem.* **29**, 2914 (1964).
[230] S. L. Friess and N. Farnham, *J. Am. Chem. Soc.* **72**, 5518 (1950).
[231] J. A. Berson and S. Suzuki, *J. Am. Chem. Soc.* **81**, 4088 (1959).

Cyclic ketones are especially reactive. The corresponding lactones are obtained in good yields. For example, cyclopentanone gives lactone **90** in 81% yield when treated with peroxytrifluoroacetic acid.[232]

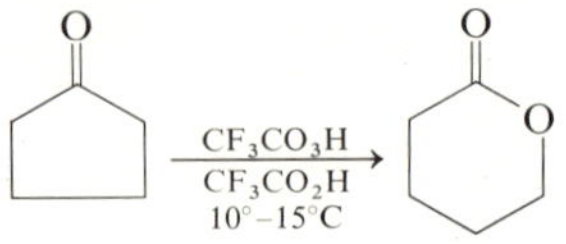

(95)

(90)

Conversion of cyclobutanones to lactones in nearly quantitative yields can be performed with alkaline hydrogen peroxide solution at room temperatures [Eq. (96)].[233]

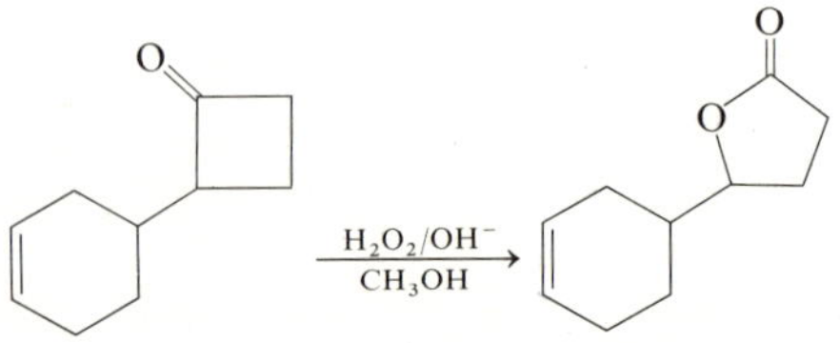

(96)

82%

Six-membered and larger ring lactones are usually equilibrated with their dimers and trimers in the presence of strong acids. Distilled solutions of peroxyacetic acid in ethyl acetate are recommended in these cases. Due to higher temperatures, which are required for the reaction to take place, further oxidation to dicarboxylic acids has been observed [Eq. (97)].[234]

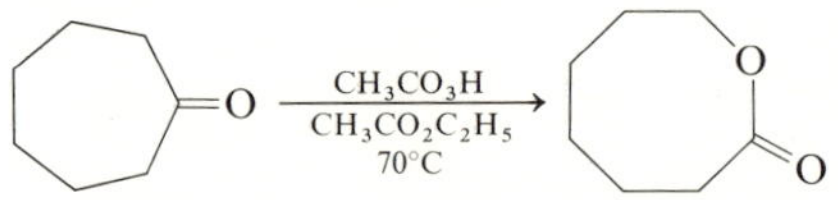

$+ HO_2C(CH_2)_5CO_2H$ (97)

29% 33%

Buffered solutions of peroxytrifluoroacetic[232] and monoperoxymaleic acid[27] can also be employed for oxidation of larger ring lactones.

The reaction of α,β-unsaturated ketones with peroxy acids normally yields unsaturated lactones [Eq. (98)][235] although both oxidation products, i.e., unsaturated lactones and epoxy lactones can be isolated in some cases. It has been found, for example, that oxidation of the carbonyl function to give **92** preceeds epoxidation in the reaction of enone ester **91** with less than 2 equivalents of *m*-chloroperoxybenzoic acid in dichloromethane. In the

232 W. F. Sager and A. Duckworth, *J. Am. Chem. Soc.* **77**, 188 (1955).
233 M. Bogdanowicz, T. Ambelang, and B. M. Trost, *Tetrahedron Lett.* p. 923 (1973).
234 P. S. Starcher and B. Phillips, *J. Am. Chem. Soc.* **80**, 4079 (1958).
235 H. M. Walton, *J. Org. Chem.* **22**, 1161 (1957).

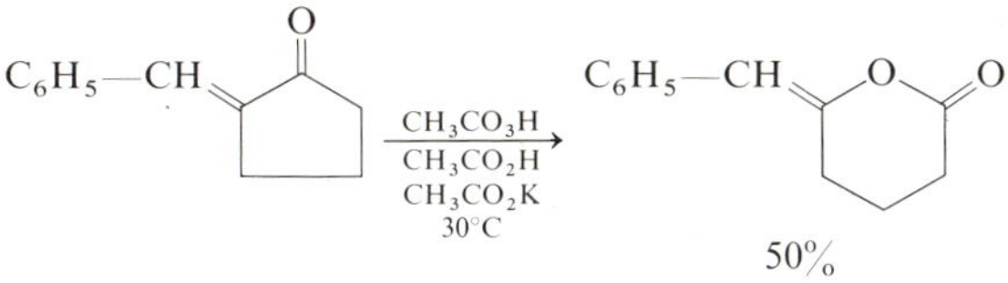

(98)

presence of excess peroxy acid, compound **93** has been found to be the sole product.[236]

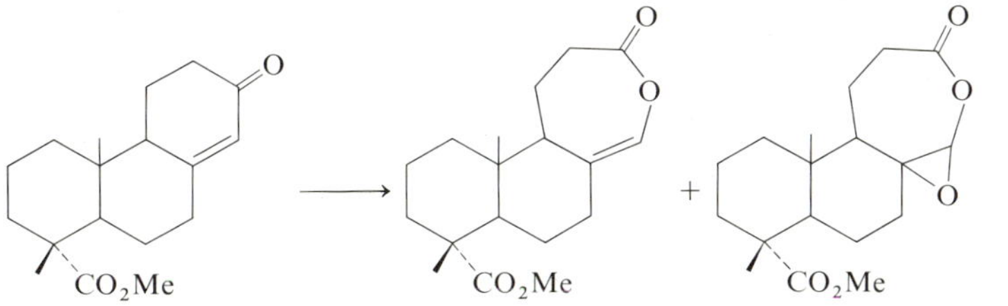

(99)

(91) **(92)** **(93)**

1. Kinetics and Mechanism

Several mechanistic studies indicate initial nucleophilic addition of peroxy acid to the carbonyl group with subsequent migration of an electron-rich group to an electrophilic oxygen.[147,178,226,226a]

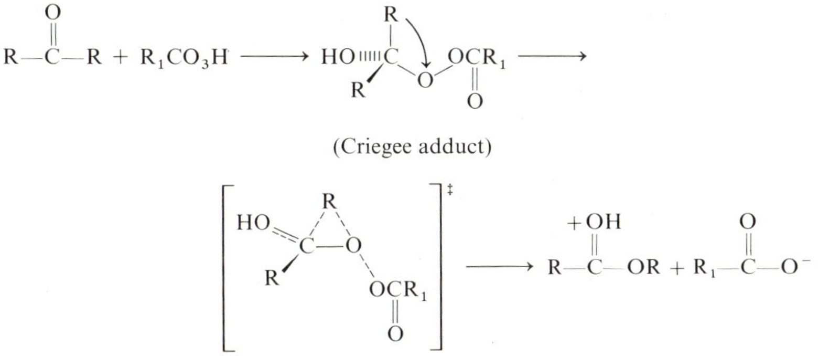

(100)

A large body of experimental evidence supports the proposed mechanism.[230,237–240] The reaction is catalyzed by acid and facilitated by electron-donating substituents in the ketone and electron-withdrawing groups in

[236] S. W. Pelletier, C. W. J. Chang, and K. N. Iyer, *J. Org. Chem.* **34**, 3477 (1969).
[237] R. Criegee, *Justus Liebigs Ann. Chem.* **560**, 127 (1948)
[238] W. von Doering and E. Dorfman, *J. Am. Chem. Soc.* **57**, 5595 (1953).
[239] S. L. Friess and A. H. Soloway, *J. Am. Chem. Soc.* **73**, 3968 (1951).
[240] S. L. Friess and R. Pinson, *J. Am. Chem. Soc.* **74**, 1302 (1952).

the peroxy acid. The rearrangement in the Criegee adduct must be intramolecular since the reaction proceeds with complete retention of configuration.[231,241–243] Benzophenone containing isotopically labeled oxygen gives phenyl benzoate with the same carbonyl oxygen isotope content as the starting material.[238] The relative migratory aptitude (the relative ease of migration) of various groups has been extensively studied and the following order has been found to be tertiary alkyl > cyclohexyl > secondary alkyl > benzyl > phenyl > primary alkyl > cyclopentyl > methyl.[227,240,244–248] Since all groups migrate in preference to methyl it is evident that all methyl ketones will give predominantly or entirely acetate esters. Electron-withdrawing groups in the para or meta position retard and electron-donating groups facilitate the migration of the phenyl group.[245] Ortho-substituted phenyl groups migrate less readily than para-substituted analogs.[245]

Steric effects in the Baeyer–Villiger reaction have received considerable attention in bicyclic systems.[249–255] It has been found that these effects can compete with or even dominate the electronic ones. For example, epicamphor (**94**) gives the electronically unfavored lactone (**95**) almost exclusively [Eq. (101)].[250] This is an example of an apparent preferential migration of a primary group over a secondary one.

The expected bridgehead migration, on the other hand, occurs almost exclusively in norcamphor (**96**) with the formation of the "normal" lactone (**98**) [Eq. (102)].[252] These results indicate strong carbonyl group steric hindrance due to a bridgehead methyl group in the former case. A chairlike transition state (**97**) has been suggested to be favored over a boatlike transition state in all these cases.[254]

[241] T. F. Gallagher and T. H. Kritchevsky, *J. Am. Chem. Soc.* **72**, 882 (1950).

[242] R. B. Turner, *J. Am. Chem. Soc.* **72**, 878 (1950).

[243] K. Mislow and J. Brenner, *J. Am. Chem. Soc.* **75**, 2318 (1953).

[244] M. F. Hawthorne, W. D. Emmons and K. S. McCallum, *J. Am. Chem. Soc.* **80**, 6393 and 6398 (1958).

[245] W. H. Saunders, *J. Am. Chem. Soc.* **77**, 4679 (1955).

[246] E. E. Smissman, J. P. Li, and Z. H. Israili, *J. Org. Chem.* **33**, 4231 (1968).

[247] R. R. Sauers and R. W. Ubersax, *J. Org. Chem.* **30**, 3938 (1965).

[248] M. A. Winnik and V. Stoute, *Can. J. Chem.* **51**, 2788 (1973).

[249] R. R. Sauers, *J. Am. Chem. Soc.* **81**, 925 (1959).

[250] R. R. Sauers and G. P. Ahearn, *J. Am. Chem. Soc.* **83**, 2759 (1961).

[251] A. Rassat and G. Ourisson, *Bull. Soc. Chim. Fr.* p. 1133 (1959).

[252] J. Meinwald and E. Frauenglass, *J. Am. Chem. Soc.* **82**, 5235 (1960).

[253] J. T. Edward and P. F. Morand, *Can. J. Chem.* **38**, 1325 (1960).

[254] M. F. Murray, B. A. Johnson, R. R. Pederson, and A. C. Ott, *J. Am. Chem. Soc.* **78**, 981 (1956).

[255] R. R. Sauers and J. A. Beisler, *J. Org. Chem.* **29**, 210 (1964).

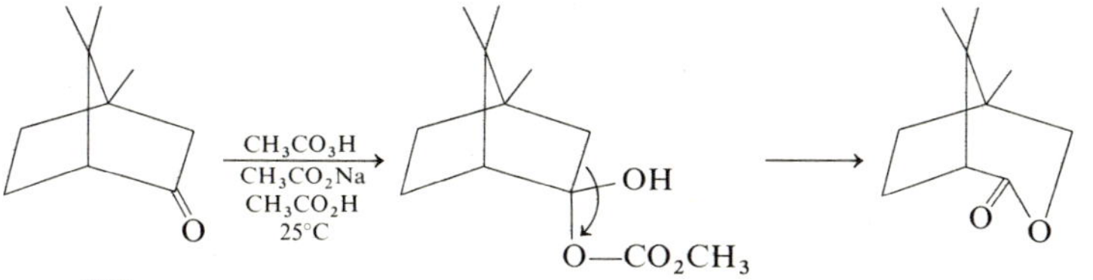

(94) **(95)** 94% (101)

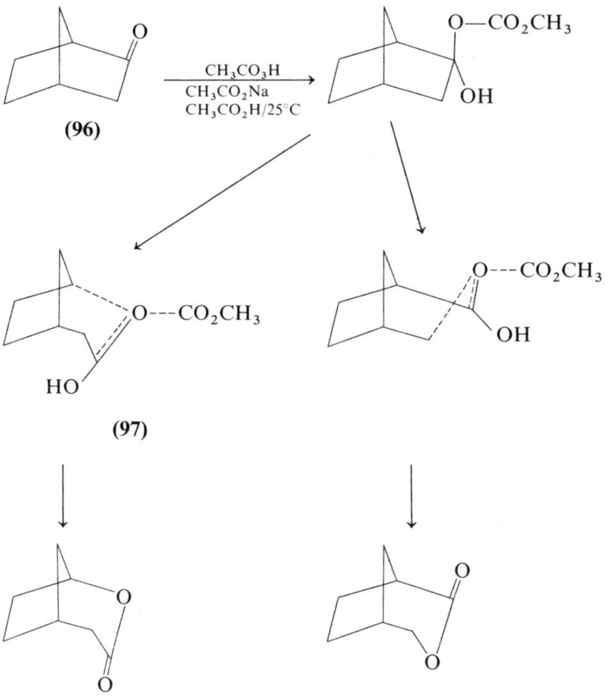

(96) **(97)**

88%

(98) (102)

The nature of the leaving group may also influence the migratory ability.[27,248] For example, peroxytrifluoroacetic acid has been found to be more sensitive to steric effects of various groups compared to monoperoxymaleic acid in the oxidation of simple ketones.[248]

All of the evidence mentioned above indicates that the size of the migrating group, its ability to support a positive charge in the transition state, and the size and reactivity of the peroxy acid, are the main factors influencing the rearrangement of the transition state. Migration of the group and loss of the leaving group are believed to proceed simultaneously.[256,257,258] Considerable carbonyl formation in the transition state (**99**) has been proposed on the basis of small α- and relatively large β-secondary isotope effect for oxidation of dueterated phenyl 2-propanone.[258] An ion pair intermediate (**100**) seems to be ruled out on the basis of recent *ab initio* and CNDO/2 studies.[259]

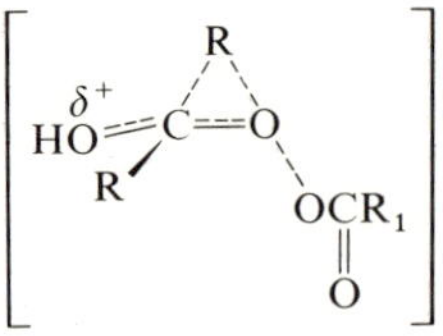

(99)

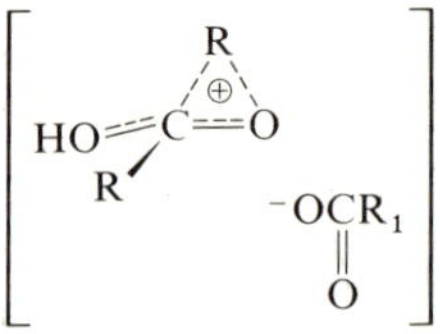

(100)

The rate for peroxy acid oxidation of ketones is, for most cases, first-order in ketone and first-order in peroxy acid.[244,256,260]

$$v = k[RCO_3H][\text{ketone}] \tag{103}$$

Rate-determining migration in the peroxy acid-ketone adduct has been shown by the negative ρ values of -1.1 $(\sigma)^{244}$ and -1.36 $(\sigma^+)^{256}$ as well as ^{14}C isotope effects for the peroxy acid oxidation of acetophenones.[256] The oxidation of *p*-methoxyacetophenone shows no ^{14}C isotope effect. This was originally interpreted as being due to the strong electron-donating power of the methoxy group. Recent evidence indicates that this phenomenon could be explained by a change of the rate-determining step to carbonyl addition as discussed below.

Detailed study of the kinetics of oxidation of acetophenones with peroxybenzoic acid in various organic solvents and 40% aqueous ethanol has been undertaken.[261] The reaction is slow in organic solvents without added catalyst. Acetic acid catalyzes only addition to the C=O function while trifluoracetic acid is strong enough to catalyze migration as well as addition. The acid catalysis appears to be general rather than specific since the rates of both steps (carbonyl addition and migration) are not correlated with the acidity function, H_0. The species operating in catalysis are presumed to be the intermolecularly hydrogen-bonded adducts **101** and **102**.

[256] B. W. Palmer and A. Fry, *J. Am. Chem. Soc.* **92**, 2580 (1970).
[257] T. Mitsuhashi, H. Miyadera, and O. Simamura, *Chem. Commun.* p. 1301 (1970).
[258] M. A. Winnik, V. Stoute, and P. Fitzgerald, *J. Am. Chem. Soc.* **96**, 1977 (1974).
[259] V. Stoute, M. A. Winnik, and I. G. Csizmadia, *J. Am. Chem. Soc.* **96**, 6388 (1974).
[260] J. L. Matheos and H. Menchaca, *J. Org. Chem.* **29**, 2026 (1964).
[261] Y. Ogata and Y. Sawaki, *J. Org. Chem.* **37**, 2953 (1972).

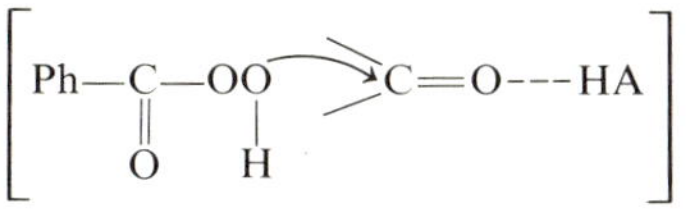

(101)

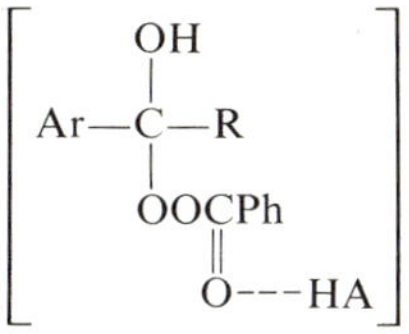

(102)

Carbonyl addition becomes the rate-determining step in ketones containing electron-donating groups. The rates of addition of peroxybenzoate ion to *o*- and *p*-hydroxyacetophenone at pH 5–12 satisfies the Taft equation to give $\rho^* = 1.3$–1.6 and $\delta = 0.6$–0.8, thus indicating the importance of steric and electronic effects. In the oxidation of *p*-methoxyacetophenone at pH 5.5, migration is the rate-determining step, while both migration and addition are important at pH 0.7–5.5. At higher acidity the acid catalyzed migration becomes predominant. The apparent rates of migration for ring substituted acetophenones in the presence of a suitable catalyst were correlated with σ^+ to give ρ values of -1.5 to -2.4.

Homolytic cleavage of the O–O bond in adduct **103** has been proposed in the reaction of peroxyacetic acid with benzophenones containing electron-withdrawing substituents.[262] These substituents were found to enhance migration of the phenyl rings to which they were attached which is contrary to the "normal" order of migration in oxidation of the same compounds with trifluoroperoxyacetic acid. Formation of the final products, according to this mechanism, would require the 1,2-shift of one of the phenyl groups from carbon to oxygen as shown below.

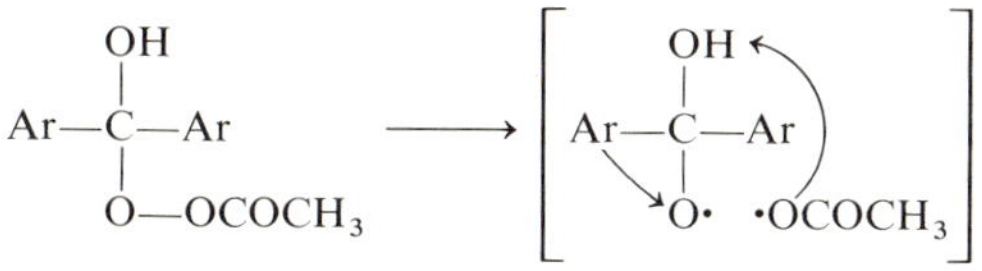

(104)

(103)

The abstraction of hydrogen by acetoxy radical has been proposed to proceed in the solvent cage since no carbon dioxide was observed. The absence of the latter seems to be the troublesome aspect of the proposed free radical mechanism.

Still another mechanistic possibility involves a 1,3-dipolar addition to the ketone of the peroxy acid molecule in its dipolar form.[176]

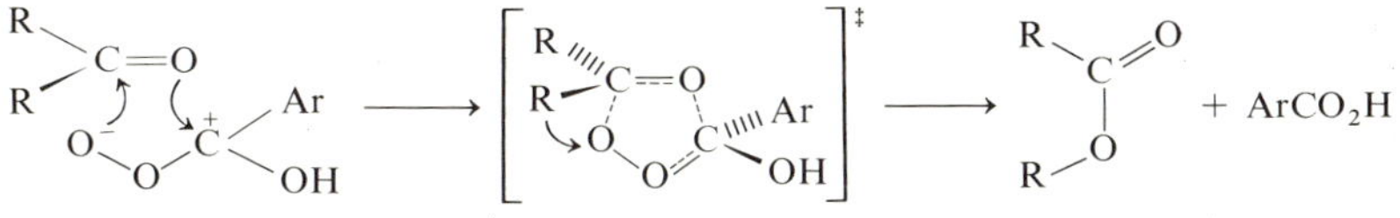

(105)

[262] J. C. Robertson and A. Swelin, *Tetrahedron Lett.* p. 2871 (1967).

No direct evidence has been found to support this mechanism but the idea seems to deserve further consideration.

The reaction of simple ketones with hydrogen peroxide[263] or alkyl hydroperoxides in alkaline solution[264] is of mechanistic rather than synthetic importance since low yields of carboxylic acids and alcohols are usually obtained. The cleavage of adduct **104** sometimes proceed differently from

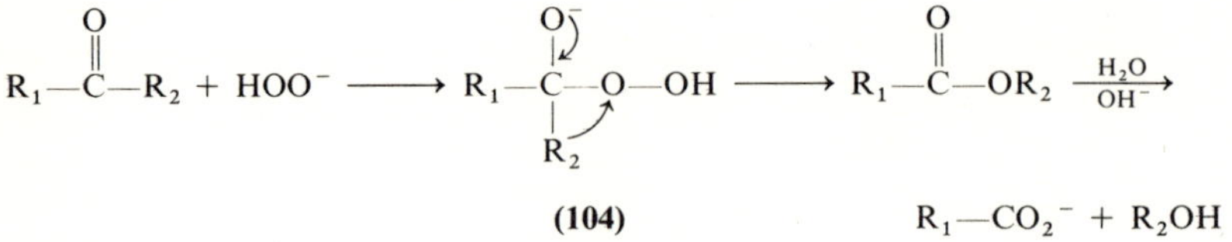

(104) (106)

that in the Baeyer–Villiger reaction.[265]

Easy conversion of cyclobutanones to γ-butyrolactones [Eq. (96)] with hydroperoxide anion has been explained by the relief of strain energy of the four-membered ring intermediate analogous to **104**.[233]

B. ALDEHYDES

Aldehydes are easily oxidized to carboxylic acids with peroxy acids.[266–268] Strong electron-donating substituents on the aryl group facilitate migration; formate esters, which are formed in these cases, hydrolize further to phenols [Eq. (107)]. Benzaldehyde, for example, reacts with peroxybenzoic acid to

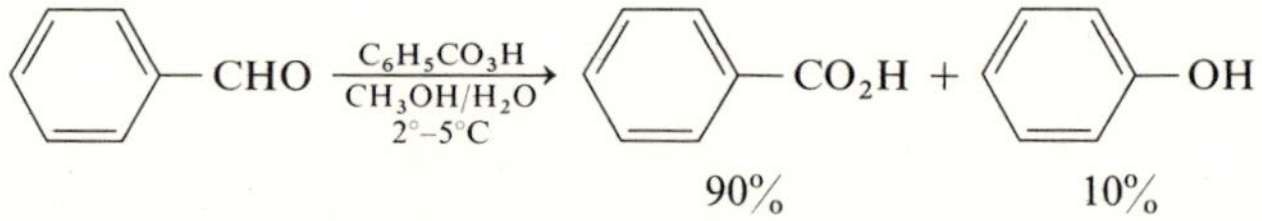

(107)

90% 10%

form benzoic acid in 90% yield, while *p*-methoxybenzaldehyde affords only 19% of the carboxylic acid and 73% of phenol, respectively.[267] A number of mono-, di-, and trimethoxybenzaldehydes have been oxidized to the corresponding phenols by *m*-chloroperoxybenzoic acid in boiling dichloromethane followed by hydrolysis of the formates.[268]

The kinetics of the reaction of benzaldehydes with peroxybenzoic acids at various pH's in 40% aqueous ethanol has been studied by Ogata and

[263] H. O. House and R. L. Wasson, *J. Org. Chem.* **22**, 1157 (1957).
[264] K. Maruyama, *Bull. Chem. Soc. Jpn.* **34**, 102 (1960).
[265] K. Maruyama, *Bull. Chem. Soc. Jpn.* **33**, 1516 (1960).
[266] Y. Ogata and Y. Sawaki, *J. Org. Chem.* **34**, 3985 (1969).
[267] Y. Ogata and Y. Sawaki, *J. Am. Chem. Soc.* **94**, 4189 (1972).
[268] I. M. Godfrey, M. V. Sargent, and J. A. Elix, *J. Chem. Soc., Perkin Trans. 1* p. 1353 (1974).

Sawaki.[267] The mechanistic scheme that has been proposed is shown in Eq. (108).

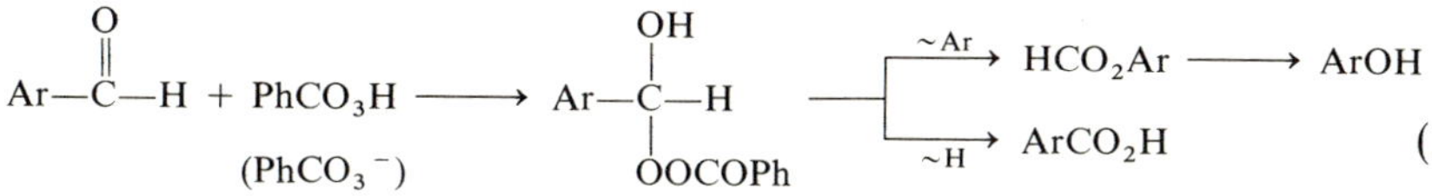

(108)

(105)

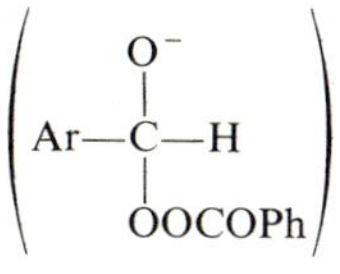

(106)

Either the aryl group or the hydrogen atom may migrate in adducts **105** and **106**. Selectivity for migration of an aryl group changes sharply with substituent and pH of the reaction medium. These migrations were found to be catalyzed by acid via a hydrogen bonded catalysis complex similar to adduct **102**. No hydroxide ion catalysis has been observed. Electron-releasing substituents in the ortho position accelerate aryl migration by a factor of approximately ten compared to those in the para position thus indicating no steric retardation by ortho substituents. This is just the opposite of the findings reported for the peroxy acid oxidation of benzophenones where retardation by ortho substituents has been found.[248] The kinetic deuterium isotope effect for aryl migration is $k_H/k_D = 1.0$. The considerable effect of solvent polarity on aryl migration ($m \leqq 0.5$) suggests significant separation of charge in the transition state for migration (**107**).

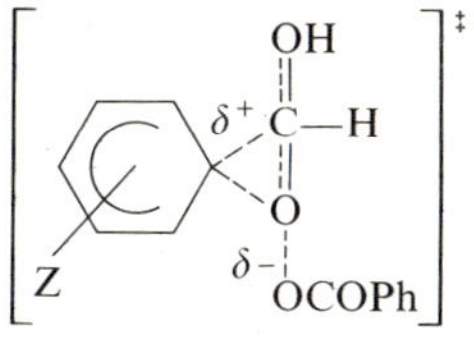

(107)

Hydride migration is characterized by a relatively large deuterium isotope effect ($k_H/k_D = 1.4-3.0$) and insensitivity to proton and hydroxide ion catalysis as well as solvent polarity ($m \leqq 0.1$). A hydride shift preferably proceeds via adduct anion **106** at pH > 8 rather than via neutral adduct **105**. The migration ratio (Ar/H) changes at pH ≈ 8.5 due to a change of the intermediate from an anion to a neutral adduct and then increases with increasing acidity of the medium since aryl migration is acid catalyzed. The relative migratory aptitude (from the product ratio $ArOH/ArCO_2H$) indicates a large negative Hammett ρ value of -4 to -5 (σ^+) for aryl migration and 1.1–1.8 for hydrogen shift, respectively.

The effect of substituents in the peroxybenzoic acids is rather small. The ρ value of 0.2–0.4 (σ) has been found for all benzaldehydes except for *o*- and *p*-hydroxybenzaldehydes where ρ is 0 at pH > 9. The evidence mentioned above led Ogata and Sawaki to the conclusion that carbonyl addition is the rate-determining step in oxidation of OH-substituted benzaldehydes while the rate-determining migration is operating in Me-, H-, Cl-, and NO_2-substituted benzaldehydes. Methoxybenzaldehyde appears to be a borderline case. *ortho*- and *para*-Hydroxy and amino aldehydes also react with alkaline hydrogen peroxide. The corresponding phenols are formed [Eq. (109)].[269] The process is known as the Dakin reaction[270] and is believed to proceed via a benzenonium-type intermediate (**108**).[271]

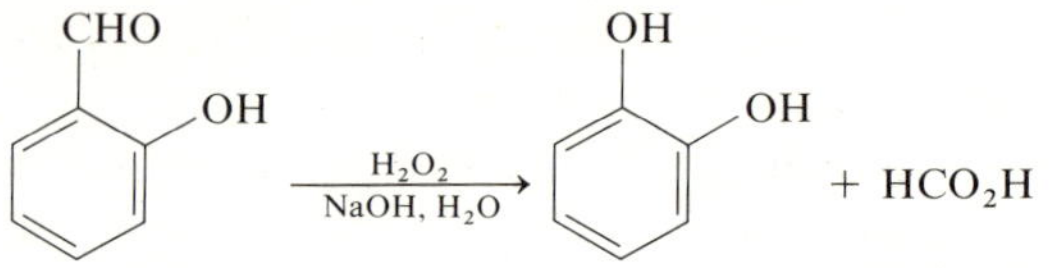

(109)

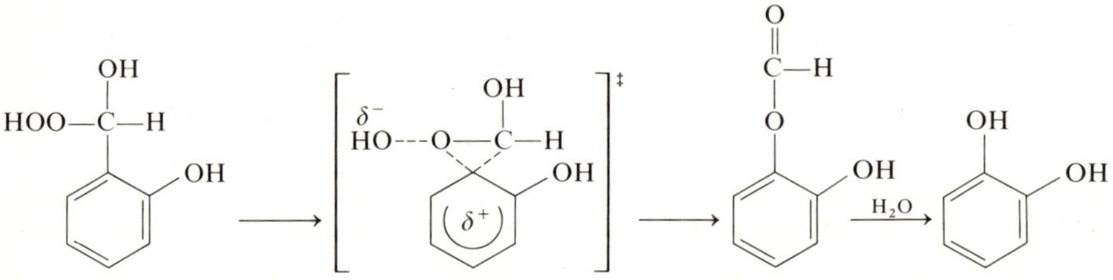

(108)

(110)

C. MISCELLANEOUS CARBONYL COMPOUNDS

α-Dicarbonyl compounds, i.e., α-diketones, *ortho*-quinones, and α-ketoesters react with peroxy acids to form the corresponding anhydrides.[272] If the Baeyer–Villiger mechanism is valid for these reactions, then the preferred migration of RCO to R is indicated [Eq. (111)]. α-Carbonyl compounds

[269] A. R. Surrey, *in* "Organic Syntheses" (E. C. Horning, ed.), Collect. Vol. III, p. 759. Wiley, New York, 1955.

[270] H. D. Dakin, *in* "Organic Syntheses" (H. Gilman and A. H. Blatt, eds.), Collect. Vol. I, 2nd ed., p. 149. Wiley, New York, 1941.

[271] Y. Ogata and A. Tabushi, *Bull. Chem. Soc. Jpn.* **32**, 108 (1959).

[272] P. Karrer and F. Haab, *Helv. Chim. Acta* **32**, 950 (1949).

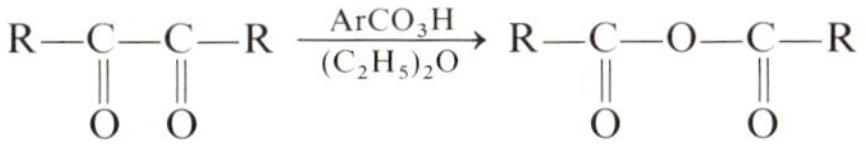

(111)

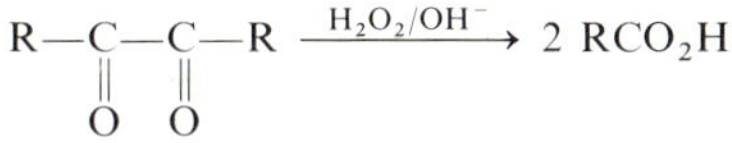

(112)

may also be oxidized with alkaline hydrogen peroxide (Eq. (112)].[273] Oxidation of α-keto acids by alkaline hydrogen peroxide presumably proceeds by the loss of carbon dioxide and hydroxide ion from an addition intermediate rather than by the Baeyer–Villiger type of rearrangement [Eq. (113)].[274]

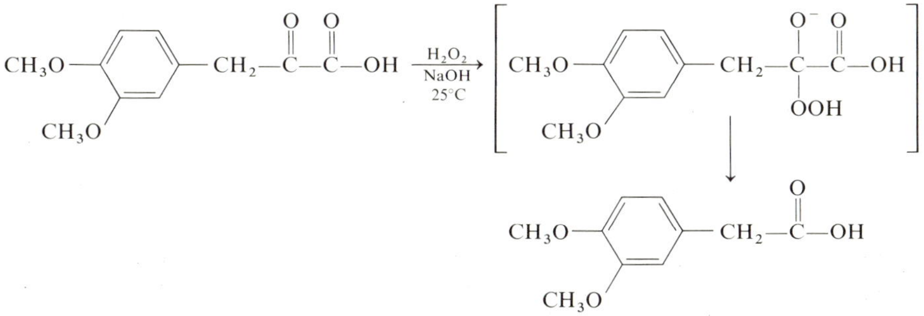

(113)

Oxidation of 1,3-dicarbonyl compounds capable of enolization with peroxy acids yields a mixture of cleavage products, presumably as a result of the epoxidation of the enol. No reaction of the Baeyer–Villiger type could be observed in these cases [Eq. (51)].[145]

The reaction of phenyl salicylates with peroxybenzoic acid in 60% (v/v) ethanol at pH 9 has recently been studied. Ethoxyphenol and catechol are formed in this reaction together with some other hydrolysis and alcoholysis products of the ester. When the reaction is carried out in aqueous methanol, *o*-methoxyphenol is formed. Although the reaction does not take place when phenyl or methyl benzoates are treated with peroxybenzoic acid under similar conditions, this novel alkoxylation reaction seems to be of some preparative value. The mechanism of this reaction most probably involves an aryl cation formed from carbonyl adduct **109**. The presence of two oxygen atoms in the latter is believed to support the electron-donating resonance stabilization in the Baeyer–Villiger type of rearrangement via transition state **110**.[275]

[273] R. P. Barnes, *J. Am. Chem. Soc.* **57**, 937 (1935).

[274] H. R. Snyder, J. S. Buck, and W. S. Ide, *in* "Organic Syntheses" (A. H. Blatt, ed.), Collect. Vol. II, p. 333. Wiley, New York, 1943.

[275] Y. Ogata, Y. Sawaki, and M. Furuta, *J. Org. Chem.* **39**, 216 (1974).

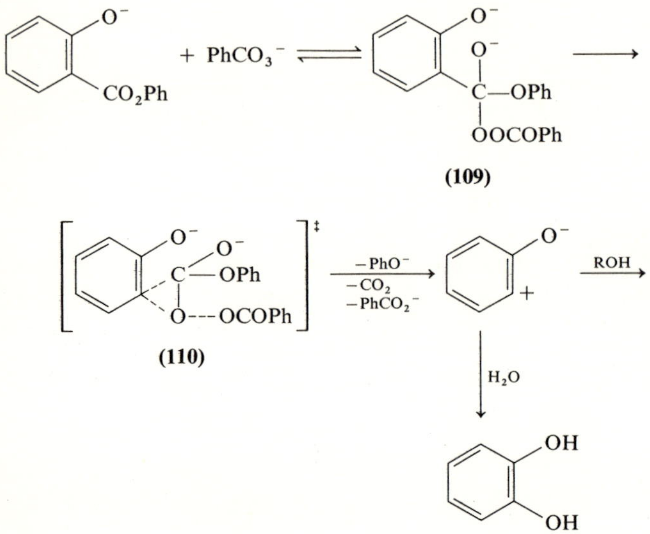

(109)

(110)

(114)

Oxidation of the β-lactam acid chlorides (**111**) with *m*-chloroperoxybenzoic acid results in a unique dechlorocarbonylation to produce *cis*-β-lactams exclusively [Eq. (115)].[276]

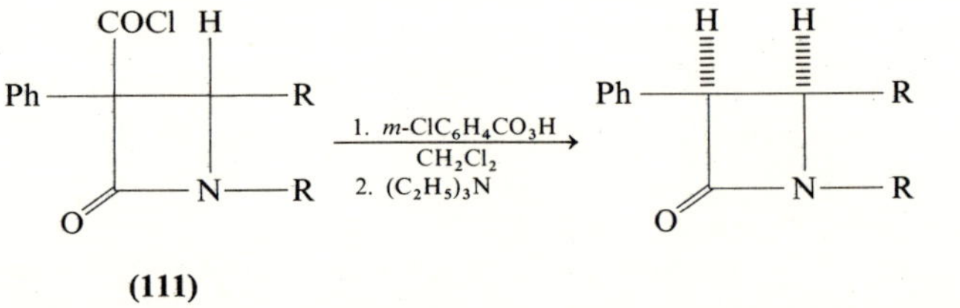

(115)

(111)

When α-hidroxy ketones are oxidized with *m*-chloroperoxybenzoic acid, a mixture of aldehydes and carboxylic acids is formed. These reactions are believed to proceed via an ene–diol intermediate which is in a rate-determining equilibrium with the starting material. This intermediate is further oxidized to a 2,3-dihydroxy epoxide which then rearranges to form the products.[277] [Eq. (116)].[277]

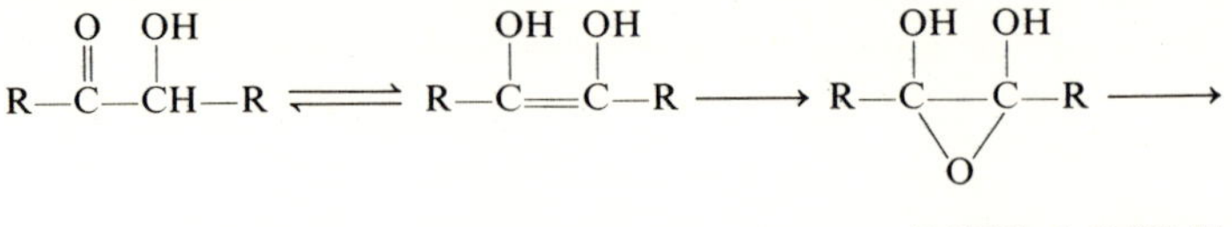

$RCHO + RCO_2H$ (116)

[276] A. K. Bose and J. C. Kapur, *Tetrahedron Lett.* p. 1811 (1973).
[277] T. Greibrokk, *Acta Chem. Scand.* **27**, 3365 (1973).

Acid chlorides can be converted to alcohols with *m*-chloroperoxybenzoic acid. Mixed peroxides, which are formed first in these reactions, rearrange to the mixed carbonates, which decompose further to give alkyl *m*-chlorobenzoates. Saponification of the latter affords alcohols in yields greater than 50% [Eq. (117)].[278]

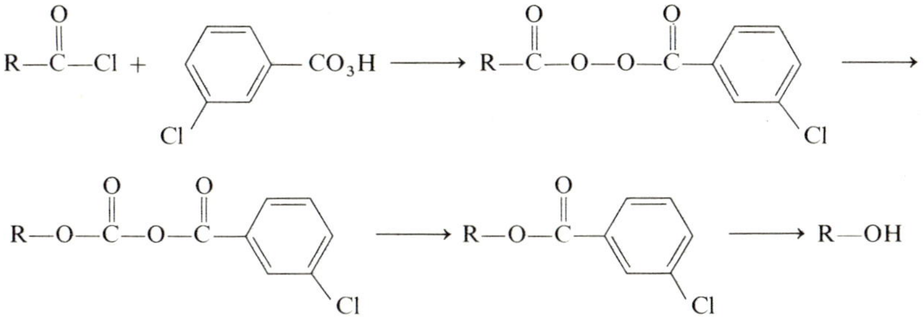

(117)

V. Oxidation of Organic Nitrogen Compounds

A. AMINES

The amino group contains nitrogen in its lowest oxidation state so it is not surprising that amines are easily oxidized with peroxy acids. Various products are obtained depending on the reaction condition used. Primary amines react with peroxy acids to yield nitroso compounds (**113**) via initially formed hydroxylamines (**112**). Nitroso compounds or their oxime tautomers (**114**) can be oxidized further to nitro derivatives (**115**) under more vigorous conditions.[11,279–285]

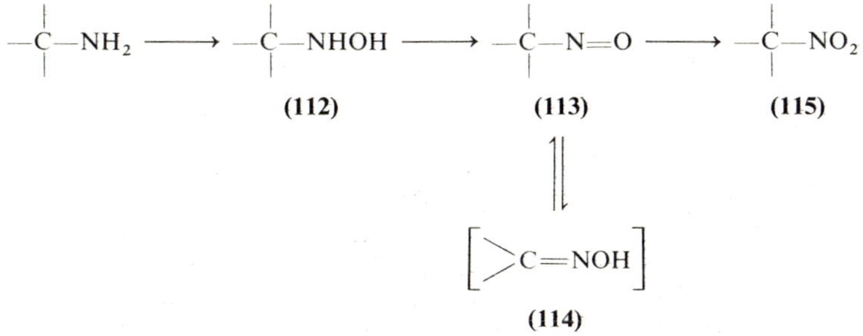

(118)

[278] D. B. Denney and N. Sherman, *J. Org. Chem.* **30**, 3760 (1965).

[279] H. Metzger, *in* "Methoden der organischen Chemie" (E. Müller, ed.), 4th ed., Vol. 10, p. 122. Thieme, Stuttgart, 1968.

[280] W. D. Emmons, *J. Am. Chem. Soc.* **79**, 5528 (1957).

[281] W. D. Emmons, *J. Am. Chem. Soc.* **79**, 6522 (1957).

[282] G. B. Payne, P. D. Deming, and P. H. Williams, *J. Org. Chem.* **26**, 659 (1961).

[283] L. DiNuno, S. Florio, and P. E. Todesco, *J. Chem. Soc. C* p. 1433 (1970).

[284] A. S. Pagano and W. D. Emmons, *Org. Synth.* **49**, 47 (1969).

[285] C. H. Robinson, L. Milewich, and P. Hofer, *J. Org. Chem.* **31**, 524 (1966).

Secondary amines react with peroxy acids to afford nitroxides (**116**)[286–289a] or nitrones (**117**) and their condensation products.

$$>NH \longrightarrow >N{-}OH \longrightarrow >N{-}O\cdot \qquad (119)$$

(116)

$$RCH_2NHR \longrightarrow RCH_2N(OH)R \longrightarrow R{-}CH{=}N(\rightarrow O)R \qquad (120)$$

(117)

Various peroxy acids have been used successfully for the oxidation of amines. Some typical examples are given below.

Oxidation of aniline with excess of cold peroxyacetic acid gives nitrosobenzene in high yields.[280] When no excess of the reagent is employed, azoxybenzene (**118**) is formed as a result of a competitive condensation of the nitroso compound with aniline [Eq. (121)].[281] Careful oxidation of cyclo-

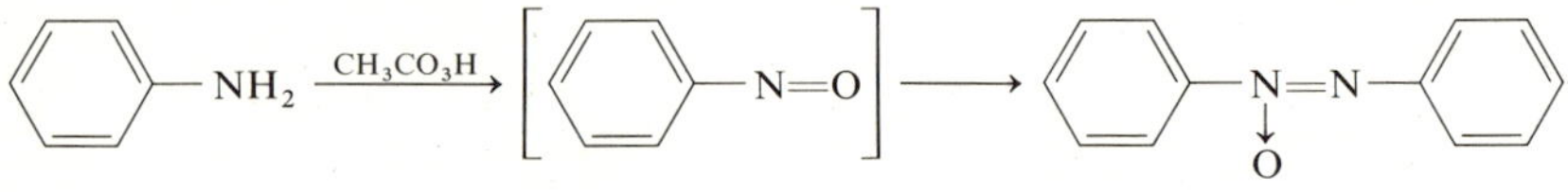

(118)

(121)

hexylamine with peroxyacetic acid affords the dimeric nitroso compound (**119**) in 41% yield [Eq. (122)].[282] When oxidation of 2,6-disubstituted ani-

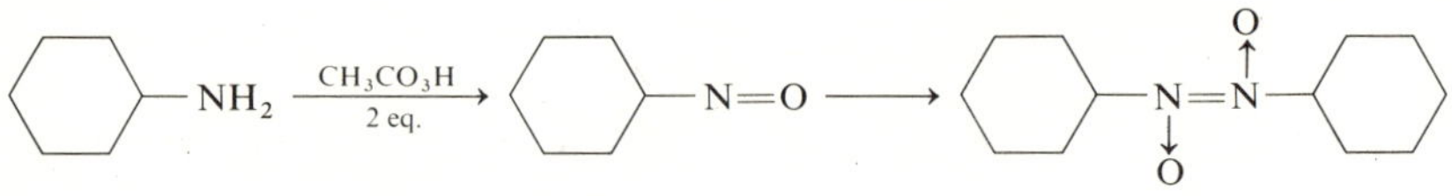

(119)

(122)

lines is performed with peroxybenzoic acid, good yields of the corresponding nitroso compounds have been reported. For example, 2,6-difluoroaniline gives the nitroso compound (**120**) in 85% yield.[283]

[286] K. Tokumaru, H. Sakuragi, and O. Simamura, *Tetrahedron Lett.* p. 3945 (1964).
[287] G. Chapelet-Letourneux, H. Lamaire, and A. Rassat, *Bull. Soc. Chim. Fr.* p. 3283 (1965).
[288] B. Zeeh and H. Metzger, *in* "Methoden der organischen Chemie" (E. Müller, ed.), 4th ed., Vol. 10, p. 1135. Thieme, Stuttgart, 1968.
[289] E. G. Rozantsev and V. D. Sholle, *Synthesis* pp. 190 and 401 (1971).
[289a] T. Yoshioka, S. Higashida, and K. Murayama, *Bull. Chem. Soc. Jpn.* **45**, 636 (1972).

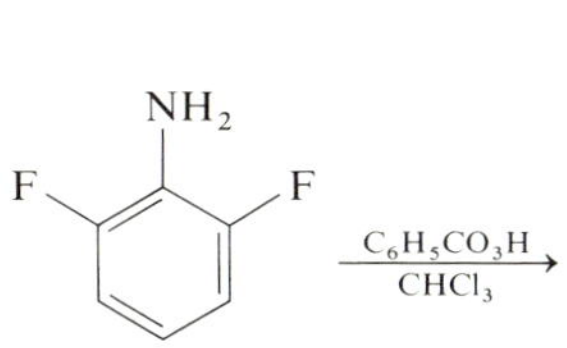

(120)

(123)

Peroxytrifluoroacetic acid is the reagent of choice for the direct oxidation of weakly basic primary aromatic amines to nitro compounds [Eq. (124)].[284]

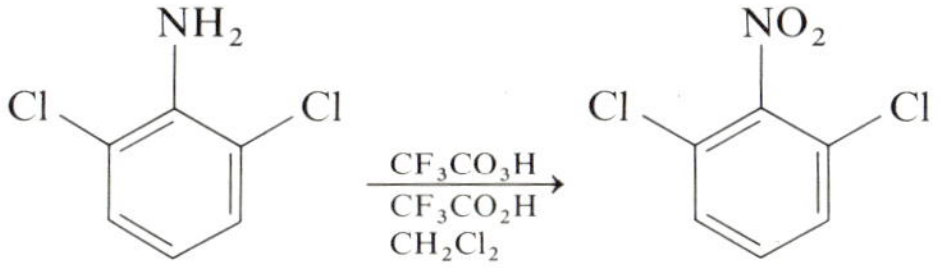

(124)

This reagent cannot be used for the oxidation of primary aromatic amines substituted with electron-donating groups. Anhydrous peroxyacetic acid is recommended in these cases.[284]

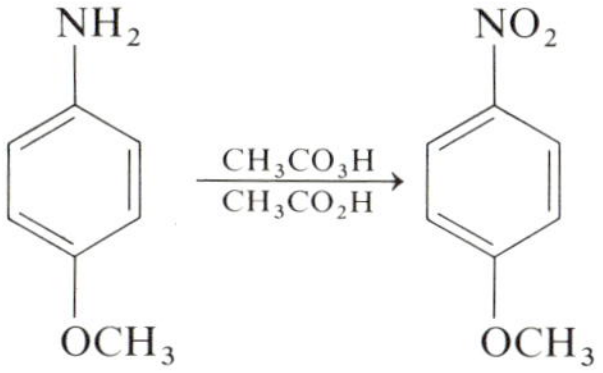

(125)

Tertiary amines react with peroxy acids to form amine oxides.[290–292]

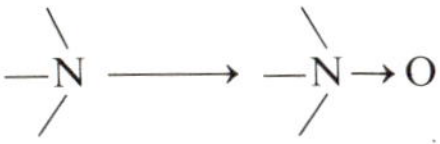

(126)

Trialkyl amines react rapidly even with hydrogen peroxide in water or alcohol [Eq. (127)],[293] and with alkyl hydroperoxides in the presence of a metal catalyst [Eq. (128)].[294]

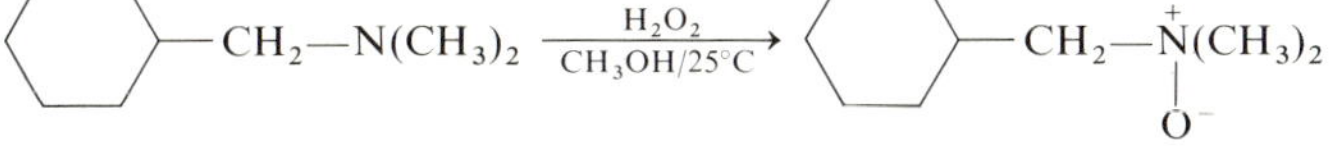

(127)

[290] A. R. Katritzky, *Q. Rev., Chem. Soc.* **10**, 395 (1956).

[291] A. C. Cope and E. R. Trumbull, *Org. React.* **11**, 317 (1960).

[292] A. R. Katritzky and J. M. Lagowski, "Chemistry of the Heterocyclic *N*-Oxides." Academic Press, New York, 1970.

[293] A. C. Cope and E. Ciganek, *in* "Organic Syntheses" (N. Rabjohn, ed.), Collect. Vol. IV, p. 612. Wiley, New York, 1963.

[294] N. N. Sheng and J. G. Zajacek, *Org. Synth.* **50**, 56 (1970).

$$CH_3(CH_2)_{11}{-}N(CH_3)_2 + (CH_3)_3COOH \xrightarrow[(CH_3)_3COH]{VO(C_5H_7O_2)_2} CH_3(CH_2)_{11}\overset{+}{N}(CH_3)_2{-}O^- \quad (128)$$

76–83%

Pyridine and other less nucleophilic heterocyclic compounds require peroxy acids and higher reaction temperatures.[295–303] For example, pyridine reacts with peroxyacetic acid to give pyridine oxide in 76–83% yield [Eq. (129)].[295] 2,6-Dibromopyridine can be converted to the corresponding *N*-oxide with peroxytrifluoroacetic acid [Eq. (130)][297]

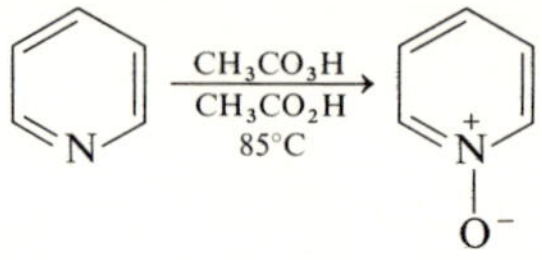

(129)

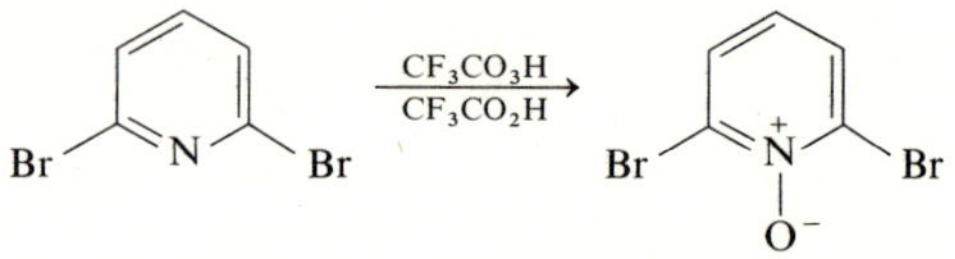

(130)

70–75%

Oxidation of ribonucleic acid components to their respective *N*-oxides has been achieved by monoperoxyphthalic acid[303–305] as well as *m*-chloroperoxybenzoic acid.[306,307]

The oxidation of tertiary amines with peroxy acids is not catalyzed by acids. Strongly acidic medium even decreases the rate by lowering the effective concentration of the free amine. Amine oxides are formed as salts of the

[295] H. S. Mosher, L. Turner, and A. Carlsmith, *in* "Organic Syntheses" (N. Rabjohn, ed.), Collect. Vol. IV, p. 828. Wiley, New York, 1963.

[296] E. C. Taylor, Jr. and A. J. Crovetty, *in*. "Organic Syntheses" (N. Rabjohn, ed.), Collect. Vol. IV, p. 704. Wiley, New York, 1963.

[297] R. F. Evans, M. van Ammers, and H. J. den Hertog, *Recl. Trav. Chim. Pays-Bas* **78**, 408 (1959).

[298] A. Pollak, B. Stanovnik, and M. Tišler, *J. Org. Chem.* **35**, 2478 (1970).

[299] G. E. Chievers and H. Suschitzky, *J. Chem. Soc. C* p. 2867 (1971)

[300] R. M. Harnden, A. G. Brown, and R. A. Vere, *J. Chem. Soc., Perkin Trans. 1* p. 333 (1973).

[301] W. Hutzenlaub, H. Yamamoto, G. B. Barlin, and W. Pfleiderer, *Chem. Ber.* **106**, 3203 (1973).

[302] J. C. Craig and K. K. Purushothaman, *J. Org. Chem.* **35**, 1721 (1970).

[303] F. Cramer, *Acc. Chem. Res.* **2**, 338 (1969).

[304] F. v. d. Haar, E. Schimme, V. A. Erdmann, and F. Cramer, *Bioorg. Chem.* **1**, 282 (1971).

[305] T. R. Fink and D. M. Krothers, *Biopolymers* **11**, 127 (1972).

[306] L. R. Subbaraman, J. Subbaraman, and E. J. Behrman, *Biochemistry* **8**, 3059 (1969).

[307] M. R. Harnden, A. G. Brown, and R. A. V. Hodge, *J. Chem. Soc., Perkin Trans. 1* p. 333 (1973).

corresponding carboxylic acids from which the free amine oxides may be obtained by passage through a column of basic alumina. Various tertiary amine oxides have been prepared in this way by oxidizing tertiary amines with *m*-chloroperoxybenzoic acid.[302]

The reaction of tertiary amines with peroxy acids is analogous to epoxidation, since an increase in both electrophilicity of the peroxy acid and nucleophilicity of the amine accelerate the reaction.[308] The nucleophilic attack of a tertiary amine on the peroxy acid is believed to proceed as shown below.

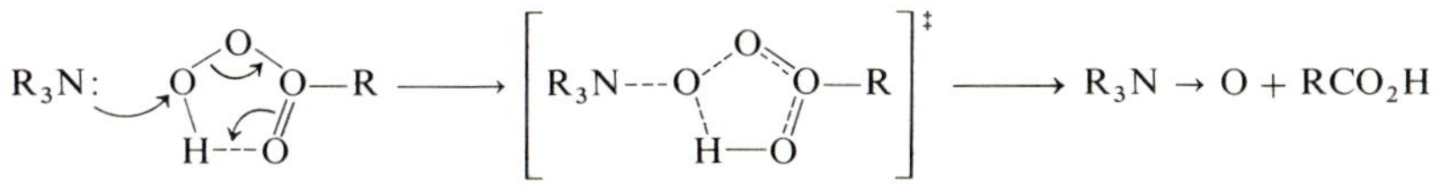

(121)

(131)

The reactions of primary and secondary amines most probably involve the same initial step with the formation of an intermediate analogous to **121**. Detailed kinetic study of the peroxyacetic acid oxidation of anilines in aqueous ethanol has shown a slow formation of phenylhydroxylamines and fast conversion of the latter into nitroso compounds.[309] Highly negative ρ values of -1.86 and -1.58 have been reported for these reactions, respectively. Oxidation of nitroso to nitro compounds presumably involves nucleophilic attack of nitrogen on the electrophilic oxygen of the peroxy acid.[310] The reaction of primary and secondary amines with *p*-nitrobenzenesulfonyl peroxide produces the corresponding imines. Hydrolysis of the latter completes the oxidative deamination of amines to produce carbonyl compounds in fair to excellent overall yields.[310a] The following mechanistic scheme has been proposed for this reaction.

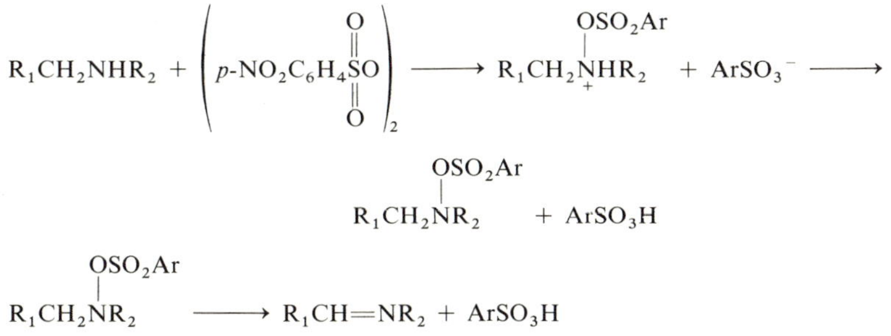

(132)

[308] Y. Ogata and I. Tabushi, *Bull. Chem. Soc. Jpn.* **31**, 969 (1958).
[309] K. B. Ibne-Rasa and J. O. Edwards, *J. Am. Chem. Soc.* **84**, 763 (1962).
[310] K. B. Ibne-Rasa, J. O. Edwards, M. T. Kost, and A. R. Gallopo, *Chem. Ind.* (*London*) p. 964 (1974).
[310a] R. V. Hoffman, *J. Am. Chem. Soc.* **98**, 6702 (1976).

B. OXIMES

Aliphatic, alicyclic, and aromatic oximes are oxidized to the corresponding nitro compounds via nitronic acids with peroxytrifluoroacetic acid in acetonitrile.[311,312]

$$R_2C{=}NOH \longrightarrow R_2C{=}NO_2H \longrightarrow R_2CHNO_2 \qquad (133)$$

A study of the stereochemistry of oxidation of a series of 2-substituted cyclohexanone oximes (**122**) shows a predominant formation of *cis*-2-substituted nitrocyclohexanes.[313]

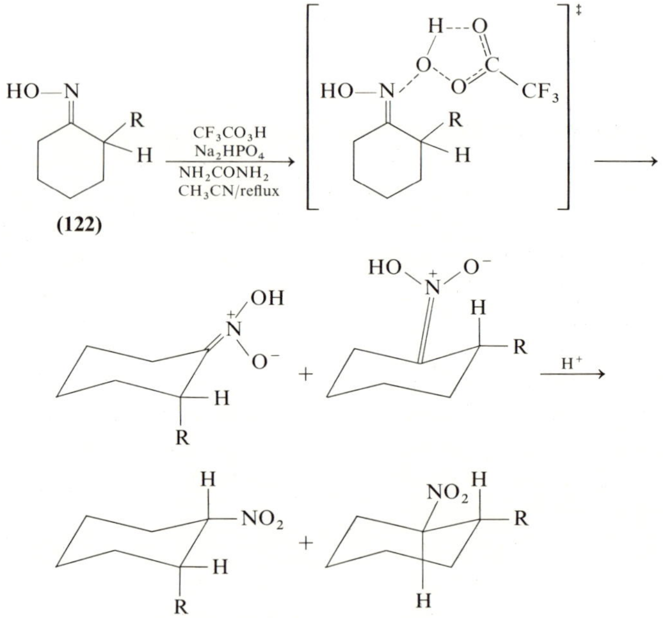

R = C_6H_5 95% 5%

R = $CH_2CO_2CH_3$ 77% 23% (134a)

The conversion of aldoximes to carboxylic acids in high yields has been accomplished with sodium peroxide. Hydroxamic acids or their conjugate bases are believed to be the intermediates which undergo oxidative cleavage to produce the corresponding carboxylic acids [Eq. (134b)].[314]

[311] W. D. Emmons and A. S. Pagano, *J. Am. Chem. Soc.* **77**, 4557 (1955).
[312] N. Kornblum, *Org. React.* **12**, 101 (1962).
[313] R. J. Sundberg and P. A. Bukowick, *J. Org. Chem.* **33**, 4098 (1968).
[314] T.-L. Ho and G. A. Olah, *Synthesis* p. 807 (1976).

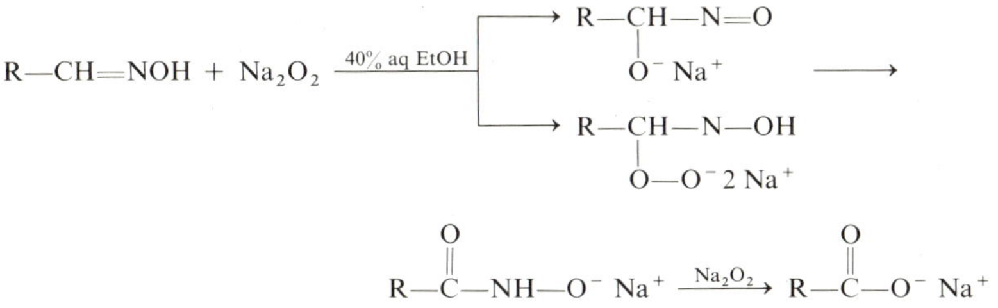

(134b)

C. AZO AND DIAZO COMPOUNDS

Azo compounds are oxidized to azoxy derivatives with peroxy acids.[11,314–320]

$$-N{=}N- \longrightarrow -\underset{\underset{O}{\downarrow}}{N}{=}N- \qquad (135)$$

The reaction most probably involves a one-step electrophilic attack of the peroxy acid at either of the lone pairs of the azo nitrogens. Analysis of substituent effects on the rate and the product isomer ratio in the reaction of *trans*-azobenzenes with peroxybenzoic acid seems to rule out the possibility of the attack at the azo double bond with the formation of an oxadiaziridine-type intermediate (**123**)[319]

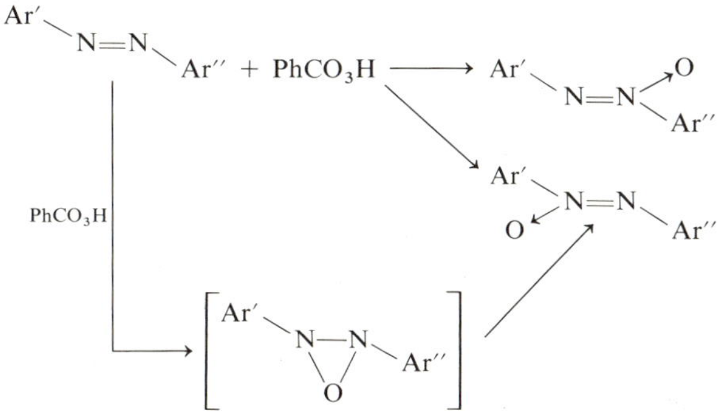

(136)

(123)

Oxidation of diazoaminobenzenes with peroxybenzoic acid also gives *N*-oxides. Electron-releasing substituents on one of the benzene rings direct

[315] G. Badger and G. E. Lewis, *J. Chem. Soc.* p. 2151 (1953).
[316] J. N. Brough, B. Lythgoe, and P. Waterhouse, *J. Chem. Soc.* p. 4069 (1954).
[317] B. T. Newbold, *J. Org. Chem.* **27**, 3919 (1962).
[318] J. M. Birchall, R. N. Haszeldine, and J. E. Kemp, *J. Chem. Soc. C* p. 1519 (1970).
[319] T. Mitsuhashi, O. Simamura, and Y. Tezuka, *Chem. Commun.* p. 1300 (1970).
[320] T. Mitsuhashi and O. Simamura, *J. Chem. Soc. B* p. 705 (1970).

the oxygen to the nearest nitrogen atom. The rate determining formation of metastable intermediate **124** which collapses to isomeric *N*-oxides **125** and **126** is believed to be the main mechanistic feature of this reaction.[320]

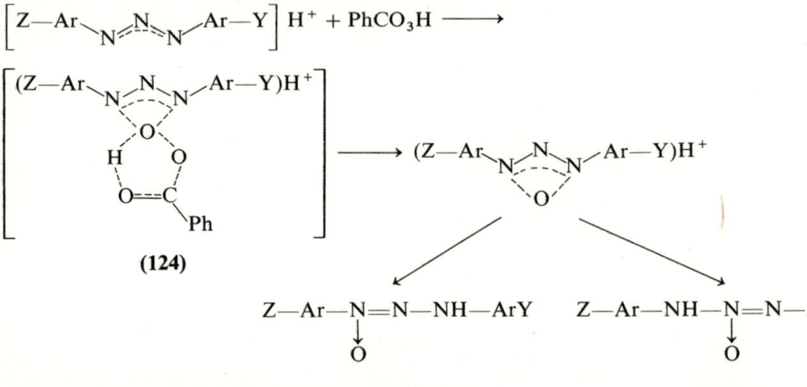

(124) (125) (126) (137)

Diazophenylmethane (**127**) reacts with peroxybenzoic acid in various solvents to yield benzophenone nearly quantitatively. The direct attack by the diazoalkane nucleophilic carbon atom on the electrophilic oxygen of the peroxy acid has been proposed.[321]

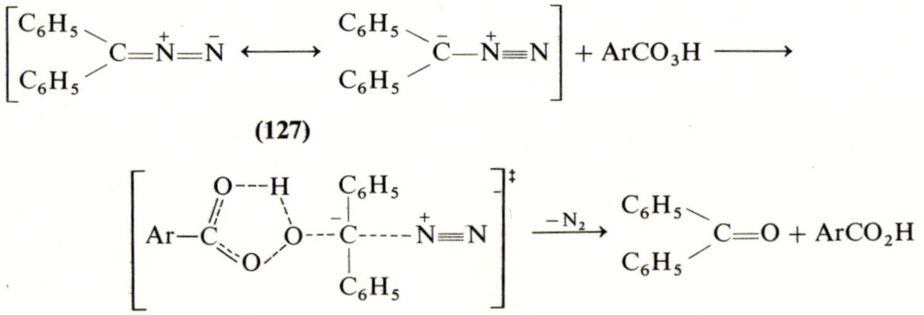

(127) (138)

The oxidation of α-diazoketones (**128**) to α-diketones with *m*-chloroperoxybenzoic acid has recently been reported.[322]

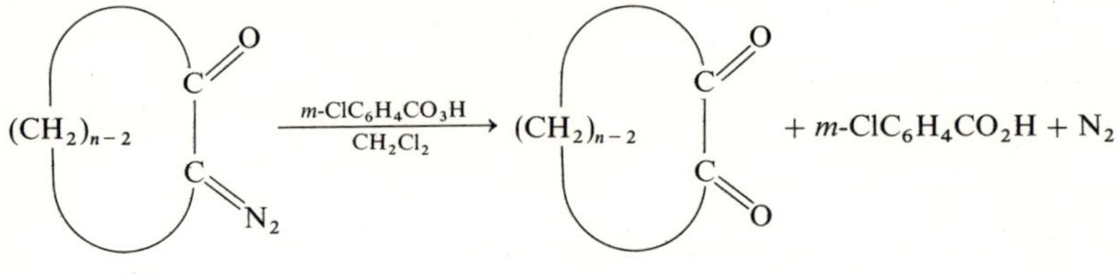

(128) (139)

[321] R. Curci, F. DiFuria, and F. Marcuzzi, *J. Org. Chem.* **36**, 3774 (1971).
[322] R. Curci, F. DiFuria, J. Ciabattoni, and P. W. Concannon, *J. Org. Chem.* **39**, 3295 (1974).

D. IMINES

The peroxy acid oxidation of imines is an important way to synthesize oxaziridines [Eq. (140)].[323–343] The main limitations of this reaction are a

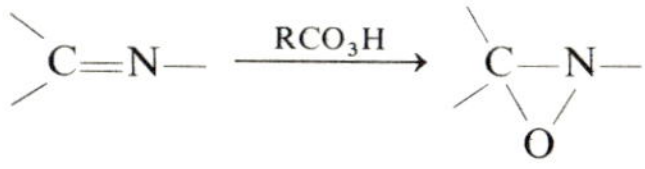

(140)

rather low stability of certain imines and some oxaziridines under acidic conditions. Carefully controlled reaction conditions are usually required.

Benzylidene-*tert*-butylamine (**129**) reacts with peroxybenzoic acid in benzene at 5°C to give oxaziridine **130** in 56–74% yield.[331]

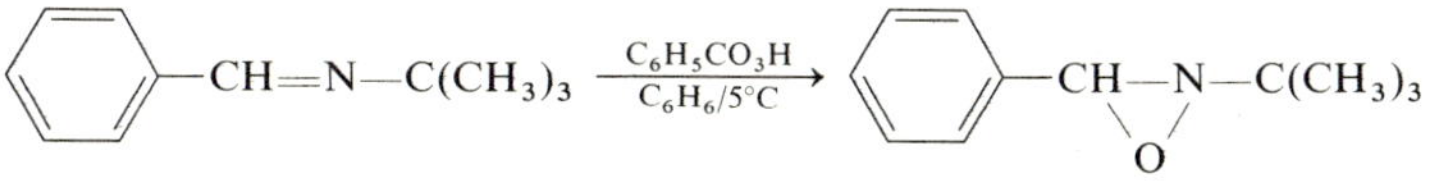

(141)

(129) **(130)**

Peroxy acid oxidation of imines is remarkably selective. Oxaziridines may be prepared in the presence of olefinic double bonds [Eq. (142)].[324]

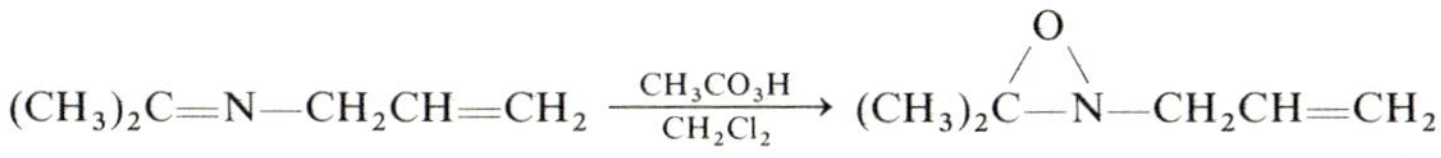

(142)

50%

[323] W. D. Emmons, *J. Am. Chem. Soc.* **79**, 5739 (1957).
[324] W. D. Emmons, *in* "Heterocyclic Chemistry" (A. Weissberger, ed.), Part 1, p. 624. Wiley (Interscience), New York, 1964.
[325] A. Padwa, *J. Am. Chem. Soc.* **87**, 4365 (1965).
[326] G. Pews, *J. Org. Chem.* **32**, 1628 (1967).
[327] J. Dupin, *Bull. Soc. Chim. Fr.* p. 3085 (1967).
[328] D. R. Boyd, *Tetrahedron Lett.* p. 4561 (1968).
[329] R. Montanari, J. Moretti, and G. Torre, *Chem. Commun.* p. 1086 (1969).
[330] D. R. Boyd and R. Graham, *J. Chem. Soc. C* p. 2648 (1969).
[331] D. W. Emmons and A. S. Pagano, *Org. Synth.* **49**, 13 (1969).
[332] D. R. Boyd, W. B. Jennings, R. Spratt, and D. M. Jerina, *Chem. Commun.* p. 745 (1970).
[333] D. R. Boyd, D. C. Neill, C. G. Watson, and W. B. Jennings, *J. Chem. Soc., Perkin Trans. 2* p. 1813 (1975).
[334] H. Kagen and I. Lillien, *J. Org. Chem.* **31**, 3729 (1966).
[335] J. K. Crandall and L. C. Crawley, *J. Org. Chem.* **39**, 489 (1974).
[336] D. H. Aue and D. Thomas, *J. Org. Chem.* **39**, 3855 (1974).
[337] J. Bjørgo, D. R. Boyd, R. S. Campbell, N. J. Thompson, and W. B. Jennings, *J. Chem. Soc., Perkin Trans. 2* p. 606 (1976).
[338] C. Bełzecki and D. Mostowicz, *J. Org. Chem.* **40**, 3878 (1975).
[339] V. Madan and L. B. Clapp, *J. Am. Chem. Soc.* **91**, 6078 (1969).
[340] V. Madan and L. B. Clapp, *J. Am. Chem. Soc.* **92**, 4902 (1970).
[341] E. Schmitz, R. Ohme, and D. Murawski, *Chem. Ber.* **98**, 2516 (1965).
[342] Y. Ogata and Y. Sawaki, *J. Am. Chem. Soc.* **95**, 4687 (1973).
[343] Y. Ogata and Y. Sawaki, *J. Am. Chem. Soc.* **95**, 4692 (1973).

C-Benzoylimines also yield oxaziridines with peroxy acids [Eq. (143)].[325]

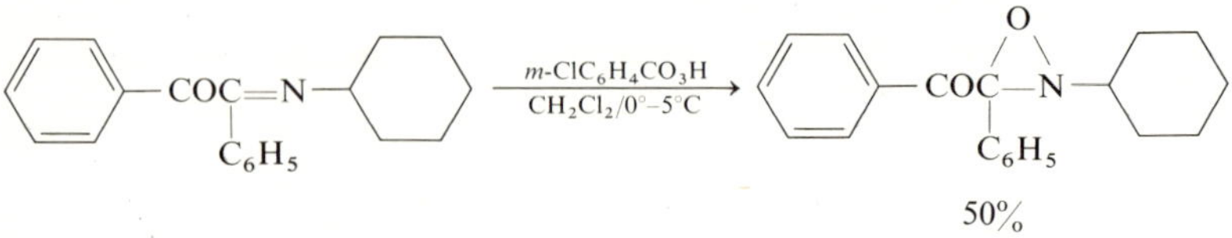

(143)

Ketenimines which possess both imine and olefinic functional groups yield various products with peroxy acids.[334,335] For example, *N*-phenyldimethylketenimine (**131**) reacts with one equivalent of *m*-chloroperoxybenzoic acid in cold dichloromethane to give a 33% yield of acetone, a 22% yield of *m*-chlorobenzoxyamide (**132**), and phenylisocyanide (**134**). The reaction is believed to proceed via an intermediate epoxyimine which decomposes to **133** and **134** or reacts with the carboxylic acid to form **132**.[335]

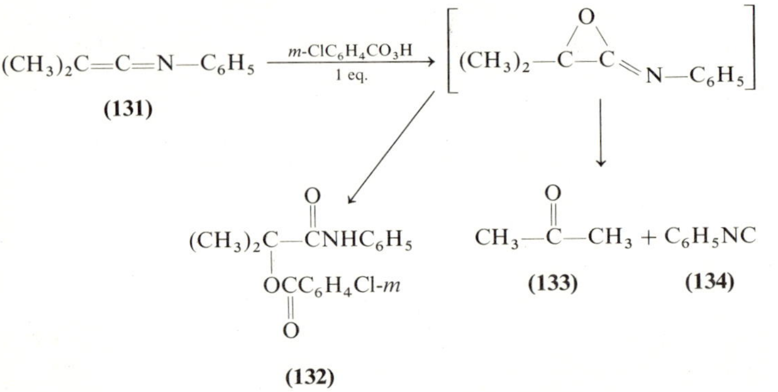

(144)

Acyclic imino ethers are oxidized to 3-alkoxyoxaziridines by *m*-chloroperoxybenzoic acid in high yields [Eq. (145)].[336]

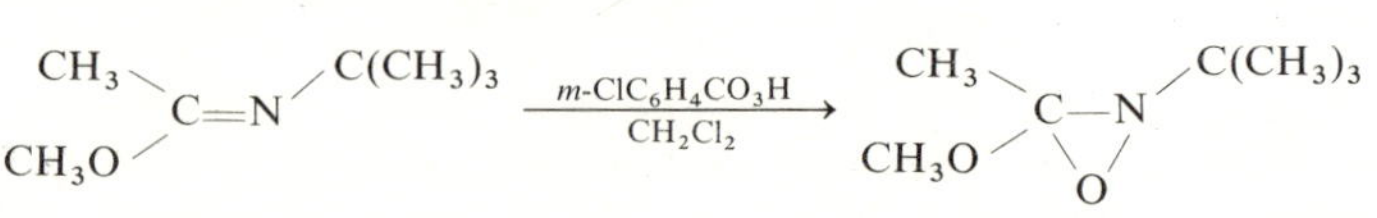

(145)

Although oxaziridines of cyclic imino ethers have been detected at −20°C, they are usually further converted to nitroso esters. For example, 2-methoxy-4-dimethyl azetine (**135**) reacts with one equivalent of *m*-chloroperoxybenzoic acid to give the corresponding nitrosobutanoate (**136**) in 50% yield.[336] Optically active oxaziridines of relatively high optical purity (ca. 50%) have been obtained by the reaction of various imines with optically active peroxy acids.[328–330,337] Oxidation of Schiff bases, prepared from (*R*)-(+)-α-phenylethylamine and a series of carbonyl compounds, with *m*-chloroperoxybenzoic acid also yields optically active oxaziridines.[338]

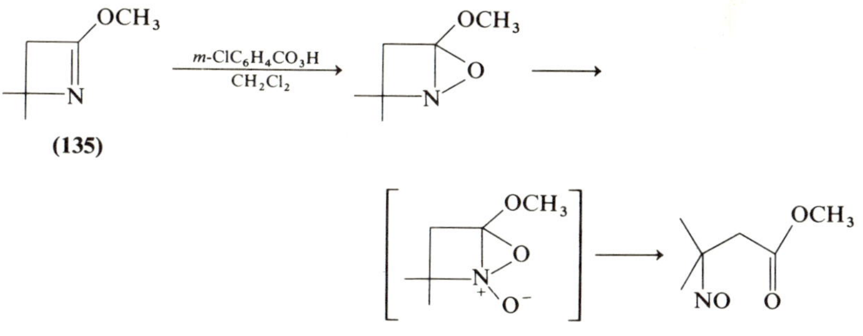

(135) (136) (146)

Two mechanisms have been proposed for the reaction of imines with peroxy acids. The first is analogous to olefin epoxidation [Eq. (147)][323,339,340] while the second resembles the Baeyer–Villiger-type reaction [Eq. (148)][341–343]

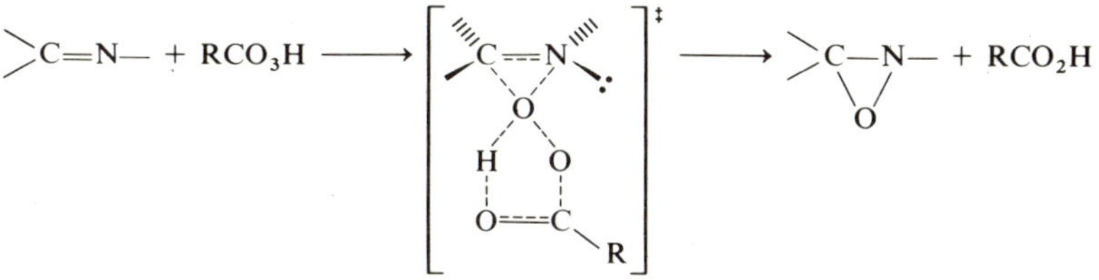

(147)

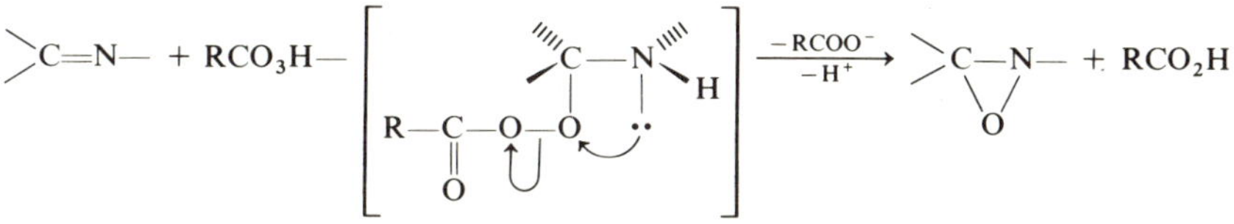

(137) (148)

Recent kinetic evidence supports the oxaziridine formation in a two-step mechanism. The reaction of acyclic imines, benzylidene-*tert*-butylimines, with peroxybenzoic acid has been studied kinetically in various solvents.[342] The reaction exhibits complex kinetics since it is catalyzed by dilute carboxylic acids and protic solvents and retarded by more concentrated carboxylic acids and basic solvents (ethers and alcohols). Alcohols are responsible for both acceleration and retardation depending on the structure of the alcohol. Ogata and Sawaki explained the acid catalysis of this reaction as being due to the formation of an intermolecularly hydrogen-bonded dimer of the imine and acid rather than the formation of a dimer between the peroxy acid and acid catalyst. Addition to the C=N bond is the rate determining step. Internal nucleophilic (S_{Ni}) reaction of the adduct then gives oxaziridine as shown in Eq. (149). Small amounts of nitrones which are formed in this

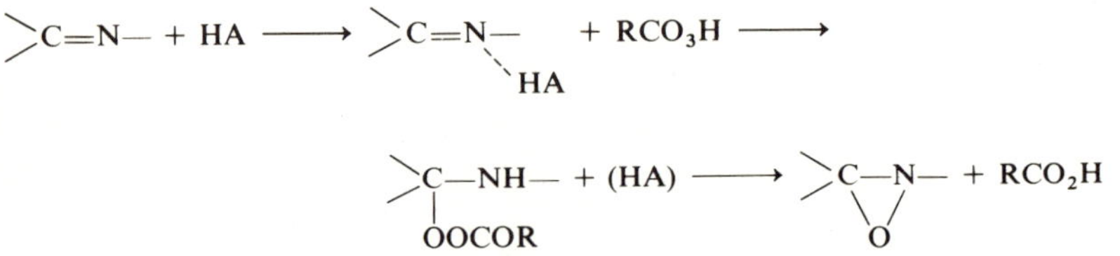

(149)

reaction, can be rationalized by the mechanism analogous to that proposed for the peroxy acid oxidation of amines. The nucleophilic site in the imine is the nitrogen lone pair of electrons of the C=N bond. Cyclic imines (**138**),

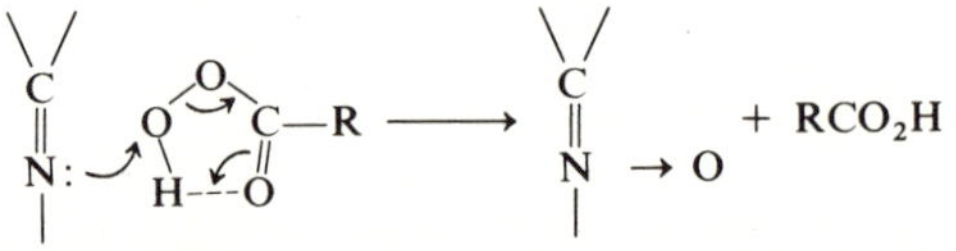

(150)

i.e., 3,4-dihydroxyisoquinoline (R = H) and 3,4-dihydro-1-methylisoquinoline (R = CH_3), react with peroxybenzoic acid to form both oxaziridine (**139**) and nitrone **140**.[343]

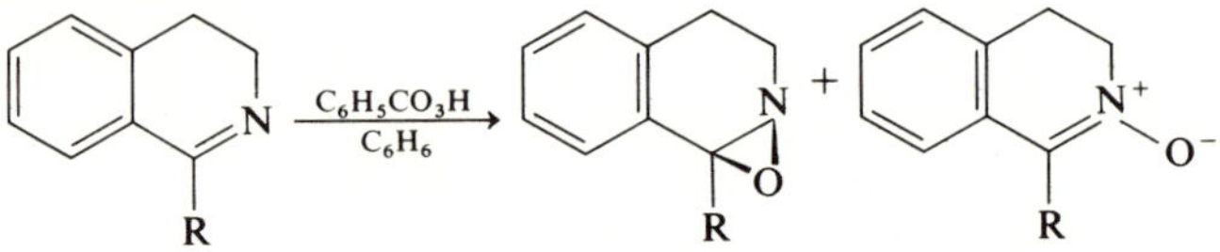

(151)

(138)	(139)	(140)
R = H	80.6%	4.8%
R = CH_3	24.0%	75.0%

The product selectivity depends on the structure of the imine and on the solvent used. In contrast to acyclic imines, the S_N1 reaction is the rate-determining step in oxidation of these cyclic imines.

Boyd and co-workers performed a detailed stereochemical study of oxidation of aldimines and ketimines with *m*-chloroperoxybenzoic acid.[332,333] They found that these reactions are not generally stereospecific. For example, the diastereoisomeric ratio **142**/**143** in oxidation of *C*-arylaldimines strongly

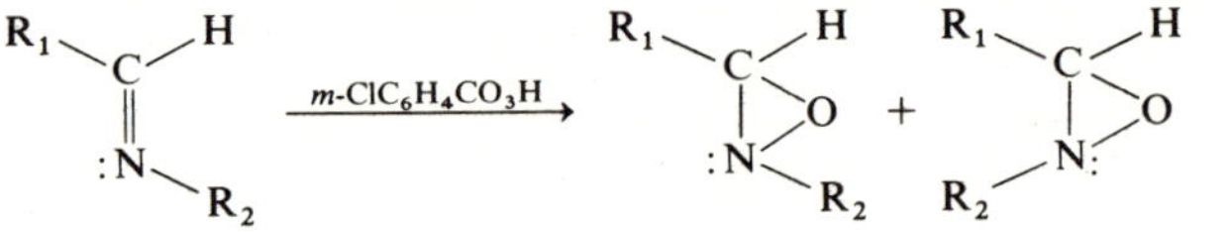

(152)

(141)	(142)	(143)
R_1 = 4-$NO_2C_6H_4$ R_2 = CH_3	46%	54% (CH_2Cl_2)
	97%	3% (C_2H_5OH)
R_1 = 4-$NO_2C_6H_4$ R_2 = $(CH_3)_3C$	>99%	<1% (CH_2Cl_2)
R_1 = C_2H_5 R_2 = CH_3	>99%	<1%
R_1 = $(CH_3)_3C$, R_2 = CH_3	>99%	<1%

depends on the solvent used. All of these results seem to favor the two-step mechanism. The prefered ground-state geometry of intermediate **137** is expected to possess a staggered conformation (**144**) on the bases of non-bonded interactions only. Steric factors alone would thus favor the eclipsed conformation (**145**) of the transition state during oxidation of *C*-arylaldimines (R_1 = Ar), which would lead to *trans*-oxaziridines. Since considerable amounts of *cis*-oxaziridines are formed in oxidation of *C*-arylaldimines, other

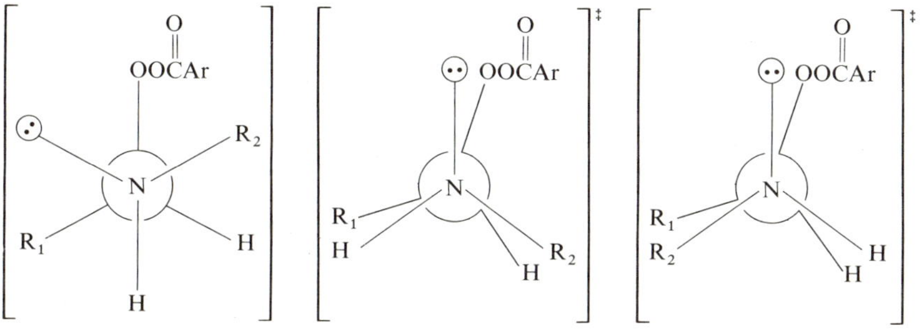

(144) **(145)** **(146)**

factors in addition to nonbonded interactions must be operative in stabilizing an alternative eclipsed conformation of the transition state (**146**) required for the production of *cis*-oxaziridines. Various effects (polar, hydrogen bonding, solvent) have been suggested to lower the energy of this otherwise energetically unfavorable transition state.

E. ALDEHYDE HYDRAZONES

Aromatic aldehyde hydrazones form azoxy compounds with peroxy acids. The nucleophilic attack of the nitrogen atom at the electrophilic oxygen of the peroxy acid is believed to be the rate-determining step in the reaction of benzaldehyde phenylhydrazone **147** with peroxybenzoic acid.[344,344a]

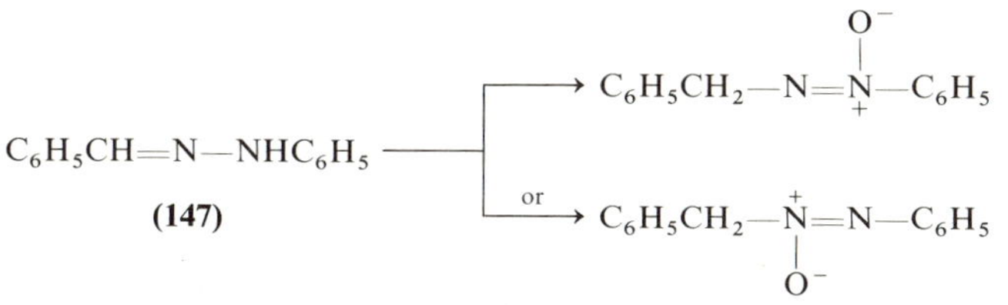

(153)

The dispute over the structure of the azoxy compound has ended in favor of the dimetric dehydro form of the azoxy compound (**148**) which is a mixture of *meso* and *dl* forms.[345]

344 B. M. Lynch and K. H. Pausacker, *J. Chem. Soc.* p. 1131 (1951); p. 3340 (1954).
344a J. N. Brough, B. Lythgoe, and P. Waterhouse, *J. Chem. Soc.* p. 4609 (1954).
345 R. B. Woodward and C. Winter, *Tetrahedron Lett.* p. 2693 (1969).

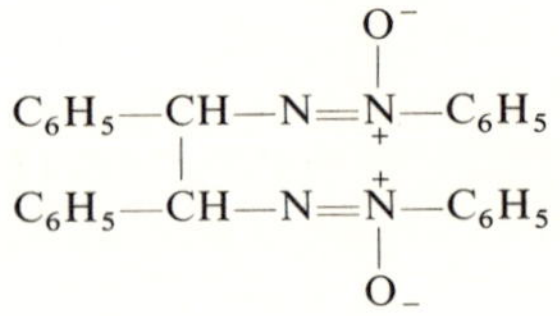

(148)

Aliphatic ketone and aldehyde alkylhydrazones have also been oxidized with peroxy acids. For example, cyclohexanone methylhydrazone reacts with one equivalent of peroxyacetic acid to give 15% of 1-(methylazo)cyclohexanol acetate (**149**) and 17.4% of azoxyacetate **150** when two equivalents of the peroxy acid are used.[346]

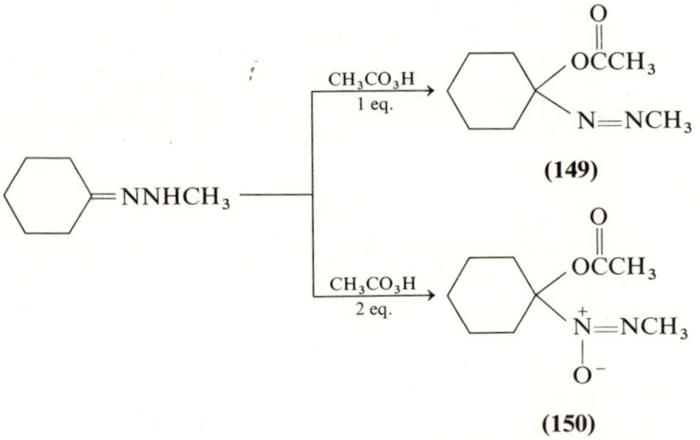

(154)

VI. Oxidation of Organic Sulfur Compounds

A. Sulfides and Sulfoxides

Peroxy acids have been used successfully for the oxidation of sulfides to sulfoxides.[11,347] Both aliphatic and aromatic sulfides react in this way. The reaction proceeds under mild conditions; at higher temperatures and prolonged reaction times, further oxidation to sulfones takes place [Eq. (155)].

$$R_2S \xrightarrow{[O]} R_2SO \xrightarrow{[O]} R_2SO_2 \qquad (155)$$

Direct oxidation to sulfones can be accomplished in most cases by two equivalents of peroxy acid [Eq. (156)].[348] Sulfides and sulfoxides are more

[346] B. T. Gillis and K. F. Schimmel, *J. Org. Chem.* **32**, 2865 (1967).
[347] D. Bernard, L. Bateman, and J. I. Cunneen, *Org. Sulfur Compd.* **1**, 229 (1961).
[348] C. R. Johnson, H. Difenbach, J. E. Keiser, and J. C. Sharp, *Tetrahedron* **25**, 5649 (1969).

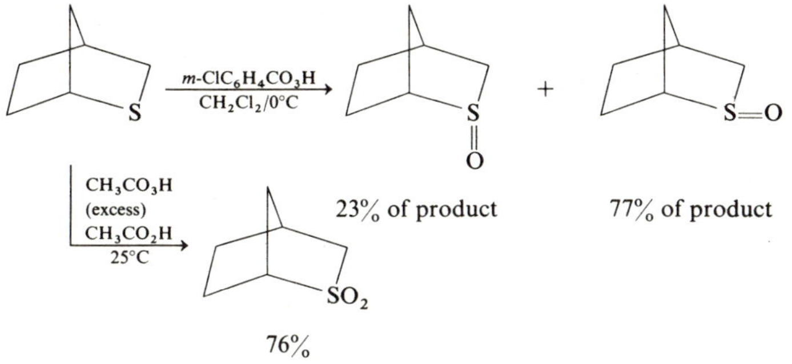

(156)

reactive than olefinic double and triple bonds and can be selectively oxidized in the presence of these functional groups. *m*-Chloroperoxybenzoic acid is most appropriate for such selective oxidations [Eq. (157)].[349] Optically

$$C_6H_5{-}C{\equiv}C{-}S{-}CH_3 \xrightarrow[CHCl_3/-23^\circ \text{ to } -10^\circ C]{m\text{-}ClC_6H_4CO_3H} C_6H_5{-}C{\equiv}C{-}\overset{\overset{\Large O}{\|}}{S}{-}CH_3 \quad (92\%)$$

$$C_6H_5{-}C{\equiv}C{-}S{-}CH_3 \xrightarrow[2\text{ eq.}]{m\text{-}ClC_6H_4CO_3H} C_6H_5{-}C{\equiv}C{-}SO_2{-}CH_3 \quad (81\%) \qquad (157)$$

active sulfoxides of a rather low optical purity (<12%) have been prepared by oxidation of sulfides with optically active (+)-peroxycamphoric acid (**151**)[350,351] and other optically active peroxy acids.[352–354]

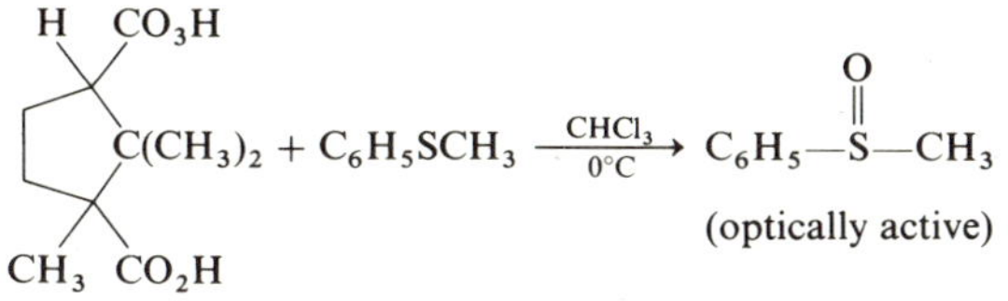

(151)

(158)

The peroxy acid oxidation of sulfides to sulfoxides is accelerated by electron-withdrawing groups in the peroxy acid and electron-donating groups in the sulfides, thus indicating that the peroxy acid is electrophilic in these reactions. A Hammett ρ value of +1.05 has been reported for the oxidation of bis(*p*-chlorobiphenyl) sulfide with substituted peroxybenzoic acids in

[349] G. A. Russell and L. A. Ochrymowycz, *J. Org. Chem.* **35**, 2106 (1970).
[350] K. Balenović, N. Bregant, and D. Francetić, *Tetrahedron Lett.* p. 20 (1960).
[351] K. Mislow, M. M. Green, and M. Raban, *J. Am. Chem. Soc.* **87**, 2761 (1965).
[352] K. Balenović, I. Bregovec, D. Francetić, I. Monković, and V. Tomasić, *Chem. Ind.* (*London*) p. 469 (1961).
[353] U. Folli, D. Iarossi, F. Montanari, and G. Torre, *J. Chem. Soc. C* p. 1317 (1968).
[354] U. Folli, D. Iarossi, and F. Montanari, *J. Chem. Soc. C* p. 1372 (1968).

isopropanol. The reaction obeys second-order kinetics (first-order in sulfide and first-order in peroxy acid), and is not subject to acid catalysis or salt effects. Kinetic evidence indicates that hydrogen is transferred intramolecularly in a process which resembles the epoxidation of olefins.[355]

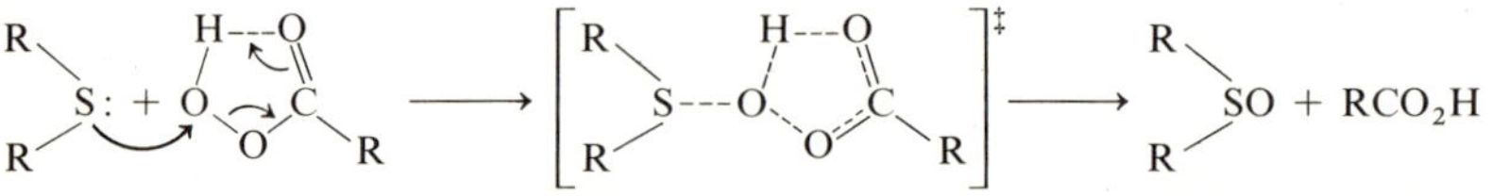

(159)

A study of the solvent effect in oxidation of *p*-nitrodiphenyl sulfide with peroxybenzoic acid shows that the rates of oxidation are not affected by the medium polarity but rather reflect specific interactions, i.e., solvation of the sulfide and the peroxy acid. The rate constants are in general larger in nonbasic aprotic solvents. Previously discussed disruption of the chelate ring of the peroxy acid with subsequent formation of the intermolecularly hydrogen-bonded adduct is believed to be the cause of such behavior.[356]

Further oxidation of sulfoxides to sulfones by peroxy acids is much slower.[356–359] Nucleophilic displacement on the peroxidic oxygen, similar to that shown in Eq. (159), has originally been suggested to be the mechanistic feature of this reaction.[359] Recent evidence seems to indicate that this oxidation involves a two-step mechanism as depicted below.[310]

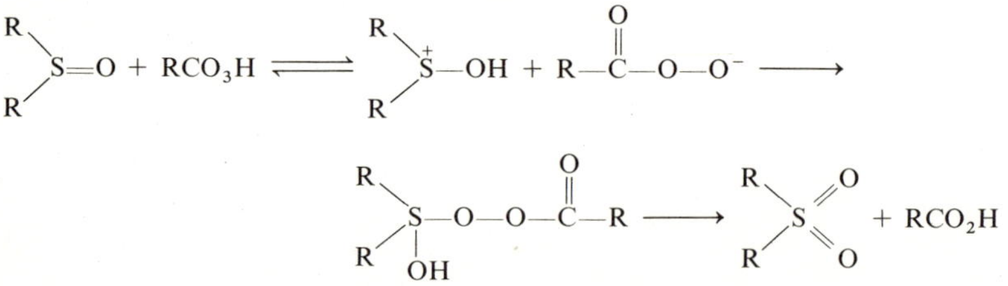

(160)

Still another mechanistic possibility would be the formation of an unstable sulfur-substituted diacyl dioxy derivative which would break down to the corresponding sulfone in an ionic or free radical process.

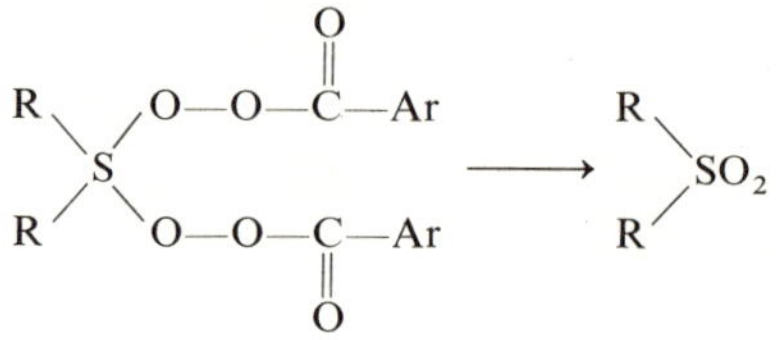

(161)

[355] C. G. Overberger and R. W. Cummins, *J. Am. Chem. Soc.* **75**, 4250 (1953).
[356] R. Curci, R. A. DiPrete, J. O. Edwards, and G. Modena, *J. Org. Chem.* **35**, 740 (1970).
[357] H. H. Szmant, H. F. Harnsberger, and F. Krahe, *J. Am. Chem. Soc.* **76**, 2185 (1954).
[358] R. Curci and P. E. Todesco, *Tetrahedron Lett.* p. 1749 (1963).
[359] R. Curci, A. Giovine, and G. Modena, *Tetrahedron* **22**, 1235 (1966).

The oxidation of sulfides to sulfoxides can also be accomplished with hydrogen peroxide and alkyl hydroperoxides. These reactions are susceptible to acid catalysis and are believed to involve nucleophilic displacement on the peroxide oxygen.[19,360]

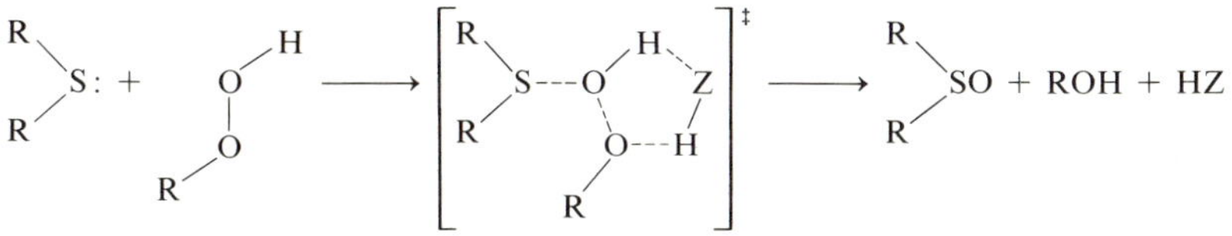

(162)

The role of the acid (HZ) is to assist the fission of the O–O bond and to facilitate the loss of the leaving group (ROH) through a transition state which allows proton transfer. Various protic solvents (alcohols, water) can be used as acids. In these solvents the reaction is second-order, first-order in peroxide, and first-order in sulfide. In the absence of any protic material, the alkyl hydroperoxide can serve as a proton donor, and the reaction becomes second-order in alkyl hydroperoxide.

A linear Hammett plot with ρ value of -0.98 has been reported for the oxidation of substituted diphenyl sulfides with hydrogen peroxide.[361]

Alkyl hydroperoxides are frequently used for oxidation of sulfides, especially in cases where over-oxidation to the sulfone is undesirable. Preparation of phenothiazine-5-oxide (**152**) is a typical example.[362]

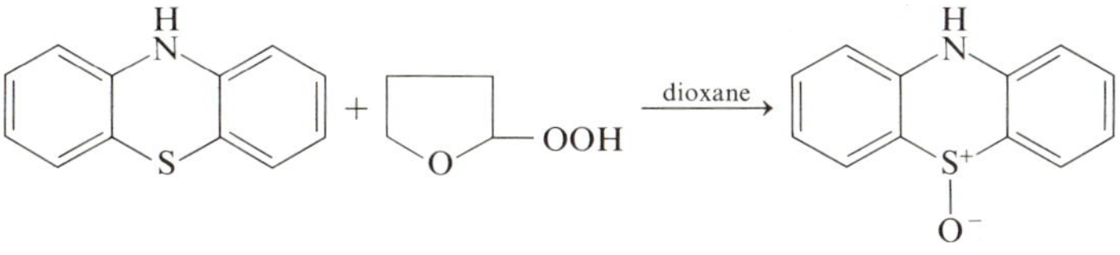

(163)

(152)

B. THIOCARBONYL COMPOUNDS

Thiocarbonyl compounds (**153**) can be converted to the corresponding S-oxides (sulfines) (**154**), and subsequently to ketones with peroxy acids in nearly quantitative yields.[363,364]

[360] E. J. Behrman and J. O. Edwards, *Prog. Phys. Org. Chem.* **4**, 93 (1967).

[361] G. Modena and L. Maioli, *Gazz. Chim. Ital.* **87**, 1306 (1957).

[362] C. Bodea and I. Silberg, *Stud. Cercet. Chim.* **14**, 317 (1963); *Chem. Abstr.* **62**, 14664 (1965).

[363] G. Opitz, *Angew. Chem., Int. Ed. Engl.* **6**, 107 (1967).

[364] B. Zwanenburg and J. Strating, *Q. Rep. Sulfur Chem.* **5**, 79 (1970).

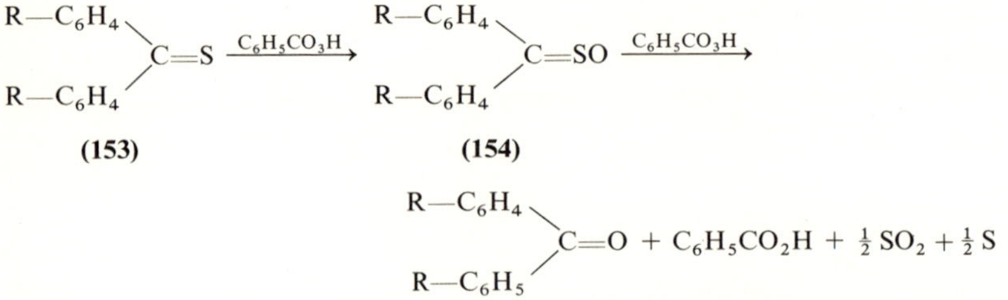

(153) (154) (164)

The first step of reaction is faster than the second one.[365] The second-order rate constants at 25°C in carbon tetrachloride for oxidation of a series of thiobenzophenones to sulfines fit the Hammett equation giving a ρ value of -0.88. Negative ρ value of -1.16 $(\sigma^+ + \sigma)$ for oxidation of sulfines to carbonyl compounds indicate that sulfines also act as nucleophiles. The effect of the solvent on these reactions correlates well with the results reported for the oxidation of sulfides and olefins, thus indicating a mechanism involving a nucleophilic attack of sulfur on the peroxidic oxygen in the peroxy acid.[366]

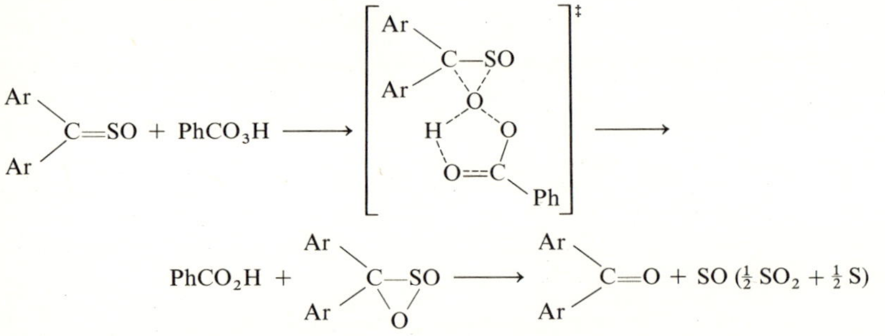

(165)

C. THIOLS AND DISULFIDES

Alkyl and aryl thiols are easily oxidized to the corresponding disulfides by hydrogen peroxide [Eq. (166)].[367,368] Further oxidation of disulfides

$$2\,RSH + H_2O_2 \longrightarrow RSSR + 2\,H_2O \qquad (166)$$

to thiosulfinates (**155**),[369,370] thiosulfonates (**156**),[369] α-disulfones (**157**),[371]

[365] A. Battaglia, A. Dondoni, P. Giorgianni, G. Maccagnani, and G. Mazzanti, *J. Chem. Soc. B* p. 1547 (1971).
[366] A. Battaglia, A. Dondoni, G. Maccagnani, and G. Mazzanti, *J. Chem. Soc., Perkin Trans. 2* p. 610 (1974).
[367] D. S. Tarbell, *Org. Sulfur Compd.* **1**, 97 (1961).
[368] J. E. Hesse and F. K. Truby, *Chem. Ind. (London)* **16**, 680 (1965).
[369] H. Bretschneider and W. Klotzer, *Monatsh. Chem.* **81**, 589 (1950).
[370] L. Sagramora, P. Koch, and A. Garbesi, and A. Fava, *Chem. Commun.* p. 985 (1967).
[371] Y. Minoura, *Nippon Gomu Kyokaishi* **32**, 177 (1959); *Chem. Abstr.* **54**, 8694 (1960).

and sulfonic acids (**158**)[372] can be performed by the controlled oxidation with either hydrogen peroxide, alkyl hydroperoxides, or peroxy acids, respectively [Eq. (167)]. Recent evidence indicates the formation of α-disulfoxides, RS(O)S(O)R, as intermediates in the peroxy acid oxidation of **155** to **156** (R = Ar).[372a]

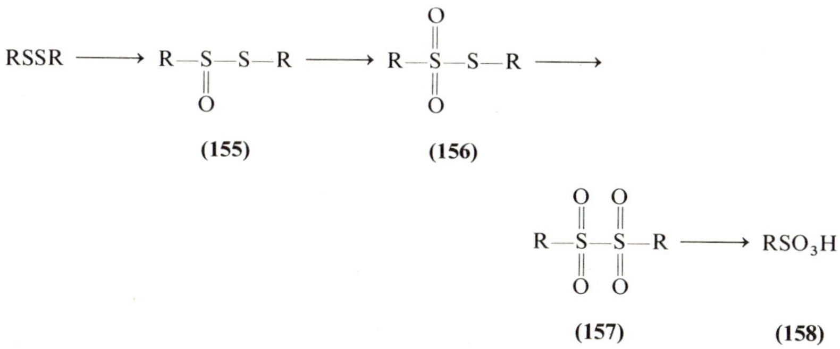

(167)

Oxidation of aliphatic thiols and thiophenols with *m*-chloroperoxybenzoic acid permits the isolation of sulfinic acids (**160**) in high yields. For example, *n*-butyl thiol reacts with this reagent at −30°C in dichloromethane to give butanesulfinic acid in 91% yield.[373] Sulfenic acids (**159**) are most probably intermediates in this reaction.

$$\text{RSH} \longrightarrow \underset{\textbf{(159)}}{[\text{RSOH}]} \longrightarrow \underset{\textbf{(160)}}{\text{RSO}_2\text{H}} \qquad (168)$$

VII. Oxidation of Organic Iodine Compounds

Oxidation of various aliphatic and aromatic iodine compounds with peroxy acids gives the corresponding iodoso compounds which can be further oxidized to iodoxy derivatives [Eq. (169)].[11,374] Iodosobenzenes

$$\text{RI} \xrightarrow{[\text{O}]} \text{RIO} \xrightarrow{[\text{O}]} \text{RIO}_2 \qquad (169)$$

are usually converted to *I,I*-diacyloxyiodobenzenes in the presence of carboxylic acids. Iodobenzene, for example, reacts with commercial (40%) peroxyacetic acid to form *I,I*-diacetyloxyiodobenzene (**161**) in yields ranging from 83–91% [Eq. (170)].[375] The use of excess peroxyacetic acid and higher

[372] H. Nawa, W. T. Brady, M. Koike, and L. J. Reed, *J. Am. Chem. Soc.* **82**, 896 (1960).
[372a] M. M. Chau and J. L. Kice, *J. Am. Chem. Soc.* **98**, 7711 (1976).
[373] W. G. Filby, K. Günther, and R. D. Penzhorn, *J. Org. Chem.* **38**, 4070 (1973).
[374] D. F. Banks, *Chem. Rev.* **66**, 243 (1966).
[375] J. G. Sharefkin and H. Saltzman, *Org. Synth.* **43**, 62 and 65 (1963).

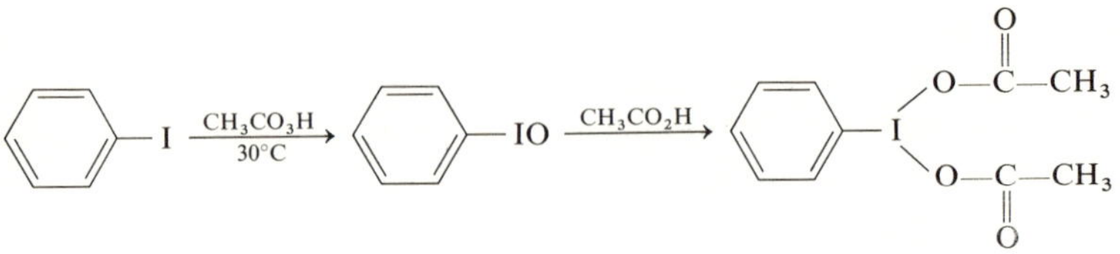

(170)

(161)

reaction temperatures result in the formation of iodoxybenzene (**162**) in 72–80% yield [Eq. (171)].[375] Peroxybenzoic acid has been found to react

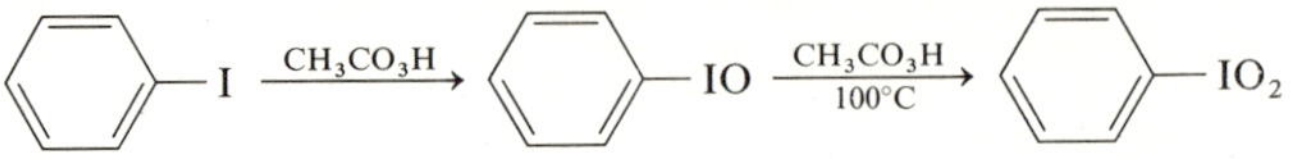

(171)

(162)

analogously.[376] The reaction of iodobenzene with peroxy acids to form iodoso compounds presumably involves the nucleophilic attack of iodine at the electrophilic oxygen according to the scheme outlined in Eq. (172).

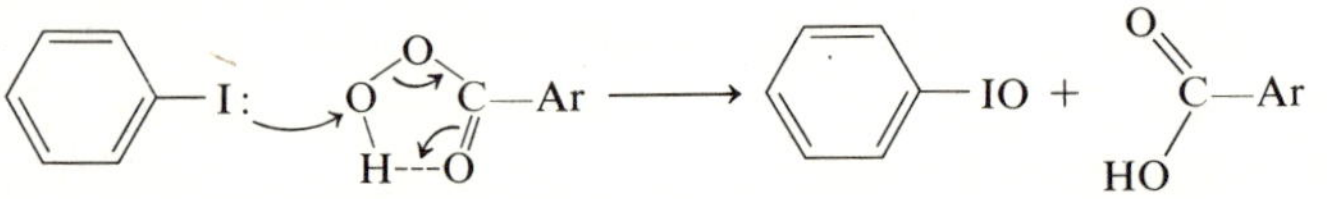

(172)

The formation of iodoxybenzene from iodosobenzene, which is the second step of the reaction, probably proceeds via a *I,I*-diacyldioxyiodobenzene. *I,I*-Dibenzoyldioxyiodobenzenes (**163**) have recently been isolated in the reaction of iodosobenzenes with substituted peroxybenzoic acids in chloroform at −5°C. These rather hazardous compounds, which contain three active oxygen atoms, decompose in the same solvent to form iodoxybenzene, benzoic acid, and small amounts of dibenzoyl peroxide in a free radical process [Eq. (173a)].[377,378]

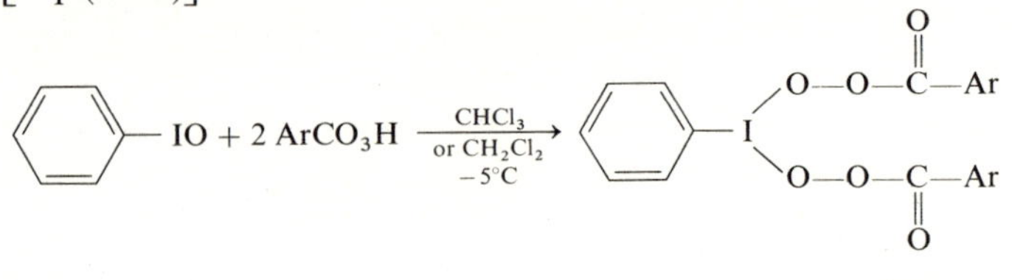

(173a)

50–70%

(163)

I,I-Dibenzoyldioxyiodobenzenes are stronger oxidizing agents compared to the corresponding peroxybenzoic acids. Various classes of compounds,

[376] J. Boeseken and E. Wicherlich, *Recl. Trav. Chem. Pays-Bas* **55**, 936 (1936).
[377] B. Plesničar and G. A. Russell, *Angew. Chem., Int. Ed. Engl.* **9**, 797 (1970).
[378] B. Plesničar, *J. Org. Chem.* **40**, 3267 (1975).

i.e., olefins, tertiary amines, iodine compounds, and sulfur compounds, have been converted in high yields to epoxides, amine oxides, iodoxy compounds, and sulfones, respectively,[377] i.e., with these reagents. The reaction is believed to involve nucleophilic attack of the substrate on electrophilic oxygen atoms which are attached to the iodine atom of the reagent since *I,I*-dibenzoyloxyiodobenzenes are the reduction products.

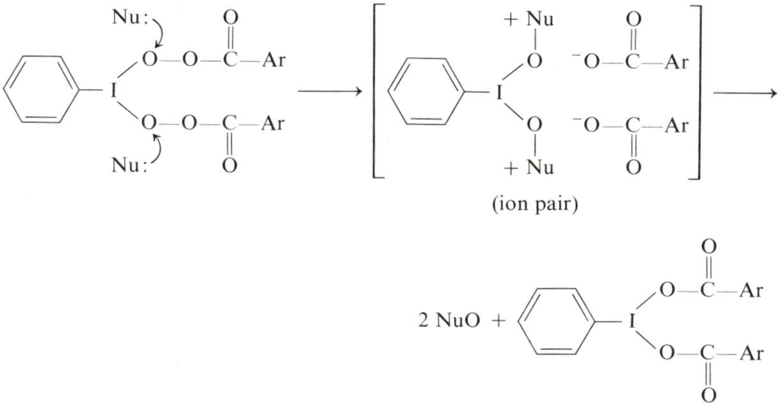

(173b)

Some interesting applications of peroxy acids in the oxidation of aromatic compounds have recently been reported. The iodination of aromatic systems by elemental iodine is rather slow. Faster reactions and higher yields have been obtained for iodination in the presence of peroxyacetic acid. Acetyl hypoiodite is believed to be the attacking electrophilic species.[379,380]

$$I_2 + CH_3CO_3H + H_2O \longrightarrow 2\,HIO + CH_3CO_2H$$

$$HIO + CH_3CO_2H + H^+ \longrightarrow (CH_3CO_2IH)^+ + H_2O \qquad (174)$$

$$(CH_3CO_2IH)^+ + ArH \longrightarrow ArI + CH_3CO_2H + H^+$$

For example, 5-iodoacenaphthene (**164**) and 2,7-diiodofluorene (**165**) were obtained in this way in 40 and 68% yields, respectively.[380] Oxidation of

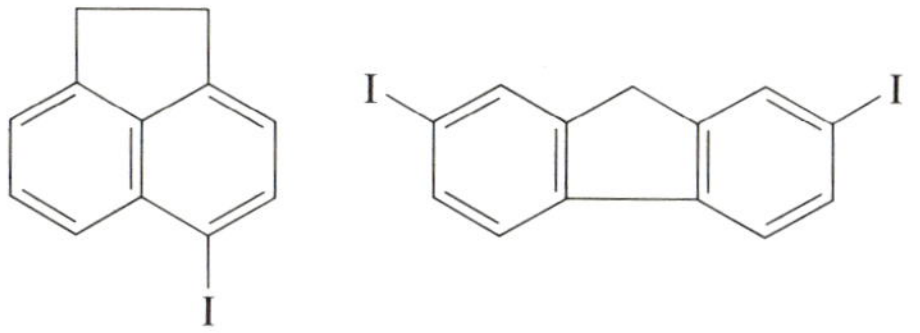

(164) **(165)**

[379] Y. Ogata and K. Nakajima, *Tetrahedron* **20**, 2751 (1964).

[380] Y. Ogata and I. Urasaki, *J. Chem. Soc. C* p. 1689 (1970).

2-iodosoisophthalic acid with peroxyacetic acid gives iodosolactone **166** in 76% yield [Eq. (175)].[381] The reaction of *o*-iodobenzamides (*o*-

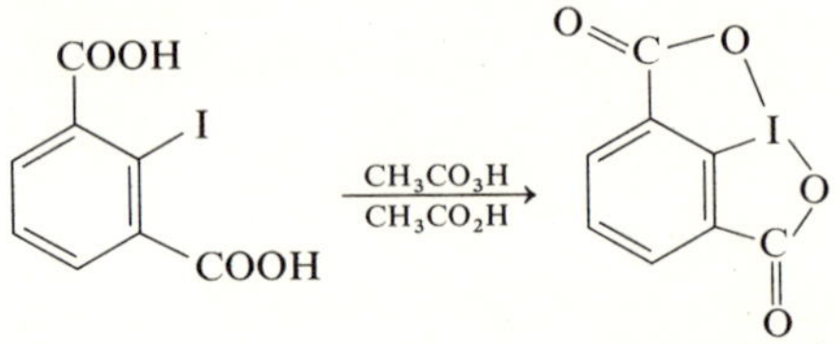

(175)

(166)

IC_6H_4CONHR) with peroxyacetic acid gives benziodazoles **167** and **168**.[382]

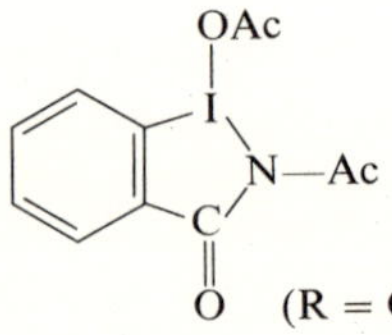

$(R = CH_3)$

(167)

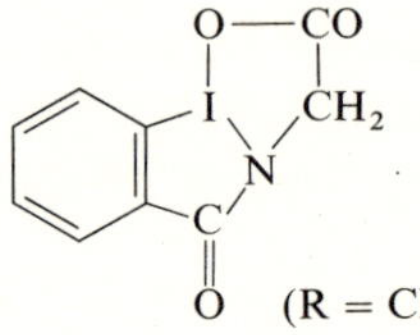

$(R = CH_2CO_2H)$

(168)

o-Iodosophenylphosphoric acid (**169**), which possesses a six-membered ring fused to the benzene ring, has been reported to be the main product of *o*-iodophenylphosphoric acid with peroxyacetic acid.[383]

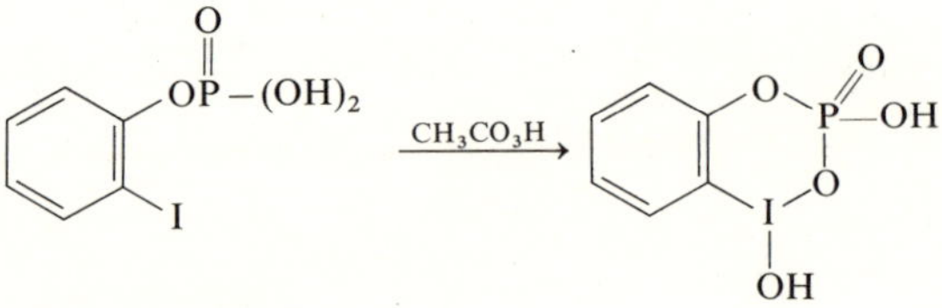

(176)

(169)

The same reagent is able to deiodinate *trans*-α-iodo-α′-acetoxystilbene (**170**) to benzil in good yields.[384]

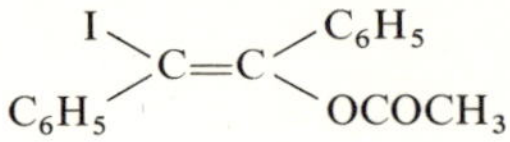 $+ CH_3CO_3H \xrightarrow{-I_2} C_6H_5COCOC_6H_5$ (177)

(170)

[381] W. C. Agosta, *Tetrahedron Lett.* p. 2681 (1965).

[382] H. J. Barber and M. A. Henderson, *J. Chem. Soc. C* p. 862 (1970).

[383] J. E. Leffler and H. Jaffe, *J. Org. Chem.* **38**, 2719 (1973).

[384] Y. Ogata and I. Urasaki, *J. Org. Chem.* **38**, 100 (1973).

VIII. Oxidation of Allylic and Other Activated Carbon–Hydrogen Bonds with Peroxy Esters

Various organic compounds can be oxidized with *tert*-butyl peroxy esters in the presence of catalytic amounts of transition metal salts.[8,9,385–388] This reaction was discovered by Kharasch and Sosnovsky in 1958 [Eq. (178)].[389]

$$RH + R{-}\overset{\displaystyle O}{\overset{\|}{C}}{-}O{-}O{-}C(CH_3)_3 \longrightarrow R{-}O{-}\overset{\displaystyle O}{\overset{\|}{C}}{-}R + (CH_3)_3COH \qquad (178)$$

Some typical examples of this type of oxidation are given below.

Copper-catalyzed reactions of peroxy esters with allylic carbon–hydrogen bonds represent one of the best methods for producing allyl alcohol derivatives. These reactions are usually quite selective. Commercially available *tert*-butyl peroxyacetate and *tert*-butyl peroxybenzoate are most commonly used for this purpose.

Cyclohexene reacts with *tert*-butyl peroxybenzoate in the presence of cuprous bromide to produce cyclohex-1-en-3-yl benzoate (**171**) in 77% yield [Eq. (179)].[389–391] The reaction of *tert*-butyl peroxyacetate and *tert*-butyl

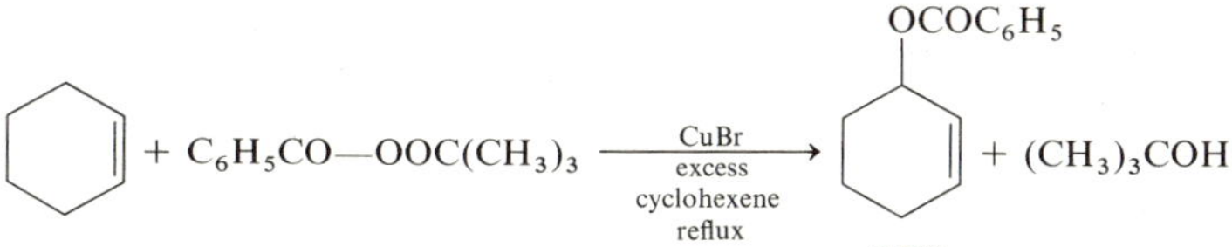

(179)

(**171**)

peroxybenzoate with cyclooctene and cyclodecene yields the corresponding acetates and benzoates, respectively.

Both peroxy esters may also be used for the preparation of other types of esters. In this case an excess of a carboxylic acid must be present in the reaction mixture.[389]

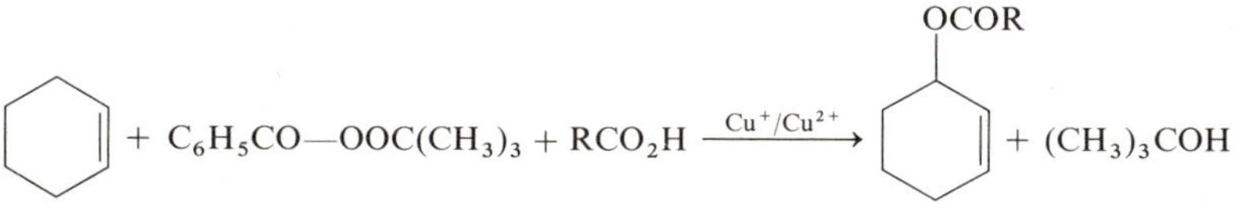

(180)

[385] G. Sosnovsky and D. J. Rawlinson, *Synthesis* **1** (1972).

[386] A. R. Doumaux, Jr., *in*. "Oxidation, Techniques and Applications in Organic Synthesis" (R. L. Augustine and D. J. Trecker, eds.), Vol. 2. p. 141. Dekker, New York, 1971.

[387] D. C. Nonhebel, *Chem. Soc. (London), Spec. Publ.* **24**, 409 (1970).

[388] J. K. Kochi, *in* "Free Radicals" (J. K. Kochi, ed.), Vol. 1, p. 591. Wiley (Interscience), New York, 1973.

[389] M. S. Kharasch and G. Sosnovsky, *J. Am. Chem. Soc.* **80**, 756 (1958).

[390] M. S. Kharasch, G. Sosnovsky, and N. C. Yang, *J. Am. Chem. Soc.* **81**, 5819 (1959).

[391] K. Pederson, P. Jakobson, and S. O. Lawesson, *Org. Synth.* **48**, 18 (1968).

Oxidation of Δ′-*p*-menthene with *tert*-butyl peroxyacetate to a mixture of acetates followed by hydrolysis and subsequent oxidation of the alcohol gave an optically inactive ketone [Eq. (181)].[392] The unusual selectivity of

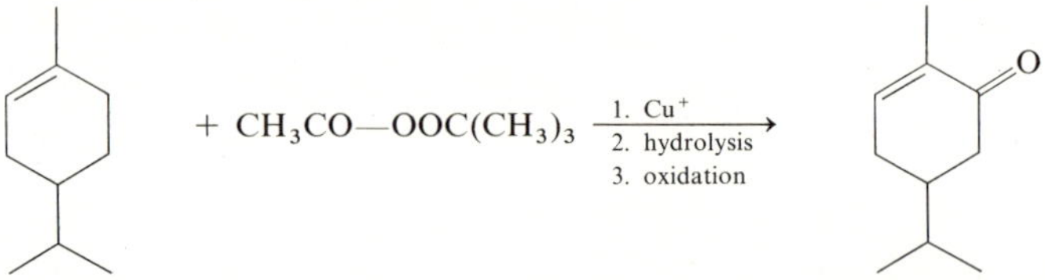

(181)

(optically active) (optically inactive)

copper ion catalyzed reactions of peroxy esters has been explained by an oxidation–reduction sequence according to the mechanistic scheme shown in Eq. (182).[393–396] Scavenging of alkoxy radicals and the absence of carbon

$$RCO—O—OC(CH_3)_3 + Cu^+ \longrightarrow RCO_2^-Cu^{2+} + (CH_3)_3CO\cdot$$

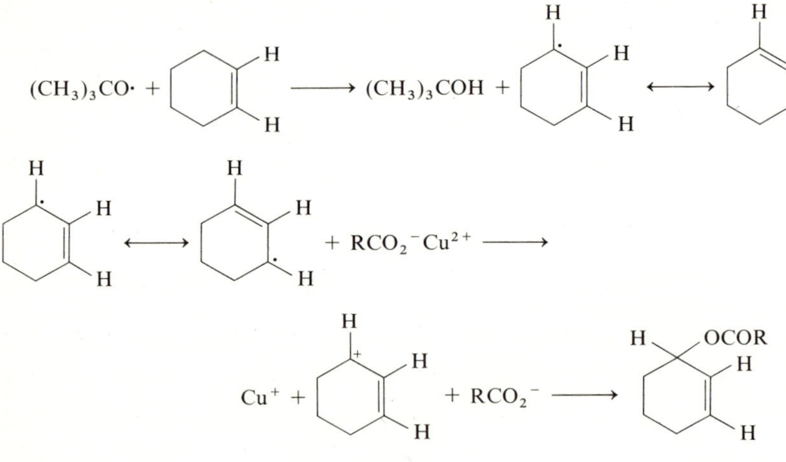

(182)

dioxide in catalytic reactions rule out the alternative fragmentation of the peroxy ester.

$$R—\underset{\underset{O}{\|}}{C}—O—O—C(CH_3)_3 \longrightarrow R—\underset{\underset{O}{\|}}{C}—O\cdot + Cu^{2+}[OC(CH_3)_3] \qquad (183)$$

In general, the selectivity of the reaction with acyclic olefins is less pronounced. Isomerization was observed in the reaction of *tert*-butyl peroxy-

[392] D. Z. Denney, A. Appelbaum, and D. B. Denney, *J. Am. Chem. Soc.* **84**, 4969 (1962).
[393] J. K. Kochi, *Tetrahedron* **18**, 483 (1962).
[394] J. K. Kochi and H. E. Mains, *J. Org. Chem.* **30**, 1862 (1965).
[395] J. K. Kochi and A. Bemis, *Tetrahedron* **24**, 5099 (1968).
[396] J. K. Kochi, *J. Am. Chem. Soc.* **84**, 3271 (1962).

acetate or benzoate with either 1-butene or *cis*- and *trans*-2-butene. A mixture of ca. 90% of 3-acyloxy-1-butene (**172**) and 10% of 1-acyloxy-2-butene (**173**) has been obtained.[397,398]

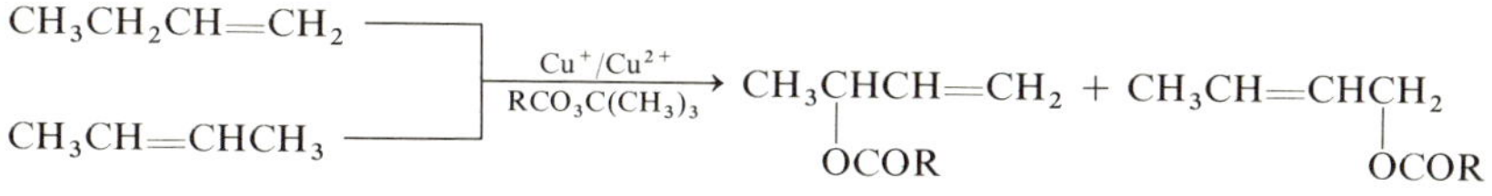

(172) **(173)**

(184)

Similar results were obtained with 1-pentene and other higher-substituted internal olefins.[398] The predominant product of the reaction of a terminal olefin is always the allylic ester with a terminal double bond. The collapse of an allylic cation–copper(I) carboxylate complex (**174**) in which the metal is bonded to the least highly substituted double bond is believed to be responsible for this product distribution.[394,396]

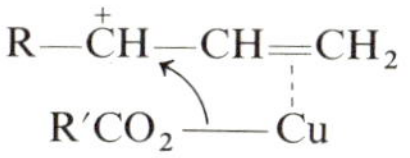

(174)

Another mechanistic possibility assumes coordination of cuprous ions with olefins. An olefin with a terminal double bond forms a more stable complex compared to the olefin with an internal double bond. This pattern is also retained in the resonance-stabilized allylic radicals. The attraction between the radicals and the paramagnetic cupric carboxylate results in the electron and ligand transfer to give the products.[8]

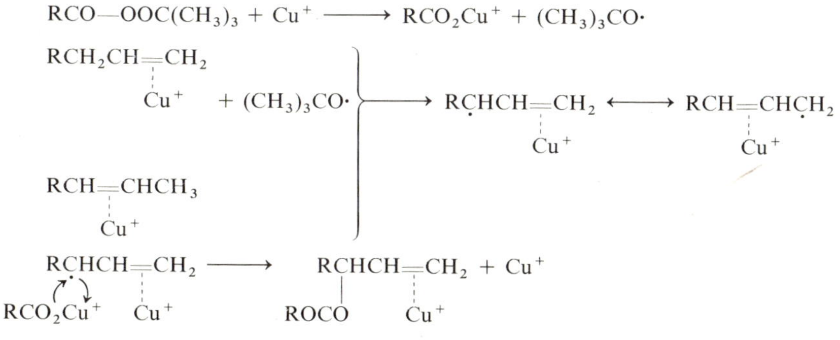

(185)

[397] J. K. Kochi, *J. Am. Chem. Soc.* **83**, 3162 (1961).
[398] J. K. Kochi, *J. Am. Chem. Soc.* **84**, 774 (1962).

A different type of reaction has apparently been observed in the case of norbornadiene (**175**). Instead of the expected benzoate, 7-*tert*-butoxynorbornadiene (**176**) was isolated in 26% yield. Detailed investigation of the mechanism of the reaction of deuterium-labeled norbornadiene with *tert*-butyl peroxybenzoate in the presence of cuprous bromide showed that 7-*tert*-butoxynorbornadiene with deuterium at all skeletal positions, in approximately statistical distribution, is formed. This indicates attack of *tert*-butoxy radicals at all olefinic carbon atoms and a subsequent rearrangement of the norbornadiene skeleton.[399–401]

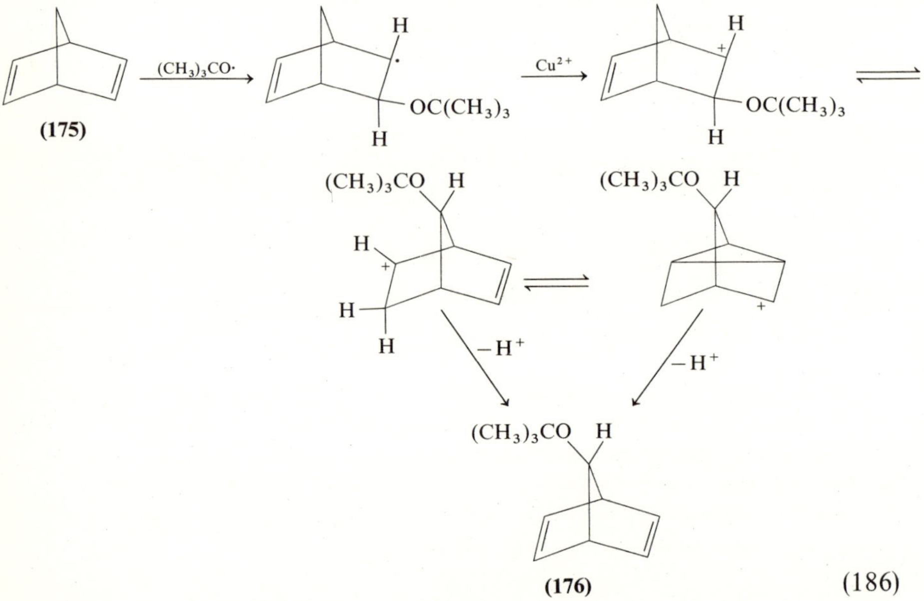

(186)

Copper catalyzed reactions of peroxy esters have also been used for the oxidation of hydrocarbons with benzylic hydrogens. For example, cumene (**177**) reacts with *tert*-butyl peroxyacetate and *tert*-butyl peroxybenzoate to produce 2-phenyl-2-acetoxypropane and 2-phenyl-2-benzoyloxypropane, respectively. In the absence of copper ions, dicumyl is the main product.[402–404]

[399] P. R. Story, *J. Am. Chem. Soc.* **82**, 2085 (1960).
[400] P. R. Story, *J. Org. Chem.* **26**, 287 (1961).
[401] P. R. Story, *Tetrahedron Lett.* p. 401 (1962).
[402] G. Sosnovsky and S. O. Lawesson, *Angew. Chem., Int. Ed. Engl.* **3**, 269 (1964).
[403] C. Walling and A. Zavitsas, *J. Am. Chem. Soc.* **85**, 2084 (1963).
[404] G. Sosnovsky, *Tetrahedron* **21**, 871 (1965).

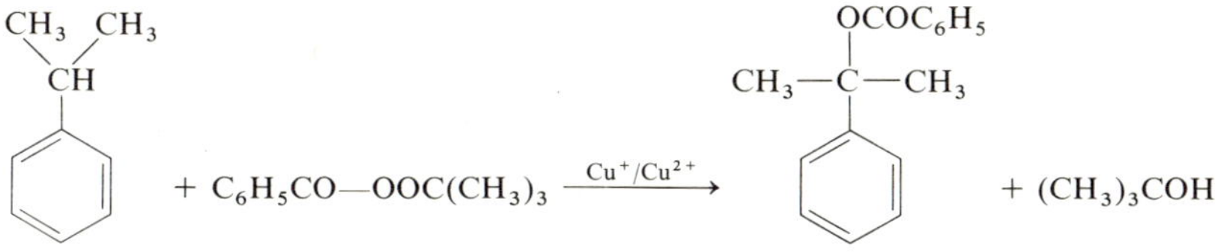

(177)

(187)

The acyloxy group is introduced at the α-position of the heteroatom in the reaction of dialkyl and aryl alkyl ethers with peroxy esters in the presence of copper salts [Eq. (188)].[405–414] Cyclic ethers react anomalously. 2-*tert*-

$$RCH_2{-}O{-}CH_2R + RCO{-}OOC(CH_3)_3 \xrightarrow{Cu^+/Cu^{2+}} R{-}\underset{\substack{|\\ OCOR}}{CH}{-}O{-}CH_2R + (CH_3)_3COH \qquad (188)$$

Butoxytetrahydrofuran has been reported to be the major product in the decomposition of *tert*-butyl peroxy esters in boiling tetrahydrofuran [Eq. (189)].[406,408,410] The 2-acyloxy derivative (**178**) is believed to be formed

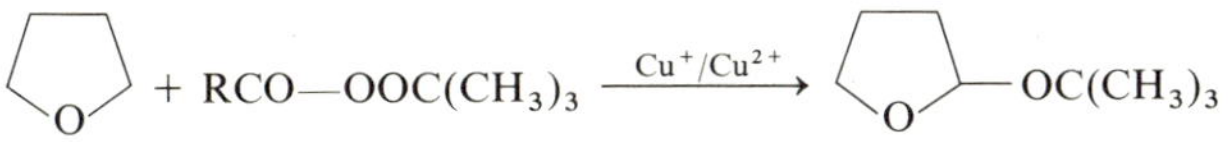

(189)

first with subsequent thermal decomposition to the corresponding carboxylic acid and 2,3-dihydrofuran (**179**). The latter adds to alcohols to give the alkoxy derivatives.[406,408,410]

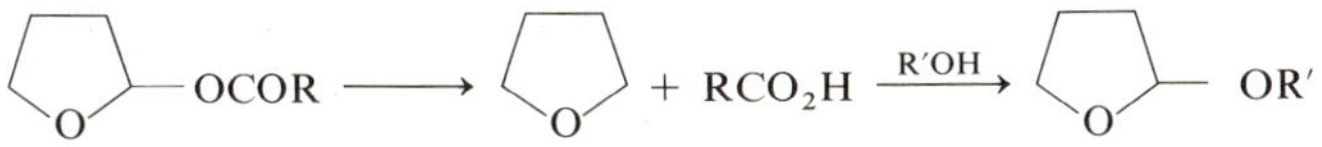

(190)

(178) **(179)**

[405] S. O. Lawesson and C. Berglund, *Acta Chem. Scand.* **15**, 36 (1961).
[406] G. Sosnovsky, *Tetrahedron* **13**, 241 (1961).
[407] S. O. Lawesson and C. Berglund, *Angew. Chem.* **73**, 65 (1961).
[408] S. O. Lawesson and C. Berglund, *Ark. Kemi* **17**, 475 (1961).
[409] S. O. Lawesson and C. Berglund, *Tetrahedron Lett. No.* **2**, p. 4 (1960).
[410] G. Sosnovsky, *J. Org. Chem.* **25**, 874 (1960).
[411] S. O. Lawesson, C. Berglund, and S. Gronwell, *Acta Chem. Scand.* **14**, 944 (1960).
[412] S. O. Lawesson, C. Berglund, and S. Gronwell, *Acta Chem. Scand.* **15**, 249 (1961).
[413] G. Sosnovsky, *Tetrahedron* **18**, 15 (1962).
[414] G. Sosnovsky and H. J. ONeill, *J. Org. Chem.* **27**, 3469 (1962).

Thioethers react analogously.[415,416] For example, tetrahydrothiophene reacts with *tert*-butyl peroxy esters in the presence of copper ions to give 2-acyloxy derivatives (**180**) in good yields. At high temperatures, **180** decomposes to 2,3-dihydrothiophene (**181**) which rather quickly reacts further to form **182**.

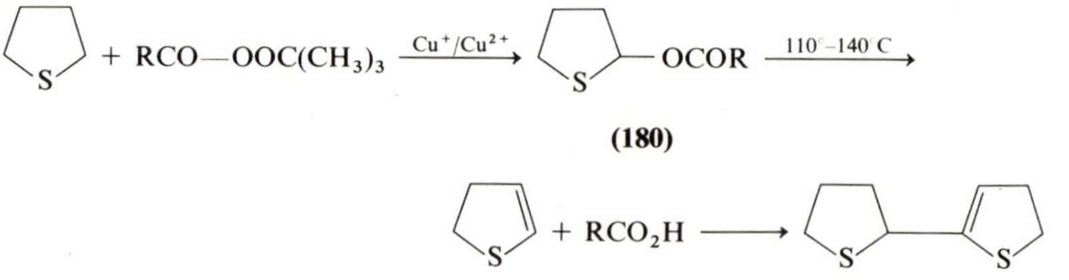

(180) (181) (182) (191)

A one-step oxidative phosphorylation has recently been achieved by the redox reaction of diisopropyl *tert*-butylperoxyphosphate (**183**)[415] with cyclohexene in the presence of cuprous bromide. Cyclohex-1-en-3-yl diisopropyl phosphate (**184**) is obtained in 40% yield by using this reagent.[416]

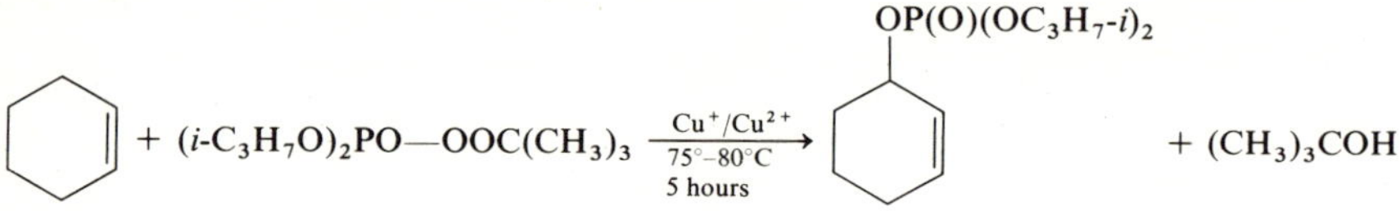

(183) (184)

ACKNOWLEDGEMENT

Support of the research from the Boris Kidrič Fund is gratefully acknowledged.

[415] G. Sosnovsky and E. H. Zaret, *Synthesis* p. 202 (1972).
[416] G. Sosnovsky, G. Karas, and D. J. Rawlinson, *Phosphorus* **3**, 87 (1973).

CHAPTER **IV**

Oxidations with Nitric Acid or Nitrogen Oxides

YOSHIRO OGATA

ISBN 0-12-697252-4

I. Introduction[1]

A. PROPERTIES OF NITRIC ACID AND NITROUS ACID

Aqueous nitric acid solutions of various concentrations are produced on fractional distillation of 68% HNO_3 (sp. gr. 1.414) which has a constant boiling point of 120.5°C (the maximum boiling point). Concentrated nitric acid (99.7%) boils at 86°C with decomposition. The boiling points of 99.7% acid are tabulated below.

Pressure (mm)	Boiling point (°C)
47	22.1
60	26.2
110	35.6

Specific gravity at 15°/4° changes with weight percentage of nitric acid as tabulated below.

HNO_3 (%)	Specific gravity
17.1	1.10
24.84	1.15
32.36	1.20
47.49	1.30
65.30	1.40
94.09	1.50
99.67	1.52

Fuming nitric acid is prepared by dissolving NO_2 in concentrated nitric acid; the amount of NO_2 in fuming nitric acid is generally 7.5% (sp. gr. 1.526)–12.7% (sp. gr. 1.544).

Ordinary commercial nitric acid is 62% HNO_3 (sp. gr. 1.383) and usually contains a small amount of NO_2 and nitrous acid. The content of nitrous acid can be estimated by titration with aqueous permanganate according to the equation below.

$$\%\ HNO_2 = 2.35\ (\text{gm}) \times (\text{ml of } 0.1\ N\ KMnO_4)/\text{sample (gm)}$$

Nitric acid may contain other nitrogen oxides such as NO, N_2O_3, N_2O_4, and N_2O_5 as well as NO_2 and HNO_2. Some of these (NO, NO_2, etc.) are

[1] For the review of nitric acid oxidation of organic compounds, see Y. Ogata, "Oxidation and Reduction of Organic Compounds," p. 417. Nankodo, Tokyo, 1963; K. Namba and T. Yoshida, *Yuki Gosei Kagaku Kyokai Shi* **24**, 958 (1966); Y. Ogata and H. Tezuka, *Kagaku* **24**, 631 (1969); H. Akabayashi, *Yuki Gosei Kagaku Kyokai Shi* **27**, 233 (1969); A. V. Topchiev, "Nitration of Hydrocarbons and Other Organic Compounds." Pergamon, Oxford, (1959).

odd-electron species and hence can initiate oxidation by abstracting a hydrogen atom from organic compounds.

The acid strengths of nitric acid and other strong acids in trifluoroacetic acid were measured spectrophotometrically.[2] Their relative acidities decrease in the order:

$$HClO_4 > H_2SO_4 > ClSO_3H > HBr > HNO_3 > H_3PO_4$$

Therefore, nitric acid can be considered to be a weak strong acid.

Nitrous acid is prepared as an aqueous solution by the addition of a strong acid to aqueous alkali nitrite, but nitrous acid is unstable and easily converted, almost quantitatively, to dinitrogen trioxide (blue) and then to a mixture of nitrogen dioxide (brown–red) and nitrogen monoxide (colorless). The color of the solution thus becomes green rapidly.

$$2\,HNO_2 \rightleftharpoons N_2O_3 + H_2O \rightleftharpoons NO_2 + NO + H_2O$$

B. CHARACTERISTICS OF NITRIC ACID OXIDATION

Nitric acid, nitrous acid, and nitrogen oxides have two features as reagents for organic reactions: (a) use as nitrating or nitrosating agents and (b) use as oxidizing agents. For substrates which possess π electrons such as aromatic compounds and olefins, they mainly act as nitrating or nitrosating agents. This is especially true when fuming nitric acid is used or nitric acid is used in the presence of a strong acid such as sulfuric acid.[3] On the other hand, for substrates which possess no π electrons such as alkanes, they act as oxidizing agents, especially in a dilute solution and in the presence of nitrogen oxides or nitrous acid.

The most typical among the nitrogen-containing oxidants, including organic oxidants such as nitrobenzene, is nitric acid. The salient features of nitric acid oxidation are listed here.

(1) The oxidation requires the presence of an initiator such as nitrous acid or NO_2. This feature is different from that in aromatic nitration, where an attacking agent is NO_2^+, and where nitrous acid or NO_3^- retards the rate.[3] Pure nitric acid, which does not contain nitrous acid, generally shows a long induction period.[4–6] Hence, for smooth nitric acid oxidations it is necessary to use nitric acid containing nitrogen oxides or sodium nitrite.

[2] J. Bessiere, *Bull. Chim. Soc. Fr.* **9**, 3356 (1969); *Chem. Abstr.* **72**, 365454 (1970); *Anal. Chim. Acta* **52**, 55 (1970); *Chem. Abstr.* **73**, 124163 (1970).

[3] P. B. D. De LaMare and J. H. Ridd, "Aromatic Substitution," p. 73. Butterworth, London, 1959; C. K. Ingold. "Structure and Mechanism in Organic Chemistry," 2nd ed., p. 321. Cornell Univ. Press, Ithaca, New York, 1969.

[4] Y. Ogata, Y. Sawaki, F. Matsunaga, and H. Tezuka, *Tetrahedron* **22**, 2655 (1966).

[5] Y. Ogata and Y. Sawaki, *J. Am. Chem. Soc.* **88**, 5832 (1966).

[6] Y. Ogata, H. Tezuka, and Y. Sawaki, *Tetrahedron* **23**, 1007 (1969).

(2) The reaction is often autocatalytic, since the reaction produces an initiator, e.g., NO_2, and sometimes proceeds violently. Since the solubility of the initiator in the solution is limited, the oxidation in an open vessel is controllable. Interestingly, ordinary radical inhibitors such as hydroquinone can be used as initiators for this reaction, since they are easily oxidized by nitric acid to give NO_2 or HNO_2, while ordinary radical initiators such as benzoyl peroxide or azobisisobutyronitrile have little effect in most cases, since the radicals which they produce would be consumed by HNO_2.

(3) Nitric acid oxidations are often catalyzed by mineral acids. Since nitric acid itself is strong, an increase in the concentration of HNO_3 accelerates the reaction by increasing the concentrations of both reagent and catalyst.

(4) Sometimes the nitric acid oxidation introduces one or more nitrogen atoms into the substrate as well as oxygen atoms depending on the substrate and the reaction conditions. For example, the conversions of alkanes to nitroalkanes, of carbonyl compounds to furoxans and of alkenes to dinitroalkanes are observed. This behavior may arise from the greater stability of the N–O bond compared with that of the Cr–O bond which is involved in chromic acid oxidations and that of the O–O bond which is involved in autoxidation or peroxide formation.

II. Alkanes and Arylalkanes

A. VAPOR PHASE NITRATION OF ALKANES

Unsubstituted alkanes resist attack by nitric acid or nitrogen oxides at moderate temperatures but can be nitrated in the vapor phase at high temperatures (350°–450°C) to give a variety of products including C–C fission products. The gas phase nitration of butane with HNO_3, for example, at 420°C in a flow system affords nitromethane, nitroethane, nitropropane, and 1- and 2-nitrobutane in the composition as shown below.[7]

$$CH_3CH_2CH_2CH_3 \xrightarrow[420^\circ C/\text{vapor}]{HNO_3} \underset{10\%}{CH_3NO_2} + \underset{13\%}{CH_3CH_2NO_2} + \underset{8\%}{CH_3CH_2CH_2NO_2} + \underset{24\%}{CH_3CH_2CH_2CH_2NO_2} + \underset{45\%}{CH_3\text{—}\underset{\substack{|\\NO_2}}{CH}\text{—}CH_2CH_3} \quad (1)$$

Addition of gaseous oxygen to the reaction system increases the total yield of nitro products but decreases the yields of nitrobutanes; i.e., addition

[7] G. B. Bachman, H. B. Hass, and L. M. Addison, *J. Org. Chem.* **17**, 906, 914, 928, 935, and 942 (1952).

of oxygen enhances cleavage of C–C bonds which forms lower nitroalkanes. Addition of molecular bromine increases both the yield and conversion evidently by facilitating the formation of carbon radicals by hydrogen atom abstraction with atomic bromine. An increase in the surface-to-volume ratio results in an increase in selectivity of products.

As shown in the nitration of butane [Eq. (1)], the reactivity of hydrogen atoms in alkanes increases only slightly in the order: primary < secondary < tertiary; the difference between the reactivities, however, is small. In another example, the nitration of pentane yields the following composition of products: 1-nitro, 18.9%; 2-nitro, 18.2%; and 3-nitro, 20.2%.

These facts suggest a mechanism for the nitration involving alkyl radicals and $\cdot NO_2$ or $\cdot OH$.

$$HONO_2 \rightleftharpoons HO\cdot + \cdot NO_2 \quad (2a)$$

$$RH + \cdot OH \longrightarrow R\cdot + H_2O \quad (2b)$$

$$RH + \cdot NO_2 \longrightarrow R\cdot + HNO_2 \quad (2c)$$

$$R\cdot + \cdot NO_2 \longrightarrow R{-}NO_2 \quad (2d)$$

$$R'CH_2\cdot + \cdot NO_2 \longrightarrow R'CH_2O\cdot + NO \quad (2e)$$

$$R'CH_2O\cdot \longrightarrow R'\cdot + CH_2O + \text{etc.} \quad (2f)$$

Thus, while nitration of arenes is a typical electrophilic substitution by NO_2^+, the nitration of aliphatic compounds in the vapor phase involves hydrogen atom abstration followed either by coupling of the formed alkyl radicals with NO_2 or further oxidation of the alkyl radicals.

B. LIQUID PHASE OXIDATION

Reaction of alkanes, cycloalkanes, and arylalkanes with nitric acid or nitrogen oxides in the liquid phase gives nitro compounds, nitrites, and carbonyl derivatives along with various oxidation and polynitration products.

Alkyl and cycloalkylbenzenes are preferentially nitrated and oxidized at benzyl position. For example, ethylbenzene and cyclohexylbenzene give mainly α-nitroethylbenzene and α-nitrocyclohexylbenzene, respectively.[8]

The ease of nitration of saturated carbon atoms is generally in the order: tertiary > secondary > primary; exceptions, however, are observed in some cases. For example, camphane (**1**) and isocamphane (**2**) are nitrated in the liquid phase to yield only secondary nitro compounds.[8]

[8] H. Shechter and D. K. Brain, *J. Am. Chem. Soc.* **85**, 1806 (1963).

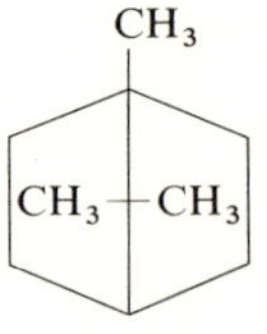

(1)

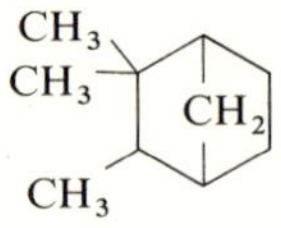

(2)

When an optically active alkane such as (+)-3-methylheptane, HCMeEt(Bu), is nitrated by 50% HNO_3 at 100°C, the product obtained, 3-methyl-3-nitroheptane, O_2NCMeEt(Bu), is racemic. Nitrations of *cis*- and *trans*-decalins and *cis*- and *trans*-hydrindanes are not also stereospecific. Therefore, a mechanism involving bimolecular substitution with a transition state such as **3**[9] is not favored, but a mechanism involving a free carbon radical which has escaped the solvent cage, such as **4**, is probable.

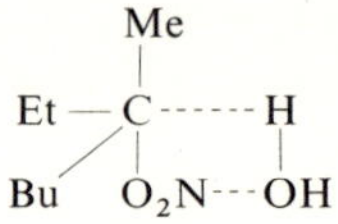

(3)

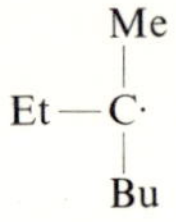

(4)

Alkyl nitrites, which are by-products, are easily decomposed from carbonyl compounds, alcohols, and carboxylic acids. Toluene is converted to benzaldehyde in the reaction below.

$$PhCH_3 + HNO_3 \longrightarrow PhCHO \longrightarrow PhCO_2H$$

Benzoic acid is converted to benzophenone in a more rapid reaction.

$$PhCH_2Ph + 2\ HNO_3 \longrightarrow PhCOPh + H_2O + 2\ HNO_2$$

The oxidation of diphenylmethane with 2.5 *M* HNO_3 in aqueous acetic acid (70% CH_3CO_2H) at 90°C gives only benzophenone (the yield of isolated product is 84%, but the yield estimated by UV spectrophotometry is ca. 100%) without appreciable amounts of nitro compounds.[10] The rate in excess HNO_3 under the same conditions was determined by measuring the UV absorption of the product and was found to be almost independent of the concentration of nitric acid and first-order with respect to diphenylmethane. It is of interest to note that (i) the rate is independent of the initiator, HNO_2, (ii) a plot of log k vs. H_0 gives a straight line with a slope of ca. −1, implying a one-proton participation at the transition state, and (iii) electron-attracting groups retard the reaction. These facts can be explained by the following

[9] C. K. Ingold, *J. Chem. Soc.* p. 244 (1935).

[10] Y. Ogata, H. Tezuka, and T. Kamei, *J. Org. Chem.* **34**, 845 (1969).

mechanism [Eqs. (3–9)], i.e., a rate-determining hydrogen atom abstraction by the radical cation $HNO_2^{+\cdot}$ (protonated NO_2).[10]

$$HNO_3 + HNO_2 \xrightleftharpoons{K_3} 2\,\cdot NO_2 + H_2O \quad (3)$$

$$NO_2 + H^+ \xrightleftharpoons{K_4} HNO_2^{+\cdot} \quad (4)$$

$$Ph_2CH_2 + HNO_2^{+\cdot} \xrightarrow[\text{slow}]{k_5} Ph_2CH\cdot + H_2NO_2^+ \quad (5)$$

$$Ph_2CH\cdot + \cdot NO_2 \longrightarrow Ph_2CHONO \quad (6)$$

$$Ph_2CHONO + H_2O \longrightarrow Ph_2CHOH + HNO_2 \quad (7)$$

$$Ph_2CHOH + HNO_2^{+\cdot} \longrightarrow Ph_2\dot{C}OH + H_2NO_2^+ \quad (8)$$

$$Ph_2\dot{C}OH + \cdot NO_2 \longrightarrow Ph_2C(OH){-}ONO \longrightarrow Ph_2CO + HNO_2 \quad (9)$$

Attack of the alkyl radical on nitric acid to form $Ph_2CH{-}O{-}\overset{+}{N}(O^-){-}OH$ which would then be oxidized to Ph_2CONO_2 is unlikely since addition to a O═N bond is unusual and the intermediate would be thermodynamically unstable.

With Eq. (5) as the rate-determining step, the rate expression is

$$v = k_5[Ph_2CH_2][HNO_2^{+\cdot}] = k_5K_4[H^+][NO_2][Ph_2CH_2]$$

and the observed apparent rate constant is

$$k = kK[H^+]\,[NO_2]$$

This rate expression is consistent with the behavior of the rate as a function of H_0 which suggests that one proton participates in the transition state.

The hydrogen abstractor could be NO^+,[11] $N_2O_3^+$,[12] or other species derived from nitrous acid, but probably not neutral species such as NO_2, since these would not explain the acidity effect.[12] The independence of rate on $[HNO_2]$ may be explained by saturation of the solution with NO_2; at $[HNO_3] > 0.3$ *M* gaseous NO_2 escapes from the system, hence, the concentration of NO_2 is held constant.

The effect of ring substituents satisfies the Hammett equation, $\log(k/k_0) = \sigma\rho$, giving a ρ value of -1.7 in 70% aqueous acid (Fig. 1) and a ρ value of -1.2 in 60% dioxane[13] (Fig. 2). One can explain this by arguing that the absolute

[11] E. J. Strojny, R. T. Iwamasa, and L. K. Frevel, *J. Am. Chem. Soc.* **93**, 1171 (1971).

[12] The formation of $N_2O_3^+$ form NO^+ and NO_2 was reported by J. D. S. Goulden and D. J. Millen, *J. Chem. Soc.* p. 2620 (1950).

[13] Y. Ogata, H. Tezuka, and T. Kamei, *Tetrahedron* **25**, 4919 (1969).

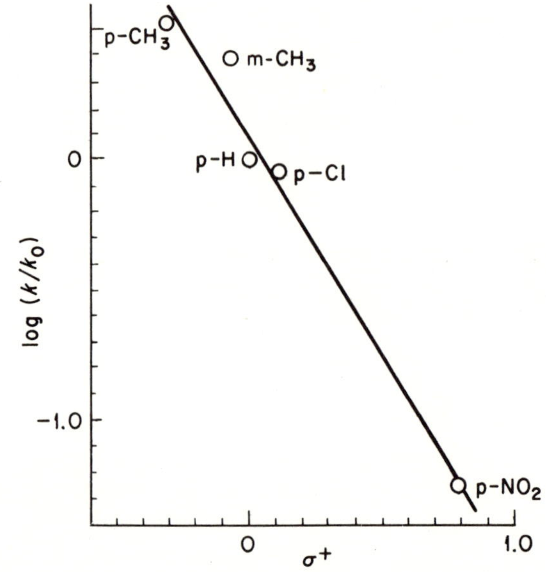

FIG. 1. Hammett plots for the relative rates of nitric acid oxidation of diphenylmethanes in 70 vol.% acetic acid. Initial conc.: [$ArCH_2Ph$] = 0.050*M*, [HNO_3] = 2.5 *M*, [$NaNO_2$] = 0.02 *M*. [Data from Ogata *et al.*[13] Courtesy of Pergamon Press]

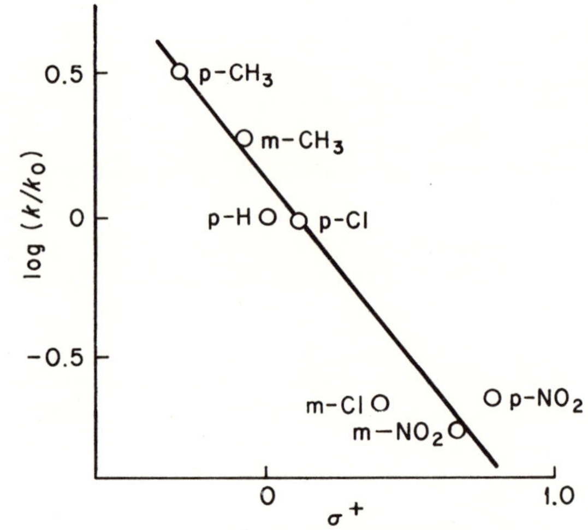

FIG. 2. Hammett plots for the relative rates of nitric acid oxidation of diphenylmethanes in 60 vol.% dioxane. Initial conc.: [$ArCH_2Ph$] = 0.05 *M*, [HNO_3] = 3.0 *M*, [$NaNO_2$] = 0.02*M*. [Data from Ogata *et al.*[13] Courtesy of Pergamon Press]

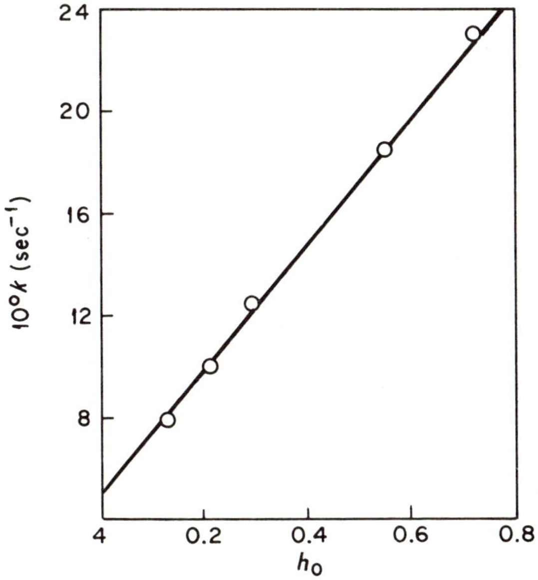

FIG. 3. Plot of 10^5 k *vs* h_0 in the nitric acid oxidation of triphenylmethane in 60% dioxane at 90°C. Initial concentration: $[HNO_3] = 1.0$ M, $[NaNO_2] = 0.01$ M, $[H_2SO_4] = 0 - 1.0$ M, $[Ph_3CH] = 0.050$ M. [Data from Ogata *et al.*[15] Courtesy of Pergamon Press]

ρ value is smaller in a solvent such as dioxane which can complex with the attacking species,[14] thus reducing its electrophilicity.

The plot of log k vs $-H_0$ for the oxidation in 60% dioxane gives a slope of 0.5–0.6, implying that the rate constant should be expressed as the sum of first and zeroth-orders in h_0, i.e., $k = k_a h_0 + k_b$, where $\log h_0 = -H_0$.

A plot of log $(k - k_b)$ vs. H_0 for 60% dioxane gives a line with a slope of ca. -1. This suggests that $\cdot NO_2$ as well as HNO_2^{+} participates in hydrogen abstraction.

In the oxidation of triphenylmethane the same solvent effect is observed, and as shown in Fig. 3, the dependency of the rate constant on h_0 is also expressed as[15]

$$k = k_a h_0 + k_b$$

Polar substituents affect mainly k_a as shown in Fig. 4. Since introduction of an electron-releasing group in the ring facilitates the nitric acid oxidation, the

[14] K. U. Ingold, *Can J. Chem.* **41**, 2816 (1963); H. Sakurai and A. Hosoi, *J. Am. Chem. Soc.* **89**, 458 (1967); B. Rubin, H. H. Sisler, and H. Shechter, *ibid.* **74**, 877 (1952); H. Shechter and F. Conrad, *ibid.* **75**, 5610 (1953).

[15] Y. Ogata, H. Tezuka, and T. Kamei, *Tetrahedron* **26**, 4313 (1970).

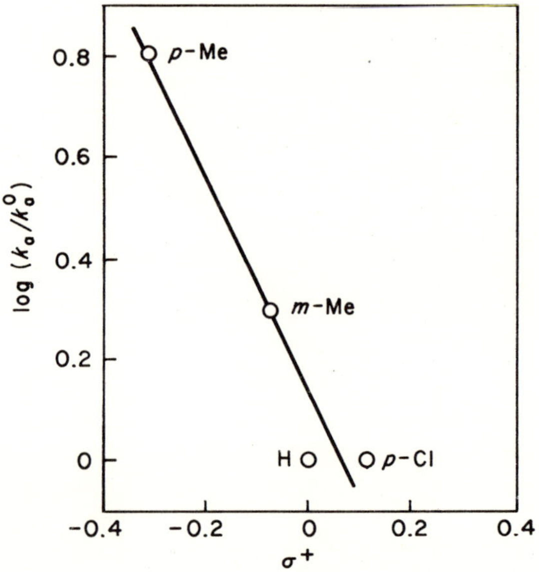

FIG. 4. Substituent effect on k_a (see equation on p. 303) in the nitric acid oxidation of diphenylmethanes in 60% dioxane at 90°C. [Data from Ogata *et al.*[15] Courtesy of Pergamon Press]

reactivity of CH_3 in the oxidation by 30% aqueous HNO_3 at 90°C increases in the order[16]: toluene < xylenes < mesitylene < 1,2,4-trimethylbenzene < 1,2,4,5-tetramethylbenzene.

The nitric acid oxidation of *o*-, *m*-, and *p*-xylenes gives the corresponding toluic acids. For example, *o*-xylene (90%) is oxidized with 69.3% HNO_3 at reflux temperature for 55 hours to yield *o*-toluic acid (53–55%).[17] Xylenes are further oxidized to the corresponding benzenedicraboxylic acids under vigorous conditions, e.g., oxidation with 30% HNO_3 at 180°C preferably under pressure.[18] A mixture of xylenes thus oxidized gives 64% dicarboxylic acids.[18a] *m*-Xylene gives a considerable amount of a nitro compound besides the dicarboxylic acid.

The effect of substituents on *p*- and *m*-xylenes are reported for the following reactions [Eqs. (10a) and (10b)]. In the oxidation of *p*-xylene **5** both toluic acids **6** and **7** are produced and **6** predominates, whereas in the oxidation of *m*-

[16] T. Yoshida, K. Yamashita, and K. Namba, *Kogyo Kagaku Zasshi* **73**, 445 (1970).
[17] E. Zaugg and T. Rapala, *Org. Synth., Collect. Vol.* **3**, 820 (1955).
[18] B. Srinivasan and S. B. Chandalia, *Ind. Chem. J.* **5**, 33 (1971); *Chem. Abstr.* **75**, 48621 (1971).
[18a] V. Kudlacek and Z. Pokorny, *Sb. Ved. Pr. Vys. Sk. Chemickotechnol., Pardubice* **16**, 41 (1967); *Chem. Abstr.* **71**, 12754 (1969).

xylene **8**, only toluic acid **9** is formed along with substituted isophthalic acid **10**.[19]

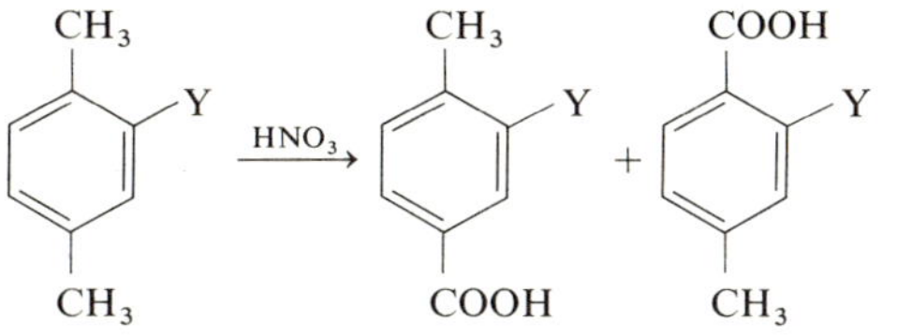

(10a)

(5) **(6)** **(7)**

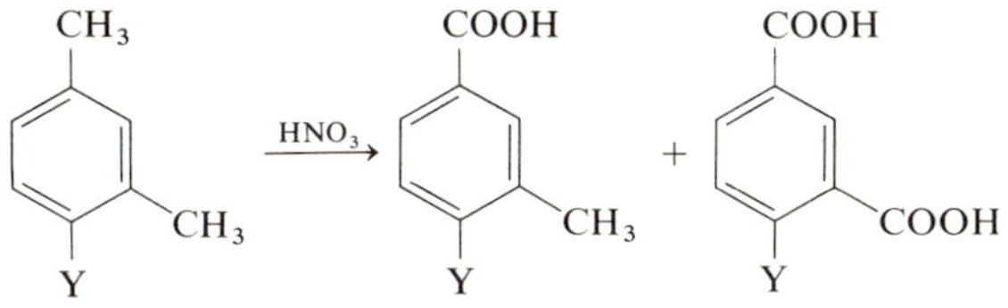

(10b)

(8) **(9)** **(10)**

Comparison of yields shows that the reactivity of the CH_3 to give **6** by oxidation of the CH_3 meta to Y decreases in the order: Br > Cl > NO_2 > F and the reactivity to give **9** by oxidation of the *p*-CH_3 decreases in the order: Br > Cl > F > NO_2. The reactivity orders of the halogens suggest the importance of an inductive effect, probably because of the steric inhibition of reasonance caused by the ortho methyl group in **5** and **8**.

The presence of an electron-attracting group on the benzene ring slows down the oxidation rate of a benzylic position, as well as simultaneously retarding the nitration of the ring to an even greater extent.[20] Hence, the preparation of some substituted benzenecarboxylic acids that possess an electron-attracting group on the ring are successful. The oxidation of methyl *p*-toluate, for example, to give terephthalic acid is an industrially important reaction. Oxidation of *p*-nitrotoluene with 20% HNO_3 at 160°C and 20–25 kg/cm^2 pressure gives *p*-nitrobenzoic acid.[21] *o*-Nitrotoluene is oxidized with 35% HNO_3 in a sealed tube to *o*-nitrobenzoic acid (45%) and picric acid (11%),[22] which most likely comes from the decarboxylative hydroxylation of nitrobenzoic acid. The cobalt-catalyzed autoxidation product of durene [1,2,4,5-$C_6H_2(CH_3)_4$] on oxidation with 17% HNO_3 at 200°C for 1.5–2 hours yields pyromellitic acid [1,2,4,5-$C_6H_2(COOH)_4$].[23]

[19] W. Crandall, R. Beasly, L. L. Lambing, and R. Moriconi, *J. Org. Chem.* **32**, 134 (1967).

[20] Cf. R. G. Coombes, R. B. Moodie, and K. Schofield, *J. Chem. Soc. B* p. 800 (1969), also see Ogata *et al.*[15]

[21] B. Srinivasan and S. B. Chandalic, *Indian J. Technol.* **9**, 274 (1971); *Chem. Abstr.* **75**, 118080 (1971).

[22] C. Hakanson and M. Nilsson, *Acta Chem. Scand.* **21**, 1978 (1967).

[23] A. G. Akhmetov *et al.*, *Neftekhimiya* **10**, 879 (1970); *Chem. Abstr.* **74**, 88316 (1971).

Oxidation of ethylbenzene with 30% HNO_3 in a molar ratio of EtPh: HNO_3 of 3:1 at 80°C produces acetophenone, benzoic acid, benzaldehyde, α-nitroethylbenzene, and a small amount of α-methylbenzyl alcohol and dibenzoylfuroxan.[24]

Cumene is oxidized by 3 *M* HNO_3 at 90°C in aqueous dioxane to give a mixture of acetophenone (70–76%) and benzoic acid (30%), which are the same products obtained from the HNO_3 oxidation of cumyl alcohol. Cumyl alcohol with HNO_3 gives acetophenone, and acetophenone with HNO_3 gives benzoic acid. Therefore, the oxidation of cumene probably proceeds via these imtermediates,[25] where α-methyl groups are eliminated stepwise.

$$\mathrm{PhCH(CH_3)_2} \xrightarrow{\mathrm{HNO_3}} \underset{\text{cumyl alcohol}}{\mathrm{PhC(CH_3)_2OH}} \xrightarrow{\mathrm{HNO_3}} \underset{\text{acetophenone}}{\mathrm{PhCOCH_3}} \xrightarrow{\mathrm{HNO_3}} \mathrm{PhCOOH} \qquad (11)$$

The oxidation of cumene in a heterogeneous system with 20% HNO_3 was reported to give acetophenone, α-nitrocumene, cumyl alcohol, and a little toluic acid.[26]

Since the isopropyl group is more likely to undergo oxidation than the methyl group, the oxidation of *p*-cymene with 69.3% HNO_3 (sp. gr. 1.42) gives *p*-toluic acid alone.[27]

$$p\text{-}\mathrm{CH_3C_6H_4CH(CH_3)_2} + \mathrm{HNO_3} \xrightarrow[\text{56–59\%}]{\text{reflux/8 hours}} p\text{-}\mathrm{CH_3C_6H_4COOH} \qquad (12)$$

Quite a different reaction, i.e., nitrative coupling of phenyl rings to form biphenyl, occurs on treatment of alkylbenzenes with 90% HNO_3 at low temperature.[28–28b] For example, reaction of *o*-xylene with 90% HNO_3 at −25°C gives nitrobixylyls (47%) consisting mainly of **11**.[28]

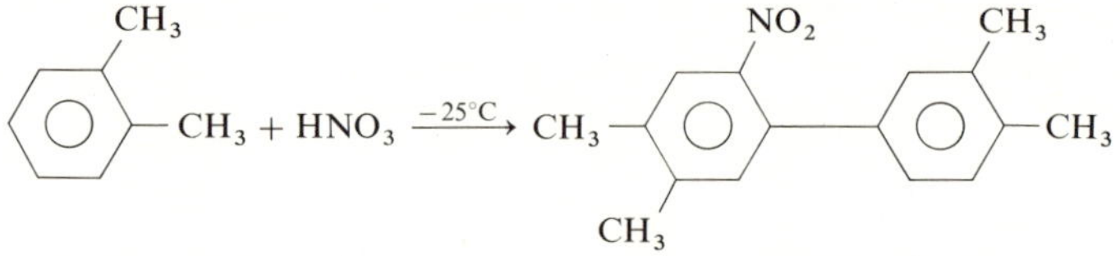

(13)

(11)

Lowering the temperature to −40°C minimizes nitration of the products.[28b] A mixture of two alkylbenzenes (RPh and R′Ph) gives crossed biaryls (R—C_6H_4—C_6H_4—R′). If DNO_3 is used as a reagent, no deuterium is

[24] M. Tamura, T. Yoshida, and K. Namba, *Kogyo Kayaku* **29**, 151 (1968).
[25] Y. Ogata and H. Tezuka, *Bull. Chem. Soc. Jpn.* **43**, 3285 (1970).
[26] M. Tamura, T. Yoshida, and K. Namba, *Kogyo Kayaku* **29**, 151 (1968).
[27] F. W. Tuley and C. S. Marvel, *Org. Synth., Collect. Vol.* **3**, 822 (1955).
[28] I. Puskas and E. K. Fields, *J. Org. Chem.* **31**, 4204 (1966).
[28a] I. Puskas and E. K. Fields, *J. Org. Chem.* **32**, 589 (1967).
[28b] I. Puskas and E. K. Fields, *J. Org. Chem.* **32**, 3924 (1967).

incorporated in the product. Therefore, it is suggested that the reaction involves attack of an electrophilic π complex between the alkylbenzene and HNO_3 on the nucleophilic alkylbenzene R′Ph, followed by deprotonation to form a biaryl.[28b]

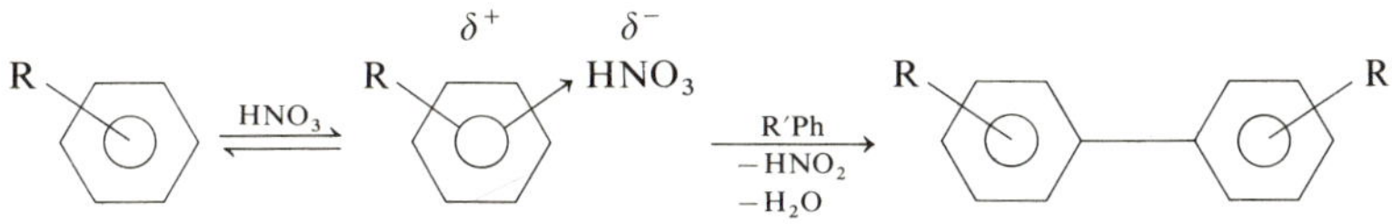

(14)

III. Carbon–Carbon Multiple Bonds

A. REACTION OF NITRIC ACID ON OLEFINS AND ACETYLENES

The multiple bonds of olefins and acetylenes are cleaved by nitric acid oxidation in the liquid phase. This type of reaction involves some important preparations. For example, propylene gives oxalic acid (71.4% as the calcium salt) by oxidation with 70% HNO_3 and oxygen at 25°–35°C.[29] 2-Butene gives propionic acid (80%), acetic acid (17.4%), and oxalic acid (2.8%) with HNO_3–O_2 at 45°–50°C in the presence of Na_3VO_4.[30] Thus, nitric acid is a convenient and inexpensive reagent for these selective oxidations.

Oleic acid, $CH_3(CH_2)_7CH{=}CH(CH_2)_7CO_2H$, is oxidized to azelaic acid, $HOOC(CH_2)_7COOH$, and suberic acid, $HOOC(CH_2)_6COOH$, together with some lower dicarboxylic acids on treatment with 64% HNO_3 at 105°C under pressure in the presence of 0.1% V_2O_5 as a catalyst.[31]

Cyclohexene is converted by 70% HNO_3 oxidation in the presence of NH_4VO_3 at 0°–60°C to adipic acid (68% yield at optimum conditions) often accompanying lower dibasic acids such as glutaric acid (10–20%) and succinic acid (3–6%).[32,33] 4-Chlorocyclohexene under similar conditions gives β-chloroadipic acid and a small amount of β-chloroglutaric and succinic acids.[34]

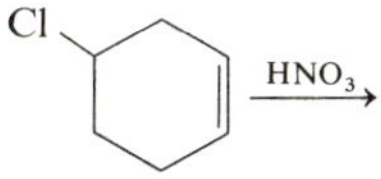

$HOOCCH_2CHClCH_2CH_2COOH$ +

$HOOCCH_2CHClCH_2COOH$ + $HOOCCH_2CH_2COOH$ (15)

29 A. Charamel, J. Descroix, J. Duroux, and S. Siquet (Rhone-Poulene), French Patent 1,501,725 (1967); *Chem. Abstr.* **70**, 3295 (1969).

30 J. Boichard *et al.*, French Patent 1,548,597 (1968); *Chem Abstr.* **71**, 90841 (1969); French Patent 1,544,793 (1968); *Chem. Abstr.* **71**, 123591 (1969).

31 G. Gut and A. Guyer, *Helv. Chim. Acta* **47**, 1673 (1963).

32 J. F. Franz, J. F. Herber, and W. S. Knowles, *J. Org. Chem.* **30**, 1488 (1965); *Chem. Ind.* (*London*) p. 250 (1961).

33 G. Gut and W. Lindenmann, *Chimia* **22**, 307 (1968); *Chem. Abstr.* **64**, 76566 (1968).

34 M. Azume and Y. Suzuki, *Kogyo Kagaku Zasshi* **74**, 397 and 390 (1971).

β-Cyanoadipic acid is obtained in a similar fashion from 4-cyanocyclohexene.[35]

Cyclohexene and other cycloolefins often give addition products of N_2O_5[33] which are cleaved to dibasic acids without a vanadium catalyst. For example, cyclohexene gives with N_2O_4 at 0°–5°C the 2-nitrosocyclohexyl nitrate dimer [Eq. (16a)], which can, in turn, be cleaved to adipic acid in the presence of vanadium ion, V(V).[32]

A mechanism for the oxidation of cyclohexene with nitric acid containing a little N_2O_4 and vanadium catalyst was postulated as follows.[32]

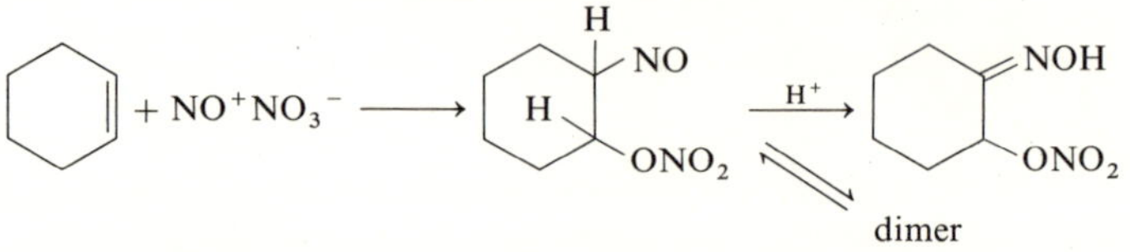

(16a)

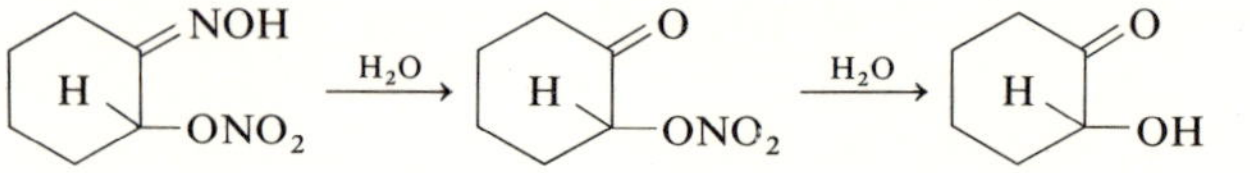

(16b)

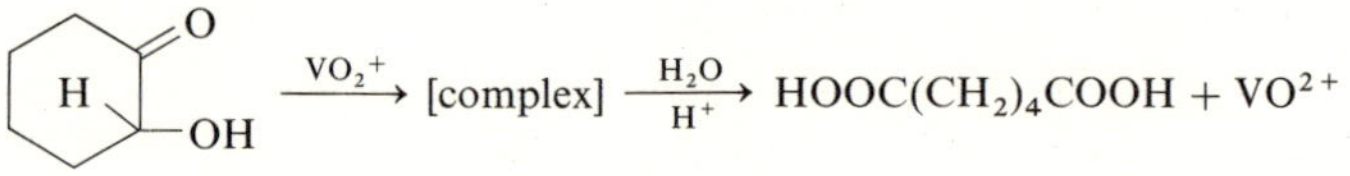

(16c)

$$VO^{2+} + HNO_3 \longrightarrow VO_2^{+} + NO_2 + H^{+} \qquad (16d)$$

Generally, the oxidation involves nitrosonitrate formation, conversion of the nitroso group to the oxime, hydrolysis of the oxime, and then hydrolysis of the nitrate to form the α-keto alcohol. The vanadium catalyst, e.g., NH_4VO_3, serves to cleave the C–C bond. When the reaction is carried out without the vanadium catalyst, the nitrosonitrate dimer precipitates.[32]

Ethylene can be transformed to glyoxal by 2.3% HNO_3 in the presence of a catalytic amount of $PdCl_3$ and sodium nitrite at 40°C, but the yield is low (4.6%).[36]

$$CH_2{=}CH_2 \xrightarrow{HNO_3} O{=}CH{-}CH = O \qquad (17)$$

Acetylene gives oxalic acid on treatment with 48% HNO_3 in the presence of Pd.[37]

$$CH{\equiv}CH + HNO_3 \xrightarrow{Pd/24°C} HOOC{-}COOH \qquad (18)$$

[35] M. Amagasa and Y. Suzuki, *Kogyo Kagaku Zasshi* **74**, 401 (1971).

[36] M. Tsutsumi and A. Yasui, Japanese Patent 7,005,522 (1970); *Chem. Abstr.* **72**, 110795(1970);

[37] Y. N. Kukushkin, V. V. Kobzev, and L. P. Morozova, *Zh. Prikl. Khim.* **43**, 2759 (1970); *Chem. Abstr.* **74**, 111357 (1971).

B. REACTION OF DINITROGEN TETROXIDE WITH OLEFINS

Dinitrogen tetroxide (N_2O_4) adds to olefins below room temperature.[38–40] The reaction, which is carried out under oxygen to eliminate NO, gives dinitro compounds (**12**) and nitro-nitrites (**13**), which are further hydrolyzed to nitro alcohols (**14**) in the presence of water or are oxidized to nitronitrates (**15**).

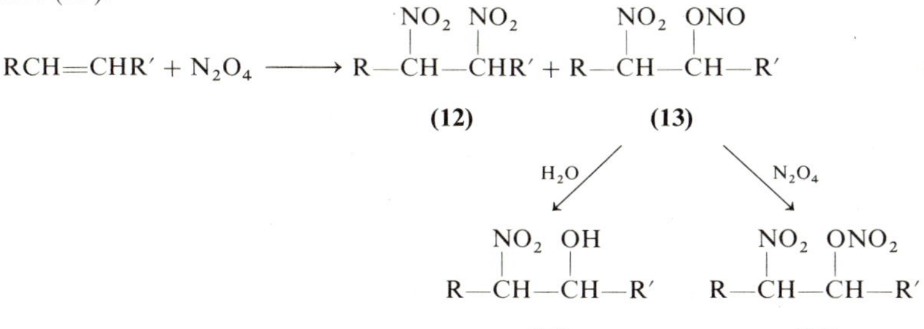

(19)

(**12**) (**13**) (**14**) (**15**)

This reaction can be used for the preparation of nitro alcohols (**14**) and nitro olefins or amino alcohols.[41,42]

The orientation of addition observed with terminal olefins [Eq. (20)] shows that the C–N bond is formed at the ethylene carbon which bears more hydrogen atoms and the C–O bond is formed at the carbon atom which bears fewer hydrogen atoms, irrespective of the polarity of the R group.[38,39]

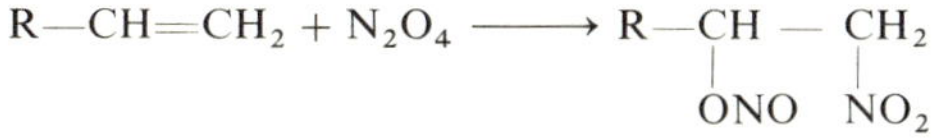

(20)

R: CH_3,[38] CF_3,[43] CO_2CH_3[44,45]

In some cases, the reaction of olefins with N_2O_4 gives nitrosonitrates.[46] Isobutene, for example, gives **16**.

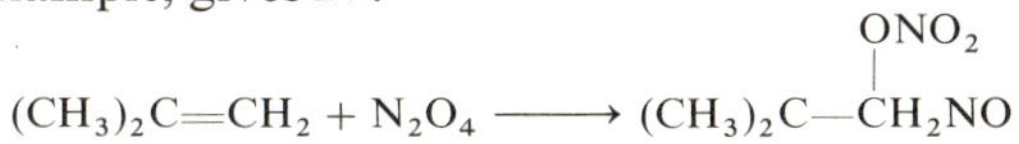

(21)

(**16**)

[38] N. Levy and C. W. Scaife, *J. Chem. Soc.* pp. 1093, 1096, and 1100 (1946).
[39] N. Levy, C. W. Scaife, and A. E. W. Smith, *J. Chem. Soc.* p. 52 (1948).
[40] H. Baldock, N. Levy, and C. W. Scaife, *J. Chem. Soc.* p. 2627 (1949).
[41] W. K. Seifert, *J. Org. Chem.* **28**, 125 (1963).
[42] G. A. Bonetti, C. B. DeSavigny, C. Michalski, and R. Rosenthal, *J. Org. Chem.* **33**, 237 (1968).
[43] A. V. Fokin, V. A. Komarov, I. N. Sorochkin, and S. M. Davydova, *Zh. Vses. Khim.* **10**, 354 (1965); *Chem. Abstr.* **63**, 8182 (1965).
[44] H. Shechter and F. Conrad, *J. Am. Chem. Soc.* **74**, 3052 (1952).
[45] H. Shechter and F. Conrad, *J. Am. Chem. Soc.* **75**, 5610 (1953).
[46] B. F. Ustavshchikov, V. A. Podgornova, N. V. Dormidontova, and M. I. Farberov, *Neftekhimiya* **5**, 873 (1965); *Chem. Abstr.* **64**, 7981 (1966).

Further oxidation and hydrolysis of product **16** can produce methacrylic acid.[45] This process for the preparation of methacrylic acid from isobutene was developed by chemists of the Escambia Chemical Corp. and is called the Escambia process.[47] A 69–74% yield of α-hydroxyisobutyric acid (**19**) is obtained when isobutene is added to excess N_2O_4 or to a HNO_3 solution of N_2O_4 at 0°C and the resulting mixture is allowed to undergo oxidation and hydrolysis. The intermediates α-nitratoisobutyraldehyde (**17**) and acid **18** have been isolated. The following mechanism [Eq. (22)] is postulated for the reaction.

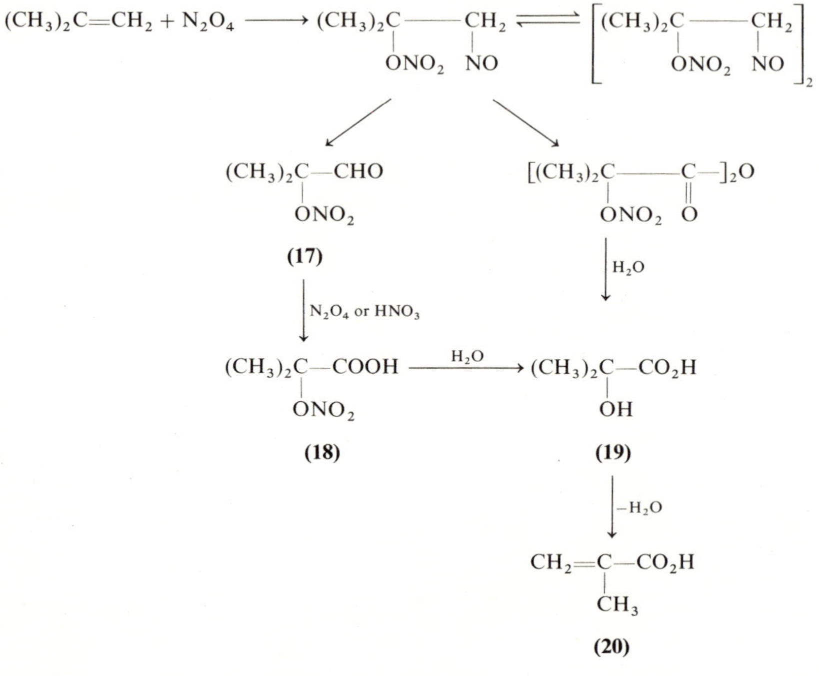

(22)

This process can be applied to some olefins for the preparation of methacrylic acid derivatives [Eq. (23)].[48,48a]

[47] E. F. Schoenbrunn and J. H. Gardner, *J. Am. Chem. Soc.* **82**, 4905 (1960).

[48] B. F. Ustavshchikov, V. A. Podgornova, N. V. Dormidontova, and M. I. Farberow, *Dokl. Akad. Nauk SSSR* **157**, 143 (1964); *Chem. Abstr.* **61**, 9393 (1964).

[48a] I. B. F. Ustavshchikov, V. A. Podgornova, N. V. Dormidontova, and M. I. Farberow, *Probl. Org. Sint.* p. 18 (1965); *Chem. Abstr.* **64**, 4928 (1966); B. F. Ustavshchikov, V. A. Podgornova, and M. I. Farberow, *Dokl. Akad. Nauk SSSR* **168**, 1335 (1966); *Chem. Abstr.* **65**, 15218 (1966).

$$RCH_2-C(R')=CH_2 \xrightarrow{N_2O_4/H_2O} RCH=C(R')-COOH \quad (23)$$

$R = H, R' = H; R = CH_3, R' = CH_3; R = C_2H_5, R' = CH_3;$

$R = H', R' = CH_3, R = H, R' = CH_3$

On the other hand, oxidation of propylene under the reaction conditions, which result in the formation of **19**, affords lactic acid.[48,49]

$$CH_3-CH=CH_2 \xrightarrow[85\,\%]{N_2O_4/15^\circ-20^\circ C} CH_3-CH(ONO_2)-COOH \xrightarrow{H_2O} CH_3-CH(OH)-COOH \quad (24)$$

α-Nitratopropionic acid, which is formed initially, is hydrolyzed to give lactic acid in the presence of an acidic catalyst (hydrochloric acid or an ion-exchange resin).[48,49] However, if either an ether or ester is used as a solvent the same reaction of propylene gives nitro compounds similar to those shown in Eq. (19);[49] the same is true for isobutene.[39]

$$CH_3-CH=CH_2 \xrightarrow[\text{in an ether or ester solvent}]{N_2O_4\ -20^\circ C \text{ to } -60^\circ C} CH_3-CH(OH)-CH_2(NO_2) + CH_3-CH(NO_2)-CH_2(NO_2) + CH_3-CH(ONO_2)-CH_2(NO_2) \quad (25)$$

The position of addition of groups without solvent or in 2,2-dichlorodiethyl ether [Eqs. (21 and 22)] suggests an initial attack of NO^+ on the terminal carbon atom of the olefin followed by addition of ONO_2^-. On the other hand, the orientation of the reaction in an ether or ester solvent [Eq. (25)] is readily explained by a radical attack of NO_2. If the solvent is an ether or an ester, it may suppress the ionization of N_2O_4 to $NO^+NO_3^-$ but does not affect its radical dissociation. An explanation for the formation of α-nitratocarboxylic acid (**18**), which involves free radical oxidation of nitro compounds,[38–40]

$$ONO-\overset{|}{\underset{|}{C}}-CH_2NO_2 \xrightarrow{N_2O_4} O_2NO-\overset{|}{\underset{|}{C}}-CO_2H \quad (26)$$

is not valid, since no carboxylic acid formation is observed with less polar alkenes such as ethylene and octene under these conditions.[38] Hence, the

[49] S. Fumasoni, G. Giacobbe, R. Martinelli, and G. Schippa, *Chim. Ind.* (*Milan*) **47**, 1064 (1965); *Chem. Abstr.* **64**, 6472 (1966).

addition of N_2O_4 may occur via either ionic addition of NO^+ or radical addition of $\cdot NO_2$ depending on the reaction conditions.

C. Reaction of Nitrogen Monoxide with Olefins

Olefins react with NO which contains a trace of NO_2 to form addition products.[50–52] The initial main reaction is the addition of NO_2 and NO with the latter adding to the carbon atom of the double bond which bears fewer hydrogen atoms.

It should be noted that a rearrangement leading to a product of the reverse orientation may occur during the workup. For example, trimethylethylene was reported to give an abnormal addition product, but this may be due to rearrangement during the workup under basic conditions.

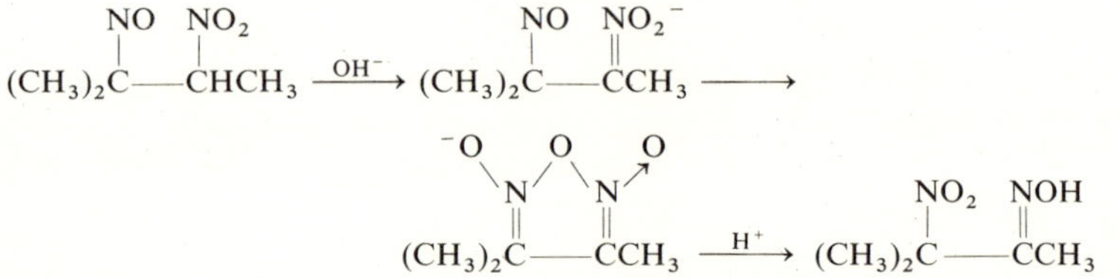
(27)

Isobutene reacts with NO to give α-nitrosobutene (**22**) and β-nitroisobutene (**23**) which are used for starting materials for the preparation of methyl methacrylate.[50,52] A mechanism involving initial addition of NO_2 (present in trace amounts) by a radical reaction (Scheme 1) was postulated.

The yields of **21** and **23** were 66% and 22%, respectively, at an early stage, but α-nitroisobutene (**22**) formed by the decomposition of **21**, raised the total yield of nitroisobutenes to 74%. β-Nitroisobutene (**23**) can be transformed to the α-isomer (**22**) by treatment with an ion-exchange resin. By treatment with hydrogen chloride, the α-isomer (**22**) is converted to $(CH_3)_2CCl{-}CCl = NOH$ which is hydrolyzed to a mixture of precursors of methacrylic acid, $(CH_3)_2C(OH)CO_2H$ and $(CH_3)_2CClCO_2H$ etc., in a high yield.[52]

Fluoro olefins also react with NO to give nitronitroso compounds and their derivatives (dinitro and ketonitro compounds) with analogous orientation.[53–55]

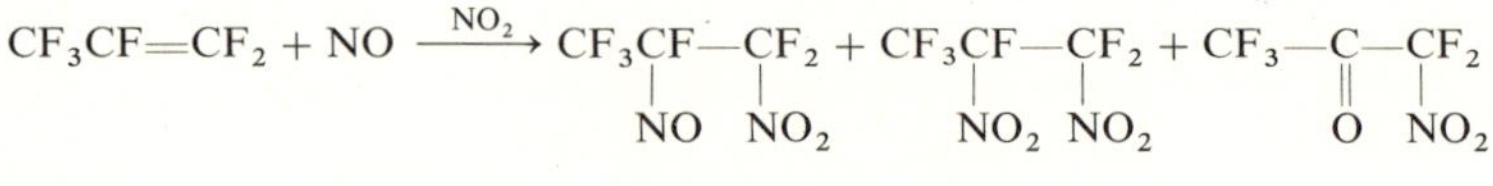
(28)

[50] J. F. Brown, *J. Am. Chem. Soc.* **79**, 2480 (1957).
[51] C. A. Burkhard and J. F. Brown, *J. Org. Chem.* **29**, 2235 (1964).
[52] L. V. Phillips and D. M. Coyne, *J. Org. Chem.* **29**, 1937 (1964).
[53] J. D. Park, A. P. Stefani, G. H. Crawford, and J. R. Lacher, *J. Org. Chem.* **26**, 3316 (1961).
[54] J. M. Birchall, A. J. Bloom, R. N. Haszeldine, and C. J. Willis, *J. Chem. Soc.* p. 3021 (1962).
[55] E. Bagley, J. M. Birchall, and R. N. Haszeldine, *J. Chem. Soc. C* p. 1232 (1966).

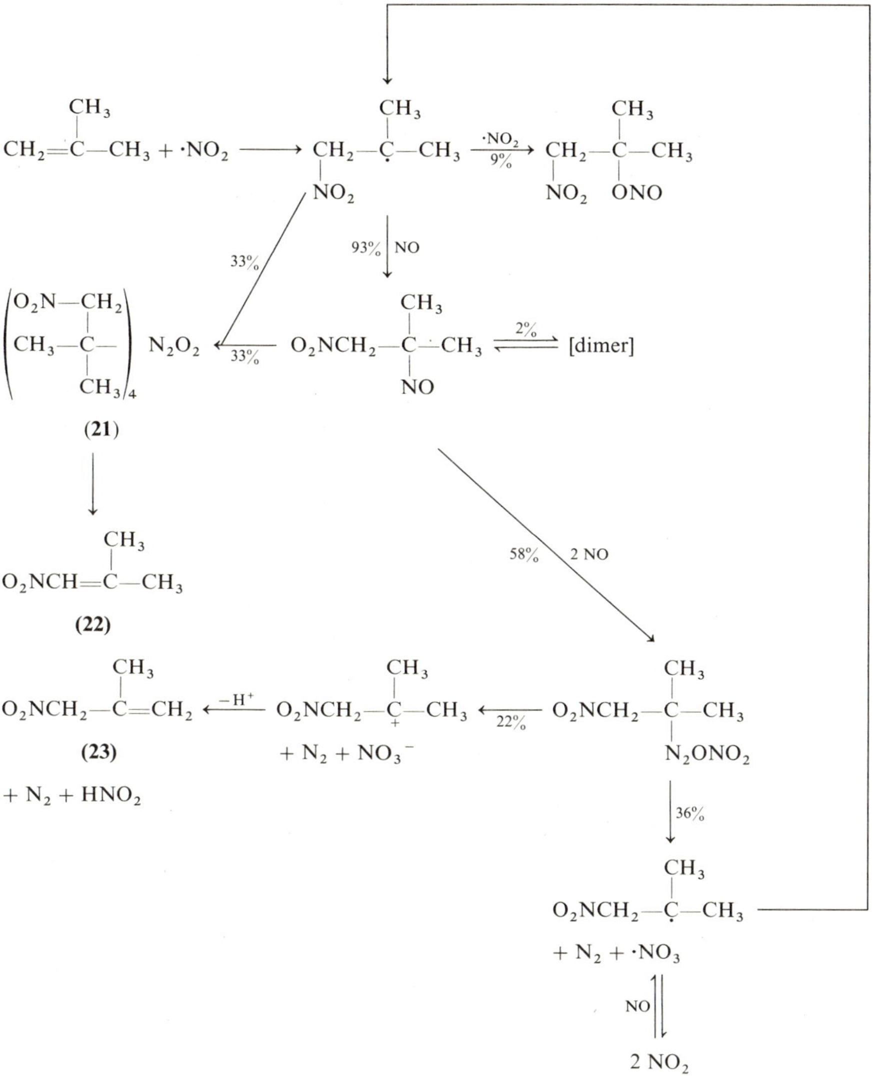

SCHEME 1

IV. Alcohols and Ethers

A. ALIPHATIC AND ALICYCLIC ALCOHOLS AND ETHERS

Alcohols are oxidized by nitric acid to aldehydes or ketones which are then oxidized to carboxylic acids. An important example of this reaction is the oxidation of cyclohexanol to cyclohexanone which is then converted to adipic acid.[56,57] Thus, cyclohexanol (5 *M*) is added slowly with stirring to 50% HNO_3 (16.6 *M*) in the presence of ammonium vanadate; the solution is maintained at 55°–60°C with ice water. After stirring for more than 1 hour the mixture is cooled to 0°C and the adipic acid precipitates. The crude adipic acid is collected by filtration and dried, yield 58–60%, mp 146°–149°C. Recrystallization from concentrated HNO_3 (sp. gr. 1.42) gives pure adipic acid, mp 151°–152°C.[58]

It is claimed that either addition of copper or iron vanadate[59,60] or passage of a 1:1 mixture of NO and NO_2[61] or NO_2 alone[62] through the reaction mixture increases the yield of adipic acid.

Other cycloalkanols can be similarly cleaved to the corresponding dicarboxylic acids, e.g., cyclododecanol gives dodecanedioic acid.[63]

Glycols give the corresponding dicarboxylic acids, e.g., ethylene glycol is oxidized by HNO_3–H_2SO_4 (20–25:30–40) in the presence of 0.001–0.1% V_2O_5 at 50°–70°C to give oxalic acid in a yield of over 93%.[64] Dicarboxylic acids (C_5–C_9) can be prepared by the oxidation of the corresponding glycols with 57% HNO_3 at 95°–105°C.[65]

For the kinetic and mechanistic study of the oxidation of aliphatic alcohols, β-methoxyethanol was chosen. The rate in 3–5 *M* aqueous sulfuric acid was followed by determination of the concentration of alcohol by NMR. The stoichiometry is as follows (no aldehyde was detected).[11]

$$3\ CH_3OCH_2CH_2OH + 4\ HNO_3 \longrightarrow 3\ CH_3OCH_2COOH + 4\ NO + 5\ H_2O \qquad (29)$$

The rate of the reaction in excess HNO_3 satisfies the equation $v = k$ [alcohol], and a plot of log *k* vs. $-H_R$[66] gives a line with a slope of 0.5, which is probably a result of the presence of basic material, e.g., alcohol or N_2O_4. Use of $-H_0$[67] gives a line with a slope of unity. The reaction has an induction period which can be reduced by the addition of NO as an initiator.

56 H. C. Godt, Jr. and J. F. Quinn, *J. Am. Chem. Soc.* **78**, 1461 (1956).
57 W. J. Ascelt and D. W. van Krevelen, *Recl. Trav. Chim. Pays-Bas* **82**, 51 (1963).
58 B. A. Ellis, *Org. Synth., Collect. Vol. 1*, 2nd ed., p. 18 (1956).
59 F. Bende, H. Voellinger, and K. Pohl, French Patent 1,527,392 (1968); *Chem. Abstr.* **71**, 12567 (1969).
60 Vickers-Zimmer A. -G., British Patent 1,145,815 (1969); *Chem. Abstr.* **70**, 105977 (1969).
61 H. Richter, K. H. Maedeback, and V. Kaufmann, East German Patent 68,897 (1969); *Chem. Abstr.* **72**, 66393 (1970).
62 G. Lartigau and H. Lemoine, German Offen. 1,966,064 (1971); *Chem. Abstr.* **75**, 77499 (1971).
63 J. O. White and D. D. Davis, German Offen. 1,912,569 (1969); *Chem. Abstr.* **72**, 21327 (1970).
64 Novomoskovsk Aniline Dye Plant, French Patent 1,479,735 (1967); *Chem. Abstr.* **68**, 21539 (1968).
65 A. S. Nekrasov, A. N. Bashkirov, and V. B. Abramovich, *Tr., Nauchnos Issled. Inst. Neftekhim. Proizvod.* **2**, 75 (1970); *Chem. Abstr.* **74**, 87276 (1971).
66 N. C. Deno, J. J. Jauruzelski, and A. Schriesheim, *J. Am. Chem. Soc.* **77**, 3044 (1955).
67 M. A. Paul and F. A. Long, *Chem. Rev.* **57**, 1 (1957).

A mechanism involving the abstraction of a hydride ion from the α-carbon of β-methoxyethanol by NO^+ is postulated as follows.

$$H_2O + 2\,NO + HNO_3 \rightleftharpoons 3\,HONO \tag{30a}$$

$$HONO + H^+ \rightleftharpoons H_2O + NO^+ \tag{30b}$$

$$CH_3OCH_2CH_2OH + NO^+ \xrightarrow{\text{slow}} CH_3OCH_2\overset{+}{C}HOH + HNO \tag{30c}$$

$$CH_3OCH_2\overset{+}{C}HOH + H_2O \rightleftharpoons CH_3OCH_2CH(OH)_2 \tag{30d}$$

$$CH_3OCH_2CH(OH)_2 \xrightarrow[-HNO]{NO^+} CH_3OCH_2\overset{+}{C}(OH)_2 \overset{-H^+}{\rightleftharpoons} CH_3OCH_2CO_2H \tag{30e}$$

An observation which supports this mechanism is that β-methoxyethanol reacts with $NO^+BF_4^-$ to give β-methoxypropionic acid.[11]

Dinitrogen tetroxide can be used for the oxidation of alcohols. For example, oxidation of fluorinated alcohols with a mixture of NO_2 and air in the gas phase at 300°–400°C produces the corresponding aldehyde hydrates with purities of 40–43% and in yields of 33–85%.[68]

$$H(CF_2)_nCH_2OH + N_2O_4 \longrightarrow H(CF_2)_nCH(OH)_2 \tag{31}$$

$n = 6, 8, \text{ and } 10$

A mechanism involving the abstraction of an α-hydrogen atom by a $\cdot NO_2$ radical is postulated for this reaction [$R_f = H(CF_2)_n$].

$$R_fCH_2OH + \cdot NO_2 \longrightarrow R_f\dot{C}HOH + HNO_2 \tag{32a}$$

$$R_f\dot{C}HOH + \cdot NO_2 \longrightarrow R_f{-}\overset{\displaystyle ONO}{\overset{|}{C}}HOH + R_3\overset{\displaystyle NO_2}{\overset{|}{C}}HOH \tag{32b}$$

$$2\,HNO_2 \longrightarrow H_2O + NO\cdot + \cdot NO_2 \tag{32c}$$

$$R_f{-}\overset{\displaystyle ONO}{\overset{|}{C}}HOH \xrightarrow{-NO\cdot} R_f{-}\overset{\displaystyle O\cdot}{\overset{|}{C}}HOH \xrightarrow{R_fCH_2OH} R_f{-}\overset{\displaystyle OH}{\overset{|}{C}}H{-}OH + R_f\dot{C}HOH \tag{32d}$$

$$R_f{-}\overset{\displaystyle NO_2}{\overset{|}{C}}HOH \xrightarrow{-HNO_2} R{-}CHO \xrightarrow{H_2O} R{-}CH(OH)_2 \tag{32e}$$

B. BENZYL ALCOHOL AND BENZYL ETHER

Benzyl alcohol is easily oxidized with 0.7 *M* (4.4%) HNO_3 in aqueous dioxane at 90°C, but practically no oxidation is observed if the concentration of HNO_3 is less than 0.1 *M* (0.63%). Benzyl alcohol is oxidized to benzaldehyde quantitatively by 0.1 *M* HNO_3 in aqueous dioxane to which a strong acid such as perchloric or sulfuric acid has been added.[4]

$$PhCH_2OH + HNO_3 \longrightarrow PhCHO + HNO_2 + H_2O \tag{33}$$

[68] R. M. Scribner, *J. Org. Chem.* **29**, 279 (1964).

Benzyl alkyl ethers react similarly to give a quantitative amount of benzaldehyde but only under more vigorous conditions; e.g., 1.5 M HNO_3.

$$PhCH_2OR + HNO_3 \longrightarrow PhCHO + HNO_2 + ROH \qquad (34)$$

The rate of this reaction [Eq. (34)] in 40% aqueous dioxane was followed by measuring the UV absorption of the aldehyde which is produced. The rate in excess HNO_3 is expressed below.[5]

$$v = k[\text{substrate}]$$

The reaction requires an initiator such as nitrous acid, but the initiator does not affect the rate constant k. The effect of acidity on k in a solution of over 0.5 M HNO_3 is expressed below.

$$k = k'h_0[HNO_3]^0$$

This means that the rate is proportional to the acidity function h_0,[67] but independent of the nitric acid concentration. (See the right-hand side of the curve in Fig. 5.) These facts suggest the mechanism below which involves α-hydrogen atom abstraction by protonated $\cdot NO_2$.

$$HNO_3 + HNO_2 \overset{K_b}{\rightleftharpoons} 2\cdot NO_2 + H_2O \qquad (35a)$$

$$\cdot NO_2 + H^+ \rightleftharpoons HNO_2^{\ddagger} \qquad (35b)$$

$$PhCH_2OR + HNO_2^{\ddagger} \xrightarrow[\text{slow}]{k_c} Ph\dot{C}HOR + HNO_2 + H^+ \qquad (35c)$$

$$Ph\dot{C}HOR + \cdot NO_2 \longrightarrow PhCH(ONO)(OR) \xrightarrow{H_2O} PhCHO + ROH + HNO_2 \qquad (35d)$$

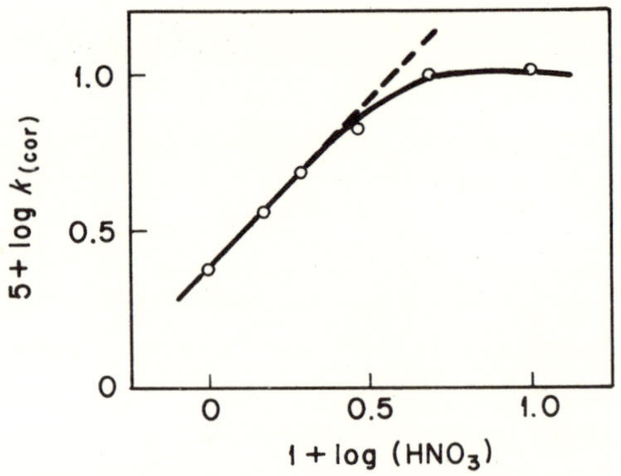

FIG. 5. Correlation between [HNO_3] and rate constant k for the HNO_3 oxidation of $PhCH_2OCH_3$ in 40% dioxane at 90°C. The values of k were corrected to the same acidity: $H_0 = 0$. The dotted line is of unit slope. [Data of Ogata and Sawaki.[5] Courtesy of *J. Am. Chem. Soc.*]

This mechanism should lead to the rate equation

$$v = k_c K_b[H^+][NO_2][PhCH_2OR]$$

but the limited solubility of the attacking species may make the rate independent of $[NO_2]$, $[HNO_2]$, and $[HNO_3]$. The rate in a sealed tube increases with $[HNO_3]$ and pressure of NO_2,[69] which indicates that the rate depends on $[NO_2]$ as shown in the above rate equation in this case.

On the other hand, the same reaction with less than 0.3 *M* HNO_3 shows first-order dependence on $[HNO_3]$ (Fig. 5), which probably means that the concentration of the attacking species in the reaction system is below its saturation point.

The effect of substituents in the benzene ring satisfies the Hammett equation giving a ρ value of -1.9 with σ^+.[5] This large negative ρ value suggests that the attacking species is positively charged, since hydrogen atom abstraction with a neutral species, e.g., ROO·, generally has a ρ value closer to zero.[70] In the same way, the electron-releasing ability of the alkyl group R in $PhCH_2OR$ accelerates the reaction in the order: Ph < Me < Et < *n*-Pr < *i*-Pr. Benzyl alcohol exhibits a rate behavior which is analogous to that of benzyl methyl ether and appears to be oxidized by a similar mechanism.[4,69]

Monoethers (and diethers in some cases) of hydroquinones and catechols are converted to the corresponding quinones by treatment with aqueous acidic nitrous acid. A mechanism involving an oxime is suggested.[71]

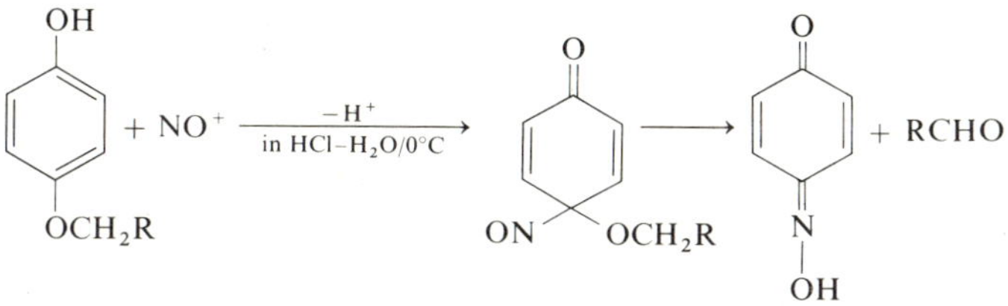

(36a)

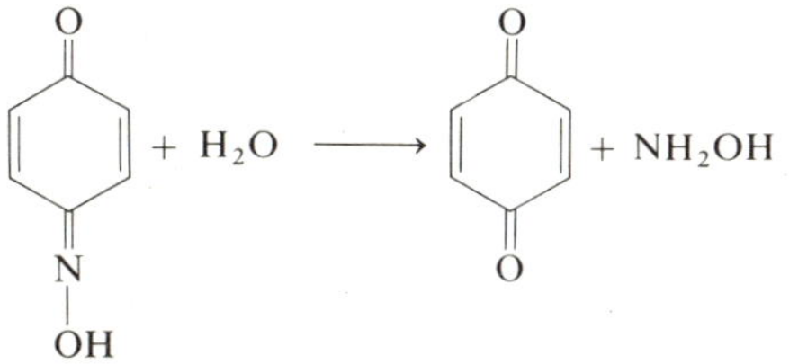

(36b)

[69] M. Tamura and T. Yoshida, *Kogyo Kayaku* **29**, 361 (1968).

[70] H. O. Pritchard, *Chem. Rev.* **52**, 529 (1953); G. A. Russell and R. C. Williamson, *J. Am. Chem. Soc.* **86**, 2357 (1964).

[71] D. H. R. Barton, P. G. Gordon, and D. G. Hewitt, *J. Chem. Soc. C* p. 1206 (1971).

V. Alkyl Halides

Alkyl halides are converted by aqueous nitric acid to the corresponding aldehyde which is then oxidized to the carboxylic acid. The initial reaction is the hydrolysis of the chloride to the alcohol.

$$RCH_2X + H_2O \longrightarrow RCH_2OH \xrightarrow{HNO_3} RCHO \xrightarrow{HNO_3} RCO_2H \qquad (37)$$

A process for the preparation of terephthalic acid uses this procedure starting from the chloromethylation of toluene to give *p*-methylbenzyl chloride which is then oxidized by HNO_3. The process is called the Grosskinsky process.[72]

$$CH_3C_6H_5 \xrightarrow{HCl + HCHO} p\text{-}CH_3C_6H_4CH_2Cl \quad (+\, o\text{-isomer}) \xrightarrow[\text{under pressure}]{HNO_3 \cdot H_2O} p\text{-}HOOCC_6H_4COOH \quad (+\, o\text{-isomer}) \qquad (38)$$

Also α,α′-dichloro-*p*-xylene can be oxidized to terephthalic acid under milder conditions, but terephthalaldehyde and 4-formylbenzoic acid are usually also produced.[73–75]

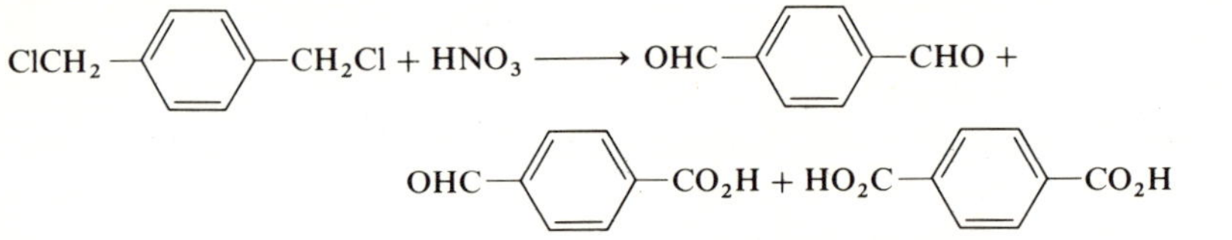

(39)

Table I shows a typical example of the oxidation products.

TABLE I

PRODUCTS FROM OXIDATION OF α,α-DICHLORO-*p*-XYLENE

$[HNO_3]$ (%)	Temp. (°C)	Time (hr)	*p*-OHC–C_6H_4–CHO (%)	*p*-OHC–C_6H_4–CO_2H (%)	*p*-$C_6H_4(CO_2H)_2$ (%)
10	106	7	56	23	15
15(V_2O_5)	106	7	36	35	16
19	104	6	70	23	5
50	106	8	2	43	49

[72] A. Benning, H. Fühbuss, and O. Grosskinsky, German Patent 1,000,801 (1957); *Chem. Abstr.* **54**, 1448 (1960); A. Benning, U. Doelling, and O. Grosskinsky, German Patent 1,102,128 (1958); *Chem. Abstr.* **56**, 421 (1962).

[73] I. S. Bengelsdorf, *J. Org. Chem.* **23**, 242 (1958).

[74] M. Masumura, T. Horie, M. Tsukayama, H. Sakuragi, and H. Shimoo, *Kogyo Kagaku Zasshi* **72**, 675 (1968).

[75] M. Masumura, T. Horie, and M. Tsukayama, *Kogyo Kagaku Zasshi* **72**, 682 (1969).

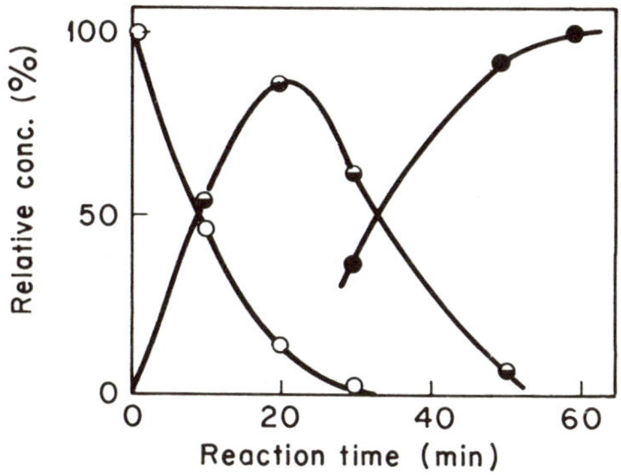

FIG. 6. Consecutive conversion of benzyl chloride with 1.34 *M* HNO_3 at 103°C. [Data of Ogata *et al.*[76] Courtesy of Chem. Soc. Japan].
—○— benzyl chloride
—◒— benzyl alcohol
—●— benzaldehyde

Similarly, 2,4-bis(chloromethyl)toluene is oxidized with 6–7% HNO_3 to 4-methylisophthalaldehyde (72%) which is further oxidized with 40–50% HNO_3 to 4-methylisophthalic acid (80%).[74] Analogous behavior is observed with 2,4-bis(chloromethyl)anisole.[75]

The kinetics of the oxidation of benzyl chloride to benzaldehyde with less than 10% HNO_3 in 40% dioxane was studied.[76] Gas chromatograph analysis of the reaction products is shown in Fig. 6. As is obvious from these results, at an early stage of the reaction, a decrease in the concentration of benzyl chloride corresponds to an increase in the concentration of benzyl alcohol. After the maximum concentration of benzyl alcohol is attained, the decrease of benzyl alcohol concentration corresponds to an increase of benzaldehyde concentration.

$$PhCH_2Cl \xrightarrow[-HCl]{H_2O} PhCH_2OH \xrightarrow[-HNO_2]{HNO_3} PhCHO \qquad (40)$$

Hence, there are two consecutive reactions with no direct conversion of benzyl chloride to benzaldehyde.

Hydrolysis of benzyl chloride is a first-order reaction with the rate being independent of acidity. Hence, the rate is faster than that of oxidation of benzyl ahcohol with less than 5% HNO_3 in 40% dioxane, but the rates of oxidation and hydrolysis are comparable at 5% HNO_3 at 90°C. At acidities higher than this, the oxidation, which is acid-catalyzed, is faster than the hydrolysis and thus the latter may be the rate-determining step in the presence of an initiator such as nitrous acid.

[76] Y. Ogata, Y. Sawaki, H. Tezuka, and F. Matsunaga, *Kogyo Kagaku Zasshi* **69**, 901 (1966).

Cyclic conjugate *vic*-dichloroolefins can be easily oxidized with NO_2 to α-diketones. For example, hexachlorocyclopentadiene gives tetrachlorocyclopentene-1,2-dione.[77]

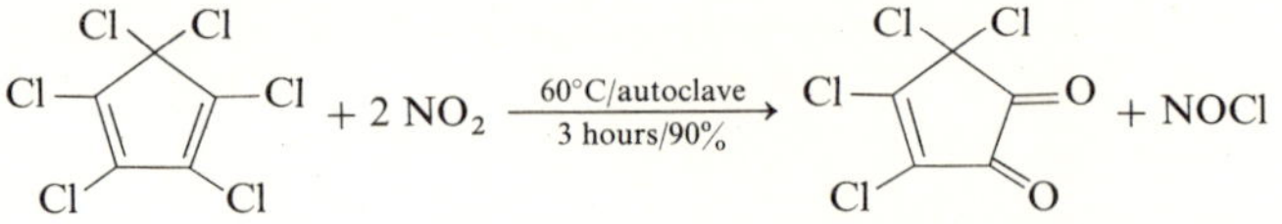

(41)

However, trichloroethylene reacts with NO_2 to give a NO_2-substituted product.

$$Cl_2C{=}CHCl + 2\,NO_2 \xrightarrow[32\%]{80°C} Cl_2C{=}C(NO_2)Cl + HNO_2 \qquad (42)$$

VI. Aldehydes

Aldehydes are oxidized by HNO_3 or N_2O_4 to the corresponding carboxylic acids. An example is the oxidation of chloral hydrate with fuming nitric acid which is an important method for the preparation of trichloroacetic acid.[78]

$$Cl_3CCH(OH)_2 + HNO_3\ (\text{sp. gr. } 1.26) \xrightarrow[57.6\%]{60°\text{–}65°C} Cl_3CCO_2H \qquad (43)$$

Oxidation of acetaldehyde with NO_2 usually gives only acetic acid by oxidation of the CHO group, but oxidation with dilute HNO_3 under appropriate conditions gives glyoxal in considerable yield along with acetic, formic, and oxalic acids.

$$CH_3CHO \nearrow OHC{-}CHO\ (\rightarrow HCO_2H + HO_2CCO_2H, \text{etc.})$$
$$CH_3CHO \searrow CH_3COOH \qquad (44)$$

The recommended procedure for the preparation of glyoxal from paraldehyde [$(CH_3CHO)_3$] involves the use of 23% HNO_3 to which formic acid has been added. The formic acid initiates the reaction and suppresses by-product formation. Furthermore, a continuous process is recommended to maintain the constant composition of the reaction system.[79] The optimum conditions are: a mixture of 97.5% paraldehyde (81 gm) and 44.4% nitric acid (178 gm) was slowly poured into a mixture of aqueous solution (2080 gm) containing 11.91% glyoxal, 7.8% formic acid, 8.11% nitric acid, and $NaNO_2$

[77] R. M. Scribner, *J. Org. Chem.* **30**, 3657 (1965).

[78] W. E. Judson, *Ber. Dtsch. Chem. Ges.* **3**, 782 (1870); Yu. V. Korshun, *Tr. Khark. Khim.-Tekhnol. Inst.* **4**, 119 (1944); *Chem. Abstr.* **42**, 6050 (1948).

[79] G. Nishikawa, T. Tanaka, and Y. Murayama, *Kogyo Kagaku Zasshi* **68**, 1873 (1965).

(0.5 gm) at 40°C for 45 minutes. Glyoxal was analyzed by acidimetry after converting it to glycolic acid by alkali.

Aromatic aldehydes are oxidized with nitric acid to acids by oxidation of the CHO group, but the reaction is much slower than the oxidation of the corresponding alcohols to aldehydes and needs higher acidities and/or temperatures. Thus, benzaldehyde in 40% dioxane can be oxidized with 4 *M* HNO_3 at 90°C to benzoic acid, while the oxidation of benzyl alcohol to benzaldehyde requires only 0.7 *M* HNO_3 under the same conditions.[6]

The reaction has an induction period which is shortened by addition of HNO_2 and prolonged by addition of urea which consumes HNO_2.[80] The reaction in a heterogeneous closed system actually occurs in the oily layer which contains more NO_2 than the aqueous layer.[81]

The oxidation rate for benzaldehydes in 40% dioxane is first-order in aldehyde ($v = k[\text{ArCHO}]$) and a plot of log k vs. H_0 results in a line with a slope of ca. -1, i.e., the rate as the following.

$$v = kh_0[\text{ArCHO}] \tag{45}$$

The rate is independent of nitric acid concentration over a wide range of concentrations, since the oxidation requires a solution saturated with nitrogen oxides.[6]

An ionic mechanism, which involves attack by NO^+, does not readily account for solvent effects. An increase in water content in aqueous dioxane should result in a decrease in the concentration of the NO^+ ion and an increase in the concentration of the $\cdot NO_2$ radical, but actually an increase in the rate is observed. A mechanism involving abstraction of the aldehyde hydrogen from the hydrated aldehyde with protonated nitrogen dioxide ($HNO_2{}^{+\cdot}$) was suggested.[6,82]

$$\text{HNO}_3 + \text{HNO}_2 \rightleftharpoons 2\,\cdot\text{NO}_2 + \text{H}_2\text{O} \tag{46a}$$

$$\text{H}^+ + \cdot\text{NO}_2 \rightleftharpoons \text{HNO}_2^{+\cdot} \tag{46b}$$

$$\text{ArCHO} + \text{H}_2\text{O} \rightleftharpoons \text{ArCH(OH)}_2 \tag{46c}$$

$$\text{ArCH(OH)}_2 + \text{HNO}_2^{+\cdot} \longrightarrow \text{Ar}\dot{\text{C}}\text{(OH)}_2 + \text{H}_2\text{NO}_2{}^+ \tag{46d}$$

$$\text{ArCHO} + \text{HNO}_2^{+\cdot} \longrightarrow \text{Ar}\dot{\text{C}}{=}\text{O} + \text{H}_2\text{NO}_2{}^+ \tag{46e}$$

$$\text{Ar}\dot{\text{C}}\text{(OH)}_2 + \cdot\text{NO}_2 \longrightarrow \text{ArC(OH)}_2\text{—ONO} \longrightarrow \text{ArCO}_2\text{H} + \text{HNO}_2 \tag{46f}$$

$$\text{Ar}\dot{\text{C}}{=}\text{O} + \cdot\text{NO}_2 \longrightarrow \text{ArC}(=\text{O})\text{—ONO} \xrightarrow{\text{H}_2\text{O}} \text{ArCO}_2\text{H} + \text{HNO}_2 \tag{46g}$$

[80] M. Tamura, T. Yoshida, and K. Namba, *Kogyo Kayaku* **30**, 32 (1969).

[81] T. Horie and M. Masumura, *Kogyo Kagaku Zasshi* **72**, 686 (1969).

[82] Y. Ogata and H. Tezuka, *Kogyo Kagaku Zasshi* **72**, 2501 (1969); for the hydration equilibrium constant of p-nitrobenzaldehyde, see J. M. Sayer, *J. Org. Chem.* **40**, 2545 (1975).

The effect of ring substituents on the rate of the oxidation of benzaldehydes is abnormal and the Hammett plot lies on crossed lines as shown in Fig. 7. The effect is explained by assuming a preliminary hydration equilibrium of aldehyde [Eq. (46c)], which is shifted to the right by electron-attracting groups which also retard hydrogen abstraction. Competition between the hydration and the hydrogen abstraction gives rise to the curved line.[6,82] Hence, when there is no hydration in the oxidation, e.g., the NO_2 oxidation in nonaqueous solvents or neat, the normal substituent effect which is observed in general oxidations is expected. In fact, oxidation of benzaldehydes with NO_2 in benzene as a solvent in a sealed tube fits the Hammett equation with a ρ value of -1.[83]

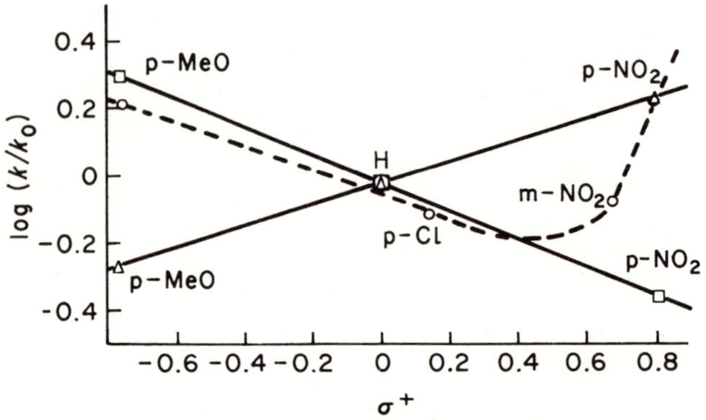

FIG. 7. Hammett plots for the relative rates of nitric acid oxidation of benzaldehydes in solvents of various dioxane contents at 90°C. Initial concn.: $[HNO_3] = 3.0\ M$, $[ArCHO] = 0.050\ M$, $[NaNO_2] = 0.01\ M$.
△ 20% dioxane
○ 40% dioxane
□ 60% dioxane
[Data from Ogata *et al.*[6] Courtesy of Pergamon Press.]

VII. Ketones

A. FORMATION OF OXIMES (REACTION OF NITROUS ACID)

Nitrous acid oxidizes ketones which possess an α-methylene group to form α-diketones and hydroxylamines.[84,85]

$$RCOCH_2R' + HNO_2 \longrightarrow RCOCOR' + NH_2OH \tag{47}$$

[83] T. Yoshida, F. Yamamoto, and K. Namba, *9th Symp. Radical React. 1968* Lecture No. 29; T. Yoshida and K. Namba, *Kogyo Kayaku* **29**, 353 (1968).
[84] R. Singer and P. A. Vamplew, *J. Chem. Soc.* p. 3052 (1957).
[85] R. Stewart, "Oxidation Mechanism," p. 111. Benjamin, New York, 1964.

The reaction involves α-nitrosation, rearrangement of the α-nitroso ketone to the oxime, and then hydrolysis of the oxime (see Section VIII, C).

The reaction of cyclohexylphenyl ketone with nitrous acid in concentrated H_2SO_4 yields benzoic acid and cyclohexanone oxime, the latter being converted to ε-caprolactam by a Beckmann rearrangement in sulfuric acid.

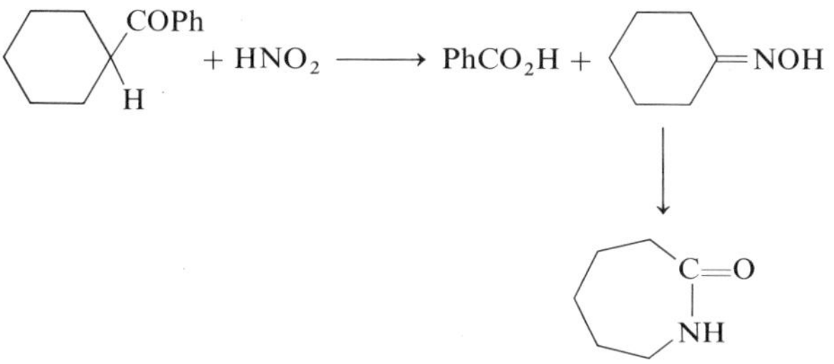

(48)

This process is analogous to the SNIA Viscosa Co. process which converts cyclohexanecarboxylic acid to ε-caprolactam with nitrosyl sulfate ($NOOSO_3H$) or a mixture nitrous and sulfuric acids. The rate of the reaction [Eq. (48)] was measured by UV spectrophotometry and was expressed as

$$v = k[\text{ketone}]$$

which is independent of [HNO_2]. The rate constant k decreases with increasing concentration of sulfuric acid, which means that the rate-determining step is deprotonation of the conjugate acid of the ketone.[86,87]

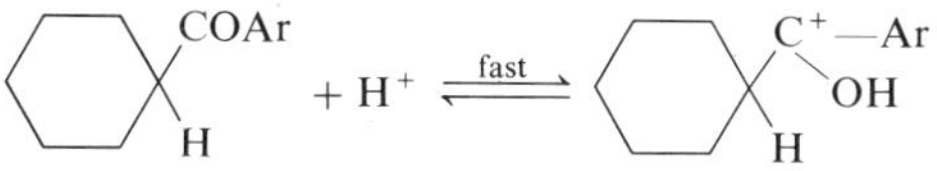

(49a)

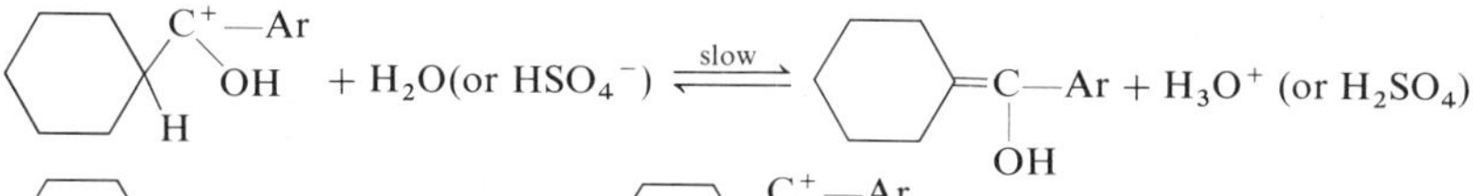

(49b)

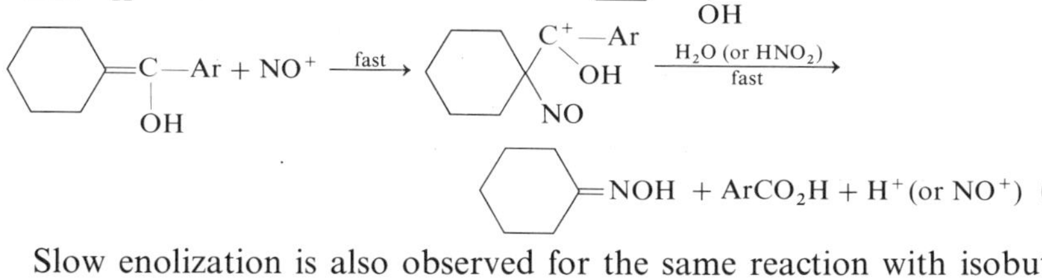

(49c)

Slow enolization is also observed for the same reaction with isobutyrophenone, but fast reversible enolization followed by a slow attack of NO^+ is observed with acetophenone and propiophenone in view of their rate expression given below.

$$v = k[\text{ketone}][NO^+] \tag{50}$$

[86] Y. Ogata, Y. Furuya, and M. Ito, *J. Am. Chem. Soc.* **85**, 3649 (1963).
[87] Y. Ogata, Y. Furuya, and M. Ito, *Bull. Chem. Soc. Jpn.* **37**, 290 (1964).

A positive ρ value of 2.6 for acetophenones may be ascribed to the favorable effect on enolization by electron-attracting groups,[87] which favor the proton transfer from protonated ketone to the base [Eq. (49b)].

B. FORMATION OF DIACYLFUROXANS

Ketones react with nitric acid to form diacylfuroxans [Eq. (51)] and C–C cleavage products such as carboxylic acids. The reaction is probably initiated by attack of NO^+. Formation of a diacylfuroxan from acetophenone was reported early in 1887.[88] This is a kind of oxidative coupling of two ketone molecules followed by cyclization with a bridging =N–O–N= [Eq. (51)]. Several examples have been reported as indicated in Eq. (51).

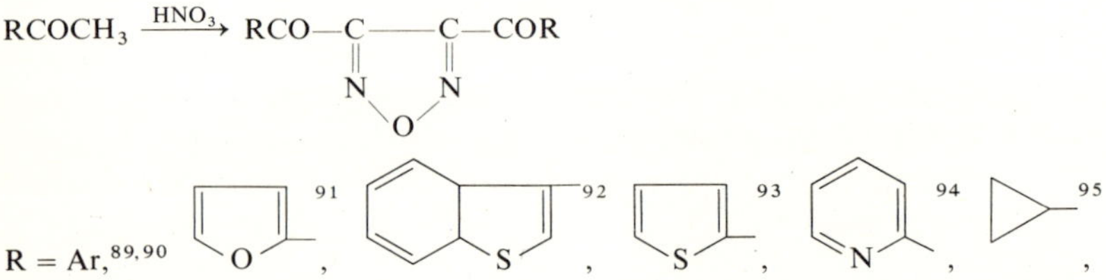

(51)

In the reaction of an acetophenone, a small amount of bis(benzoylformaldoximino)furoxan (**29** in Scheme 2) accompanies the formation of dibenzoylfuroxan (**28**).[90]

The presence of nitous acid is required for a smooth reaction. A postulate assuming N_2O_3 as an attacking species is questionable, since N_2O_3 is neither a radical nor an ion. Hence, it is more probable that the reaction is initiated by hydrogen abstraction by NO_2 or addition of NO^+ to an enol. In the HNO_3 oxidation of benzyl phenyl ketone, no evidence for hydrogen atom abstraction is observed as will be stated below,[96] and thus attack by NO^+ on the keto[84] or enol form[86,96] of the ketone is reasonable.

As to the mechanism for the formation of diacylfuroxans, Scheme 2 which goes by way of the oxime and nitrolic acid (**26**) is probable. The N_2O_4 oxidation of benzaldoxime, which is described in Section VIII,C, is analogous to this reaction and supports this mechanism.

[88] A. F. Holleman, *Ber. Dtsch. Chem. Ges.* **20**, 3360 (1887).
[89] H. R. Snyder, *J. Am. Chem. Soc.* **77**, 4233 (1955).
[90] J. H. Boyer and M. S. Chang, *Chem. Ind.* (*London*) p. 730 (1959).
[91] K. Hayes and C. O'Keefe, *J. Org. Chem.* **19**, 1897 (1954).
[92] D. A. Shirley, B. H. Gross, and M. J. Danzig, *J. Org. Chem.* **23**, 1024 (1958).
[93] M. S. Chang and A. J. Matuszko, *J. Org. Chem.* **26**, 5239 (1961).
[94] M. S. Chang, *J. Org. Chem.* **28**, 3542 (1963).
[95] M. S. Chang and J. U. Lowe, *J. Org. Chem.* **33**, 866 (1968).
[96] Y. Ogata and H. Tezuka, *Tetrahedron* **25**, 4797 (1969).

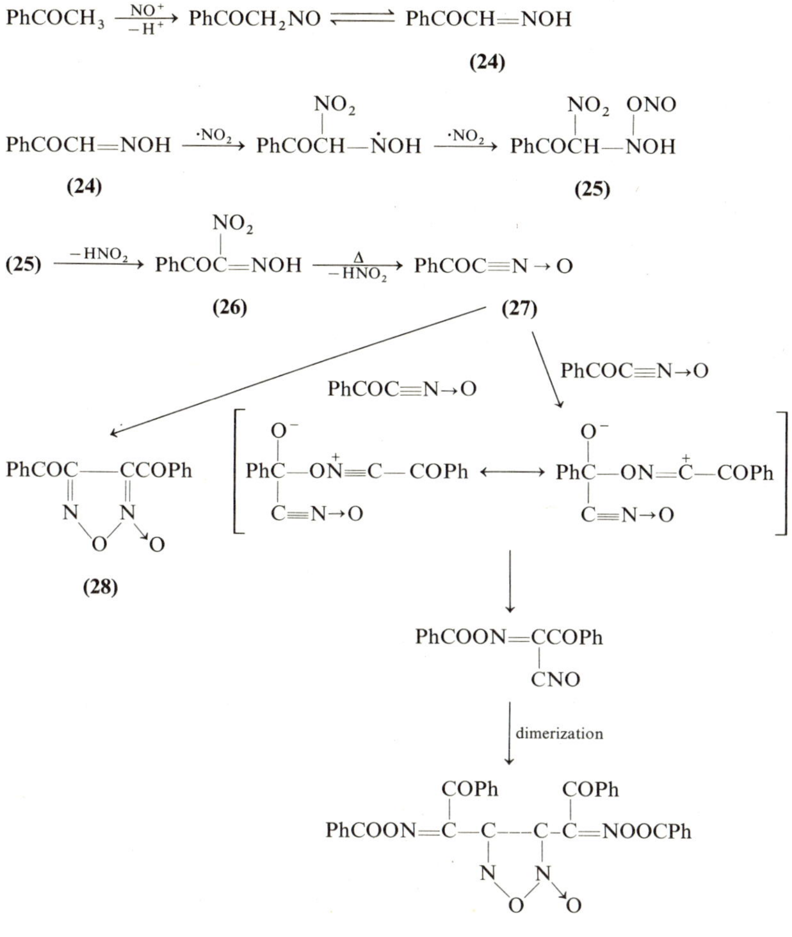

(29)

SCHEME 2

The diacylfuroxans are also produced by the reaction of N_2O_4 with ketones. Thus, acetone reacts with N_2O_4 at 0°–5°C to yield an unstable intermediate which on heating to 50°C gives diacetylfuroxan (**30**). The mechanism for the reaction may involve enolization of acetone followed by an attack of NO^+ from N_2O_4.

Higher temperatures are unfavorable for furoxan formation, since homolysis of N_2O_4 to give $\cdot NO_2$ (but not heterolysis to give NO^+) is favored by higher temperatures.[97]

[97] P. Gray and A. D. Yoffe, *Chem. Rev.* **55**, 1069 (1955).

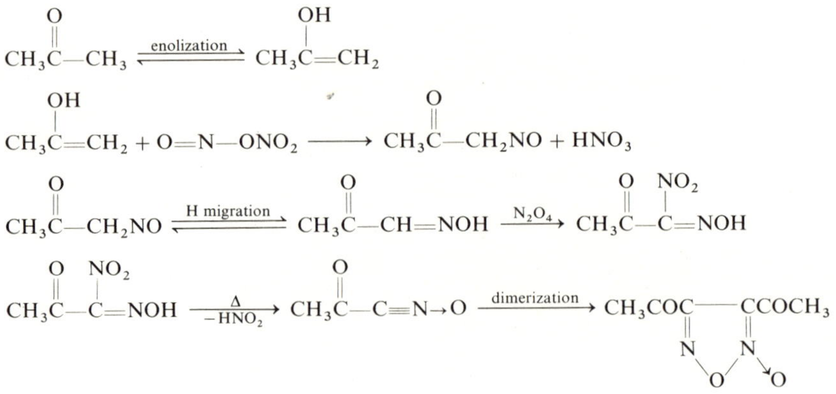

(30) (52)

C. FORMATION OF CARBOXYLIC ACIDS

Another reaction of ketones with nitric acid is cleavage of the bond between the carbonyl carbon and an α-carbon atom (α-cleavage). For example, acetophenone is oxidized by 1 *M* HNO_3 in 70% aqueous acetic acid in the presence of 0.5 *M* H_2SO_4 to benzoic acid (ca. 75%).

$$PhCOCH_3 + HNO_3 \longrightarrow PhCO_2H + CH_3CO_2H \qquad (53)$$

Similarly, *p*-(*p*-nitrophenyl)benzoic acid is prepared by the HNO_3 oxidation of *p*-(*p*-nitrophenyl)acetophenone.[98]

The carboxy group formed in oxidation of aliphatic ketones does not always come from the carbonyl group, but may come from one of the α-carbon atoms. For example, in the oxidative cleavage of 5,14-octadecadione (**31**) statistical attack and cleavage at either side of the CO groups of the diketone should give sebacic acid (**32**), azelaic acid (**33**) and suberic acid (**34**) in a molar ratio of 1:2:1. These acids are obtained but the observed ratio for oxidation by 50% HNO_3 at 55°C is 1.06:0.6:1. Hence, the yield of azelaic acid (**33**) is far less than expected, which suggests that it is more susceptible to oxidative degradation than the other two acids (**32** and **34**).[99]

$$CH_3(CH_2)_3\underset{\underset{O}{\|}}{C}(CH_2)_8\underset{\underset{O}{\|}}{C}(CH_2)_3CH_3 \longrightarrow$$

(31)

$$HOOC(CH_2)_8COOH + HOOC(CH_2)_7COOH + HOOC(CH_2)_6COOH \qquad (54)$$

(32)	(33)	(34)
25.8%	14.6%	24.2%

[98] R. C. Franklin, U. S. Patent 2,742,491 (1956); *Chem. Abstr.* **51**, 1280 (1957).
[99] C. M. Selwitz and A. C. Whitaker, *J. Org. Chem.* **22**, 116 (1957).

Benzyl phenyl ketone (deoxybenzoin **35**) is oxidized by nitric acid to benzoin (**36**), benzil (**37**), and C–C cleavage products, i.e., benzoic and *p*-nitrobenzoic acids and benzaldehyde.[96]

$$\underset{\textbf{(35)}}{PhCH_2COPh} \longrightarrow \underset{\textbf{(36)}}{PhCH(OH)COPh} + PhCOCOPh + PhCO_2H$$

$$+\ p\text{-}NO_2C_6H_4CO_2H + PhCHO \qquad (55)$$

The ratio of products changes with changing reaction conditions. For example, yields for the reaction with 1 *M* HNO_3 in 40% aqueous dioxane at 60°C are: benzoin, 25%; benzil 12% and ($ArCO_2H$ + PhCHO), 62%. An electron-attracting group Z in Z–$C_6H_4CH_2COPh$ increases the rate, suggesting rate-determining enolization. The rate constant is proportional to $[H_3O^+]$. Nitrous acid alone cannot give the fission products. These facts are accounted for by the following mechanism.[96]

$$PhCH_2COPh \underset{}{\overset{H^+}{\rightleftharpoons}} PhCH_2\text{—}\overset{+}{C}(OH)\text{—}Ph \xrightarrow[\text{slow}]{-H^+(H_2O)} PhCH{=}C(OH)Ph \qquad (56a)$$

$$PhCH{=}C(OH)Ph + NO^+ \overset{-H^+}{\rightleftharpoons} PhC(NO){=}C(OH)Ph \xrightarrow{NO_2/H_2O}$$

$$Ph\text{—}CH(OH)\text{—}C({=}O)\text{—}Ph + Ph\text{—}C({=}O)\text{—}C({=}O)\text{—}Ph + PhCHO + PhCO_2H \qquad (56b)$$

Oxidation of cyclohexanone to adipic acid is an intermediate step in the oxidation of cyclohexanol as described above. Since this reaction is of technical importance, the reaction conditions, catalysts, and mechanism have been well studied. This reaction also seems to go through the enol[100,101] (see Scheme 3.)

6-Nitro-6-hydroxyiminohexanoic acid (**37**) and the hemihydrate of cyclohexanedione (**38**) have been isolated from the reaction mixture. This oxidation needs nitrous acid as an initiator and a vanadium and/or copper catalyst for the ring opening. The vanadium catalyst may convert **38** to adipic acid with glutaric acid produced as a by-product. Succinic acid and lower decomposition products are mainly formed without the catalyst. The copper catalyst is effective notably at higher temperature, where it combines with nitrogen oxides and prevents the further breakdown of unstable intermediates.[101]

[100] H. C. Godt and J. F. Quinn, *J. Am. Chem. Soc.* **78**, 1461 (1956).
[101] W. J. van Asselt and D. W. van Krevelen, *Recl. Trav. Chim. Pays-Bas* **82**, 51 (1963).

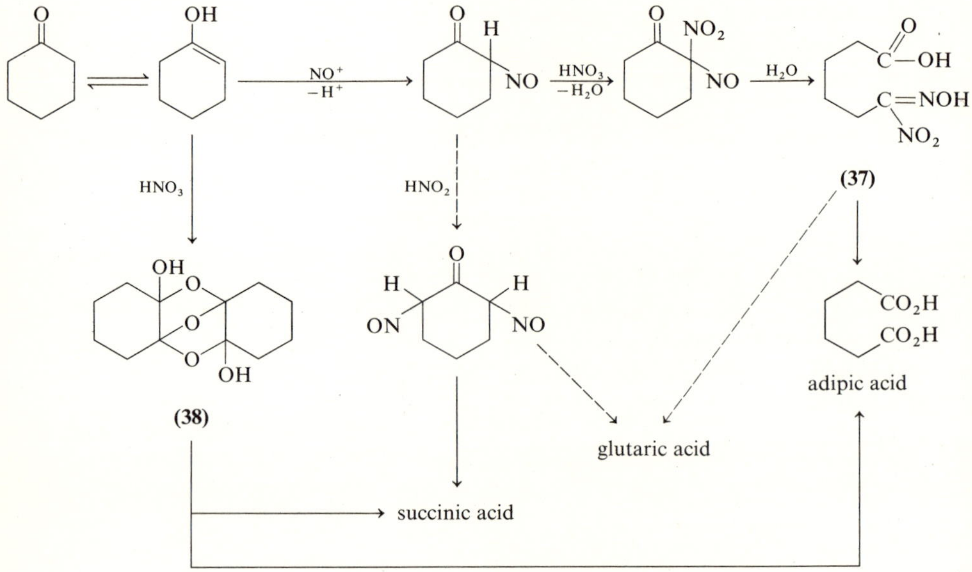

SCHEME 3

VIII. Organonitrogen Compounds

A. AMINES

It is well known that primary and secondary aromatic amines react with concentrated HNO_3 to give ortho- and para- (and in strong acidic media meta-) nitro compounds and in some cases, *N*-nitro and *N*-nitroso compounds and nitrosobenzene.[102, 103] It is also well known that primary aromatic amines are converted by HNO_2 to diazonium salts and secondary amines to *N*-nitroso compounds.[104] Tertiary aromatic amines are nitrated on the ring, and often give products cleaved at N–C, C–H, and/or C–C bonds.[105]

Oxidation of trialkylamines with 63% HNO_3 in a mixture of acetic acid and acetic anhydride at 30°–50°C gives the following results. Triethylamine gives mainly *N*,*N*-diethlynitrosoamine and *N*,*N*-diethylacetamine as well as a little *N*,*N*-diethylformamide.[106]

[102] P. B. D. De La Mare and J. H. Ridd, "Aromatic Substitution." Butterworth, London, 1959.

[103] K. Kishore and S. N. Upadhyaya, *J. Indian Chem. Soc.* **47**, 727 (1970); *Chem. Abstr.* **74**, 52841 (1971).

[104] T. E. Corrigan and K. A. Kobe, *in* "Unit Processes in Organic Syntheses" (P. H. Groggins, ed.), p. 82. McGraw-Hill, New York, 1958.

[105] J. Glazer, E. D. Hughes, C. K. Ingold, A. T. James, and E. Roberts, *J. Chem. Soc.* p. 2657 (1950).

[106] Y. Ogata, Y. Sawaki, and Y. Kuriyama, *Tetrahedron* **24**, 3425 (1968).

$$Et_3N + HNO_3 \xrightarrow[50^\circ C/85\text{ minutes}]{\text{in AcOH–Ac}_2\text{O}} \underset{23\%}{Et_2N\text{–}NO} + \underset{13\%}{Et_2N\text{–}COCH_3} + \underset{2\%}{Et_2N\text{–}CHO} \quad (57)$$

Tri-*n*-butylamine, however, gives mainly *N,N*-dibutylformamide and a little *N,N*-dibutylnitrosoamine, *N,N*-dibutylacetamide, and *N,N*-dibutylbutyramide.

$$n\text{-}Bu_3N + HNO_3 \xrightarrow[50^\circ C/1\text{ hour}]{\text{in AcOH-Ac}_2\text{O}} n\text{-}Bu_2N\text{–}CHO + n\text{-}Bu_2N\text{–}NO + n\text{-}Bu_2N\text{–}COCH_3 + n\text{-}Bu_2N\text{–}COC_3H_7 \quad (58)$$

These reactions have an induction period which can be eliminated by the addition of NO_2. The addition of HC1 or HBr accelerates the reaction and changes the main product to *N,N*-dialkylnitrosoamine, $R_2N\text{–}NO$ (see Scheme 4).

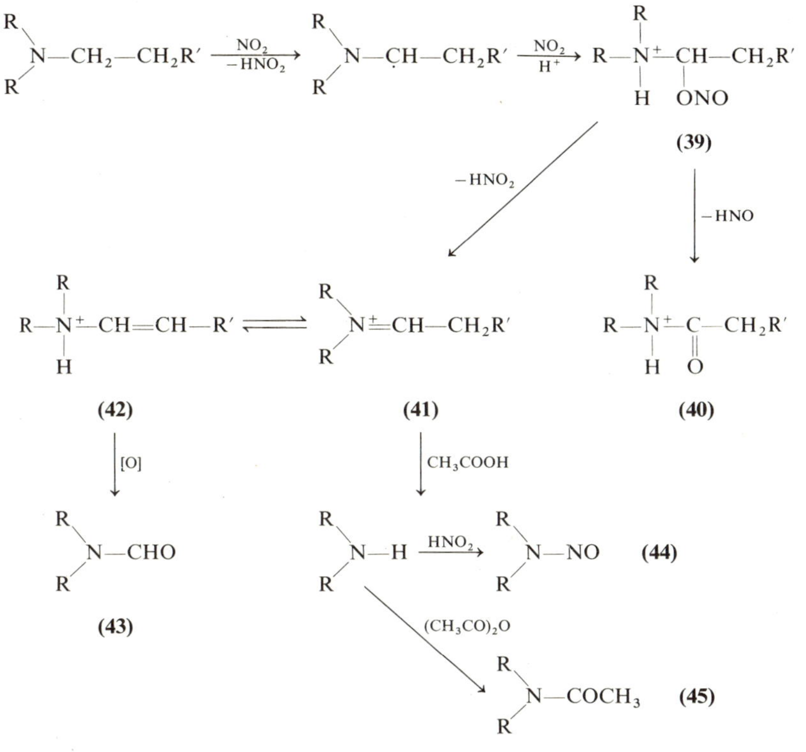

SCHEME 4

The initial reaction probably forms nitrite **39** which then loses HNO or HNO_2 to give **40** or **41**, respectively. Immonium ion **41** is converted to a mixture of nitrosoamine **44** and acetamide **45**, while **42** is oxidized to formamide **43**. Since **41** is actually more stable than **42** for triethylamine (R = Et, R′ = H), the N=C cleavage products, **44** and **45**, are favored over **43**. On the other hand, formamide **43** is favored for tri-*n*-butylamine (R = *n*-Bu, R′ = Et), since **42** is more stable than **41** in this case. As evidence for the intermediacy of **41** and **42**, oxidation of the enamine $n\text{-}Bu_2NCH{=}CHEt$ also gives the same products, **43**, **44**, and **45**, and the content of **43** is increased by acceleration of oxidation on addition of HNO_2 to the reaction system.[106] The formation of nitroso compound **44** is favored by the presence of excess hydrogen halide, probably because molecular halogen, formed by oxidation with HNO_3, can oxidize amines faster than NO_2, resulting in the carbon–nitrogen bond fission.[106] Amines which possess a bulky alkyl group such as diethylisopropylamine or diethylisobutylamine are relatively stable, and give mainly **44** and **45**.

B. ALKYLPYRIDINES, QUINOLINES, AND ACRIDINES

Alkylpyridines are oxidized by dilute (ca. 30%) HNO_3 to the corresponding pyridinecarboxylic acids (R = alkyl).

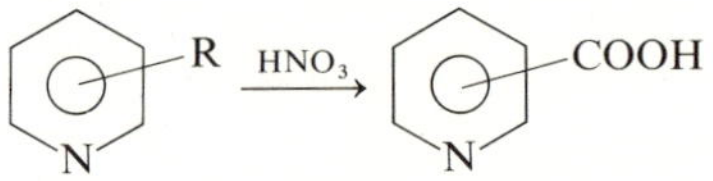

(59)

The oxidation rate is first-order with respect to the pyridine and the rate constants decrease in the order:[107,108] 2-Me-5-Et > 2-Me > 4-Me > 3-Me and 2-Me-5-Et > 2, 5-Me_2 > 3-Me. The ethyl group is oxidized more easily than the methyl group because of the presence of a secondary hydrogen atom.

Quinoline is oxidized with 28.1% HNO_3 at 231°C under 56 atm of pressure[109] or with 63% HNO_3 in H_2SO_4 at 180°–240°C under unknown pressure,[110] yielding nicotinic acid via the pyridinedicarboxylic acid, since its α-carboxyl group is easily decarboxylated.[111] A yield of 79% of nicotinic acid was reported.[110]

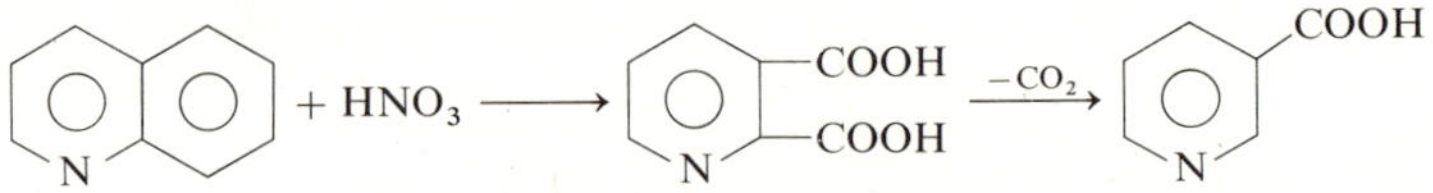

(60)

[107] B. F. Ustavshchikov *et al.*, *Dokl. Akad. Nauk SSSR* **118**, 1311 (1969).

[108] T. S. Titova, B. F. Ustavshchikov, M. I. Farberov, and E. V. Degtyarev, *Zh. Prikl. Khim.* (*Leningrad*) **42**, 910 (1969).

[109] A. Stocker, O. Marti, T. Pfammatter, and G. Schreiner, German Offen. 1,956,117 (1970); *Chem. Abstr.* **73**, 45352 (1970).

[110] J. Bialek, *Przem. Chem.* **46**, 526 (1967); *Chem. Abstr.* **68**, 114393 (1968).

[111] M. R. F. Ashworth, R. P. Daffern, and D. L. Hammick, *J. Chem. Soc.* p. 809 (1939).

Similar procedures for the preparation of nicotinic acid or its derivatives by the oxidation of dialkylpridines followed by α-decarboxylation are reported.[107,109,112]

Acridinium ion reacts with nitric acid in two different ways.[113] One is N–C fission or cleavage at the central ring to form 2-(2-carboxy-4-nitrobenzoyl)pyridine

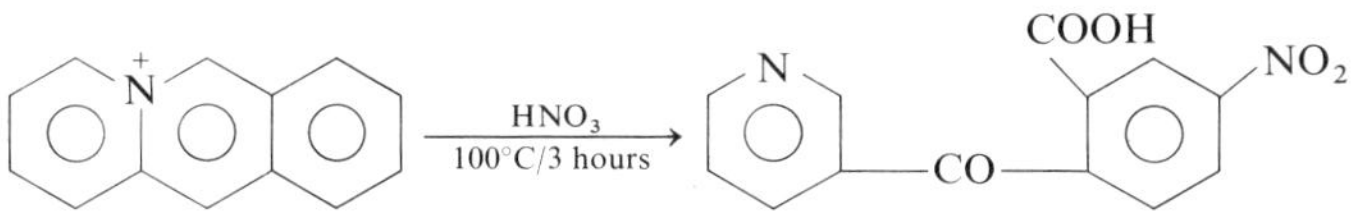

(61)

and the other is C–C fission or cleavage at the terminal ring leading to 2,3-dicarboxyquinolizinium ion [Eq. (62)].

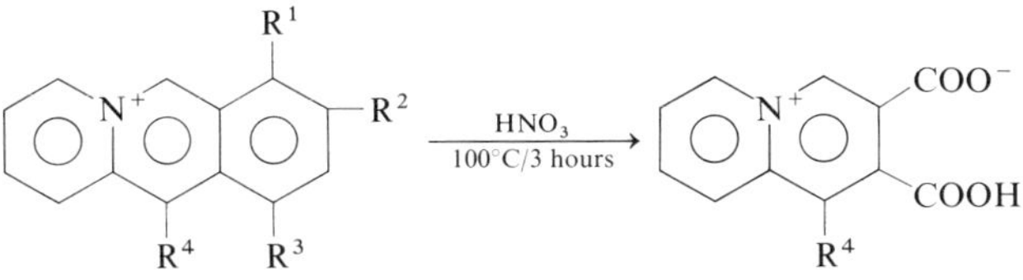

(62)

R^1, R^2, R^3, R^4 = H, OH, or MeO

The latter reaction occurs, when the terminal ring carries one or more hydroxy or methoxy groups since the ring in this case is more susceptible to oxidation.[113]

C. OXIMES

The reaction of oximes with N_2O_4 resembles the reaction of enolizable ketones. For example, benzaldoxime gives initially the *gem*-dinitro compound (**46**) and the *gem*-trinitro compound (**47**) via addition of N_2O_4 to the C=N double bond. Besides these, there are obtained other oxidation products which are indicated and accounted for in Scheme 5.[114–118]

[112] B. F. Ustavshchikov, L. L. Zalygin, T. S. Titova, E. V. Degtyaev, and M. I. Garberov, *Izv. Vyssh. Uchebn. Zaved., Khim. Khim. Tekhnol.* **11**, 896 (1968); *Chem. Abstr.* **70**, 46604 (1969).

[113] C. K. Bradsher and M. W. Barker, *J. Org. Chem.* **29**, 452 (1964).

[114] J. H. Boyer and H. Alul, *J. Am. Chem. Soc.* **81**, 4237 (1959).

[115] L. F. Fieser and W. E. Doering, *J. Am. Chem. Soc.* **68**, 2252 (1946).

[116] S. S. Novikov, L. I. Khmel'nitskii, and O. V. Lebedeva, *Zh. Obshch. Khim.* **28**, 2296 (1958); *Chem. Abstr.* **53**, 3111 (1959).

[117] L. I. Khmel'nitskii, S. S. Novikov, and O. V. Levedev, *Izv. Akad. Nauk SSSR, Otd. Khim. Nauk* p. 477 (1961); *Chem. Abstr.* **55**, 23389 (1961).

[118] P. A. S. Smith, "The Chemistry of Open Chain Organic Nitrogen Compounds," Vol. II, p. 56. Benjamin, New York, 1966.

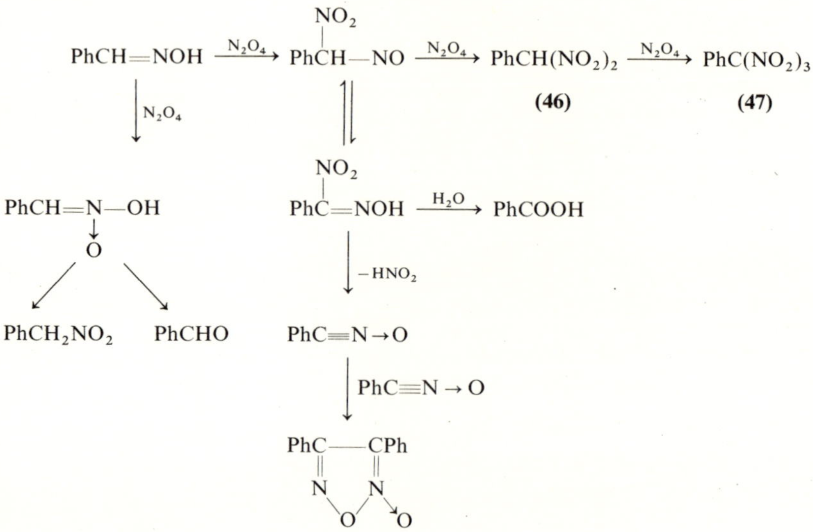

SCHEME 5

Oxidation of oximes with nitrous acid affords carbonyl compounds along with gaseous products containing N_2O, NO, and N_2.[119] It was revealed by a study using $H^{15}NO_2$ that the nitrogen atoms of N_2O and N_2 come from both the oxime and HNO_2 in equal amounts, while the nitrogen atom of NO comes from HNO_2. The gas which is evolved in the presence of mineral acid consists mostly of N_2O (77.9%) and N_2 (20%), while that in the absence of strong acid contains 67% NO and 22% N_2O. These facts suggest the following mechanism for the HNO_2 oxidation of oximes in strong acidic media.

$$HNO_2 + H^+ \rightleftharpoons H_2NO_2^+ \rightleftharpoons H_2O + NO^+ \quad (63)$$

$NO^+ + R{-}CH{=}NOH \longrightarrow$

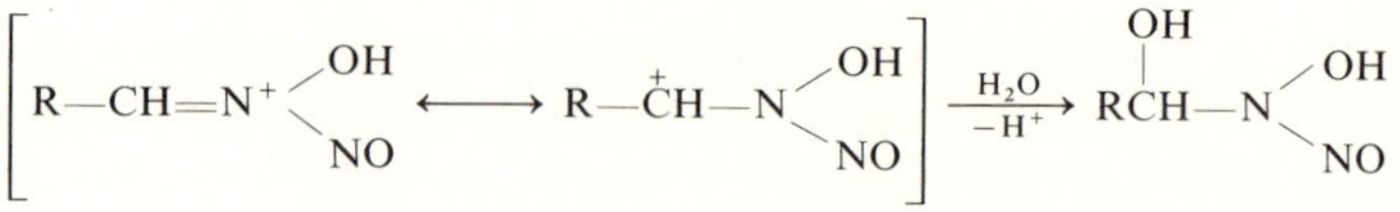

(64)

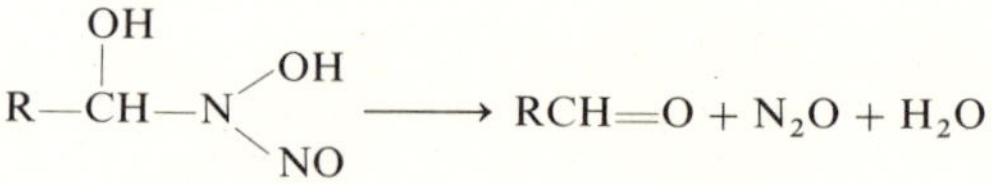

(65)

For N_2 formation the following steps are proposed.

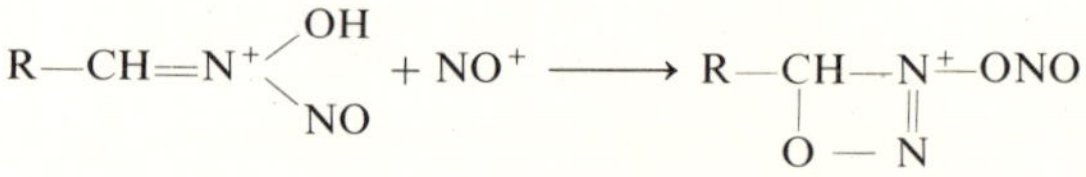

(66a)

[119] J. M. Kliegman and R. K. Barnes, *J. Org. Chem.* **37**, 4223 (1972).

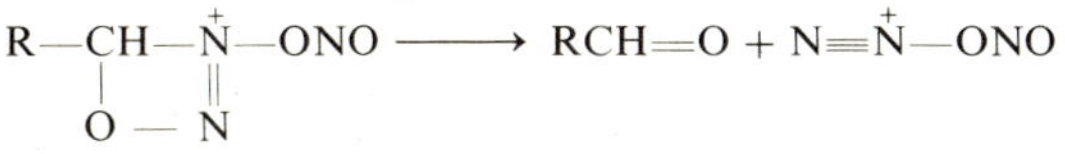

(66b)

$$N{\equiv}\overset{+}{N}{-}ONO \longrightarrow N_2 + NO_2^{+} \qquad (66c)$$

$$\overset{-}{N}{=}\overset{+}{N}{=}\overset{+}{O}NO \longrightarrow N_2O + NO^{+} \qquad (66d)$$

In the absence of mineral acid the attacking species may be HNO_2 instead of NO^+.

$$RCH{=}NOH + HNO_2 \longrightarrow RCH{=}NO\cdot + NO + H_2O \qquad (67a)$$

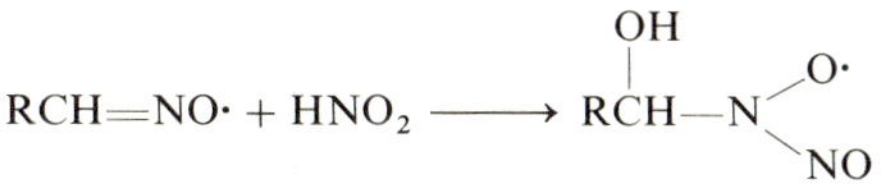

(67b)

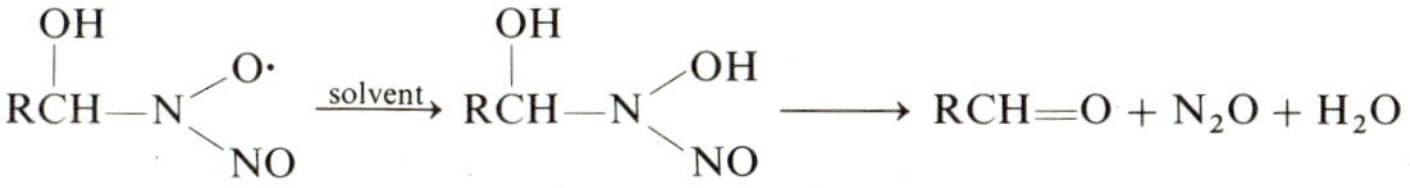

(67c)

D. IMINES

N-Benzylideneanilines react with a nitrosonium salt in anhydrous media to produce benzaldehydes and benzenediazonium ions.

$$ArCH{=}NAr' + NO^{+}BF_4^{-} \xrightarrow[90\%]{} ArCHO + Ar'N_2^{+} + BF_4^{-} \qquad (68)$$

Since aliphatic diazonium ions decompose easily to give molecular nitrogen and carbonium ions, the similar reaction with *N*-benzylidenealkylamines gives rise to either a carbonium ion, if it is fairly stable (e.g., Ph_3C^+), or a dialkyloxonium ion [Eq. (70b)] by reaction with a product that contains a carbonyl group.[120] For example, *N*-benzylidenetriphenylamine gives benzaldehyde (over 95%), gaseous nitrogen, and triphenylmethyl cation.

$$PhCH{=}NCPh_3 + NO^{+}BF_4^{-} \xrightarrow[25°C]{in\ CH_3CN} PhCHO + N_2 + Ph_3C^{+}BF_4^{-} \qquad (69)$$

However, the reaction of *N*-benzylidenebenzylamine gives benzaldehyde (28%) and N-protonated imine ($PhCH{=}\overset{+}{N}HCH_2Ph$) (54%) along with gaseous products (54%). When the reaction solution is quenched by reduction with a silane, the products contain dibenzyl ether, which is a product produced from an oxonium ion [Eq. (70b)].

$$PhCH{=}NCH_2Ph + NO^{+} \xrightarrow[25\%]{in\ CH_3CN} PhCHO + N_2 + PhCH_2^{+} \qquad (70a)$$

$$PhCH_2^{+} + PhCH{=}O \longrightarrow Ph\overset{+}{C}H{-}O{-}CH_2Ph \xrightarrow{Et_3SiH} (PhCH_2)_2O \qquad (70b)$$

This scheme is supported by the fact that $PhCH{=}NCD_2Ph$ yields dibenzyl ether containing two deuterium atoms, $PhCH_2OCD_2Ph$. The mechanism for the cleavage of the C=N bond may involve an attack of NO^+.

[120] M. P. Doyle, M. A. Zaleta, J. E. DeBoer, and W. Wierengs, *J. Org. Chem.* **38**, 1663 (1973).

The reaction of *N*-benzylidenebenzhydrylamine, gives the following products.

$$\mathrm{PhCH{=}NCHPh_2 + NO^+ \xrightarrow[25^\circ C]{\text{in } CH_3CN} \underset{33\%}{PhCHO} + \underset{\mathbf{(48)}\ 15\%}{Ph_2CO} + \underset{\mathbf{(49)}\ 57\text{–}67\%}{PhCH{=}\overset{+}{N}HCHPh_2} + N_2 + NO + N_2O} \quad (71)$$

Here, the formation of benzophenone (**48**) suggests a hydrogen-transfer mechanism [Eq. (72)]. The proton source for formation of the protonated imine (**49**) may be HNO formed from the electrophilic attack of NO^+ on an organic species.

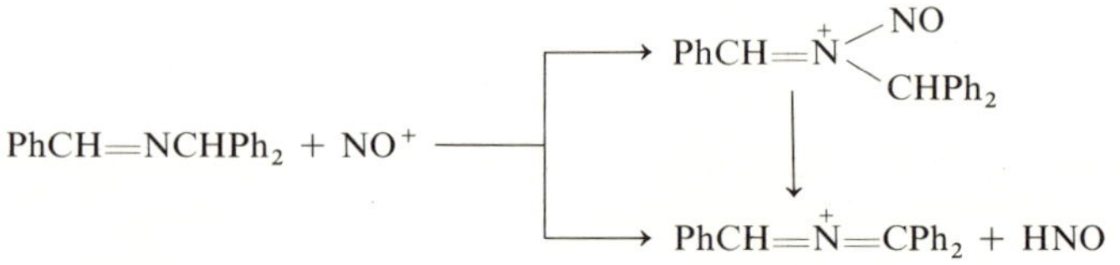

(72)

E. NITROSO COMPOUNDS

Aromatic compounds having a strong electron-releasing group are nitrosated by nitrous acid and the nitroso compounds which are produced can be readily oxidized with nitric acid to nitro compounds. In some cases, aromatic nitration may go by way of preliminary nitrosation. Acceleration of nitration by addition of nitrous acid is therefore observed with phenol and dimethylaniline, but not with benzene which appears to be nitrated directly.[121] Mercury-catalyzed nitration also seems to go via nitrosation.[122]

Oxidation of nitrosobenzene requires an initiator such as nitrous acid and proceeds autocatalytically, since the concentration of NO_2 increases as the reaction proceeds. Ordinary radical sources such as benzoyl peroxide and azobisisobutyronitrile, which can produce an appreciable amount of nitrous acid from nitric acid, are also effective initiators.[123]

An increase of dioxane content in aqueous dioxane solvent results in an increase in the oxidation rate and the yield of nitrosobenzene. For example, the yield of 37% in 20% dioxane rises to 78% in 60% dioxane.[123] This fact suggests that an increase of dioxane content favors an attack by a radical species such as $\cdot NO_2$ rather than an ionized species such as NO^+ and

[121] C. A. Burton, E. D. Hughes, C. K. Ingold, D. I. H. Jacobs, M. H. Jones, G. J. Minkoff, and R. I. Reed, *J. Chem. Soc.* p. 2628 (1950).

[122] F. H. Westheimer, E. Segel, and R. Schramm, *J. Am. Chem. Soc.* **69**, 773 (1947); R. M. Schramm and F. H. Westheimer, *J. Am. Chem. Soc.* **70**, 1782 (1948).

[123] Y. Ogata and H. Tezuka, *J. Am. Chem. Soc.* **89**, 5428 (1967). The rate was measured by S. Veibel, *Z. Phys. Chem., Abt B* **10**, 22 (1930).

suppresses acid-catalyzed side reactions.[124] The rate of oxidation with dilute ($\leq$ 1 *M*) HNO_3 containing a little HNO_2 in aqueous dioxane depends on the nitrate ion and nitrous acid concentrations.[123]

$$v = k[NO_3^-]^{0.5}[HNO_2]^{0.5}[PhNO] \tag{73}$$

An increase in acidity with H_2SO_4 raises the rate constant k, which fits the following equation, where h_0 is the acidity function.

$$k = k'k_0^{0.5}(1 + 0.5h_0) \tag{74}$$

The reaction is facilitated by electron-releasing groups (Table II).

TABLE II

SUBSTITUENT EFFECT ON THE NITRIC ACID OXIDATION OF SUBSTITUTED NITROSOBENZENES[a]

Substituent	$10^5\ k_0$ (sec^{-1})	Substituent	$10^5\ k_0$ (sec^{-1})
p-Me	9.5	*p*-Cl	4.8
m-Me	11.7	*m*-Cl	5.1
H	8.8	*m*-NO_2	4.2

[a] In 70% dioxane at 70°C, initial concentrations: [ArNO] = 0.048 *M*, [HNO_2] = 0.40 *M*, [$NaNO_2$] = 0.005 *M*. (Data of Ogata and Tezuka.[123] Courtesy of *J. Am. Chem. Soc.*)

These facts are accounted for by the following mechanism which involves attack of both $\cdot NO_2$ and $HNO_2^{\ddagger}$ on the double bond of the nitroso group.

$$ArN{=}O + \cdot NO_2 \longrightarrow Ar\dot{N}{-}O{-}NO_2 \tag{75a}$$

$$ArN{=}O + HNO_2^{\ddagger} \longrightarrow Ar\dot{N}{-}O{-}NO_2H^+ \tag{75b}$$

$$Ar\dot{N}{-}O{-}NO_2 + \cdot NO_2 \xrightarrow{\text{slow}} ArN(ONO){-}ONO_2 \tag{75c}$$

$$Ar\dot{N}{-}O{-}NO_2H^+ + \cdot NO_2 \xrightarrow{\text{slow}} Ar{-}N(ONO){-}ONO_2H^+ \tag{75d}$$

$$ArN(ONO){-}ONO_2 \longrightarrow Ar\overset{+}{N}(O^-){=}O + N_2O_3 \tag{75e}$$

$$Ar{-}N(ONO){-}ONO_2H^+ \longrightarrow Ar{-}\overset{+}{N}(O^-){=}O + N_2O_3 + H^+ \tag{75f}$$

$$N_2O_3 + H_2O \rightleftharpoons 2\ HNO_2 \tag{75g}$$

[124] E. Bamberger, H. Büsdorf and H. Sand, *Ber. Dtsch. Chem. Ges.* **31**, 1513 (1898).

An attack of NO_2 on the nitrogen atom of a nitroso group gives another radical

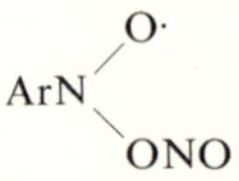

but the observed substituent effect suggests electrophilic behavior of the attacking species. Hence, attack on the oxygen atom is more probable.

The rate equation derived from the above mechanism with rate-determining steps of Eqs. (75c and 75d) is consistent with the following observation

$$v = k'h_0^{0.5}(1 + 0.5h_0)[NO_3^-]^{0.5}[HNO_2]^{0.5}[PhNO]$$

where the first term in the parentheses corresponds to the attack of $\cdot NO_2$ and the second term to the attack of HNO_2^+. Since 0.5 h_0 is much smaller than 1, HNO_2^+ contributes very little to this oxidation probably because nitrosobenzene, which is highly reactive toward radicals,[125] can readily react with $\cdot NO_2$.

p-Nitrosoanisole is demethylated by nitric acid to eventually form *p*-nitrophenol,[126] since *p*-nitrosoanisole is readily subject to acid-catalyzed hydrolysis,[127] but no hydrolysis occurs with *p*-nitroanisole.

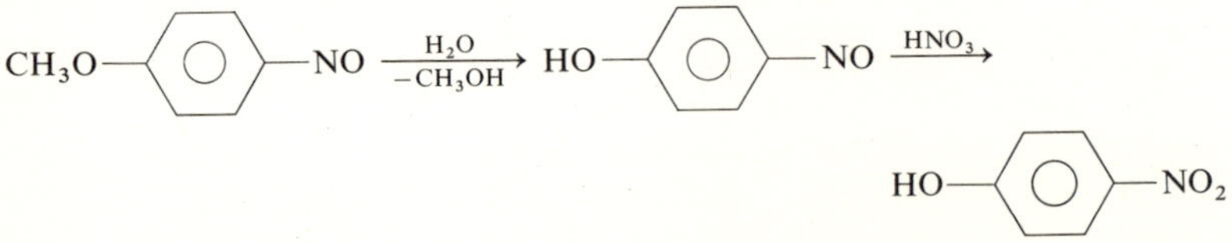

(76)

The preparation of *p*-nitrophenol from phenol via *p*-nitrosophenol is a useful procedure since the para isomer is not contaminated with the ortho isomer which is always formed in direct nitration. The nitric acid oxidation of *p*-nitrosophenol is a fast reaction and occurs easily even in 20% aqueous dioxane at 40°C, which is different from other nitrosoarenes. The rate of nitration is much higher than that expected from the Hammett equation. The rate equation in 20% dioxane measured by UV spectrophotometry is given as Eq. (77).[123]

$$v = k[HNO_3][HOC_6H_4NO] \quad (77)$$

[125] P. A. S. Smith, "The Chemistry of Open Chain Nitrogen Compounds," Vol. II, p. 373. Benjamin, New York, 1966.

[126] Y. Ogata and H. Tezuka, *J. Org. Chem.* **33**, 3179 (1968).

[127] J. T. Hays, E. H. Butts, and H. L. Young, *J. Org. Chem.* **32**, 153 (1967); Y. Furuya, D. Morikawa, and K. Itoh, *Bull. Chem. Soc. Jpn.* **40**, 1425 (1967).

The rate is independent of [HNO_2] and this behavior also differs from that of the other nitrosoarenes.

The suggested mechanism is given in Eqs. (78a-d). The greater stability of radical **50**, which is due to resonance between $\dot{N}$ and the OH group, is a reason for its different behavior from the other nitrosoarenes. The rate is determined by the second step [Eq. (78b)] for the other nitrosoarenes, while the rate-determining step for nitrosophenol is probably the last step [Eq. (78d)].[123]

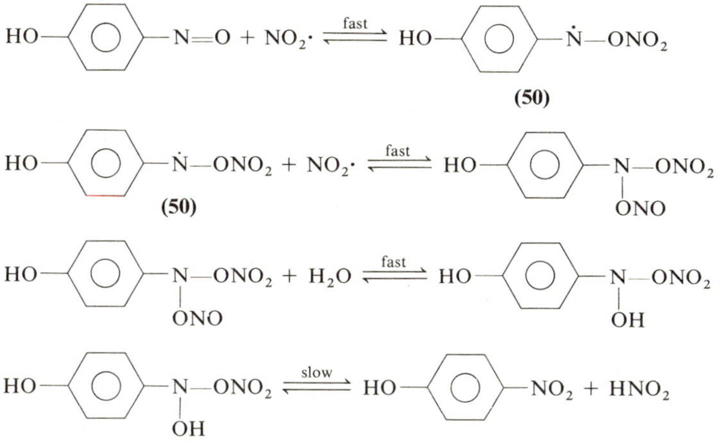

(78a) (78b) (78c) (78d)

The observed rate law can be explained by assuming that the last step is rate-determining. It is a γ-elimination instead of the more common β-elimination, but in this case β-elimination produces nitrosophenol, the starting material. This behavior may be due to the presence of a lone pair of electrons on nitrogen. A similar γ-elimination is reported for the conversion of nitrolic acid to nirile oxide.

$$PhC(NO_2){=}NOH \xrightarrow{\Delta} PhC{\equiv}N \rightarrow O + HNO_2 \qquad (79)$$

On the other hand, the acid-catalyzed oxidation of *p*-nitrosophenol by HNO_3–HNO_2 in aqueous sulfuric acid in the acidity region of $-1 > H_0 > -6$ has the following rate expression.

$$v = k[HOC_6H_4NO][NO_3^-][HNO_2]$$

The kinetic data suggest a mechanism involving an attack of N_2O_4.[128]

$$HNO_3 + HNO_2 \rightleftharpoons N_2O_4 + H_2O \qquad (80a)$$

$$N_2O_4 + HOC_6H_4NO \longrightarrow HOC_6H_4NO_2 + N_2O_3 \qquad (80b)$$

$$N_2O_3 + H_2O \rightleftharpoons 2\,HNO_2 \qquad (80c)$$

[128] A. Granzow and A. Wilson, *J. Org. Chem.* **37**, 1189 (1972).

At higher acidity, however, where most *p*-nitrosophenol is protonated, the protonated form may reacts with N_2O_4.[128]

A kinetic study[129] of the N_2O_4 oxidation of 2,5-dimethylnitrosobenzene in CCl_4 shows that the rate equation is expressed by the following equation.

$$v = k[\text{substrate}][N_2O_4] \qquad (81)$$

The reaction is not acid-catalyzed (and is even retarded by $SbCl_5$) and has a large negative entropy of activation ($\Delta S^{\ddagger} = -32.4$ eu) which implies that the number of degrees of freedom in the transition state are small. The use of $^{15}N_2O_4$ shows no incorporation of ^{15}N into the nitro group of the product. These facts suggest a cyclic transition state formed via a radical attack of $\cdot NO_2$.[129]

(82a)

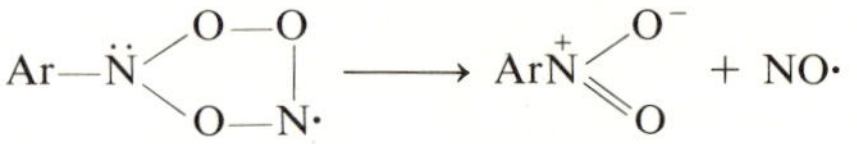

(82b)

$$NO\cdot + \cdot NO_2 \rightleftharpoons N_2O_3 \qquad (82c)$$

Ar = 2,5-$(CH_3)_2C_6H_3$

IX. Organosulfur Compounds

A. SULFIDES

In general, the oxidation of dialkyl,[130–134] diaryl,[135] and alicyclic sulfides[136] (e.g., $C_5H_{10}S$) with nitric acid or N_2O_4 gives the corresponding sulfoxides. For example, oxidation of neat dimethyl sulfide with N_2O_4 at 0°C gives ca. 90% yield of dimethyl sulfoxide.[133]

$$R{-}S{-}R + HNO_3 \text{ (or } N_2O_4) \longrightarrow R{-}\underset{\downarrow \atop O}{S}{-}R \qquad (83)$$

R = Me, Et, *n*-Pr, *i*-Pr

[129] T. G. Bonner and R. A. Hancock, *J. Chem. Soc. B* p. 519 (1970).
[130] T. P. Hilditch, *J. Chem. Soc.* p. 1618 (1908).
[131] A. Saytzeff, *Justus Liebigs Ann. Chem.* **139**, 354 (1866).
[132] L. Horner and F. Hübenett, *Justus Liebigs Ann. Chem.* **579**, 193 (1953).
[133] C. C. Addison and J. C. Sheldon, *J. Chem. Soc.* p. 2705 (1956).
[134] G. Laurence, *C. R. Hebd. Seances Acad. Sci., Ser. C* **269**, 352 (1969); *Chem. Abstr.* **71**, 90731 (1969).
[135] S. Oae, N. Kunieda, and W. Takagi, *Chem. Ind. (London)* p. 1790 (1965).
[136] C. R. Johnson and D. McCants, Jr., *J. Am. Chem. Soc.* **86**, 2935 (1964).

Oxidation of a diaryl sulfide to the sulfoxide often accompanies the nitration of one of the aryl rings of the sulfide.[137,138] *p*-Acetamino- and *p*-dimethylaminodiphenyl sulfides give *S*-phenyl fission-products during the HNO_3 oxidation in nitroethane.[137]

The oxidation of sulfides generally stops at the sulfoxide stage and does not produce sulfones because sulfoxides form oxonium salts with nitric acid, e.g., $[ArS(\overset{+}{O}H)Me][NO_3^-]$, which resist further oxidation.[139] Dialkyl sulfides can, however, be converted to the corresponding sulfones under more vigorous conditions. Thus, dialkyl sulfides are oxidized to dialkyl sulfones with concentrated nitric acid at 120°–150°C for 43–109 minutes.[140]

$$3\,R{-}S{-}R + 8\,HNO_3 \longrightarrow 3\,R{-}SO_2{-}R + 6\,NO_2 + 2\,NO + 4\,H_2O \qquad (84)$$

The yields change with R = CH_3, 88%; Et, 97%; *n*-Pr, 69%; *n*-Bu, 69%; *n*-octyl, 83%. For dimethyl sulfoxide the yield was 86%.

In order to elucidate the kinetics and mechanism for the oxidation of diaryl sulfides to sulfoxides, *p,p'*-dichlorodiphenyl sulfide was chosen, since ring nitration does not occur.[140] Nitric acid oxidation in 90% aqueous acetic acid at 40°C gives a quantitative yield of only *p,p'*-dichlorodiphenyl sulfoxide. Surprisingly, oxidation under air induces autoxidation of the sulfide, while oxidation under a nitrogen atmosphere with complete elimination of oxygen (using the freeze-thaw technique) corresponds to the stoichiometry indicated in Eq. (85) (See Table III).[138]

$$Ar{-}S{-}Ar + HNO_3 \longrightarrow Ar{-}\underset{\substack{\downarrow \\ O}}{S}{-}Ar + HNO_2 \qquad (85)$$

Measurement of the rate of this oxidation under N_2 in 90% acetic acid by UV spectrophotometry is summarized as follows: (i) The rate is expressed below where h_0 is acidity function ($-\log h_0 = H_0$).

$$v = kh_0^2[Ar_2S][HNO_3][HNO_2] \qquad (86)$$

(ii) Nitrous acid is an excellent initiator and the reaction is stopped by addition of urea. (iii) The rate of the HNO_3 oxidation is ca. 1400-fold faster than the HNO_2 oxidation. (iv) The reaction is retarded by an electron-attracting group on the benzene ring. (v) The N_2O_4 oxidation is accelerated by increasing the solvent polarity and retarded by solvation of N_2O_4; thus $CHCl_3 > CCl_4 \approx Ac_2O > AcOH \gg$ dioxane.

[137] A. Compagnini, M. Santagali, N. Marziano, and R. Passerini, *Ann. Chim. (Rome)* **60**, 527 (1970); *Chem. Abstr.* **74**, 53202 (1971).

[138] Y. Ogata and T. Kamei, *Tetrahedron* **26**, 5667 (1970).

[139] F. G. Bordwell and P. J. Boultan, *J. Am. Chem. Soc.* **79**, 717 (1957).

[140] D. W. Goheen and C. F. Bunnett, *J. Org. Chem.* **26**, 1331 (1961).

TABLE III

STOICHIOMETRY OF NITRIC ACID OXIDATION OF *p,p'*-DICHLORODIPHENYL SULFIDE TO THE SULFOXIDE[a]

[(*p*-Cl-C_6H_4-)$_2$S]	[HNO_3] (*M*)	[(*p*-Cl-C_6H_4-)$_2$SO] (*M*)	[(*p*-Cl-C_6H_4-)$_2$SO]/[HNO_3] (*M*)
Under air			
0.06	0.04	0.06	1.5
0.08	0.04	0.08	2.0
0.06	0.02	0.06	3.0
Under N_2			
0.03	0.02	0.020	1.0
0.05	0.02	0.023	1.1

[a] In 90% (vol) Acetic Acid at 40°C; [H_2SO_4] = 0.56 *M*. (Data of Ogata and Kamei.[138] Courtesy of Pagamon Press).

These findings suggest the following mechanism.[138]

$$H^+ + NO_3^- \underset{}{\overset{K_a}{\rightleftharpoons}} HNO_3 \qquad (87a)$$

$$HNO_3 + HNO_2 \overset{K_b}{\rightleftharpoons} N_2O_4 + H_2O \qquad (87b)$$

$$N_2O_4 + H^+ \overset{K_c}{\rightleftharpoons} N_2O_4H^+ \qquad (87c)$$

$$ArSAr + N_2O_4H^+ \xrightarrow[\text{slow}]{k_d} ArSOAr + N_2O_3 + H^+ \qquad (87d)$$

$$N_2O_3 + H_2O \rightleftharpoons 2\,HNO_2 \qquad (87e)$$

This mechanism leads to the rate equation consistent with the observed one, where *k* and *K* are rate and equilibrium constants of the subscripted steps, respectively.[138]

$$v = k_d K_a K_b K_c [H^+]^2 [NO_3^-][HNO_2][Ar_2S]/[H_2O] \qquad (88)$$

The oxidation is much faster in chloroform than in acetic acid as stated above. This phenomenon may be explained by the facile solvation of N_2O_4 in acetic acid which suppresses the homolysis of N_2O_4. A kinetic study of the oxidation of diphenyl sulfide is also reported in nitromethane or nitroethane as a solvent.[141]

B. MISCELLANEOUS SULFUR COMPOUNDS

Dialkyl and diaryl disulfides react with N_2O_4 in CCl_4 to afford good yields of sulfonic acid anhydrides.[142]

[141] N. Marziano *et al.*, *Ann. Chim. (Rome)* **59**, 565 and 573 (1969); *Chem. Abstr.* **71**, 90512 and 90513 (1969); *Boll. Sedute Accad. Gioenia Sci. Nat. Catania* [4] **9**, 579 (1969); *Chem. Abstr.* **72**, 66169 (1970).

[142] N. Kunieda and S. Oae, *Bull. Chem. Soc. Jpn.* **41**, 233 (1968).

$$R{-}S{-}S{-}R + N_2O_4 \xrightarrow[\substack{0^\circ C\ (2\ hours)\\ 30^\circ{-}50^\circ C\ (2\ hours)}]{in\ CCl_4} R{-}SO_2{-}O{-}SO_2{-}R \quad (89)$$

For various R groups the yields are Me, 85%; *n*-Pr, 74%; Ph, 81%; *p*-MeC_6H_4, 92%; *p*-ClC_6H_4, 90%.

Under the same reaction conditions, thiosulfinates also produce sulfonic acid anhydrides.[142]

$$R{-}SO{-}S{-}R + N_2O_4 \xrightarrow[0^\circ{-}50^\circ C/4\ hours]{in\ CCl_4} R{-}SO_2{-}O{-}SO_2{-}R \quad (90)$$

(R = *p*-$CH_3C_6H_4$; 94% yield)

However, no thiosulfinate is isolated during the oxidation of disulfide.

Sodium thiocyanate is oxidized by nitric acid containing nitrous acid to hypothiocyanous acid which is converted to thiocyanogen, hydrogen cyanide, and sulfur dicyanide.[143,144] At high concentrations of SCN^- the rate is expressed as

$$v = k[NO_3^-][HNO_2][H^+] \quad (91a)$$

while at low concentrations of SCN^- as the following.

$$v = k'[NO_3^-][HNO_2][SCN^-][H^+]^2 \quad (91b)$$

These facts suggest the mechanism below.

$$NO_3^- + HNO_2 + H^+ \rightleftharpoons N_2O_4 + H_2O \quad (92a)$$

$$N_2O_4 + HNCS \longrightarrow HOSCN \rightleftharpoons (SCN)_2 + H_2O \quad (92b)$$

$$2\ HOSCN \longrightarrow HOOSCN + H^+ + SCN^- \quad (92c)$$

$$HOSCN + HO_2SCN \longrightarrow HO_3SCN + H^+ + SCN^- \quad (92d)$$

$$H_2O + HO_3SCN \longrightarrow H_2SO_4 + HCN \quad (92e)$$

$$CN^- + (SCN)_2 \longrightarrow S(CN)_2 + SCN^- \quad (92f)$$

$$H^+ + HNO_2 + SCN^- \rightleftharpoons NOSCN + H_2O \quad (92g)$$

The rate equation at high [SCN^-] [Eq. (91a)] which does not involve an [SCN^-] term corresponds to a mechanism with the rate-determining step [Eq. (92a)], while rate equation (91b) corresponds to the rate-determining step indicated in Eq. (92b).

Thiophenes are nitrated by fuming HNO_3 to yield 2- and then 2,4- and 2,5-dinitrothiophenes but no sulfoxide.[145] Tetrachlorothiophene is oxidized

[143] G. Stedman and P. A. E. Whincup, *J. Chem. Soc. A* p. 1145 (1969).

[144] M. N. Hughes, E. D. Phillips, G. Stedman, and P. A. E. Whincup, *J. Chem. Soc. A* p. 1148 (1969).

[145] F. F. Blicke, *Heterocycl. Compd.* **1**, 227 (1950).

by concentrated HNO_3 at 80°–100°C with elimination of chloride ion to form 3,4-dichlorothiacyclopentene-2,5-diene.[146]

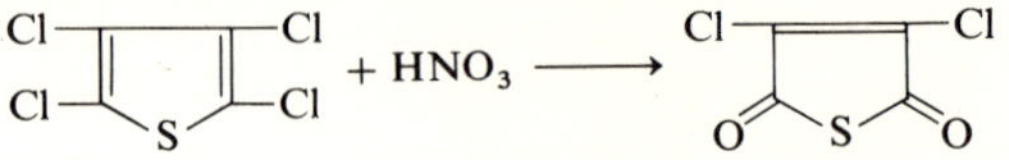

(93)

ACKNOWLEDGMENTS

The author wishes to thank Dr. H. Tezuka for his assistance in finding literature references for this chapter and Mr. M. Mineno, Mr. T. Ikejiri, and Mr. M. Inaishi for typing the manuscript.

[146] O. Scherer and F. Kluge, German Patent 1,290,551 (1969); *Chem. Abstr.* **71**, 3262 (1969).

CHAPTER **V**

Alkaline Hypohalite Oxidations

SUJIT K. CHAKRABARTTY

I. Introduction

Oxidation of methyl ketones by means of alkaline hypohalite solution was first reported by Lieben in 1822. Since then, the process has been known as the haloform reaction, and the subject was reviewed in 1939 by Fuson and Bull.[1] Most of the recent work with hypohalite involves direct oxidation

[1] R. C. Fuson and B. A. Bull, *Chem. Rev.* **15**, 275 (1934).

ISBN 0-12-697252-4

of the organic compounds by alkaline hypohalite for preparative purposes. Some of these reactions were documented by Fieser and Fieser[2] in 1967 and 1969.

As an aid to understanding the scope of hypohalite oxidation, a survey of the properties of the oxidant will be given. The subject matter will be organized according to functional elements in organic structure which undergo or take part in oxidation by hypohalite. Whenever possible the reaction mechanism will be explored and considerable emphasis will be given to revealing the selective nature of the oxidant.

II. The Chemical Nature of the Oxidant

A. REDOX POTENTIALS IN AQUEOUS SOLUTION[3]

In a saturated solution of a halogen in water at 25°C, species other than solvated halogen molecules occur in equilibrium with the free halogen and its products. The nature of the solution can be defined in terms of the following two equilibria.

$$X_2(g,l,s) \rightleftharpoons X_2(aq)$$

$$K_1(Cl_2) = 0.062$$
$$K_1(Br_2) = 0.21$$
$$K_1(I_2) = 0.0013$$

$$X_2(aq) \rightleftharpoons H^+ + X^- + HOX$$

$$K_2(Cl_2) = 4.2 \times 10^{-4}$$
$$K_2(Br_2) = 7.2 \times 10^{-9}$$
$$K_2(I_2) = 2.0 \times 10^{-13}$$

The equilibrium concentration[4] of the various species are given below in Table I.

An appreciable concentration of hypochlorous acid can be generated in a saturated aqueous solution of chlorine but a smaller concentration of HOBr, and only a negligible concentration of HOI, can be obtained in this fashion. The reaction of halogen with water does not, therefore, constitute a suitable method for preparing an aqueous solution of the hypohalous acids. An alternative method of preparation, and one which increases the yield of desired product by removing the hydrohalic acid formed by hydrolysis, is the following reaction, which is carried out by passing the halogen into a well-agitated suspension of mercuric oxide.

[2] L. F. Fieser and M. Fieser, "Reagents for Organic Synthesis," Vols. I and II. Wiley, New York, 1967 and 1969 resp.

[3] W. M. Latimer, "Oxidation States of the Elements and Their Potentials in Aqueous Solution," 2nd ed., pp. 54–59. Prentice-Hall, Englewood Cliffs, New Jersey, 1952.

[4] F. A. Cotton and G. Wilkinson, "Advanced Inorganic Chemistry," 2nd ed., pp. 567–571. Wiley (Interscience), New York, 1966.

TABLE I

EQUILIBRIUM CONCENTRATION OF HALOGENS IN AQUEOUS SOLUTION[a]

	Cl_2	Br_2	I_2
Total solubility	0.091	0.21	0.0013
$[X_2(aq.)]$ (moles/liter)	0.061	0.21	0.0013
$[H^+] = [X^-] = [HOX]$	0.030	1.15×10^{-3}	6.4×10^{-6}

[a] At 25°C, (moles/liter).

$$2\,X_2 + 2\,HgO + H_2O \rightleftharpoons HgO \cdot HgX_2 + 2\,HOX$$

In principle, all hypohalite anions can be produced by dissolving halogens in base according to the general reaction

$$X_2 + 2OH^- \rightleftharpoons X^- + XO^- + H_2O$$

where the equilibrium constants are quite favorable, e.g., 7.5×10^{15} for Cl_2, 2×10^6 for Br_2, and 30 for I_2. The tendency of the hypohalite anions to disproportionate in basic solutions ($3\,XO^- \rightleftharpoons 2\,X^- + XO_3^-$), however, creates complications. These reactions have large equilibrium constants, i.e., 10^{27} for ClO^-, 10^{15} for BrO^-, and 10^{20} for IO^- but, fortunately, the reaction of ClO^- is slow at and below room temperature, and reasonably pure solutions of Cl^- and ClO^- free from ClO_3^- can be prepared. The disproportionation of BrO^- is moderately fast at room temperature, while that of IO^- is very fast at all temperatures giving IO_3^- quantitatively according to the following equation.

$$3\,I_2 + 6\,OH^- \rightleftharpoons 5\,I^- + IO_3^- + 3\,H_2O$$

The close proximity of the standard potentials of halogens in various oxidation states and the ease of disproportionation of hypohalite ions in basic solution are responsible for generating members of species of a particular halogen with varying degrees of oxidizing power which may interfere in any oxidation reaction. For this reason, the basic thermodynamic data,[4] as given in Table II, are important for defining the oxidizing power of the species that can be encountered in the course of any reaction with hypohalite. It is seen that so far as reduction to the free halogen in aqueous acid solution is concerned, the ability of the oxyhalogen acids to function as oxidizing agents decreases with increasing oxidation state. A similar situation exists for reduction to halide in basic solution. Further, the oxyhalogen acids are much stronger oxidizing agents in acid solution than are the corresponding anions in basic medium.

The hypohalous acids with their small dissociation constants: HOCl, 3.4×10^{-8}; HOBr, 2×10^{-9}; HOI, 1×10^{-11}, are all weak acids. Thus, depending on the pH of the solution, the active oxidant may be the

TABLE II

STANDARD POTENTIALS (IN VOLTS) FOR REACTION OF HALOGENS WITH H_2O

Reaction	Cl	Br	I
$H^+ + HOX + e \rightleftharpoons \frac{1}{2} X_2 + H_2O$	1.63	1.59	1.45
$3\,H^+ + HOX_2 + 3\,e \rightleftharpoons \frac{1}{2} X_2 + 2\,H_2O$	1.64	—	—
$6\,H^+ + XO_3^- + 5\,e \rightleftharpoons \frac{1}{2} X_2 + 3\,H_2O$	1.47	1.52	1.20
$8\,H^+ + XO_4^- + 7\,e \rightleftharpoons \frac{1}{2} X_2 + 4\,H_2O$	1.36	—	1.34
$XO^- + H_2O + 2\,e \rightleftharpoons X^- + 2\,OH^-$	0.89	0.76	0.49
$XO_2^- + 2\,H_2O + 4\,e \rightleftharpoons X^- + 4\,OH^-$	0.78	—	—
$XO_3^- + 3\,H_2O + 6\,e \rightleftharpoons X^- + 6\,OH^-$	0.63	0.61	0.26
$XO_4^- + 4\,H_2O + 8\,e \rightleftharpoons X^- + 8\,OH^-$	0.56	—	0.39

TABLE III

PERCENTAGE OF TOTAL CHLORINE IN HYPOCHLORITE SOLUTION PRESENT AS HYPOCHLOROUS ACID AT VARIOUS pH LEVELS, 25°C

	pH							
	5.0	6.5	7.5	8.0	8.5	9.0	10	12
%HOCl	99.7	91.0	50.0	24.0	9.1	3.1	0.31	0.003

hypohalous acids or their anions. The composition of saturated solutions of hypochlorite are given in Table III. At pH 7.5, half of the available chlorine will be present as hypochlorous acid and half as OCl^-. At pH 10, only 0.3% of the available chlorine will be present as HOCl; the remainder is OCl^-.

B. STABILITY: ROLE OF pH AND CATALYST IN DECOMPOSITION OF HYPOCHLORITE

While hypochlorite ion undergoes minor decomposition to oxygen and chloride ion, its main reaction is disproportionation to chlorate and chloride. Several comprehensive studies of this reaction[5–15] have been made and the

[5] F. Foerster and P. Dolch, *Z. Elektrochem.* **23**, 137 (1917).

[6] O. R. Howell, *Proc. R. Soc. London, Ser. A* **104**, 134 (1923).

[7] E. Chirnoaga, *J. Chem. Soc.* p. 1693 (1926).

[8] J. R. Lewis, *J. Phys. Chem.* **32**, 243 and 1808 (1928); **35**, 915 (1931).

[9] C. G. Fink, *Trans. Electrochem. Soc.* **71**, 487 (1937).

[10] M. W. Lister, *Can. J. Chem.* **30**, 879 (1952); **34**, 465 (1956); **40**, 729 (1962).

[11] G. H. Ayres and M. H. Booth, *J. Am. Chem. Soc.* **77**, 825 and 828 (1955).

[12] R. E. Connick and Y. T. Chia, *J. Am. Chem. Soc.* **81**, 1280 (1959).

[13] T. Yokoyama and O. Takayasu, *Kogyo Kagaku Zasshi* **70**, 1619 (1967).

[14] A. Prokopcikas, J. Butkevicius, and L. Matuliauskiene, *Kinet. Katal.* **11**, 795 (1970).

[15] B. P. Nikol'skii, V. G. Ktunchak, T. V. L'vova, V. V. Pal'chevskii, and R. I. Sosnovskii, *Dokl. Akad. Nauk SSSR* **191**, 1324 (1970); **197**, 140 (1971).

overall second-order kinetics are attributed to the following reaction

$$2\ NaOCl \rightarrow NaClO_2 + NaCl$$
$$NaOCl + NaClO_2 \rightarrow NaClO_3 + NaCl$$

(a slow bimolecular reaction to produce chlorite followed by faster oxidation of chlorite with OCl^-). The activation energies for the two steps are 24.8 and 20.8 kcal/gm-mole, respectively, and the rates are such that at 40°C, a solution of sodium hypochlorite will contain about 1% as much chlorite as hypochlorite. The rate of the overall reaction increases dramatically at ionic strength above 0.8, but there is no catalytic effect of the specifics ions OH^-, Cl^-, or $CO_3{}^{2-}$.

The decomposition to chloride and oxygen is a bimolecular reaction

$$OCl^- \rightarrow Cl^- + \tfrac{1}{2}\,O_2$$

where the rate of oxygen evolution is proportional to $[OCl^-]^2$; the rate constant is 7.5×10^{-6} gm-mole(liter)$^{-1}$(min)$^{-1}$ at 60°C and ionic strength 3.5. The activation energy is 26.6 kcal/gm-mole. Since kinetic data obtained from the reactions of carefully purified hypochlorite solution, using coprecipitation of Cu, Ni, and Co by $CaCO_3$, $Mg(OH)_2$, $La(OH)_3$, or sodium periodate, did not differ, the decomposition to oxygen was assumed to be uncatalyzed. However, the evolution of oxygen becomes a major reaction when trace amounts of metals or their oxides are present. The catalytic decomposition of sodium hypochlorite with high initial pH (~10) showed negligible change in pH during reaction. In the initial pH range 7–8, the first part of the reaction showed a slow decrease, followed by a sharp decrease to the constant region of 3–4, and some chlorine appeared as a reaction product simultaneously with the rapid drop in pH. With a higher concentration of hypochlorite having initial pH in the 7–8 range, a sharp decrease in pH during reaction was observed; the lower the initial pH, the more rapid was the drop in pH. The possible reactions in the system are listed below.

$$2\ OCl^- \rightarrow O_2 + 2\ Cl^- \qquad E^0 = +0.49\ V$$
$$3\ OCl^- \rightarrow ClO_3{}^- + 2\ Cl^- \qquad E^0 = +0.39\ V$$
$$2\ HOCl \rightarrow O_2 + 2\ Cl^- + 2\ H^+ \qquad E^0 = +0.26\ V$$
$$3\ HOCl \rightarrow ClO_3{}^- + 2\ Cl^- + 3\ H^+ \qquad E^0 = +0.06\ V$$
$$HOCl + Cl^- + H^+ \rightarrow Cl_2 + H_2O \qquad E^0 = +0.13\ V$$

The reaction rate depends on several variables, such as hypochlorite concentration, amount of catalyst, rate of stirring, initial pH, buffer constituents, ionic strength, and temperature. The energy of activation for decomposition of hypochlorite solution catalyzed by colloidal hydrous iridium dioxide was calculated to be 16.4 kcal.

By using radioactive chlorine as a tracer, Taube and Dodgen[16] followed the hydrolysis of chlorine and its oxidation to various oxidation states in

[16] H. Taube and H. Dodgen, *J. Am. Chem. Soc.* **71**, 3330 (1949).

acid medium. The species ClO_2 is frequently a product which slowly disproportionates and reacts with other oxidation states of chlorine in acid at room temperature. This species, ClO_2, also appears in the disproportionation of ClO_3^-, in the reduction of ClO_3^- by Cl^-, in the oxidation of ClO_2 by Cl_2 or HOCl, and in the disproportionation of ClO_2 in acid.

III. The Haloform Reaction

A. OXIDATION OF METHYL KETONE

The haloform reaction is the process whereby the haloforms are derived from organic compounds by the action of hypohalites. An oxidizable terminal methyl group is a prerequisite for haloform formation. The process involves stepwise halogenation followed by chain cleavage of the resulting trihalomethyl derivatives. Thus, methyl ketones, acetaldehyde, or other compounds containing CH_3CO, CH_2XCO, or CHX_2CO undergo the haloform reaction. The literature published after Lieben (1870) and reviewed by Fuson and Bull,[1] noted that certain alcohols, amines, β-diketones, olefins, oximes, esters, and acetylides were also oxidized by hypohalite. In all these cases, however, processes such as oxidation, hydrolysis, and halogenation occur prior to haloform reaction of the resulting methyl ketone or acetaldehyde derivatives.

1. Kinetics and Mechanism of Haloform Reactions

The mechanism of the haloform reaction plays a very important role in the evaluation of the potential of a hypohalite oxidant. Starting with the simplest case of the haloform reaction, the reaction of acetone with sodium hypohalite, it was shown by Bartlett[17] that with hypobromite and hypoiodite solutions having high concentrations of hydroxyl ions, the reaction velocity is independent of the concentration of halogenating agent and has the same value for iodine and bromine. The same behavior is found in the base-catalyzed halogenation of acetone in alkaline solutions of low concentration, where the product is tribromo- or triiodoacetone. The reaction of acetone with a strongly alkaline solution of hydrochlorite is several hundred times slower than the rate of enolization at the employed concentration. The rate of this reaction is controlled by second-order kinetics between the enol (which is present at substantially its equilibrium concentration at all times) and the hypochlorite ion. The intermediates that are formed in haloform reaction are products of unsymmetrical halogenation.

Bartlett's observations were confirmed by Bell and Longuet-Higgins[18] from another set of kinetic data obtained by extended measurement of the

[17] P. D. Bartlett, *J. Am. Chem. Soc.* **56**, 967 (1934).

[18] R. P. Bell and H. C. Longuet-Higgins, *J. Chem. Soc.* p. 636 (1946).

same reaction over a range of temperature. Thus, it was established that the initial step of the haloform reaction is the enolization of the ketone. The mechanistic path can be described according to Scheme 1. Since the halogen

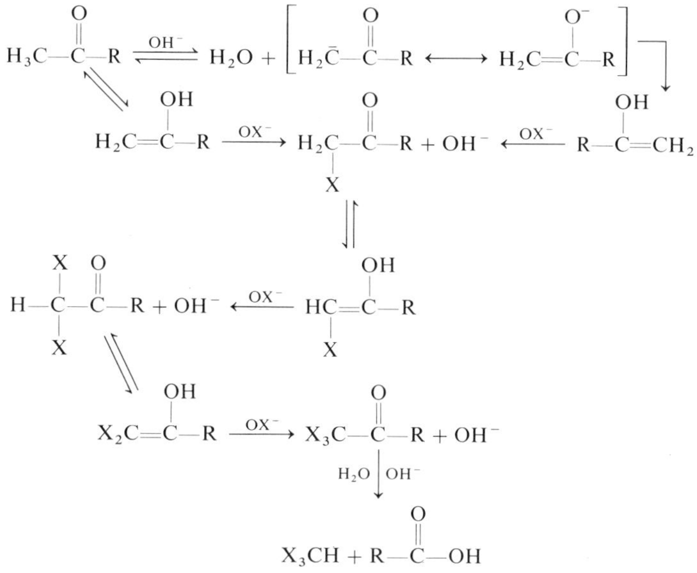

SCHEME 1

in hypohalite anion has a partial positive character, one may speculate that the halogenation of the "enol" may involve a cyclic change.

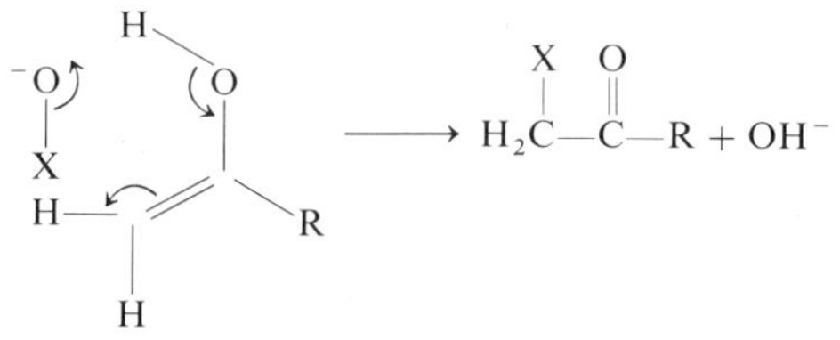

However, no attempt has yet been made to establish this step in the haloform reaction.

A typical procedure for oxidation of methyl ketone is cited below from Newman and Holmes.[19]

β-Naphthoic Acid from Methyl β-Naphthyl Ketone. The desired amount of hypochlorite solution was prepared by passing chlorine in a cold solution of alkali (in which a suitable amount of ice is added) to the proper gain in

[19] M. S. Newman and H. L. Holmes, *in* "Organic Syntheses" (A. H. Blatt, ed.), Collect. Vol. II, p. 428. Wiley, New York, 1943.

weight, at which point a solution at 0°C was obtained. The flask was fitted with a thermometer and an efficient stirrer. The solution was warmed to 55°C, the ketone was added, and a temperature of 60°–70°C was maintained by occasional cooling for 30 minutes when the exothermic reaction was over. Bisulfite solution was added to destroy excess hypochlorite and the mixture was cooled and acidified. Usual separation and crystallization from 95% ethanol gave pure β-naphthoic acid in 88% yield.

B. OXIDATION OF ENOLIZABLE KETONES

Levine and Stephens[20,21] showed that hypohalite oxidation of ketones to acids is not limited to methyl ketones. Both cyclic[23] and noncyclic ketones are oxidized to carboxylic acids using sodium hypobromite or sodium hypochlorite solutions Table IV.[20–22]

TABLE IV

OXIDATION OF ENOLIZABLE KETONES

Substrate	Product	Yield (%)	Ref.
Propiophenone	Benzoic acid	64–96	20,21
Propyl 2-thienyl ketone	2-Thiophenic acid	—	20
5-Methyl-2-propionylthiophene	5-Methyl-2-thiophenic acid	67	20
Ethyl 2-furyl ketone	2-Furoic acid	59	20
5,5-Dimethyl-1,3-cyclohexanedione (dimedon)	β,β-Dimethylglutaric acid	90	21
Cyclohexanone	Adipic acid	82	22
Cyclopentanone	Glutaric acid	87	22

Neiswender, Moniz, and Dixon[24] found that a methylene or methyl group attached to an aromatic ring can be oxidized by NaOCl to a carboxy group provided that the ring also contains an acetyl group, which is similarly oxidized to a carboxy group. For example, *p*-ethylacetophenone is oxidized to terephthalic acid in 95% yield. The explanation advanced is that in the alkaline medium the ketone (**1**) forms a resonance-stabilized anion (**2** ↔ **3**) which undergoes chlorination and this, in turn, favors further enolization.

[20] R. Levine and M. W. Farrar, *J. Am. Chem. Soc.* **71**, 1496 (1949).
[21] R. Levine and J. R. Stephens, *J. Am. Chem. Soc.* **72**, 1642 (1950).
[22] W. T. Smith and G. L. McLeod, *in* "Organic Synthesis" (N. Rabjohn, ed.), Collect. Vol. IV, p. 345. Wiley, New York, 1963.
[23] M. W. Farrar, *J. Org. Chem.* **22**, 1708 (1957).
[24] D. D. Neiswender, Jr., W. B. Moniz, and J. A. Dixon, *J. Am. Chem. Soc.* **82**, 2876 (1960).

A second chlorination results in **4** which would be hydrolyzed readily to diketone **5**. From this point a normal haloform reaction could occur.

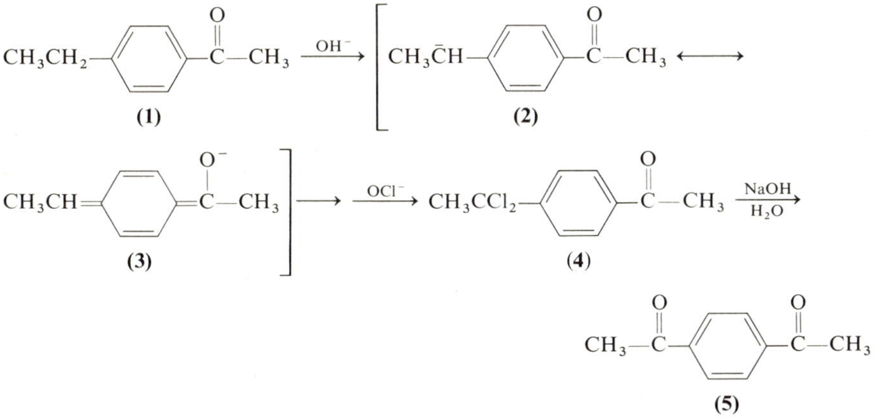

In another example, 2-acetyl-9,10-dihydrophenanthrene on oxidation with NaOCl at pH 12–13 afforded diphenyl-2,2,4-tricarboxylic acid. Phenanthrenequinone-2-carboxylic acid is postulated to be an intermediate and indeed, oxidation of phenanthrenequinone with sodium hypochlorite affords diphenic acid in 96% yield. It is to be noted here that the oxidation of methylene groups to carbonyl systems by hypohalite was first reported by Schiessler and Eldred[25] who obtained 9-oxofluorene-2-carboxylic acid as an "anomalous" oxidation product from 2-acetylfluorene. The role of the acetyl group in activating the methyl or methylene group becomes evident from the fact that ethylbenzene or fluorene is unreactive under the exact conditions used to oxidize *p*-ethylacetophenone or acetylfluorene. That both the resonance and inductive effect of the acetyl group might conceivably contribute to the oxidation of the alkyl side chain was demonstrated by oxidation of *m*-methylacetophenone to isophthalic acid in 13% yield, where activation is limited to the inductive effect.

IV. Oxidation of Methyl and Methylene Groups

It was shown in the preceding section that methyl and methylene groups activated by a carbonyl group are readily oxidized with hypohalite anion. Chakrabartty and Kretschmer[26] have recently shown that any electronegative group (e.g., nitro, cyano, carboxy, alkoxy, amino, thioalkoxy, or a heterocyclic ring) present in a structure makes the compound susceptible to hypohalite oxidation (Table V).

[25] R. W. Schiessler and N. R. Eldred, *J. Am. Chem. Soc.* **70**, 3958 (1948).

[26] S. K. Chakrabartty and H. O. Kretschmer, *J. Chem. Soc., Perkin Trans. 1* p. 222 (1974).

TABLE V

OXIDATION DATA FOR MODEL COMPOUNDS[a]

Substrate	Molar ratio of substrate to NaOCl	Product	Yield (mole %)
o-Nitrotoluene	1:22	*p*-Nitrobenzoic acid	60.7
		1-Chloromethyl-2-nitrobenzene	8.0
p-Nitrotoluene	1:22	*p*-Nitrobenzoic acid	81.7
		1-Chloromethyl-4-nitrobenzene	10.6
m-Nitrotoluene	1:22	*m*-Nitrobenzoic acid	70.0
		1-Chloromethyl-3-nitrobenzene	40.4
1-Ethyl-4-nitrobenzene	1:20	*p*-Nitrobenzoic acid	90.0
		1-(1-Chloromethyl)-4-nitrobenzene	7.3
		1-(1-Hydroxyethyl)-4-nitrobenzene	Trace
1-Butyl-4-nitrobenzene	1:20	*p*-Nitrobenzoic acid	25.0
		Acetic acid	
		Propionic acid	
3-Nitro-*o*-xylene	1:20	2-Methyl-6-nitrobenzoic acid	33.2
		2-Methyl-3-nitrobenzoic acid	43.1
		3-Nitro-*o*-phthalic acid	11.0
4-Nitro-*o*-xylene	1:20	2-Methyl-5-nitrobenzoic acid	35.8
		2-Methyl-3-nitrobenzoic acid	35.8
		4-Nitro-*o*-phthalic acid	9.6
5-Nitro-*m*-xylene	1:20	3-Methyl-5-nitrobenzoic acid	77.7
		5-Nitroisophthalic acid	7.9
p-Nitrobibenzyl	1:25	*p*-Nitrobenzil	72.0
	1:25	*p*-Nitrobenzoic acid	62.0
		Benzoic acid	63.3
Benzil	1:4	Benzoic acid	138.0
Benzoin	1:4	Benzoic acid	160.0
o-Toluonitrile	1:20	*o*-Phthalic acid	21.0
p-Toluonitrile	1:20	*p*-Cyanobenzoic acid	92.2
		p-Chloromethylbenzonitrile	Trace
m-Toluonitrile	1:20	*m*-Cyanobenzoic acid	72.6
		m-Chloromethylbenzonitrile	Trace
o-Toluoic acid	1:15	*o*-Phthalic acid	36.0
		o-Chloromethylbenzoic acid	33.8
p-Toluoic acid	1:8	Terephthalic acid	6.4
		p-Chloromethylbenzoic acid	43.5
	1:16	Terephthalic acid	32.6
		p-Chloromethylbenzoic acid	11.1
m-Toluic acid	1:16	Isophthalic acid	37.4
		m-Chloromethylbenzoic acid	21.4
		m-Hydroxymethylbenzoic acid	11.8
2-Methylpyridine	1:7	Pyridine-2-carboxylic acid	57.9
2,6-Dimethylpyridine	1:4	6-Methylpyridine-2-carboxylic acid	48.4
		6-Chloromethylpyridine-2-carboxylic acid	7.0
		Other chloro-product (non-acid)	Trace

TABLE V (*cont.*)

Substrate	Molar ratio of substrate to NaOCl	Product	Yield (mole %)
Quinoline	1:10	Isophthalic acid	22.2
		3,4-Dichloroquinoline-2-one	34.4
		Trichloroquinoline-2-one	39.0
Isoquinoline	1:10	*o*-Phthalic acid	58.3
Nitrocyclohexane	1:10	Chlorocyclohexane	
		Chlorocyclohexene	
		Dichlorocyclohexane	
Cyclohexanecarboxylic acid	1:10	Succinic acid	
		Glutaric acid	
		Adipic acid	
2-Methylcyclohexane	1:10	Succinic acid	
		Glutaric acid	
		Adipic acid	
3-Methylcyclohexanone	1:10	Succinic acid	
		Glutaric acid	
		Methyladipic acid	
Norborane	1:10	Norcamphor	
		Bicyclo[2.2.1]hept-2-ene	
		Chloronorboranes	
Norcamphor	1:10	Carbon dioxide	216
		Succinic acid	
		Cyclopentanedicarboxylic acid	
		Chloropentanedicarboxylic acid	
		Chloronorcamphor	
		Dichloronorcamphor	
Triptycene	1:10	Monochlorinated	
		Dichlorinated	
Adamantane	1:10	Carbon dioxide	143
		Monochloroadamantane	
		Dichloroadamantane	
		Trichloroadamantane	
Adamantane-2-one	1:10	Carbon dioxide	200
		Succinic acid	
		Chlorocyclopentanedicarboxylic acid	
		Dichloro- and trichloro-adamantanone	

[a] This table was taken from S. K. Chakrabartty and H. O. Kretschmer, *J. Chem. Soc., Perkin Tran. 1* p. 222 (1974).

A. ACTIVATION BY NITRO SUBSTITUENTS

Nitro derivatives of toluene, ethylbenzene, butylbenzene, and xylenes have been oxidized to the corresponding carboxylic acids in moderate to excellent yield. Using an approximate 1:20 *M* substrate:hypohalite ratio and a 1-hour reaction time, the extent of substrate oxidation was 33.8% for *ortho*-, 46.3% for *para*-, and 47.0% for *meta*-nitrotoluene. The extent of oxidation for 1-ethyl-4-nitrobenzene was 90%. Besides the complete oxidation products, namely, benzoic acids from nitrotoluenes, chloromethyl derivatives (ranging from 8% in case of the ortho and 20% in case of the meta isomer) were also isolated.

Oxidation of nitro derivatives of *o*- and *m*-xylene indicates that both methyl groups can be oxidized. 3-Nitro-1, 2-xylene is oxidized to the extent of 75% and gave 2-methyl-3-nitrobenzoic acid, 2-methyl-6-nitrobenzoic acid, and 3-nitrophthalic acid in relative mole ratios of 3:6:1. 4-Nitro-1,2-xylene, reacting similarly, gives 2-methyl-5-nitrobenzoic acid, 2-methyl-4-nitrobenzoic acid, and 4-nitrophthalic acid with mole ratios of 5:5:1, respectively. 5-Nitro-1,3-xylene gives 3-methyl-5-nitrobenzoic acid and 5-nitroisophthalic acid in a 10:1 mole ratio from 80% of the substrate.

p-Nitrobenzoic acid, propionic acid, and acetic acid have been isolated as oxidation products of *p*-nitrobutylbenzene. Since the propionic acid cannot be further oxidized with hypohalite, isolation of acetic acid from the oxidation products of *p*-nitrobutylbenzene suggests that an α-triketone

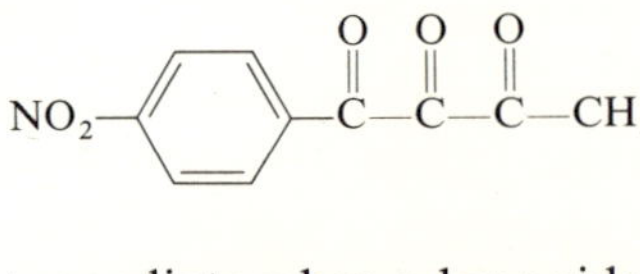

may be formed as an intermediate when a long side chain exists.

Oxidation of *p*-nitrobibenzyl using 1:20 *M* ratio of the reactants, gives *p*-nitrobenzoic acid, benzoic acid, and 1-(*p*-nitro) benzil. The major product is the diketone, which on further oxidation is converted to *p*-nitrobenzoic acid and benzoic acid.

The isolation of chloro products and ketonic products in the above reactions is consistent with the general mechanism of hypohalite oxidation presented in Section III,B.

B. ACTIVATION BY CYANO SUBSTITUENTS

Oxidation of a methyl group attached to an aromatic ring and activated by a cyano group is achieved by reacting methylbenzonitriles with sodium hypohalite using a 1:20 *M* ratio of the reactants. Phthalic acid has been obtained (21% yield) from *o*-methylbenzonitrile. It is not known whether the hydrolysis of the cyano group precedes oxidation of the methyl group.

Hydrolysis of the cyano group has not been observed for the para and meta isomers. *p*-Methylbenzonitrile gives *p*-cyanobenzoic acid (92% yield), and *m*-methylbenzonitrile is oxidized to *m*-cyanobenzoic acid (72% yield). In both cases trace amounts of chloromethyl derivatives were obtained.

C. ACTIVATION BY CARBOXY SUBSTITUENTS

The carboxylic function can also activate methyl groups attached to an aromatic ring for hypohalite oxidation. Treatment of *o*-toluic acid with 16 moles of sodium hypohalite gives phthalic acid in 36% yield and an equimolar quantity of 2-(chloromethyl)benzoic acid. *p*-Toluic acid, using an *M* ratio of 1:8 of the reactants, yields 8% terephthalic acid and 54% 4-(chloromethyl)-benzoic acid. When a 1:16 *M* ratio is used, the yield of terephthalic acid increases six times and the yield of chlorinated product decreases by a factor of four. *m*-Toluic acid gives 37.4% isophthalic acid, 21.4% 3-(chloromethyl)-benzoic acid, and 11.8% 3-(hydroxymethyl)benzoic acid. Hydrolysis of the chlorinated product to give the hydroxy derivative has not been observed in other cases.

The contribution of the inductive effect of the acetyl group[6] to the oxidation of an alkyl side chain has been reported earlier by Neiswender *et al.* A similar yield of meta and para products and a lesser yield of the ortho product from nitro, cyano, and carboxy substrates seems to indicate that inductive and steric effects jointly predominate over enolate ion resonance stabilization.

D. ACTIVATION BY PYRIDINE RINGS

Oxidative cleavage of pyridine rings usually requires very drastic conditions, but sodium hypohalite, which is a mild oxidant, can cleave certain pyridine rings. Pyridine reacts very readily with sodium hypohalite giving CO_2 and NH_3 as reaction products. No other product could be recovered and characterized. Combined GLC-mass spectral analysis of the rapidly quenched reaction mixture gave two peaks, one for nonreacted pyridine (a major component) and a minor component having the characteristic mass spectrum of 2-pyridone (molecular ion, *m/e* 95, as base peak; major fragments, *m/e* 67 and 66; dipositive ions, *m/e* 47.5, 47, and 46.5; and a minor fragment, *m/e* 39, respectively).

The ring of α-picoline has been found to be cleaved, but the cleavage reaction is quite insignificant. Using an *M* ratio of the reactants of 1:8, the oxidation of the methyl group gives a 57–60% yield of 2-picolinic acid. In the case of 2,6-lutidine, oxidation is confined to the methyl group only. Using an *M* ratio of 1:5, a yield of 39.7% of 5-methyl-2-picolinic acid was obtained along with traces of a chloromethyl product.

The heterocyclic ring of quinoline is cleaved to give an unusual product, isophthalic acid. Using an *M* ratio of 1:10 of the reactants, yields of 21%

isophthalic acid, 28.7% dichloroquinoline-2-one, and 38.8% trichloroquinoline-2-one are obtained. Isoquinoline is converted to phthalic acid (85.8% yield) using an *M* ratio of 1:10 for the substrate and sodium hypohalite. No other product was formed in the reaction.

Thus, it is quite evident that the 2-unsubstituted pyridine ring is not stable upon hypohalite oxidation. The structures which can undergo an azomethine-type addition will be attacked by hypohalite on the unsubstituted carbon alpha to the heteroatom and thereby undergo ring cleavage. Alkyl substitution on this α-carbon alters the phenomenon and oxidation then takes the normal course, i.e., formation of a "carbanion" on the side chain followed by further oxidation.

V. Oxidation of Saturated Carbocyclic Compounds

Neiswender *et al.* have reported that when 6-acetyltetraline was oxidized[24] with sodium hypochlorite, an 18% yield of the normal haloform reaction product, tetralin-6-carboxylic acid, was obtained along with 56% yield of a mixture of tricarboxylic acids, 3-(2,4-dicarboxylphenyl)propanoic acid, and 3-(2,5-dicarboxyphenyl)propanoic acid, which are products resulting from cleavage of the alicyclic ring. Mechanistically, the reaction is comparable to that of *p*-nitrobibenzyl, nitroxylene, or acetylbutylbenzene with sodium hypohalite. The cleavage of cyclohexanone to adipic acid and of cyclopentanone to glutaric acid follows the same path.

Chakrabartty and Kretschmer's work[26] provides further information about the fate of saturated carbocyclic systems. A nitro group on an alicyclic ring seems to be too labile to activate adjacent carbon–hydrogen bonds for hypohalite oxidation. Nitrocyclohexane, after reacting with sodium hypohalite (1:20 *M* ratio), gave a mixture of chlorocyclohexane, chlorocyclohexene, and dichlorocyclohexane as products. Replacement of a nitro group with halogen has been reported earlier[27] for certain aromatic systems.

On the other hand, carboxy groups on an alicyclic ring participate in the oxidative cleavage of the cyclohexane ring. Cyclohexanecarboxylic acid, when treated with sodium hypochlorite (1:20 *M* ratio), gave a mixture of succinic, glutaric, and adipic acids as well as CO_2. The relative abundance of each acid in the reaction mixture corresponds to $C_4 > C_5 \gg C_6$ acid. From the product distribution the following reaction path may be envisioned (Scheme 2). The halogenation of structure **6** with OCl^- may be a concerted cyclic process (see Section III, A, 1).

Cyclic ketones are known to undergo haloform-type reactions. Methyl substituents on such ketones were found to affect the nature and yield of cleavage products. From 2-methylcyclohexanone, a mixture of glutaric,

[27] S. Marmor, *J. Org. Chem.* **28**, 1656 (1963).

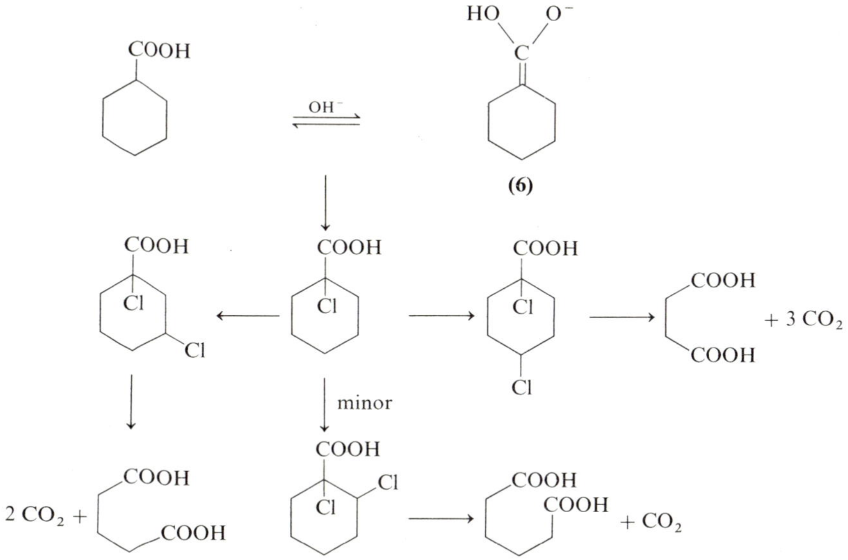

SCHEME 2

adipic, succinic, 2-methyladipic, and 2-methylglutaric acids was obtained ($C_4 > C_6 \sim C_5 \gg CH_3$—$C_5 \sim CH_3$—$C_6$ acids). The relatively insignificant amount of methyl-substituted dicarboxylic acids in the product mixture suggests that chlorination of a methine carbon atom alpha to a carbonyl group (cf. oxidation of a cyclohexane carboxylic acid) initiates the oxidation and reaction paths (Scheme 3). For the 3-methylcyclohexanone oxidation,

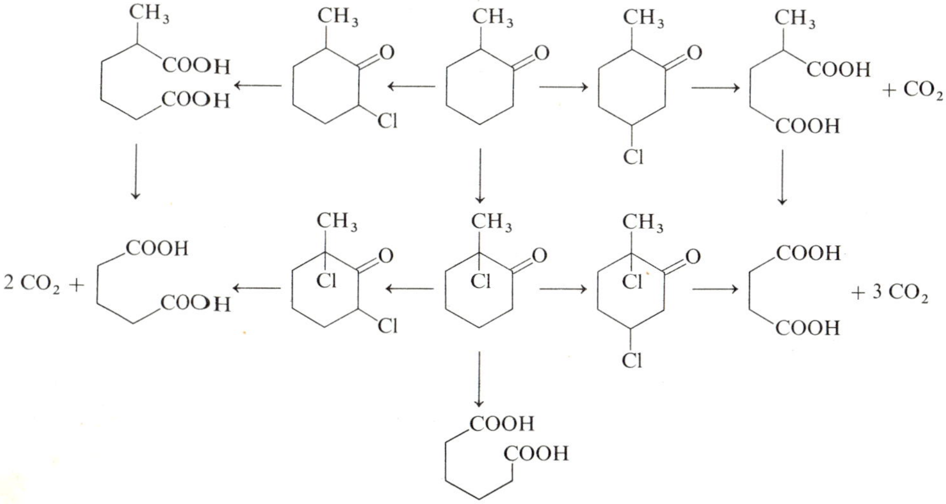

SCHEME 3

dicarboxylic acids were formed in the relative amounts $C_5 > C_4 > CH_3$–C_6 Formation of both glutaric and succinic acids, and the predominance of the former over the others, again indicates the role of methine carbon in the reaction* (Scheme 4).

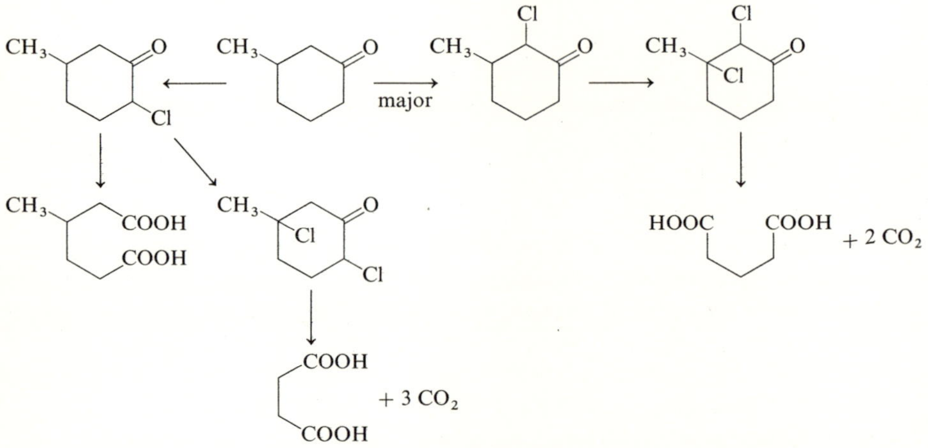

SCHEME 4

The reactivity of saturated carbocyclic systems toward hypohalite increases as the system becomes more complex. Cyclohexane is inert toward hypochlorite; norbornane is oxidized to norcamphor, but chloronorbornane and bicyclo[2.2.1]hept-2-ene are the major products; adamantane undergoes extensive chlorination followed by ring cleavage to give CO_2 as the end product. Incorporating a heterogroup on the bridge of the carbocyclic system facilitates ring cleavage. Norcamphor was oxidized to cyclopentanedicarboxylic acid, succinic acid, and CO_2. Adamantane-2-one gave cyclopentanedicarboxylic acid and CO_2. Adamantane-1-carboxylic acid gave extensively chlorinated products and CO_2. If the yield of carbon dioxide from different substrates undergoing oxidation under identical conditions can be taken as a measure of ring cleavage, the reactivity of these substrates is in the order: adamantane < adamantane-2-one < norcamphor < adamantane-1-carbocyclic acid. Incorporating a benzene ring, on the other hand, decreases the reactivity of the system. Triptycene, when oxidized with sodium hypochlorite, gives only 9-chloro- and 9,10-dichlorotriptycene.

* The reaction path for oxidation of the cyclohexane ring described in this section is purely conjecture and it needs to be elaborated upon by further study.

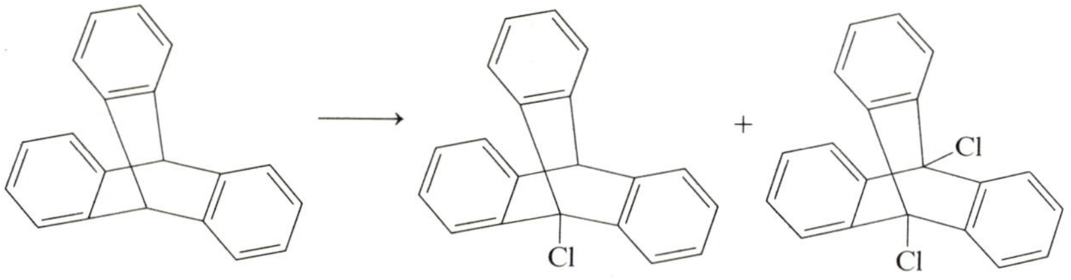

VI. Ruthenium-Catalyzed Oxidation of Carbocyclic Compounds[28]

Oxidation of organic compounds can be achieved by using catalytic amounts of $RuCl_3$ in combination with sodium hypochlorite. In this procedure, RuO_4 is the actual oxidant but 5.25% aqueous sodium hypochlorite is the effective reagent. In a typical oxidation, a solution of cyclohexanol (10 mmole) in water (40 ml) containing 0.5 ml of 2% aqueous $RuCl_3$ can be titrated at 0°C with 1.51 *N* sodium hypochlorite. Each drop of hypochlorite causes a color change and after 13.8 ml (20.8 mEq) is added, a reasonably stable yellow end point is reached. A 90–95% yield of cyclohexanone can be obtained. Preparative experiments with other substrates are conveniently done by stirring a methylene chloride solution of the substrate with an aqueous solution containing the catalyst and the desired amount of hypochlorite. Cyclohexene is oxidized to adipaldehyde and adipic acid by this method. Oxidation of 1,2-cyclohexanediol gives 2-hydroxycyclohexanone and adipic acid. Oxidation of 1,2-cyclohexanedione gives adipic, glutaric, and succinic acids.

Degradation of the benzene ring can also be achieved with this reagent. At room temperature, 3-phenylpropionic acid is oxidized to succinic acid (94%) and benzoic acid (6%). In CCl_4–water suspension, treatment of phenylcyclohexane with this reagent yields cyclohexanecarboxylic acid (25%).

Wolfe, Hasan, and Campbell[28] have suggested that OsO_4, $RhCl_3$, and $IrCl_3$ can also be employed in combination with aqueous hypochlorite for this type of oxidation.

VII. Reaction with Other Active Hydrogen Compounds

A. MOX AS A SOURCE OF X^+ ION

In the Hofmann preparation of primary amines an amide is decarbonylated with the intermediate formation of an isocyanate by treatment with bromine and alkali or with a hypohalite solution to give a primary amine with one less carbon atom.

[28] S. Wolfe, S. K. Hasan, and J. R. Campbell, *J. Chem. Soc. D* No 21, p. 1420 (1970).

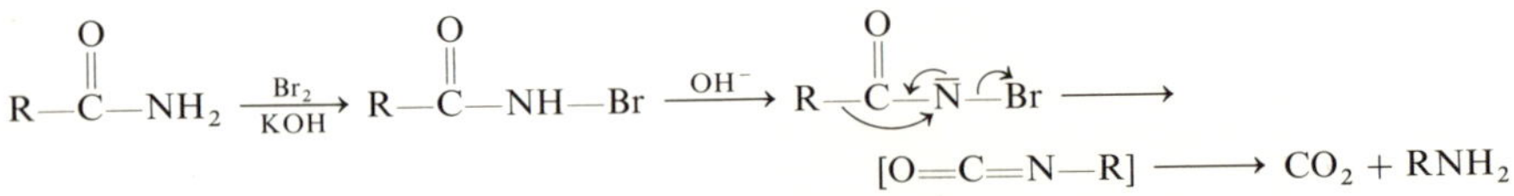

The preparation of 3-aminopyridine from nicotinamide in moderate yield (60–65%) has been reported[29] using the above process. The commercial preparation of anthranilic acid from phthalimide, also uses the Hofmann route.

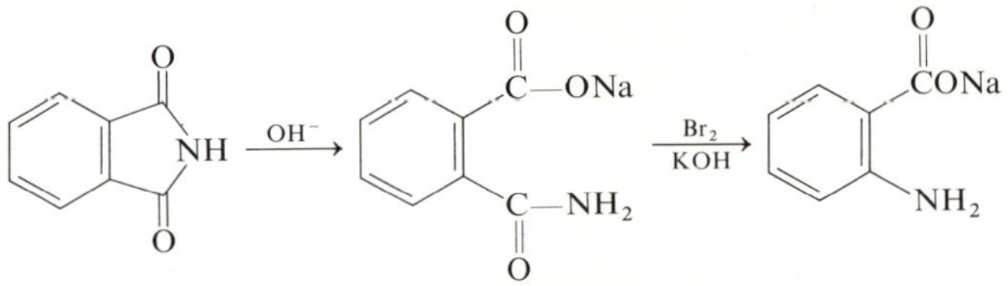

The preparation of hydrazine by oxidation of ammonia with hypochlorite anion is well known.[30] A similar oxidation path may occur in the preparation of benzofurazone oxide[31] from *o*-nitroaniline in good yield

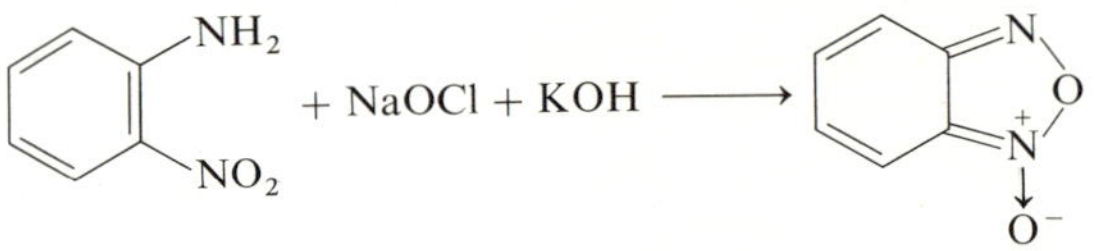

and diaziridine[32] from methylenediamine sulfate.

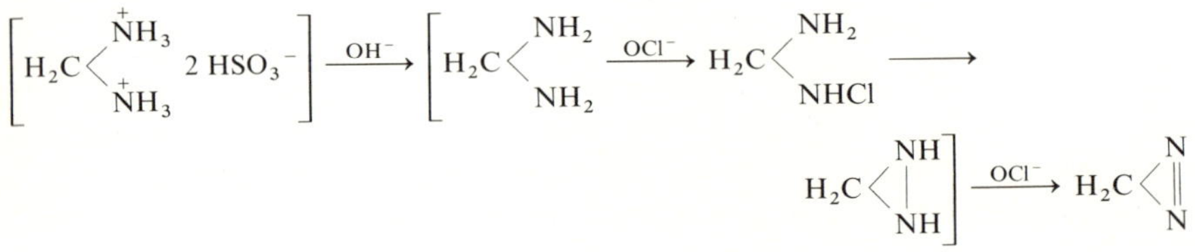

Extensive use of a hypohalite oxidant for a N-halogenation reaction has been reported in the literature. 2,6-Dibromoquinone-4-chloroimide was prepared[33] by oxidative chlorination of the stannous salt of 2,6-dibromoaminophenol with sodium hypochlorite in 85% yield.

[29] C. F. H. Allen and C. N. Wolf, *in* "Organic Syntheses" (N. Rabjohn, ed.), Collect. Vol. IV, p. 45. Wiley, New York, 1963.

[30] R. Adams and B. K. Brown, *in* "Organic Syntheses" (A. H. Blatt, ed.), Collect. Vol. I, 1st ed., p. 309. Wiley, New York, 1941.

[31] F. B. Mallory, *in* "Organic Syntheses" (N. Rabjohn, ed.), Collect. Vol. IV, p. 74. Wiley, New York, 1963.

[32] R. Ohme and E. Schmitz, *Chem. Ber.* **97**, 297 (1964).

[33] W. E. Bachmann, M. P. Cava, and A. S. Dreiding, *J. Am. Chem. Soc.* **76**, 5554 (1954).

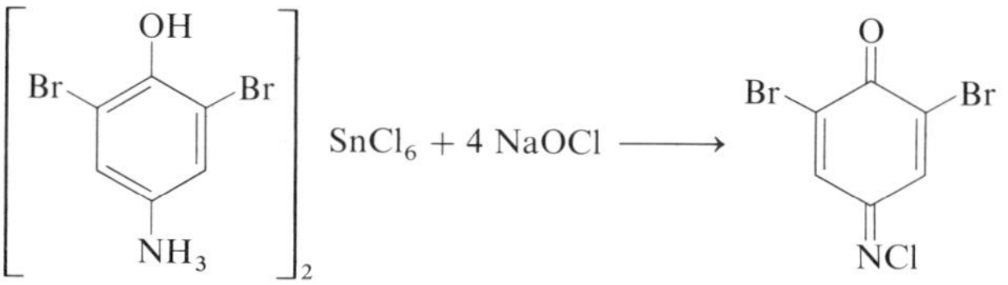

An alkyl hypohalite, particularly *tert*-butyl hypochlorite, has been found to be a much more satisfactory reagent than hypochlorous acid for the N-chlorination of amines. This is illustrated in the degradation[33] of β-acetoxy-20-amino-Δ^5-pregene to pregnenolone

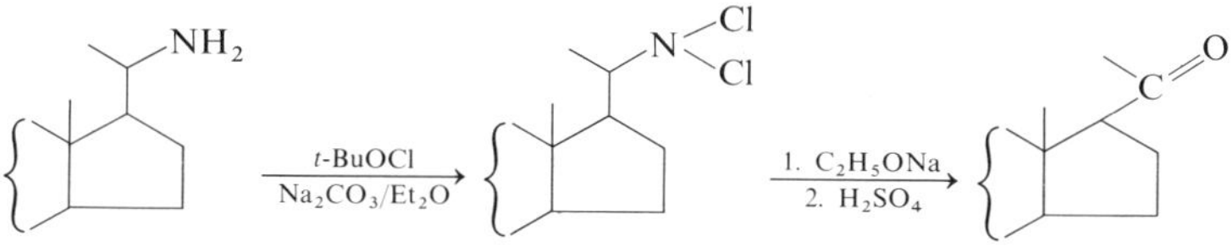

or in the synthesis[34] of phenacylamine and other α-aminoketones,

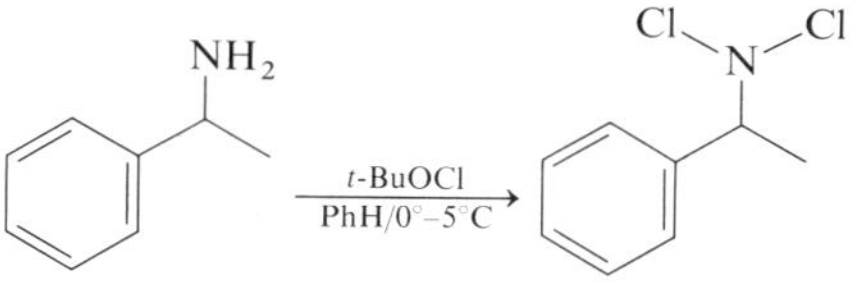

and in preparation[35] of *N*-chlorocyclohexylideneimine from cyclohexylamine.

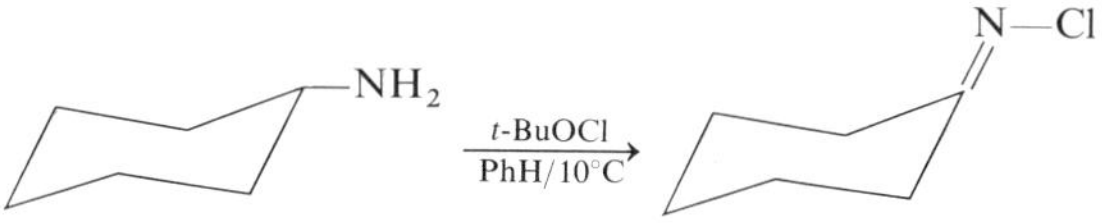

N-Halogenation with alkaline hypochlorite has recently been used in the preparation[36] of *N*-monochlorocarboxamides and of *N*-monobromocarbamate and *N*-monobromocarboxamide. The method consists of preparing the sodium salt of the monohalo derivative of the substrate by treating with a 5–6% alkaline hypohalite solution at ca. 0°C followed by careful neutralization of the salt with dilute sulfuric acid. A typical procedure is exemplified in the preparation of ethyl *N*-chlorocarbamate. To 35.6 gm (400 mmole) of ethyl carbamate in a 2-liter flask cooled in an ice bath was

[34] H. E. Baumgarten and F. A. Bower, *J. Am. Chem. Soc.* **76**, 4561 (1954); **82**, 459 (1960).
[35] G. H. Alt and W. S. Knowles, *J. Org. Chem.* **25**, 2047 (1960).
[36] C. Bachand, H. Driquez, J. M. Paton, D. Touchard, and J. Lessard, *J. Org. Chem.* **39**, 3136 (1974).

added 545 ml (0.73 mmole/ml) of sodium hypochlorite solution. The mixture was stirred until it became colorless (15 minutes), 300 ml of dichloromethane was added, and then with vigorous stirring over 2 hours, 482 mEq of 2 *N* sulfuric acid was added dropwise. The organic phase was decanted and the aqueous layer was further extracted with dichloromethane. Removal of the solvent from the combined extracts under reduced pressure at 20°–25°C yielded 49.0 gm (98%) of ethyl *N*-chlorocarbamate.

The method is not convenient for the preparation of water-soluble *N*-chloroformamide, nor is it successful with sterically hindered 2,2-dimethylpropionamide. Table VI records the yield of 28 different *N*-haloamides, as reported by Bachand, Dreguez, Paton, Touchard, and Lessard.[36]

TABLE VI

PREPARATION OF *N*-HALOAMIDES

Compound	Product	Yield (%)	Product	Yield (%)
$CH_3CH_2OCONH_2$	$CH_3CH_2OCONHCl$	98	$CH_3CH_2OCONHBr$	85
$CH_3CH_2CH_2OCONH_2$	$CH_3CH_2CH_2OCONHCl$	86	—	—
$CH_3OCH_2CH_2OCONH_2$	$CH_3OCH_2CH_2OCONHCl$	93	—	—
$ClCH_2CH_2OCONH_2$	$ClCH_2CH_2OCONHCl$	91	—	—
$Cl_3CCH_2OCONH_2$	$Cl_3CCH_2OCONHCl$	92	$Cl_3CCH_2OCONHBr$	92
$PhCH_2OCONH_2$	$PhCH_2OCONHCl$	98	$PhCH_2OCONHBr$	79
$HCONH_2$	$HCONHCl$	43	—	—
CH_3CONH_2	$CH_3CONHCl$	70	—	—
$CH_3CH_2CONH_2$	$CH_3CH_2CONHCl$	78	$CH_3CH_2CONHBr$	87
$ClCH_2CH_2CONH_2$	$ClCH_2CH_2CONHCl$	88	$ClCH_2CH_2CONHBr$	89
$BrCH_2CONH_2$	$BrCH_2CONHCl$	80	$BrCH_2CONHBr$	78
$ClCH_2CONH_2$	$ClCH_2CONHCl$	83	$ClCH_2CONHBr$	79
FCH_2CONH_2	$FCH_2CONHCl$	79	$FCH_2CONHBr$	65
$Cl_2CHCONH_2$	$Cl_2CHCONCHCl$	81	$Cl_2CHCONHBr$	57
Cl_3CCONH_2	$Cl_3CCONHCl$	85	$Cl_3CCONHBr$	45
F_3CCONH_2	$F_3CCONHCl$	63	$F_3CCONHBr$	19

Another application of the N-chlorination reaction is found in the regeneration[37] of ketones from tosylhydrazones. Carbonyl compounds can be conveniently recovered from their tosylhydrazones by reacting with sodium hypochlorite which furnishes both Cl^+ for N-chlorination and OH^- for deprotonation and nucleophilic attack.

[37] Tse-Lok-Ho and Chiu Ming Wong, *J. Org. Chem.* **39**, 3453 (1974).

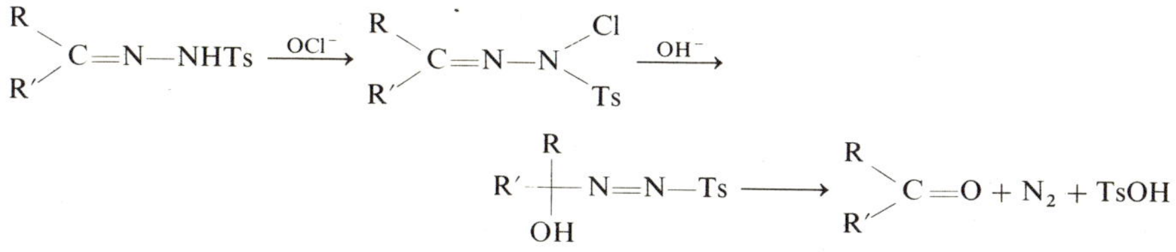

The tosylhydrazone (1.0 gm) dissolved or suspended in chloroform (30 ml) is shaken with 5% NaOCl (20 ml) for five minutes to give the desired ketone. Between 60–85% yields of simple ketones, namely, cyclohexanone, norcamphor, 2-methylcyclohexanone, acetophenone, and benzophenone can be obtained. Unfortunately, the procedure is not very suitable for cleavage of aldehyde derivatives.

Hypohalite can be used in the halogenation of a terminal acetylene.[38] Thus, phenylacetylene is brominated in 73–83% yield by reacting with aqueous alkali and bromine at room temperature. Alkyl hypohalite can be used instead of inorganic hypohalite for halogenation of acetylenes.

Alcohols can be converted to alkyl hypohalites by reacting an alkaline alcohol with halogen. The stability of the alkyl hypohalites depends on the alkyl group and the halogen atom. A tertiary alkyl hypohalite is most stable; hypochlorite is more stable than hypobromite or hypoiodite. *tert*-Butyl hypochlorite has been found to be a useful reagent[39] for selective chlorination,[40–43] oxidation,[44–46] and other reactions.

B. EPOXIDATION OF AN α,β-UNSATURATED COMPOUND

Marmor[47] reported that sodium hypochlorite reacts with 1,4-naphthoquinone in dioxane to produce the 2,3-epoxide in 71.5% yield. Similarly, the epoxide from benzalacetophenone can be obtained in high yield by reacting it with sodium hypochlorite in pyridine.

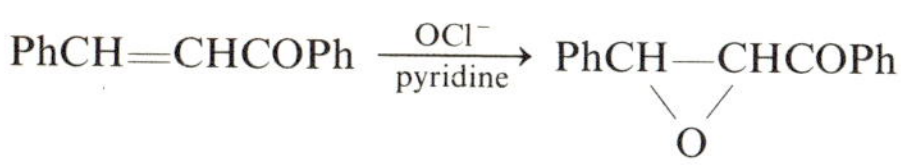

[38] S. I. Miller, G. R. Ziegler, and R. Wieleseck, *Org. Synth.* **45**, 86 (1965).
[39] M. Anbar and D. Ginsberg, *Chem. Rev.* **54**, 425 (1954).
[40] D. Ginsberg, *J. Am. Chem. Soc.* **73**, 2723 (1951).
[41] C. Walling and B. B. Jacknow, *J. Am. Chem. Soc.* **82**, 6108 and 6113 (1960).
[42] C. Walling and W. Thaler, *J. Am. Chem. Soc.* **83**, 3877 (1961).
[43] D. H. R. Barton, A. L. J. Deckwith, and A. Goosen, *J. Chem. Soc.* p. 181 (1965).
[44] C. A. Grob and H. J. Schmid, *Helv. Chim Acta* **36**, 1763 (1953).
[45] J. J. Beereboon, C. Djerassi, D. Ginsburg, and L. F. Fieser, *J. Am. Chem. Soc.* **75**, 3500 (1953).
[46] G. S. Fonken, J. L. Thompson, and R. H. Levin, *J. Am. Chem. Soc.* **77**, 172 (1955).
[47] S. Marmor, *J. Org. Chem.* **28**, 250 (1963).

Since pyridine, dioxane, and the epoxide are all susceptible to oxidation by the hypohalite anion, precaution is necessary to limit the reaction conditions with respect to the amount of oxidant, temperature, and pH of the reaction medium. Instead of epoxidation, the HOBr addition product can also be obtained by manipulation of the reaction condition,[48] as illustrated below.

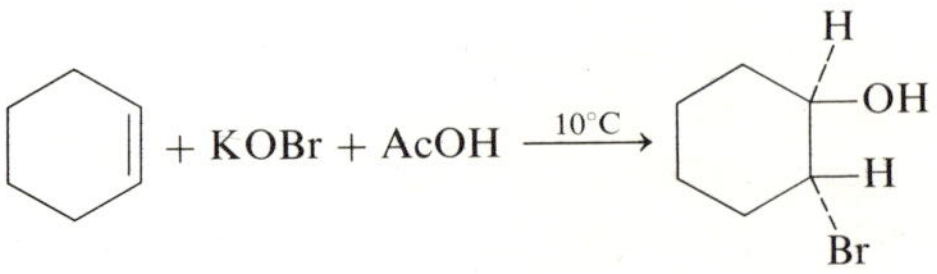

C. SHORTENING OF CHAIN LENGTH BY OXIDATIVE DECARBOXYLATION

Whistler and Schweiger[49] described the preparation of D-arabinose from D-glucose by a two-stage hypohalite oxidation.

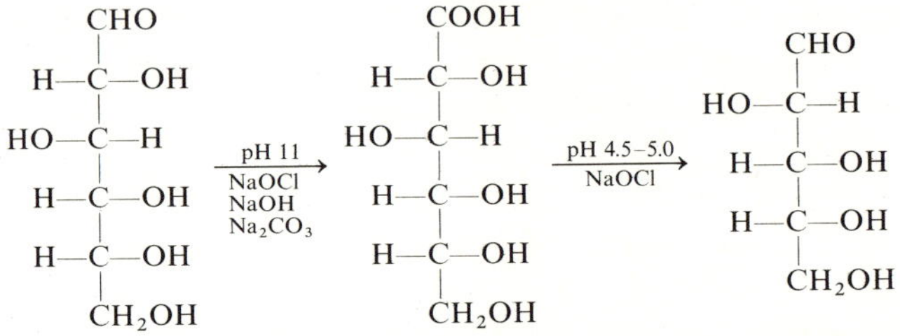

The procedure was found to be very useful for reducing oligosaccharides[50] where other chain shortening procedures are tedious or lead to a low overall yield. Thus, β-maltose monohydrate was converted to 3-*O*-α-D-glucopyranosyl-α-D-arabinose in 32.6% yield and α-lactose monohydrate to 3-*O*-β-D-galactopyranosyl-α-D-arabinose in 38.1% yield.

Inorganic hypohalites are being used for conversion of primary or secondary α-amino acids to aldehydes or ketones. Slates, Taub, Kuo, and Wendler[51] reported the preparation of 3,4-dimethoxyphenylacetone by the oxidation of α-methyl-3,4-dimethoxyphenylalanine with sodium hypochlorite. *N*-Chloramines are stated to be formed initially, which on decarboxylation produces iminium ions as intermediate products; hydrolysis of the iminium ion would give the carbonyl compound.

48 A. Kergomard, *Bull. Soc. Chim Fr.* p. 2360 (1961).
49 R. L. Whistler and R. Schweiger, *J. Am. Chem. Soc.* **81**, 5190 (1959).
50 R. L. Whistler and K. Yagi, *J. Org. Chem.* **26**, 1050 (1961).
51 H. L. Slates, D. Taub, C. H. Kuo, and N. L. Wendler, *J. Org. Chem.* **29**, 1424 (1964).

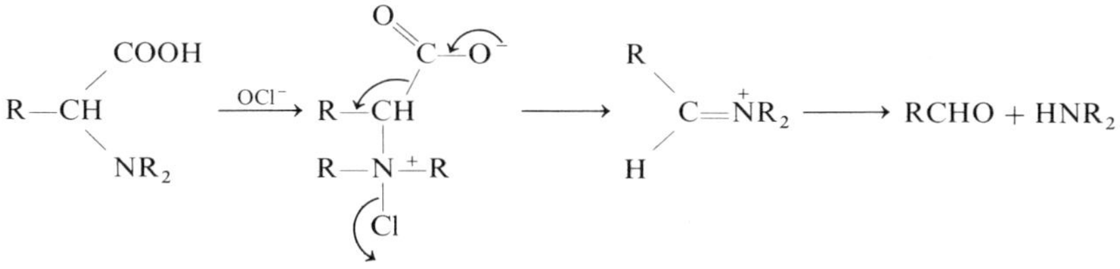

The hypohalite induced oxidative decarboxylation of various primary, secondary, and tertiary α-amino acids was studied by van Tamelen, Haarstad, and Orvis.[52] Two equivalents of hypochlorite for each mole of amino acid gave optimum decarboxylation. Maximum decarboxylation occurred at pH 1.5 with a definite trend to a lower rate as the solution pH was increased. At a lower pH the basic product, probably in a protonated iminium ion form, is less vulnerable to further reaction with any unreacted hypohalite. *N*,*N*-Dimethyglycine gave *N*-chlorodimethylamine. 2-Methyltryptophan gave 4-acetylquinoline (20%). The following reaction course (Scheme 5) was suggested for formation of a quinoline derivative.

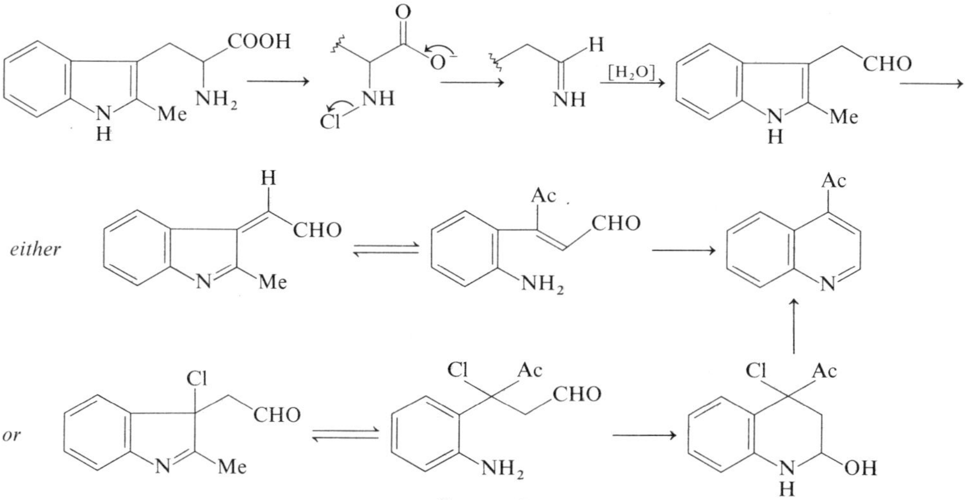

SCHEME 5

Within the tetrahydro-β-carboline category, both the secondary and tertiary amino acids were subjected to the action of alkaline hypochlorite. In neither instance could a simple oxidative decarboxylated product be isolated. The secondary amino acid, 2,3,4,5-tetrahydro-β-carboline-4-carboxylic acid, gave norharman as the product in poor yield.

[52] E. E. van Tamelen, V. B. Haarstad, and R. L. Orvis, *Tetrahedron* **24**, 687 (1968).

R=H

The *N*-methyl variant was similarly prone to over-oxidation and the action of hypohalite was found to be complex.

D. OXIDATIVE COUPLING REACTION

In the preparation of a bimolecular product from 2,4,6-trinitrobenzyl chloride in the presence of alcoholic potassium hydroxide, 2,4,6-trinitrobenzyl anion was postulated as the active intermediate. Alkaline hypochlorite oxidation of 2,4,6-trinitrotoluene should also generate the same anion, and under favorable conditions should produce bimolecular products.

Shipp and Kaplan[53] prepared 2,4,6-trinitrobenzyl chloride (85% yield) by treating 2,4,6-trinitrotoluene in tetrahydrofuran (THF)–MeOH solution with 5% aqueous sodium hypochlorite at 0°C for one minute; with sodium hypobromite, a 31.5% yield of corresponding bromide could be obtained. If the temperature of the reaction mixture was allowed to rise to 35°C, the bimolecular product, 2,2′,4,4′,6,6′-hexanitrobibenzyl was formed (79% yield). Variation of reaction conditions gave the unsaturated product, 2,2′,4,4′,6,6′-hexanitrostilbene.

Recently, Ogata and Nagura[54] studied the oxidative coupling reaction of benzyl cyanides with iodine or *tert*-butyl hypohalite and a strong base.

$$2\,PhCH_2CN + 2\,I_2 + 4\,RONa \longrightarrow PhC(CN){=}C(CN)Ph + 4\,ROH + 4\,NaI$$

The rate law, observed in this study, suggests a mechanism which involves the deprotonation of α-halobenzyl cyanide, followed by nucleophillic attack of the formed carbanion on another molecule of α-halobenzyl cyanide to give 2-halo-2,3-diphenylsuccinnonitrile. Subsequent elimination of hydrohalic acid would provide the desired stilbene derivative.

E. HALOOXY SUBSTITUTION REACTION

Green and Rowe[55] reported displacement of the nitro group by halogen when 2,4-dinitroaniline in alkaline methanol was treated with sodium hypochlorite. The true structure of the product has been established by Mallory and Varimbi[56] as 5-chloro-4-methoxybenzofurazane-1-oxide (in

[53] K. G. Shipp and L. A. Kaplan, *J. Org. Chem.* **31**, 857 (1966).
[54] Y. Ogata and K. Nagura, *J. Org. Chem.* **39**, 394 (1974).
[55] A. G. Green and R. M. Rowe, *J. Chem. Soc.* **101**, 2452 (1912).
[56] F. B. Mallory and S. P. Varimbi, *J. Org. Chem.* **28**, 1656 (1963).

equilibrium with the less stable 3-oxide). They postulate the following mechanism for this unusual reaction.

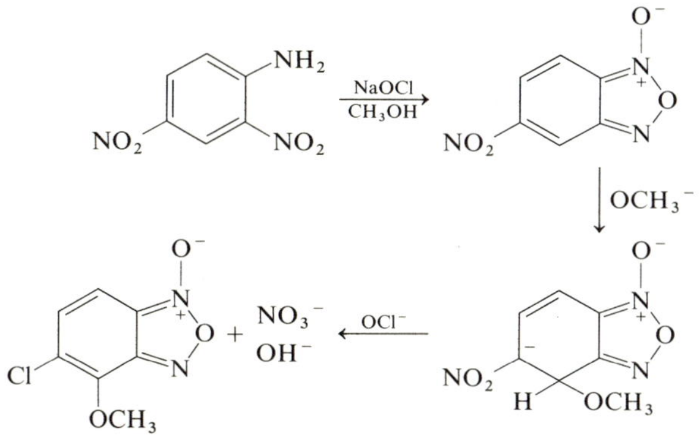

If the reaction is carried out in ethanolic KOH, the products are the ethoxy derivatives. Additional examples have been provided by Mallory, Wood, and Hurwitz[57] where, by treatment of certain aromatic nitro compounds with aqueous sodium hypochlorite in alkaline alcohol, the aromatic nitro group and the adjacent ring hydrogen are displaced by a chlorine and alkoxy group. The yield in each case is excellent (64–82%) for the reactions of 2,3-dinitroaniline, benzotriazole, and diphenylquinoxaline.

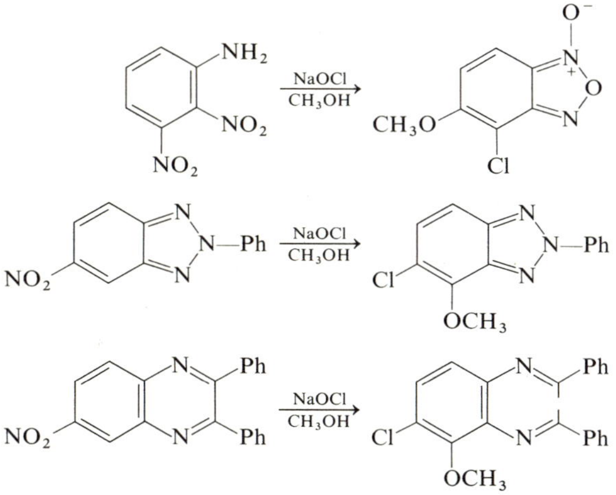

Use of sodium hypobromite in the above reaction gave bromomethoxy products.

[57] F. B. Mallory, C. S. Wood, and B. M. Hurwitz, *J. Org. Chem.* **29**, 2605 (1964).

VIII. Cleavage of Aromatic Rings

A. PHENOL DEGRADATION

Phenol[58] and naphthols[59,60] are known to react with alkaline hypochlorite solution at moderate temperature. The reaction of phenol gives variable yields of 3,5,5-trichloro-1,4-dihydroxycyclopent-2-ene carboxylic acid, which on treatment with concentrated sulfuric acid produces 3,5-dichloro-2-hydroxy-4-oxocyclopent-2-enecarboxylic acid. Dry distillation of this product under reduced pressure gives the monoenol of 2,4-dichlorocyclopentane-1,3-dione. It was concluded that the alkaline hypochlorite oxidation of phenol involves the initial formation of the normal maximum substitution product followed by addition of HOCl across the double bonds leading to the formation of an intermediate which undergoes a pseudo-Favorskii or benzilic acid rearrangement. Such a sequence is illustrated by the following scheme.

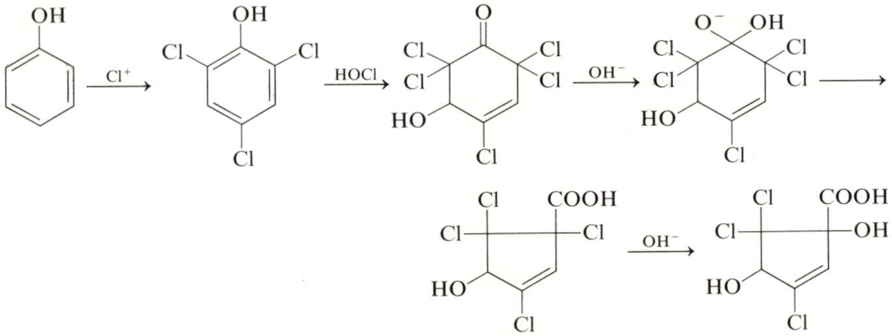

The sodium salt of 2-naphthol reacts exothermally with sodium hypochlorite at 65°C to give phthalic acid (22%) and carbon dioxide as products. From 1-naphthol, a 90% yield of phthalic acid could be obtained. 1-Hydroxyanthracene and 2,3-naphthalenedicarboxylic acid also react with sodium hypochlorite at 60°–70°C to yield phthalic and pyromellitic acids.

From these experiments, Mayo[60] and Landolt[59] concluded that some condensed aromatic systems, activated for oxidation by "strategically placed" carboxyl and hydroxyl functionalities would undergo aromatic carbon–carbon bond cleavage by sodium hypochlorite at 60°–70°C. This oxidation may sometimes be followed by decarboxylation of the primary oxidation products.

[58] C. J. Moye and S. Sternhell, *Aust. J. Chem.* **19**, 2107 (1966).

[59] R. G. Landolt, *Fuel* **54**, 299 (1975).

[60] F. R. Mayo, *Fuel* **54**, 273 (1975).

B. BROMOPICRIN REACTION

The reaction of picric acid with hypohalite to give halopicrin is remarkable in the sense that it results in the complete fragmentation of a benzene ring into six single carbon units under very mild conditions. The formation of bromopicrin by this reaction has been reported by Stenhouse in 1854. Birch and co-workers[61,62] have used this reaction as a means of locating labelled carbon atoms in biosynthetic investigations. They have indicated that a variety of polynitro compounds, e.g., 2,4,6-trinitrophloroglucinol, 2,4,6-trinitroorcinol, 2,5-dihydroxy-3,6-dinitro-1,4-dienone, and 2-bromo-2-nitropropane-1,3-diol, can undergo this reaction. A hydroxyl group on the molecule is essential and the reaction occurs more readily as the number of nitro groups is increased.

Recently, Butler and co-workers[63–65] have reported an extensive study of the bromopicrin reaction. The data obtained from the kinetic study are explained in terms of the rate-determining attack of hypobromite on an intermediate formed from hydroxide and picrate anions in an equilibrium process

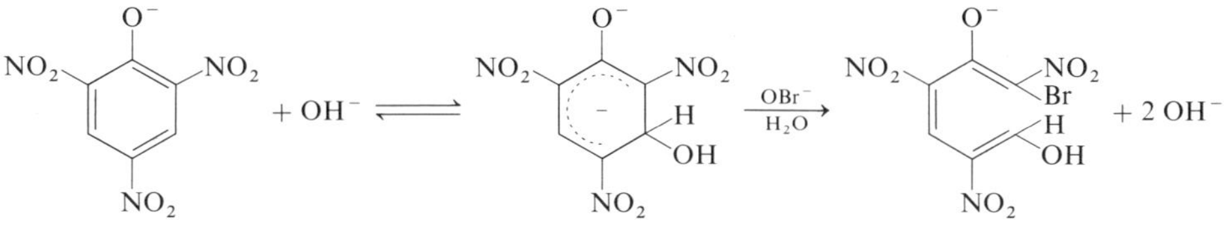

(7)

and the overall reaction is represented as

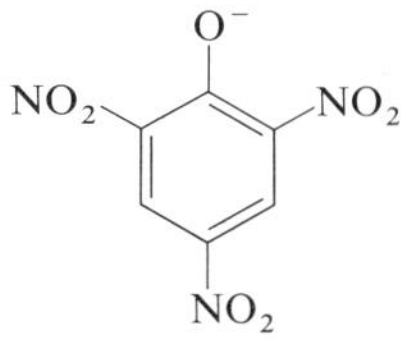

$$+\ 9[OBr^-] + 4\ H_2O \longrightarrow 3\ CBr_3NO_2 + 2\ HCO_2^- + CO_2 + 8\ OH^-$$

(8)

Elucidation of the mechanism of fragmentation of **7** to give **8** is difficult. A simpler system related to **7** is 2-nitroethanol which also reacts with hypo-

[61] A. J. Birch, R. A. Massy-Westropp, R. W. Rickards, and H. Smith, *J. Chem. Soc.* p. 360 (1958).

[62] A. J. Birch, C. J. Moye, R. W. Rickards, and Z. Vanek, *J. Chem. Soc.* p. 3586 (1962).

[63] A. R. Butler and H. F. Wallace, *J. Chem. Soc. B* p. 1758 (1970).

[64] R. I. Aylott, A. R. Butler, D. S. B. Grace, and H. McNab, *J. Chem. Soc., Perkin Trans. 2* p. 1107 (1973).

[65] S. P. Avery and A. R. Butler, *J. Chem. Soc., Perkin Trans. 2* p. 1110 (1973).

bromite to yield bromopicrin. Butler and co-workers[64] have elaborated the mechanism of the reaction as follows.

$$HOCH_2CH_2NO_2 \xrightarrow{OH^-} HOCH_2\bar{C}HNO_2 \xrightarrow{OX^-} HOCH_2CBrHNO_2 + OH^- \longrightarrow$$

$$HOCH_2\bar{C}BrNO_2 \xrightarrow{OX^-} HOCH_2CBr_2NO_2 + OH^-$$

$$\downarrow$$

$$CBr_3NO_2$$

The rate-determining step in this reaction is the first ionization (cf. haloform reaction). Subsequent halogenations, ionization, and fission of the carbon–carbon bond are all very fast.

It is also observed that under similar conditions nitroethane gives 1,1-dibromo-1-nitroethane and not the fisson product; methyl 2-nitroethyl ether is also converted to 2,2-dibromo-2-nitroethyl methyl ether without showing any tendency to cleave. α,α-Dibromo-α-nitrotoluene (made by the action of sodium hypobromite on α-nitrotoluene)[65] does not react further with sodium hypobromite to form bromopicrin. Even the *p*-nitro derivative of this compound does not possess the necessary activation to displace CBr_2NO_2 from the benzene ring under bromopicrin reaction conditions.

On these premises it was suggested that the fission of the 2,2-dibromo-2-nitroethanol molecule in an alkaline medium is a reverse aldol-type reaction

$$HO^- \; H{-}O{-}CH_2{-}CBr_2NO_2 \longrightarrow H_2O + CH_2O + \bar{C}Br_2NO_2$$

the dibromonitromethyl anion further reacting with Br_2(or OBr^- ?) to yield bromopicrin. It was also suggested that bromonitrocarbene may be the intermediate species in the final step of the overall reaction.

Index

N

Q

R

S

T

U

V

ORGANIC CHEMISTRY

A SERIES OF MONOGRAPHS

EDITOR

1. Wolfgang Kirmse. CARBENE CHEMISTRY, 1964; 2nd Edition, 1971
2. Brandes H. Smith. BRIDGED AROMATIC COMPOUNDS, 1964
3. Michael Hanack. CONFORMATION THEORY, 1965
4. Donald J. Cram. FUNDAMENTALS OF CARBANION CHEMISTRY, 1965
5. Kenneth B. Wiberg (Editor). OXIDATION IN ORGANIC CHEMISTRY, PART A, 1965; Walter S. Trahanovsky (Editor). OXIDATION IN ORGANIC CHEMISTRY, PART B, 1973; PART C, 1978
6. R. F. Hudson. STRUCTURE AND MECHANISM IN ORGANO-PHOSPHORUS CHEMISTRY, 1965
7. A. William Johnson. YLID CHEMISTRY, 1966
8. Jan Hamer (Editor). 1,4-CYCLOADDITION REACTIONS, 1967
9. Henri Ulrich. CYCLOADDITION REACTIONS OF HETEROCUMULENES, 1967
10. M. P. Cava and M. J. Mitchell. CYCLOBUTADIENE AND RELATED COMPOUNDS, 1967
11. Reinhard W. Hoffmann. DEHYDROBENZENE AND CYCLOALKYNES, 1967
12. Stanley R. Sandler and Wolf Karo. ORGANIC FUNCTIONAL GROUP PREPARATIONS, VOLUME I, 1968; VOLUME II, 1971; VOLUME III, 1972
13. Robert J. Cotter and Markus Matzner. RING-FORMING POLYMERIZATIONS, PART A, 1969; PART B, 1; B, 2, 1972
14. R. H. DeWolfe, CARBOXYLIC ORTHO ACID DERIVATIVES, 1970
15. R. Foster. ORGANIC CHARGE-TRANSFER COMPLEXES, 1969
16. James P. Snyder (Editor). NONBENZENOID AROMATICS, VOLUME I, 1969; VOLUME II, 1971
17. C. H. Rochester. ACIDITY FUNCTIONS, 1970
18. Richard J. Sundberg. THE CHEMISTRY OF INDOLES, 1970
19. A. R. Katritzky and J. M. Lagowski. CHEMISTRY OF THE HETEROCYCLIC *N*-OXIDES, 1970
20. Ivar Ugi (Editor). ISONITRILE CHEMISTRY, 1971

21. G. Chiurdoglu (Editor). CONFORMATIONAL ANALYSIS, 1971

22. Gottfried Schill. CATENANES, ROTAXANES, AND KNOTS, 1971

23. M. Liler. REACTION MECHANISMS IN SULPHURIC ACID AND OTHER STRONG ACID SOLUTIONS, 1971

24. J. B. Stothers. CARBON-13 NMR SPECTROSCOPY, 1972

25. Maurice Shamma. THE ISOQUINOLINE ALKALOIDS: CHEMISTRY AND PHARMACOLOGY, 1972

26. Samuel P. McManus (Editor). ORGANIC REACTIVE INTERMEDIATES, 1973

27. H. C. Van der Plas. RING TRANSFORMATIONS OF HETEROCYCLES, VOLUMES 1 AND 2, 1973

28. Paul N. Rylander. ORGANIC SYNTHESES WITH NOBLE CATALYSTS, 1973

29. Stanley R. Sandler and Wolf Karo. POLYMER SYNTHESES, VOLUME I, 1974; VOLUME II, 1977

30. Robert T. Blickenstaff, Anil C. Ghosh, and Gordon C. Wolf. TOTAL SYNTHESIS OF STEROIDS, 1974

31. Barry M. Trost and Lawrence S. Melvin, Jr. SULFUR YLIDES: EMERGING SYNTHETIC INTERMEDIATES, 1975

32. Sidney D. Ross, Manuel Finkelstein, and Eric J. Rudd. ANODIC OXIDATION, 1975

33. Howard Alper (Editor). TRANSITION METAL ORGANOMETALLICS IN ORGANIC SYNTHESIS, VOLUME I, 1976; VOLUME II, 1978

34. R. A. Jones and G. P. Bean. THE CHEMISTRY OF PYRROLES, 1976

35. Alan P. Marchand and Roland E. Lehr (Editors). PERICYCLIC REACTIONS, VOLUME I, 1977; VOLUME II, 1977

36. Pierre Crabbé (Editor). PROSTAGLANDIN RESEARCH, 1977

37. Eric Block. REACTIONS OF ORGANOSULFUR COMPOUNDS, 1978

38. Arthur Greenberg and Joel F. Liebman, STRAINED ORGANIC MOLECULES, 1978

39. Philip S. Bailey. OZONATION IN ORGANIC CHEMISTRY, VOL. I, 1978

40. Harry H. Wasserman and Robert W. Murray (Editors). SINGLET OXYGEN, 1978

A
B 8
C 9
D 0
E 1
F 2
G 3
H 4
I 5
J 6